10th Product Design, Robotics, Advanced Mechanical and Mechatronic Systems and Innovation Conference (PRASIC 2018)

IOP Conference Series: Materials Science and Engineering
Volume 514

Brasnov, Romania
8 – 9 November 2018

ISBN: 978-1-5108-9082-4
ISSN: 1757-8981

Printed from e-media with permission by:

Curran Associates, Inc.
57 Morehouse Lane
Red Hook, NY 12571

Some format issues inherent in the e-media version may also appear in this print version.

No changes have been made to the content of these proceedings. There may be changes to pagination and minor adjustments for aesthetics.

Printed by Curran Associates, Inc. (2019)

For permission requests, please contact the Institute of Physics at the address below.

Institute of Physics
Dirac House, Temple Back
Bristol BS1 6BE UK

Phone: 44 1 17 929 7481
Fax: 44 1 17 920 0979

techtracking@iop.org

Additional copies of this publication are available from:

Curran Associates, Inc.
57 Morehouse Lane
Red Hook, NY 12571 USA
Phone: 845-758-0400
Fax: 845-758-2633
Email: curran@proceedings.com
Web: www.proceedings.com

Table of contents

Volume 514

Product Design, Robotics, Advanced Mechanical and Mechatronic Systems and Innovation Conference (PRASIC)

8–9 November 2018, Brasov, Romania

Accepted papers received: 11 March 2019
Published online: 26 June 2019

Preface

Papers

Tribology

Management

Biomechanics

IOP Conf. Series: Materials Science and Engineering **514** (2019) 011001 doi:10.1088/1757-899X/514/1/011001

PRASIC 2018, 8 – 9 November 2018

x[th] **P**roduct Design, **R**obotics, **A**dvanced Mechanical & Mechatronic **S**ystems and **I**nnovation **C**onference

The 10th edition of the PRASIC Conference was hosted in Brasov on 8-9 November 2018.
The PRASIC'18 Conference is dedicated to scientific research in the field of mechanics, machine elements, industrial robots, advanced mechanics and mechatronic systems, tribology, and also in education, innovation and technology transfer, industrial management, mechanics and machine history.
The beginning was in 1982 under the name of PRASIC (Computer Aided Design), at the initiative of Professor Ioan Drăghici. At the beginning only the scientific researches related to Mechanisms and Machine Elements were treated, from 1996 Robotics was added, and then other scientific concerns related to them.
In 2006, it was the last edition having as sections mechanisms, Machine Elements and Robotics. In 2016 PRASIC was renamed, being **P**roduct Design, **R**obotics, **A**dvanced Mechanics & Mechatronic **S**ystems, and **I**nnovation **C**onference, keeping the brand PRASIC, well known in Romania and also outside, having an important history, and also keeping a high quality standard for scientific papers.
At the 2018 edition, 48 papers were selected on 9 sections, with the participation of colleagues from university centers such as: Politehnica University of Bucharest, Gh. Asachi Iasi University, Technical University of Cluj Napoca, Stefan cel Mare University of Suceava, University Constantin Brancusi Tg. Jiu, Dunarea de Jos University of Galati, Politehnica University of Timisoara, University of Craiova, Petroleum and Gas University of Ploiesti, Technical University of Moldova Chisinau.
The scientific opening of the Conference was provided by Professor Alberto Canen of Coppe University, Rio de Janeiro, Brazil and Professor Valeriu Dulgheru of the Technical University of Moldova.
Conference papers were submitted for publication at the IOP Publishing Conference Series - Materials Science and Engineering.

Organizing Committee,
Professor Dan SĂVESCU

Peer review statement

All papers published in this volume of *IOP Conference Series: Materials Science and Engineering* have been peer reviewed through processes administered by the proceedings Editors. Reviews were conducted by expert referees to the professional and scientific standards expected of a proceedings journal published by IOP Publishing.

IOP Conf. Series: Materials Science and Engineering **514** (2019) 012001 doi:10.1088/1757-899X/514/1/012001

Influence of the roughness on surface geometry of rolling contacts

M Benchea[1] and S Crețu[1]

[1]Mechanical Engineering, Mechatronics and Robotics Department, "Gheorghe Asachi" Technical University of Iaşi, Iaşi, Romania

E-mail: marcelin.benchea@tuiasi.ro

Abstract. The topography of the active surfaces in rolling contact depends on the surface roughness obtained by different manufacturing process technology (turning, grinding, polishing, lapping, etc.). The aim of this paper work is to evidence the influence of the surface roughness on surface geometry of rolling contacts. The experimental tests were carried out on the AMSLER machine and an experimental setup for pure rolling motion. The surfaces of the rolling samples were analyzed with SEM (scanning electron microscopy) and EDS (energy dispersive X-ray spectroscopy) techniques for structural and chemical analysis and measured with the Taylor Hobson profilometer before and after the experimental tests for surface topography.

1. Introduction

The topography of the active surfaces in rolling contact depends on the surface manufacturing process technology. In the field of engineering the exact degree of roughness can be of considerable importance, affecting the function of a component or its cost [1, 2]. The presence of nano-asperities on contact surfaces influence the running-in process of important machine parts [3], such as rolling bearings, gears, cam-cam followers unit, wheel-rail subassembly, etc.

The roughness of the surfaces that interacts has an important role on its tribological behaviour. The solutions of the models used in order to describe it [4, 5] require the input of a three-dimensional rough profile. Since the experimental acquisition of such profiles it does not allow a sufficiently large sample of profiles to be used, an essential requirement for any numerical algorithm is to generate arbitrarily rough surfaces with the same or similar proprieties of the real surfaces [6, 7].

When two engineering surfaces are loaded for the first time changes in the topography of both surfaces generally occur. These changes appear between start-up and steady state and are associated with running-in process [8, 9].

An analysis model was used to model the nonlinear strain rate dependent deformation of materials stressed in elastic-plastic domain [10, 11]. The model is developed in the frame of incremental theory of plasticity using the von Mises yield criterion and Prandtl-Reuss equations. Considering the isotropic and non-linear kinematic hardening laws of Lemaitre-Chaboche [12, 13] the model accounts for the cyclic hardening phenomena. The cyclic evaluation process of both the plastic strains and residual stresses is performed until the material shakedowns [10, 11].

Recently, to characterize the chemical and structural modifications of superficial layer on the atomic or molecular level of materials, many researchers use different techniques such as X-ray photoelectron spectroscopy (XPS), energy dispersive X-ray spectroscopy (EDS), transmission electron microscopy (TEM) or scanning electron microscopy (SEM), [14, 15]. In this paper we used SEM and EDS techniques for structural and chemical analysis of the rollers surface before and after experimental tests.

2. Experimental setup

To evaluate the running-in process and the surface roughness modification for the case of two steel rollers, were carried out experimental tests on two rollers machine (AMSLER), figure 1.

Figure 1. The AMSLER machine.

The two rollers running against each other with different rotational speed, having the constant rotational speed ratio $k = n_2/n_1 = 0.906$, where n_1 and n_2 are rotational speed of the lower and upper roller respectively. The slide-to-roll ratio ξ is determined by the following relation:

$$\xi = \frac{2\cdot(v_1 - v_2)}{v_1 + v_2} = 2\cdot\frac{(1 - k\cdot D_2/D_1)}{(1 + k\cdot D_2/D_1)} \qquad (1)$$

where D_1 and D_2 are the lower and upper roller diameters, respectively.

Two rollers with diameters of *53.5 mm* and *59 mm* made from 42CrMo4 steel were used for pure rolling motion ($\xi = 0$). The lower roller with diameter of *53.5 mm* it is cylindrical and the upper roller with diameter of *59 mm* it is crowned with radius of *300 mm*, both rollers have *10 mm* in length.

Testing conditions: normal load $F = 140\ N$, rotational speed test $n_1 = 181\ rpm$ and running time *10 min*.

3. Numerical rough surface generation

Every surface has some form of texture topography which varies according to its structure and the way it has been manufactured. According to [2] there are three categories of roughness profile parameters: amplitude parameters (the vertical characteristics of the surface deviations), spacing parameters (the horizontal characteristics of the surface deviations) and hybrid parameters. Definition of all roughness parameters are given in ISO standards [16, 17]. The most used roughness parameters in technical drawing sheets for a given workpiece are R_a (arithmetic mean height) or R_z (mean height for the five highest peaks and the five lower valleys from roughness profile); i. e. of other amplitude parameters: R_q - root mean square deviation, R_t - the maximum peak to valley height.

It was developed a numerical method for the generation of arbitrarily defined three-dimensional rough surface, [18]. The surface roughness heights can be generated according to the following relation:

$$z_{ij} = \sum_{k=1}^{n} \sum_{l=1}^{m} \alpha_{kl} \cdot \eta_{i+k,j+l} \tag{2}$$

where α_{kl} are coefficients that provide the desired autocorrelation function and $\eta_{i+k,j+l}$ are independent random numbers and have unit variance [6, 18].

$$E(\eta_{ij}\eta_{kl}) = \begin{cases} 1 \text{ if } & i = k, j = l \\ 0 \text{ if } & i \neq k, j \neq l \end{cases} \tag{3}$$

Using eqs. (3) the autocorrelation function is obtained as:

$$R_{pq} = \sum_{k=1}^{n-p} \sum_{l=1}^{m-q} \alpha_{kl} \cdot \alpha_{k+p,l+q} \tag{4}$$

Eqs. (4) represent a non-linear system of *nxm* equations with the coefficients α_{kl} as unknowns, [6]. An iterative technique has been used to solve the non-linear equations system (4) and the obtained solutions were used further in eq. (2) to obtain the roughness heights with the desired autocorrelation function.

When two rough surfaces are into contact it can be generated an equivalent rough surface [18], obtained by summing corresponding values of the roughness heights:

$$z = z_1 + z_2 \tag{5}$$

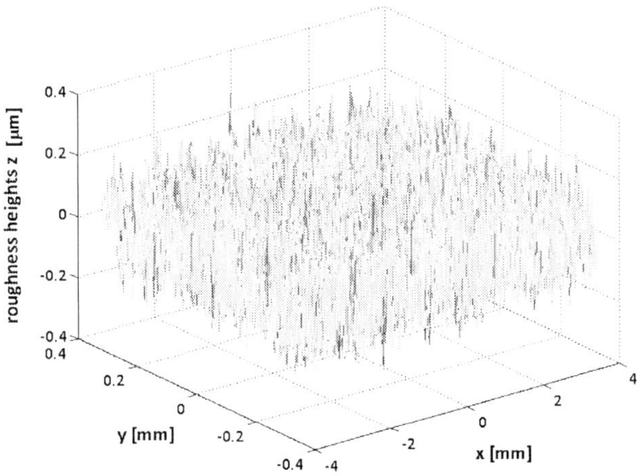

Figure 2. Arbitrarily generated rough surface [6, 18].

4. Experimental and numerical results

For structural and chemical analysis of the rollers surface before and after experimental tests it was used the scanning electron microscope with EDAX detector. In figure 3 is presented a typical EDS spectrum obtained for D_2 surface roller after experimental test.

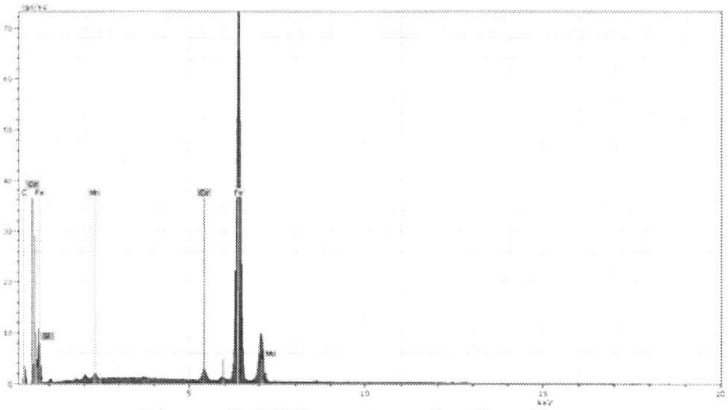

Figure 3. EDS spectrum for D_2 roller.

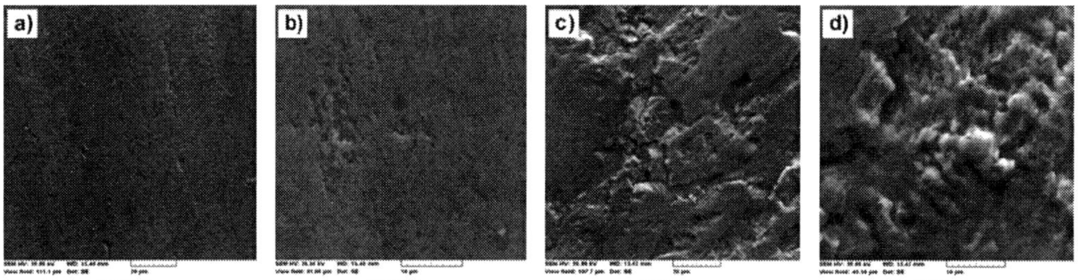

2000x 5000x 2000x 5000x

Figure 4. Surface SEM images after experimental test: D_2 roller (a, b) and D_1 roller (c, d).

Figure 4 shows microstructure of the running-in track into the rollers surface after tests on AMSLER machine, were can be observed a smoother surface for the upper roller D_2 with minor changes and a grossly rough surface for the lower roller D_1 with the flattened roughness peaks due to the rolling process.

Table 1. Chemical composition of the running-in track into the rollers surface (wt. %).

42CrMo4 steel rollers	Si	Mn	Cr	Mo	C	Fe
D_1 before	*0.385*	*0.891*	*1.199*	*0.298*	*0.415*	*balance*
D_1 after	*0.378*	*0.909*	*1.218*	*0.305*	*0.421*	*balance*
D_2 before	*0.386*	*0.889*	*1.198*	*0.299*	*0.418*	*balance*
D_2 after	*0.391*	*0.872*	*1.188*	*0.287*	*0.409*	*balance*

Table 1 presents the EDS results, showing the minor changes in the atomic composition of the running-in track into the rollers surface. These were measured before and after the rolling tests on AMSLER machine.

In figure 5 are presented the roughness profiles for the two rollers tested on the AMSLER machine, before and after the running test. The initial roughness of the rollers was $R_{a1} = 460\ nm$ and $R_{a2} = 220\ nm$, respectively $R_{q1} = 610\ nm$ and $R_{q2} = 280\ nm$.

The measurements of the rollers roughness before and after the tests evidenced that important modifications were produced for the lower roller, with a grossly roughness, for the approximate length of *1 mm* in the contact area. After the running test the average roughness height of the lower roller decreases to $R_{a1} = 390\ nm$ and root mean square deviation to $R_{q1} = 550\ nm$.

IOP Conf. Series: Materials Science and Engineering **514** (2019) 012001 doi:10.1088/1757-899X/514/1/012001

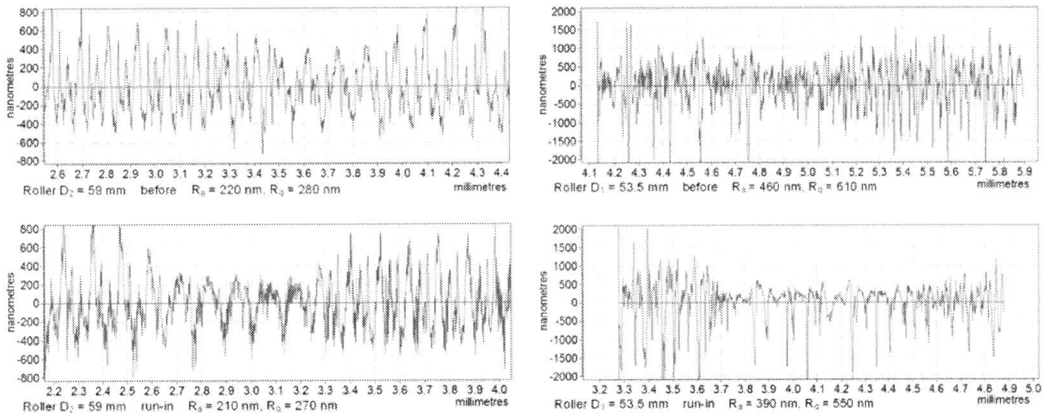

Figure 5. Experimental 2D roughness profiles before and after running test.

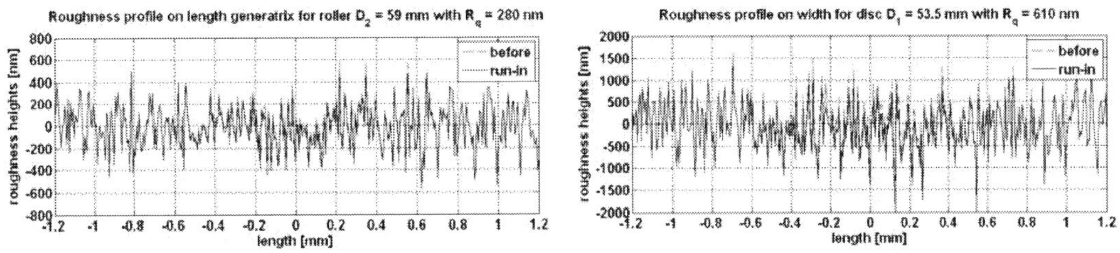

Figure 6. Numerical 2D roughness profiles before and after *300* running cycles.

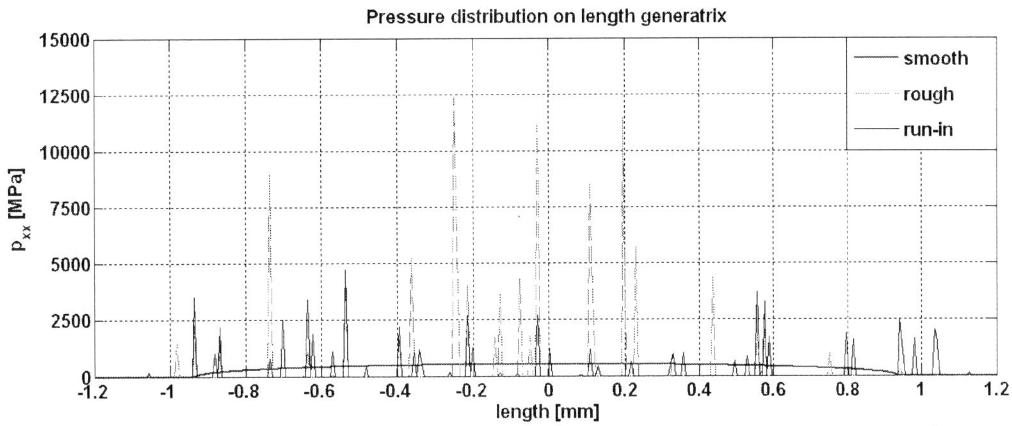

Figure 7. Pressure distribution on length generatrix of the rollers.

The same tendency was also observed with the elastic-plastic model after *300* running cycles, figure 6. The numerical results are in very good concordance with experimental test and with numerical results provided by running-in model presented in [8, 9].

The plastic modification of the surface roughness is given by the fact that the nano-asperities on the surface generates high peaks in the pressure distribution on the contact area between the two rollers, figure 7, pressure peaks which are a few times higher than the yield limit and locally produces plastic deformation of the nano-asperities. Also, if the surfaces will have a grossly roughness the changes will be more pronounced than for the surfaces with finer roughness and will diminish the higher peaks from the pressure distribution.

5. Conclusions

The topography of the active surfaces in rolling contact depends on the surface manufacturing process technology. Experimental and numerical results presented in this paper evidence the fact that the presence of nano-asperities on contact surfaces influence the running-in process and durability of a given application [3].

It was used a numerical method to generate an arbitrarily defined three-dimensional rough surface and an elastic-plastic model to take into account the running-in process.

The experimental results obtained on AMSLER machine present a very good concordance with the numerical results for two rollers case and with numerical results provided by running-in model presented in [8, 19].

The presence of nano-asperities on contact surfaces generates peaks in pressure distributions, higher peaks for surfaces with grossly roughness. For surfaces with grossly roughness the running-in process is more pronounced than for surfaces with a finer roughness.

6. References

[1] Robbe-Valloire F 2001 Statistical analysis of asperities on a rough surface *Wear* **249** pp 401-408
[2] Taylor Hobson Precision 2003 *Exploring surface texture* (TH Ltd)
[3] Stawicki T, Sedlak P, Koniuszy A 2010 The testing of the influence of the roughness of the crankshaft journal upon the durability of the crankshaft bearing in engines of agricultural machines *Sci. probl. of mach. op. and maint.* **4** pp 7-17
[4] Allwood J, Ciftci H 2005 An incremental solution method for rough contact problems *Wear* **258** pp 1601-1615
[5] Allwood J 2005 Survey and performance assessment of solution methods for elastic rough contact problems *J. Tribol.* **127** pp 10-23
[6] Bakolas V 2003 Numerical generation of arbitrarily oriented non-gaussian three-dimensional rough surfaces *Wear* **254** pp 546-554
[7] Patrikar RM 2004 Modeling and simulation of surface roughness *App. Surf. Sci.* **228** pp 213-220
[8] Jamari J 2006 *Running-in of rolling contacts* (PhD Thesis Twente)
[9] Jamari J, de Rooij MB and Schipper DJ 2007 Plastic deterministic contact of rough surfaces J. *Tribol.* **129** pp 957-962
[10] Benchea M, Iovan-Dragomir A, Creţu S 2014 Misalignment effects in cylindrical roller bearings *App. Mech. and Mat.* **658** pp 277-282
[11] Creţu S, Benchea M, Iovan-Dragomir A 2016 On basic reference rating life of cylindrical roller bearings. Part II - Elastic-Plastic Analysis *J. Balk. Tribol. Assoc.* **22/1** pp 272-280
[12] Chaboche JL 2008 A review of some plasticity and viscoplasticity constitutive theories *Int. J. Plasticity* **24** pp 1642-1693
[13] Besson J, Cailletaud G, Chaboche JL, Forest S and Blétry M 2010 *Non-linear mechanics of materials* (Springer)
[14] Guo T, Kong C, Li X, Guo P, Wang Z, Wang A 2017 Microstructure and mechanical properties of Ti/Al co-doped DLC films: Dependence on sputtering current, source gas, and substrate bias *App. Surf. Sci.* **410** pp 51-59
[15] Yuan Y, Li Z 2017 Microstructure and tribology behaviors of in-situ WC/Fe carbide coating fabricated by plasma transferred arc metallurgic reaction *App. Surf. Sci.* **423** pp 13-24
[16] ISO 4287:1996, Geometrical Product Specifications (GPS) - Surface texture: Profile method - Terms, definitions and surface texture parameters
[17] ISO 4288:1996, Geometrical Product Specifications (GPS) - Surface texture: Profile method - Rules and procedures for the assessment of surface texture
[18] Creţu S 2009 *Elastic-plastic concentrated contact* (Politehnium)
[19] Jamari J, Schipper DJ 2007 Deformation due to contact between a rough surface and a smooth ball *Wear* **262** pp 138-145

IOP Conf. Series: Materials Science and Engineering **514** (2019) 012002 doi:10.1088/1757-899X/514/1/012002

Employment of hyper-cycloidal oscillatory motion for finding the coefficient of rolling friction. Part 1: Theoretical model

S Alaci[1], R-D Pentiuc[2], F C Ciornei[1], I C Romanu[1], L Irimescu[1] and F Buium[3]

[1]Mechanics and Technologies Department, Stefan cel Mare University of Suceava, Romania
[2]Electrotechnics Department, Stefan cel Mare University of Suceava, Romania
[3]Mechanical Engineering, Mechatronics and Robotics Department, Gheorghe Asachi Technical University of Iasi, Romania

E-mail: alaci@fim.usv.ro , florina@fim.usv.ro

Abstract. The oscillation motions of physical pendula constructed with contacts between ball and cylindrical external surfaces for materializing the theoretical frictionless pivot point were applied for determining the coefficient of rolling friction (occurring as a characteristic of the relative motion between the contacting surfaces). The condition imposed for estimation of the coefficient of rolling friction consists in the requirement that the theoretical curve and the experimental data describing the decrease of angular amplitude should superpose each other. To be mentioned that specialized software for dynamic analysis characterizes the rolling friction by the internal damping of the contacting materials instead of applying the coefficient of rolling friction between the materials.

1. Introduction

The mechanisms met in applications can be categorized into three main classes [1]: linkage, cam and gear mechanisms. One of the major advantages brought by the presence of a higher pair is the significant decrease of the number of structural elements. To illustrate this affirmation, the problem of transmission of motion between two shafts with arbitrary directions is considered. This goal may be accomplished by designing a RCCC mechanism, as example [2] that contains cylindrical pairs and rotation and/or translation pairs. When for finding the solution, only fifth class pairs are preferred, the simplest structural solution includes seven binary elements joined by seven pairs of fifth class. From structural point of view, the problem is solved simpler if the higher pairs are accepted in the structure of the transmission mechanism. The rotation motion can be transmitted between two shafts of random direction by means of a higher pair of first class, as it is exposed in [3-4]. Another structural solution for motion transmission between the two shafts consists in using a first class pair of surface-surface type, met in the case of cylindrical gear mechanisms. The relative motion in a higher pair is complex and its effect is the presence of a friction torsor that has all components nonzero. For plane kinematic chains, the components of the friction torsor are: the friction force oriented along the tangent in the contact point and the moment of rolling friction, with the direction normal to the plane of motion.

Even for the plane kinematic chains, the motion between the generatrix curves of the contacting surfaces is extremely complex, depending directly on the nature of the contacting curves. As principle, the curves with best technological constraints are the circle and the straight line. In [5] is presented a technique for replacing the uphill and downhill profiles of a rotating cam with circular arcs and the effect if this substitution upon the law of motion of a translating knife-edge follower. The use only of

IOP Conf. Series: Materials Science and Engineering **514** (2019) 012002 doi:10.1088/1757-899X/514/1/012002

circles and lines as generatrix curves for the contacting surfaces substantially diminishes the possible relative motions between these. Practically, when one of the curves is maintained immobile, the points from the other one may describe five types of curves, resumed in [6]. One of the feasible situations is the outer contact between a fixed circle and a mobile circle that rolls without sliding over the first. The points attached to the mobile circle describe a family of hyper-cycloids whose shapes depend on the relative position of the point with respect to the mobile circle. In a series of recent works, [7-9] different types of pendula for which the contacting surfaces are combinations of cylindrical and/or plane surfaces were studied, with the objective of finding the value of the rolling friction torque. All these models are based on the principle of equality between the experimental damping and the theoretical decrease of amplitude.

2. The hyper-cycloidal pendulum. Characteristics and equation of motion

The hyper-cycloidal pendulum is presented in figure 1. A mobile circle (γ) of r radius is brought into contact with a fixed circle (Γ) of R radius, on the outer side.

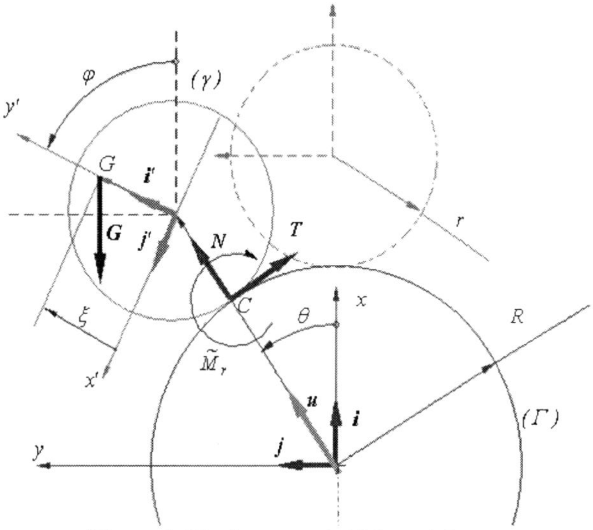

Figure 1. The hyper-cycloidal pendulum.

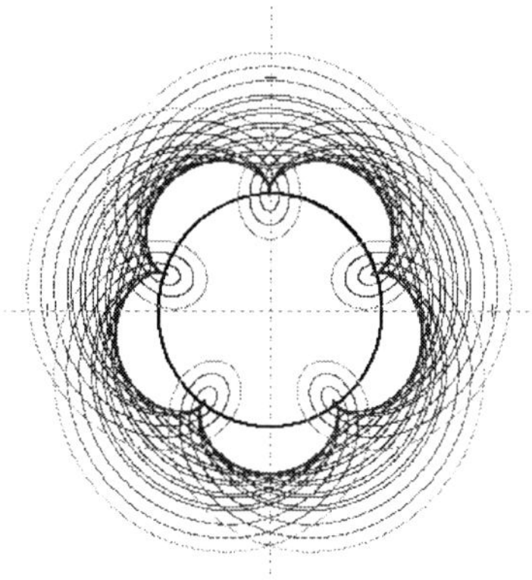

Figure 2. Family of hyper-cycloids.

The position of the mobile circle is characterized using a Cartesian coordinate system, attached to the circle, with the origin in the center of it. The $O'x'$ axis passes through the point C - the center of mass of the pendulum, positioned at a distance ξ with respect to the center of the mobile circle. For different ξ values, a family of hyper-cycloids can be generated, as in figure 2. The position of the mobile circle is described by two parameters: the angle θ of the vector radius of the center of the mobile circle with respect to the fixed frame and the angle φ made by the axes of the mobile system with the axes of the fixed frame.To conclude, for the general case, the pendulum has two degrees of freedom. Concerning the external forces, the single force is the own weight $\boldsymbol{G}$. The torsor of the reactions in the point C contains the normal reaction $\boldsymbol{N}$ oriented along the radius of the contact point, the friction force $\boldsymbol{T}$, tangent to the two circles in the contact point and the moment of rolling friction, normal to the plane of motion and oriented as to oppose to the relative rotation between the two circles. The versors of the mobile frame are denoted by $\boldsymbol{i}',\boldsymbol{j}'$ and the versor of the vector radius of the contact point is $\boldsymbol{u}$, as seen in figure 1. In order to establish the motion of the pendulum, there are applied: the theorem of motion of the center of mass:

$$M\ddot{\boldsymbol{r}}_G = \boldsymbol{G} + \boldsymbol{N} + \boldsymbol{T} \tag{1}$$

and the moment of momentum theorem with respect to the center of mass:

$$J_z \ddot{\varphi} \boldsymbol{k} = \overline{GC} \times (\boldsymbol{N} + \boldsymbol{T}) + \widetilde{M}_r \boldsymbol{k} \tag{2}$$

From the equations (1) and (2) it result two and one projection equations, respectively and thus, three scalar equations are available. The unknowns of the problem are the angles θ and φ, necessary for the characterization of motion and the values of the components of the reaction torsor in the contact point C, denoted N, T and M_r. So, two supplementary scalar equations are needed. One of these equations is get assuming that the relation between the moment of rolling friction and the normal reaction N is known. Though recent researches, based on the theory of elasticity [10] demonstrate a power law dependency, for the simplicity of the calculus, a linear dependence is accepted:

$$M_r = s_r N \tag{3}$$

The above relation is accepted in most of the cases [11-13]. Concerning the last equation necessary for compatibility reasons, it depends on the hypothesis assumed for the relative motion between the two circles, in the point C. Consequently, if $v_{relC} = 0$ is considered, situation characteristic to pure rolling, there is a relation between the angles θ and φ, and the system has one degree of freedom. The force T is unknown as value and the pure rolling condition must be fulfilled:

$$T / N < \mu_d \tag{4}$$

where μ_d is the dynamic coefficient of sliding friction. On contrary case, when sliding exists between the two surfaces, the friction force attains the limit value:

$$T = \mu_d N \tag{5}$$

and there is no connection between the two angles θ and φ any more.

3. Deducing the equation of motion of the pendulum for the pure rolling case

With the circle Γ considered immobile, the condition of pure rolling between the two circles is expressed by cancelling the velocity of the point of contact C from the mobile circle:

$$\boldsymbol{v}_{C_\gamma} = \dot{\theta} \boldsymbol{k} \times (R + r) \boldsymbol{u} + \dot{\varphi} \boldsymbol{k} \times (-r \boldsymbol{u}) = 0 \tag{6}$$

From equation (6) it results:

$$\dot{\theta} = \dot{\varphi}[r / (R + r)] \tag{7}$$

The equation (7) is integrated with the hypothesis that at the initial moment the orientations of the two axes were identical and it results:

$$\theta(\varphi) = r\varphi / (R + r), \tag{8}$$

that is the equation for the relationship between the two angles. The parametric equations of the trajectories of the center of mass of the pendulum are expressed by:

$$\begin{cases} x = (R + r)\cos\theta(\varphi) + \xi\cos\theta \\ y = (R + r)\sin\theta(\varphi) + \xi\sin\theta \end{cases} \tag{9}$$

For different values of the parameter ξ the equations (9) represent a family of hyper-cycloids. By differentiating twice with respect to time the relations (9) the components of the acceleration of the center of mass with respect to the fixed system are obtained:

$$a_{Gx} = -\left[sin\theta + \frac{\xi}{r} sin\left(\frac{r+R}{r}\theta \right) \right] \ddot{\theta}(R+r) - \left[cos\theta + \frac{\xi}{r} cos\left(\frac{r+R}{r}\theta \right) \right] \dot{\theta}^2 (R+r) \tag{10a}$$

$$a_{Gy} = \left[cos\theta + \frac{\xi}{r} cos\left(\frac{r+R}{r}\theta \right) \right] \ddot{\theta}(R+r) - \left[sin\theta + \frac{\xi}{r} sin\left(\frac{r+R}{r}\theta \right) \right] \dot{\theta}^2 (R+r) \tag{10b}$$

In order to apply the equation (1), the vectors from the left member of the vector equation (1) are expressed via the projections on the axes of the fixed frame:

$$\boldsymbol{G} + \boldsymbol{N} + \boldsymbol{T} = \begin{bmatrix} -Mg \\ 0 \\ 0 \end{bmatrix} + N \begin{bmatrix} cos\theta \\ sin\theta \\ 0 \end{bmatrix} + \begin{bmatrix} sin\theta \\ -cos(\theta) \\ 0 \end{bmatrix} \tag{11}$$

So as to apply the moment of momentum theorem (2), in addition to vector expressions of the normal force $\boldsymbol{N}$ and friction force $\boldsymbol{T}$, the expression of the vector $\overline{GC}$ that specifies the position of the contact point C with respect to the centre of mass of the pendulum G, is necessary:

$$\overline{GC} = \boldsymbol{r}_C - \boldsymbol{r}_G = \begin{bmatrix} R\,cos\theta \\ R\,sin\theta \\ 0 \end{bmatrix} - \begin{bmatrix} (R+r)cos\theta + \xi\, cos[(R+r)\theta / r] \\ (R+r)sin\theta + \xi\, sin[(R+r)\theta / r] \\ 0 \end{bmatrix} \tag{12}$$

together to the expression of the vector of rolling friction torque:

$$\boldsymbol{M}_r = -\boldsymbol{k}\, s_r N\, sgn(\dot{\theta}) \tag{13}$$

where $sgn(x)$ represents the signum function. The equation (13) shows that the moment of friction is proportional to the value of the normal reaction N and always has opposite sense to the vector of angular velocity. The above relations permit finding the differential equation of motion of the pendulum:

$$\ddot{\theta} = \frac{\left(1 + \frac{R+r}{r} \cdot \frac{\xi}{r} cos\frac{R\theta}{r} \right) s_r\, sgn(\dot{\theta}) + \frac{\xi}{r} R\, sin\frac{R\theta}{r}}{\frac{J_z + Mr^2}{r} - M\frac{\xi}{r} s_r\, sgn(\dot{\theta}) sin\frac{R\theta}{r} + M\xi\left(\frac{\xi}{r} + 2cos\frac{R\theta}{r} \right)} M\dot{\theta}^2 + \frac{\left(r\,sin\theta + \xi\, sin\frac{(R+r)\theta}{r} - s_r\, sgn\dot{\theta}\, cos\theta \right)}{\left[\frac{J_z + Mr^2}{r} - M\frac{\xi}{r} s_r\, sgn(\dot{\theta}) sin\frac{R\theta}{r} + M\xi\left(\frac{\xi}{r} + 2cos\frac{R\theta}{r} \right) \right](R+r)} Mg \tag{14}$$

The equation (14) is a nonlinear differential equation. After the numerical integration of the equation for stipulated initial conditions, the friction force T and the normal reaction N must be found, since they are necessary for confirming the pure rolling condition (4).

$$T = -M(R+r)\left[1 + \frac{\xi}{r} cos\left(\frac{R}{r}\theta \right) \right] \ddot{\theta} + M(R+r)^2 \frac{\xi}{r^2} \dot{\theta}^2\, sin\left(\frac{R}{r}\theta \right) + Mg\, sin\theta \tag{15}$$

$$N = -M(R+r)\frac{\xi}{r}\ddot{\theta}\sin\left(\frac{R}{r}\theta\right) - M(R+r)\left[1+\frac{\xi}{r}\cos\left(\frac{R}{r}\theta\right)\right]\dot{\theta}^2 + Mg\cos\theta \tag{16}$$

In [14-15] it is underlined that for the case of damped oscillations under dry friction condition, a condition necessary for validation of the obtained equation is the linear decrease of amplitude. To test the fulfilment of this stipulation, the equation (14) was integrated numerically using the Runge-Kutta IV, considering that the pendulum is launched from resting at an initial angular amplitude. For the parameters occurring in equation (14) a set of values of the parameters were considered: $R = 0.5m; r = 0.1m; \xi = -0.033m$; $M = 36.757\,Kg; J_z = 0.461\,Kg \cdot m^2$; $s_r = 5\mu m$. In figures 3 and 4 there are presented the variation of the angular elongation and the variation of the coefficient of static friction for two values of initial amplitude. From these figures it can be observed that for small oscillations, the ratio T/N takes values lower than 0.2 that is the limit value of the coefficient of sliding friction and as consequence, the angular amplitude presents linear decrease. When the launching amplitude has appreciable values, in the initial phase of the motion the pure rolling condition is disobeyed and as result, the angular amplitude presents a nonlinear diminishment. An interesting aspect is that towards the end of the motion, the pure rolling condition is satisfied and therefore the decrease of amplitude becomes linear. This aspect could be used in estimating the limit value of the dynamic coefficient of sliding friction.

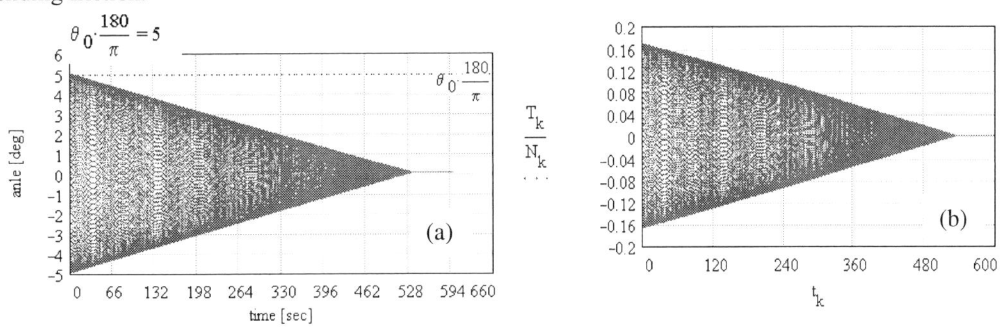

Figure 3. Variation of angular elongation (a) and variation of the coefficient of static friction (b) for the initial amplitude $\theta_0 = 5^o$.

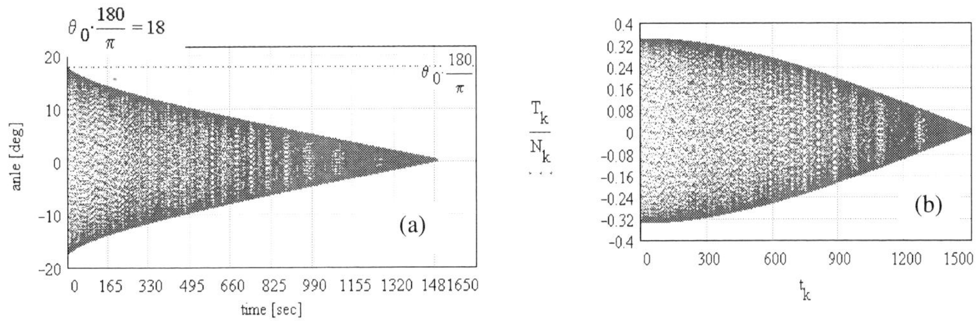

Figure 4. Variation of angular elongation (a) and of the coefficient of static friction (b) for $\theta_0 = 18^o$.

4. Conclusions

The paper presents a method and the related device used for establishing the coefficient of rolling friction. The instrument is a pendulum with external cylindrical surface as contact surface. It contacts a fixed cylindrical surface and thus the two cylinders make contact over a generatrix. When there is no sliding between the two cylindrical bodies, the points from the mobile one describe a family of hyper-

IOP Conf. Series: Materials Science and Engineering **514** (2019) 012002 doi:10.1088/1757-899X/514/1/012002

cycloids and from hereby the name of the technique. The fundamental theorems of vector mechanics are applied under the assumption of pure rolling conditions and the differential characteristic equation of motion of the pendulum is obtained. Once this law established, the characteristics of the reaction torsor in the contact between the two cylinders can be found. The normal reaction force and the tangential one permit testing if the condition of pure rolling between the two cylinders is valid. For a set of initial parameters, the equation of motion was numerically integrated and a linear damping of the amplitude of the pendulum was observed. The condition that the theoretical model and the experimental pendulum should have identical decreases of angular amplitude is the base of finding the coefficient of rolling friction. The comparison between the results of the theoretical model and the experimental data are the subject of the second part of the paper.

Acknowledgment: This work was supported by a grant of the Romanian Ministry of Research and Innovation, CCCDI – UEFISCDI, project number PN-III-P1-1.2-PCCDI-2017-0404 / 31PCCDI / 2018, within PNCDI III.

References

[1] Uicker J, Pennock G and Shigley J 2010 *Theory of Machines and Mechanisms* (Oxford Univerity Press)

[2] Yang A T and Freudenstein F 1964 Application of dual-number quaternion algebra to the analysis of spatial mechanism *ASME Journal of Applied Mechanics* **86** pp 300–308

[3] Alaci S, Ciornei F-C, Amarandei D, Băeşu M and Mihăilă D 2017 On direct coupling of two shafts. Part 1: Structural consideration Tehnomus Journal **24** pp 277-280

[4] Ciornei F C, Doroftei I, Alaci S, Prelipcean G, Dulucheanu C 2018 Analytical kinematics for direct coupled shafts using a point-surface contact, *Proceedings of ACME 2018 TU Iasi International Conference*, (to be published)

[5] Ciornei F-C and Alaci S 2014 The Use of Circular Arc Cams for the Command of a Robotic System. Part I: Theoretical Considerations, Robotica and Management **19**(2) pp 3-6

[6] Handra-Luca V, Stoica I A 1983 *Introducere în teoria mecanismelor* **Vol. 1** (Editura Dacia Cluj Napoca)

[7] Ciornei M C, Alaci S, Ciornei F C, Romanu I C 2017 A method for the determination of the coefficient of rolling friction using cycloidal pendulum *IOP Conf. Series Mater. Sci. Eng.* **227** – 012027

[8] Alaci S, Ciornei F C, Ciogole A, Ciornei M C 2017 Estimation of coefficient of rolling friction by the evolvent pendulum method, *IOP Conf. Series Mater. Sci. Eng.* **200** – 012005

[9] Alaci S, Ciornei F C, Bujoreanu C, Ciornei M C and Acsinte I L 2018 Finding the coefficient of rolling friction using a pericycloidal pendulum, *Proceedings of ACME 2018 TU Iasi International Conference*, (to be published)

[10] Cherepanov G P 2014 Theory of rolling: solution of the Coulomb problem *Journal of Applied Mechanics and Technical Physics,* **55(1)** 182-189

[11] Voinea R, Voiculescu D and Ceauşu V 1989 *Introducere în mecanica solidului cu aplicaţii în inginerie* (Editura Academiei Republicii Socialiste România Bucureşti)

[12] Kane T R and Levinson D 1985 *Dynamics: Theory and Applications* (McGraw-Hill New York.)

[13] Hibbler R C 2012 *Engineering Mechanics: Dynamics* (Prentice Hall)

[14] Torzo G and Pearnzoni P 2009 The real pendulum: theory, simulation, experiment *Lat. Am. J. Phys. Educ.* **3**(2) pp 221-228

[15] Zonetti L F C, Camargo S S, Sartori J, de Sousa D F and Nunes L A O 1999 A demonstration of dry and viscous damping of an oscillating pendulum, *Eur. J. Phys.* **20** pp 85–88

IOP Conf. Series: Materials Science and Engineering **514** (2019) 012003 doi:10.1088/1757-899X/514/1/012003

Employment of hyper-cycloidal oscillatory motion for finding the coefficient of rolling friction. Part 2: Experimental investigation

F C Ciornei[1], R-D Pentiuc[2], S Alaci[1], I C Romanu[1], L Irimescu[1] and F Buium[3]

[1]Mechanics and Technologies Department, Stefan cel Mare University of Suceava, Romania
[2]Electrotechnics Department, Stefan cel Mare University of Suceava, Romania
[3]Mechanical Engineering, Mechatronics and Robotics Department, Gheorghe Asachi Technical University of Iasi, Romania

E-mail: florina@fim.usv.ro

Abstract. The paper presents a method for finding the characteristics of rolling friction between two cylindrical external surfaces. By choosing the motion between two parallel cylinders it results that a point from one of the cylinders describes a hyper-cycloidal curve when rolling over the other cylinder. The study is focused on the friction in ball bearings. Parts of the ball bearings to be studied are directly involved in materializing the hyper-cycloidal contact. Thus, the immobile circles are materialized by the outer surface of the external ring and the mobile circles are materialized by the spherical surfaces of the balls. The motion of a laser stick attached to the pendulum is filmed. Analyzing the movie, the numerical time dependence of angular amplitude decrease is obtained. The concordance between the experimental and theoretical angular amplitude damping directs to the value of coefficient of rolling friction.

1. Introduction

In the first part of the paper the differential equation of motion of a hyper-cycloidal pendulum was obtained. The pendulum has a plane-parallel motion, the contacting surfaces being two cylinders: a fixed one with radius R and a mobile one with radius r having in common the generatrix normal to the plane of motion. Accepting the hypothesis of pure rolling between the two cylinders, the differential equation of motion was deduced. In order to obtain the moment of rolling friction M_r and implicitly the coefficient of rolling friction s_r it is necessary, as for other types of pendula studied [1-2], the application of the theoretical model results to a real pendulum for which all the parameters needed to characterize the motion are specified. In this case, as shown in figure 1, these parameters are: the radii R and r of the two cylinders, the mass M and the principal and central moment of inertia J_z of pendulum and a parameter ξ characterizing the position of the pendulum center of mass with respect to the center of the mobile circle. One of the most difficult problems is to provide a pure rolling motion between two real cylinders, tangent external in the present case. The condition that the two cylinders make a linear contact along the generatrix is a difficult task to achieve because any deviation from parallelism of the two axes changes the linear contact into a theoretical point contact, [3], actually of an elliptical contact area. In conclusion, ensuring the parallelism between the two cylinders is a very difficult task.

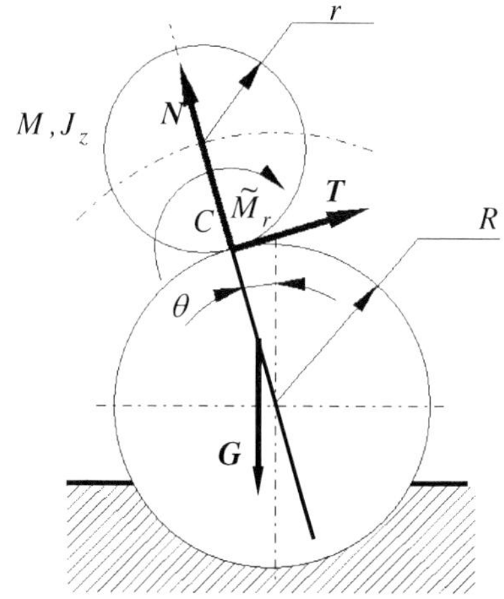

Figure 1. The characteristic parameters of the system.

2. Construction of actual pendulum and obtaining the experimental data

A first solution would be to replace the cylinder - plane contact with the contact between a fixed cylinder and a mobile sphere with the same radius as the mobile cylinder. In this way the contact area dimensions remain the same regardless of the position of the contact point. But another problem arises. In sphere-plane contact, the number of relative motions is greater than in the case of two-cylinders contact. Consequently, the additionally motions have to be canceled by joints that can introduce errors in the accuracy of the measured parameters. As a final solution, a cylinder similar to that used in [4] was applied, the mobile cylinder is materialized by two identical balls interlocked with a body of revolution coaxial with the axis of centers of the two balls, figure 2. The pendulum used in the experiments is shown in figure 2.

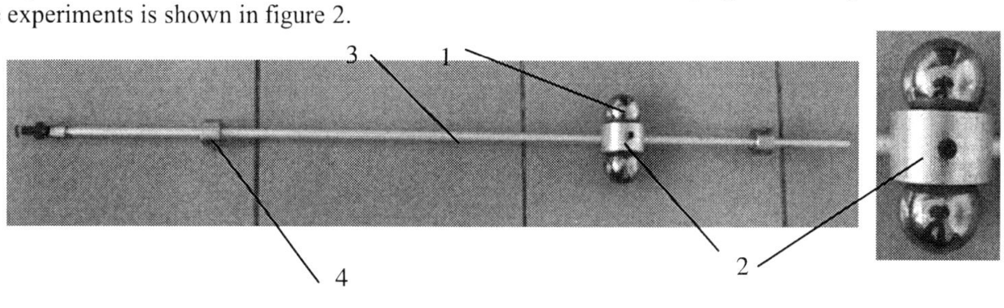

Figure 2. The pendulum used.

As can be seen in figure 2, the two balls *1* are locked by means of a revolution body *2*. In this body is also fixed the pendulum rod formed by an aluminum cylindrical rod *3*. Also, two identical cylindrical bodies *4* are coaxially mounted on the rod. Both the parts *4* and the ball-holder body *2* can be fixed in various positions, thus changing the inertial characteristics and implicitly the pendulum oscillation period. The fixed cylinder has been materialized by two identical outer bearings rings, figure 3. These are placed on the rectilinear edges of two square cross-section aluminum pipes, placed parallel in an horizontal plane, figure 2. Before putting into contact the balls with the bearing rings, it must be checked that the points of maximum height from the two rings are in the same horizontal plane. This is done using a dial gauge. After the contact between the pendulum and the two rings is made, the horizontality of the pendulum axis of oscillation should be checked in the same way. Figure 2c shows the overall view of the test-rig. At the lower end of the pendulum a laser stick is placed, the spot moving along a graduated ruler.

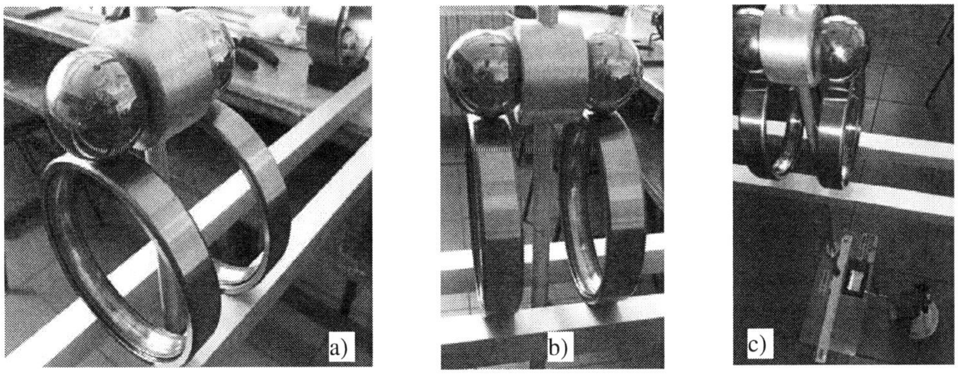

Figure 3. Materialization of the external contact by a two-cylinders generatrix.

The motion of the laser spot is recorded with a camera. The movie is split into frames, and the dependency is numerically registered.

$$x_k = x_k(t_k) \tag{1}$$

The experimental dependence is shown in figure 4. Since the theoretical model was deduced considering that the initial moment is $t = 0$ and that the reference position is even the pendulum equilibrium position, it is required the position x_0 of the spot around which the oscillation takes place.

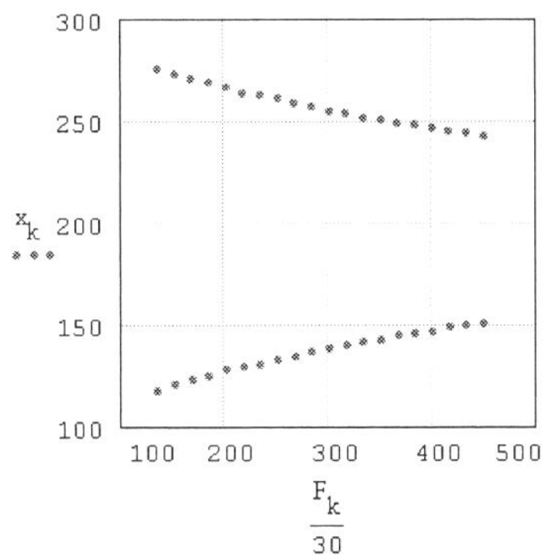

Figure 4. The variation with time of the displacement of the spot.

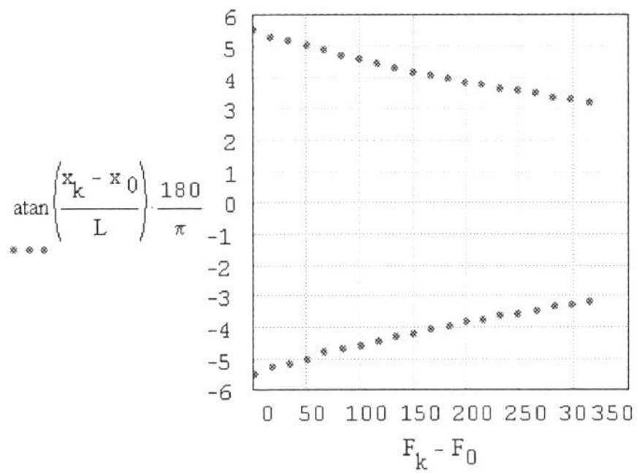

Figure 5. Determination of pendulum equilibrium position.

In order to find the reference position of the pendulum (spot), the condition of minimizing the derivative of the function is imposed, the sum being extended to all the experimental points.

$$H(x) = \sum_{k=1}^{n} \left| x - x_k \right|^2 \tag{2}$$

Figure 5 shows the time variation of the angular amplitude of the pendulum. It is noted with L the distance from the ruler and the point of contact (which in the hypothesis of the small oscillations can be considered constant). As in other work where the pendulum method was used to determine the rolling friction [5], a deviation of the experimental variation curve from a straight line was observed, as the theoretical model in the first part of the paper provides. This creates problems regarding the imposing the initial conditions for the theoretical model. Specifically, *the experimental data should be interpolated with a straight line whose slope must be the same as the slope of amplitude decrease of the theoretical model.* The approximation with a straight line can be done by optimizing both slope

IOP Conf. Series: Materials Science and Engineering **514** (2019) 012003 doi:10.1088/1757-899X/514/1/012003

and ordinate at the origin, figure 6a or only the slope figure 6b, the last interpolation ensuring that the theoretical and experimental models start from the same initial amplitude.

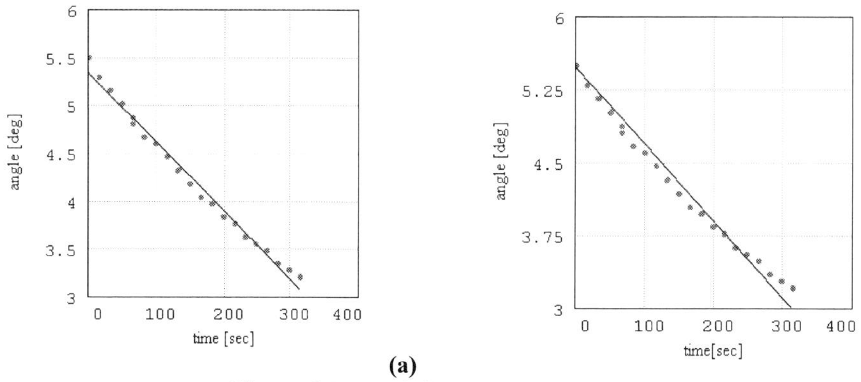

(a) **(b)**

Figure 6. Interpolation with slope optimization:
(a) with intersection point; (b) with imposed starting point.

If the geometric characteristics of the rings and balls, for example the pendulum mass, can be obtained by simple weighing, the central inertia moment and the center of mass position can not be obtained directly by measurement. For their determination, Kater's pendulum method is used as in [6].

3. Exemplifying the method

For the experimental pendulum, two balls with diameters $2r = 0.0421m$ were used, the final mass of the pendulum being $M = 1.1kg$. The outer diameter of the bearing rings has the value $2R = 0.11m$. The distance between the axes of the holes is $d = 0.3m$. The oscillation periods obtained are $T_1 = 1.465\,sec$ and $T_2 = 1.265\,sec$, the corresponding distances $d_1 = 0.446m$, $d_2 = 0.146m$ and a moment of central inertia $J_z = 0.041kg \cdot m^2$. The distance from the center of gravity to the center balls axis is $\xi \cong -0.063m$ (the equation describing the theoretical model were obtained considering $\xi > 0$). The oscillation period of the hyper-cycloidal pendulum (figure 6) is $T = 1.67\,sec$. Figure 7 presents the experimental data, the interpolation line of these in case of imposed starting point figure 6b, together with the solution of the differential equation deduced in the first part of the paper. The concordance between the experimental and theoretical data was obtained for a value of the coefficient of rolling friction $s_r = 12\mu m$. In the detail (figure 7b) the full concordance between the oscillation period of the real pendulum and the theoretical pendulum can be seen.

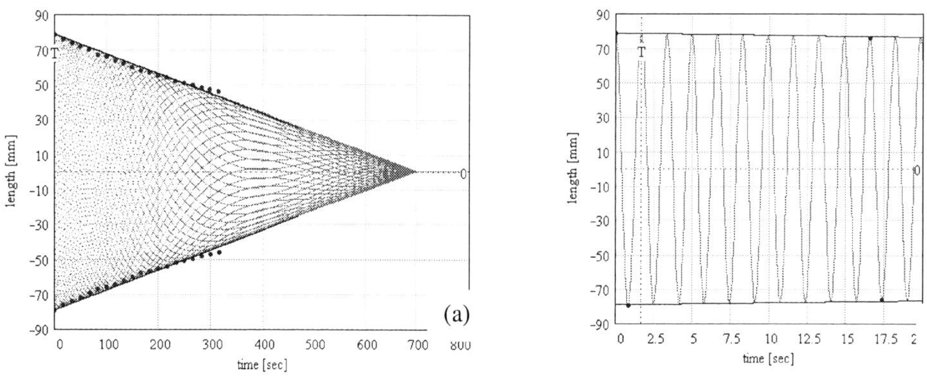

(b)

Figure 7. Experimental data and theoretical signal.

Knowing the law of pendulum motion, we can determine the dimensions of the reaction components from the contact point:

$$T = -M(R+r)\left(1+\frac{\xi}{r}\cos\frac{R\theta}{r}\right)\ddot{\theta} + M(R+r)^2\frac{\xi}{r^2}\sin\frac{R\theta}{r}\dot{\theta}^2 + Mg\sin\theta \tag{3}$$

$$N = -M(R+r)\frac{\xi}{r}\sin\frac{R\theta}{r}\ddot{\theta} - M(R+r)\left(1+\frac{\xi}{r}\cos\frac{R\theta}{r}\right)\dot{\theta}^2 + Mg\cos\theta \tag{4}$$

Figures 8 and 9 represent the variations in time of the frictional force T and normal reaction N, considered dimensionless by dividing to the weight Mg of the pendulum. The dependencies of these two forces allow the estimation of the static friction coefficient in the contact point; the variation is represented in figure 10. From this figure it can be seen that the condition necessary to have pure rolling is fulfilled. An immediate application of the current work is the possibility of modeling the rolling friction contacts with simulation software. For example, the MSC.ADAMS software, one of the most powerful dynamic analysis software, does not use the concept of coefficient of rolling friction.

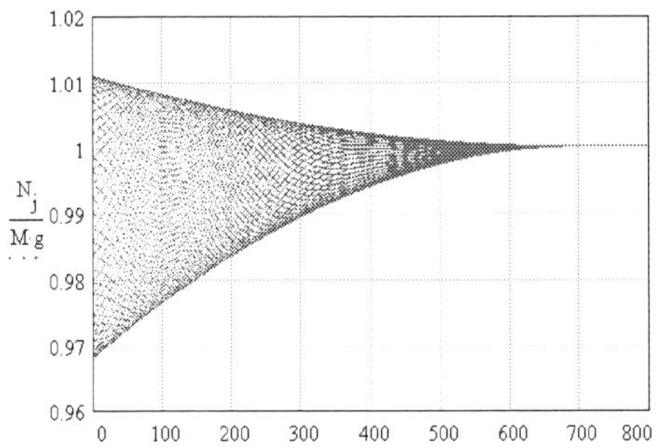

Figure 8. Variation of dimensionless friction force.

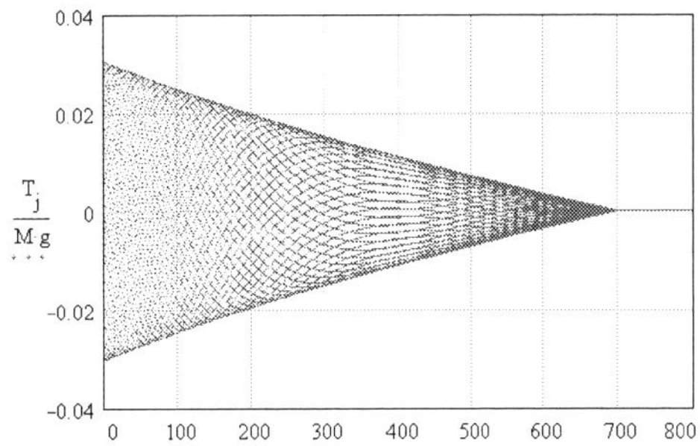

Figure 9. Variation of dimensionless normal force.

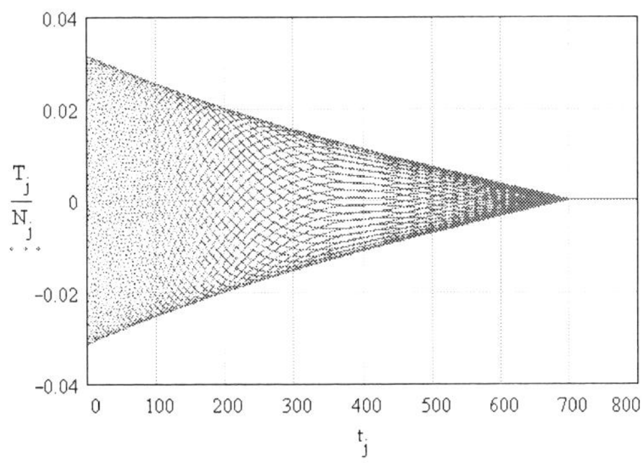

Figure 10. Variation of coefficient of friction in the contact.

In figure 11 there is presented a hyper-cycloidal pendulum modeled with this software (a), the decrease of the pendulum angular acceleration (b) and setting the parameters for contact modeling (c). The decrease of the angular acceleration is caused by the "damping" parameter without making any reference to the coefficient of rolling friction. Using the theoretical model developed in the first part of the paper and describing how to obtain the rolling coefficient of friction using experimental data, one can determine the relationship between the "damping" parameter of the software and the coefficient of rolling friction, thereby expanding the possibilities of using the software.

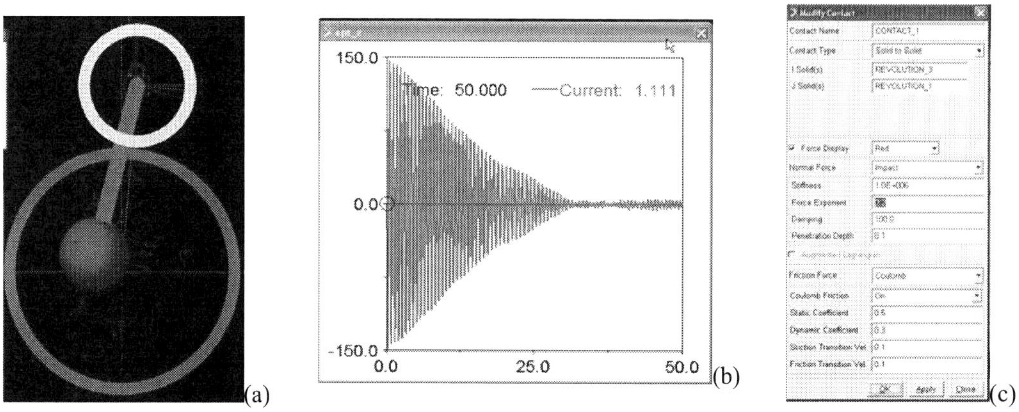

Figure 11. Modeling a hyper-cycloidal pendulum with the MSC.ADAMS software.

4. Conclusions

The work presents the method of obtaining the experimental data and how to obtain the value of the coefficient of rolling friction. Thus, the hyper-cycloidal pendulum was obtained using two identical ball bearings, connected to each other by a body of revolution. Perpendicular to the axis of the balls, at the middle of the distance between their centers, a rod that represents the body of the pendulum is fixed. In order to change the inertial characteristics of the pendulum, two cylindrical bodies are mounted on the rod, which can be fixed at different distances from the axis of the balls. The hyper-cycloidal motion of the pendulum is achieved by contacting the two balls with the outer surfaces of two identical outer bearing rings. This ensures that the axis of oscillation maintains a single direction, providing the pendulum with a plane-parallel motion. The analytical model requires the specification of the inertial characteristics of the pendulum. mass, mass center position and central moment of inertia. The last two parameters are determined using Kater's pendulum method, completing for this purpose two physical pendulums with oscillation axes parallel to the axis of the balls. Imposing the condition of identical linear damping rates of the angular amplitude of the theoretical pendulum and the actual pendulum, the value of the coefficient of rolling friction is obtained. Finally, with this value, the condition of pure rolling in the ball-ring contacts is verified. In addition to the direct application of obtaining the coefficient of rolling friction, the work may be useful to find the equivalence between the magnitude of the coefficient of rolling friction and the characteristic parameters of damping in the concentrated contacts when the damping is defined by other parameters, other than the coefficient of rolling friction.

Acknowledgement. This work was supported by a grant of the Romanian Ministry of Research and Innovation, CCCDI – UE FISCDI, project number PN-III-P 1-1.2-PCCDI-2017-0404 / 3 1PCCDI / 2018, within PNCDI III.

5. References

[1] Ciornei M C, Alaci S, Ciornei F C and Romanu I C 2017 A method for the determination of the coefficient of rolling friction using cycloidal pendulum *IOP Conf. Ser. MSE* **227** - 012027
[2] Alaci S, Ciornei F C, Ciogole A and Ciornei M C 2017 Estimation of coefficient of rolling friction by the evolvent pendulum method *IOP Conf. Ser. MSE* **200** - 012005
[3] Johnson K L 1985 *Contact Mechanics* (Cambridge University Press)
[4] Ciornei MC, Alaci S, Ciornei F C and Romanu IC 2017 A method for the determination of the coefficient of rolling friction using cycloidal pendulum *IOP Conf. Ser. MSE* **227** 012027
[5] Alaci S, Ciornei F C, Bujoreanu C Ciornei M C and Acsinte I L Finding the coefficient of rolling friction using a pericycloidal pendulum, *ACME 2018 Int. Conference, to be published*
[6] Baker G and Blackburn J 2005 *The Pendulum* (Oxford University Press)

IOP Conf. Series: Materials Science and Engineering **514** (2019) 012004 doi:10.1088/1757-899X/514/1/012004

An improved technique of finding the coefficient of rolling friction by inclined plane method

F C Ciornei[1], R D Pentiuc[2], S Alaci[1], I C Romanu[1], S T Siretean[1] and M C Ciornei[3]

[1]Mechanics and Technologies Department, Stefan cel Mare University of Suceava, Romania
[2]Electrotechnics Department, Stefan cel Mare University of Suceava, Romania
[3]Department 2 Physiology I, Carol Davila University of Medicine and Pharmacy, Bucuresti, Romania

E-mail: florina@fim.usv.ro

Abstract. During the design process the mechanical engineer faces a challenge: the choice of the tribological parameters. Even most influential technical references offer ranges for the frictional characteristics. An emblematic example is the coefficient of rolling friction (CRF). The handiest solution in establishing the correct value of the CRF for a certain contact is to find it experimentally. One of the methods is the inclined plane manner. The weak point of the method consists in the particularly small values of the slope required to evidence the effect of rolling friction. The present work shows a technique for finding the CRF. The principle of the method consists in considering both the slope of the inclined plane and the torque of rolling friction as unknowns. The two parameters can be established by experimental finding of the acceleration of an axi-symmetric body, both for ascending and descending motions.

1. Introduction

The rolling friction occurs in all nonconforming mechanical contacts [1]. The classical examples to take into account are the cam mechanisms and the rolling bearings. In Hertzian point contacts, besides the rolling friction the spinning friction appears, concretized by a spinning torque directed along the normal in the theoretical point of contact [2]. When the motion from the higher pair where friction occurs is a spatial one (the relative velocity has components both on the normal and contained in the tangent plane) the two types of friction are simultaneously present and when the experimental estimation of these frictions is aimed, difficult situations arise.

Though there are works concerning the rolling friction [3-6], in the recent technical literature there are met numerous papers in which new hypothesis and experimental techniques are presented. Perhaps the most significant aspect to be mentioned is the transition from the model that considers the proportionality between the rolling friction and the normal force [7], model created based on the analogy to the dry sliding friction, to the model based on hypothesis of the theory of elasticity which considers nonlinear dependency [8].

From experimental point of view, two main classes of methods are applied for the estimation of the rolling friction: methods where an oscillatory motion exists in contact and different types of pendula are used, and methods where the velocity from the contact point is monotonous and here the inclined plane technique can be mentioned. In the first case the coefficient of rolling friction is found as reveal by the damping of the oscillation of the pendulum while for the second case, the coefficient of rolling friction is estimated based on the acceleration of an axi-symmetric body. Both methods present advantages and drawbacks - the option for one of the other of the experimental techniques should take into account the particular conditions. For instance, if the two contacting bodies are made of materials with an important spatial gradient of the mechanical properties, then the method of oscillations is recommended since the dimensions of the analyzed region are reduced.

2. The principle of the method

The method proposed in the present work aims to establish the coefficient of rolling friction by analyzing the motion of an axi-symmetric body on an inclined plane. The technique was used in [9] where the evaluation of the coefficient of rolling friction is based on the measurements upon the acceleration of a revolute body moving downwards on an inclined plane, for diverse tilting angles. The main disadvantage of the method is the fact that the torque of rolling friction is noticed for small values of the tilting angle of the plane. Thus, the rolling friction torque influences the motion of the body when the angle of inclination of the plane is at least:

$$\alpha \cong s_r / R \tag{1}$$

where s_r is the coefficient of rolling friction and R is the radius of the contact point with respect to the axis of the rolling body. Considering that for the metallic bodies in contact, $s_r = (10^{-5} \div 10^{-4})m$ [10], it results that for a body with the radius of the contact point $R = 0.01m$ the tilting angle of the plane must take values within the range:

$$\alpha \cong (10^{-3} \div 10^{-2}) = 0.057^o \div 0.57^o \tag{2}$$

These are extremely small values, inferior to the precision of the measurement instruments. To avoid this aspect, the present paper proposes a method for estimation of the coefficient of rolling friction that also accepts as unknown the value of the angle of inclination of the plane. The rolling body used in the present study is a body made of two identical bearing balls of radius R, connected by a cylindrical part of radius r as seen in figure 1.

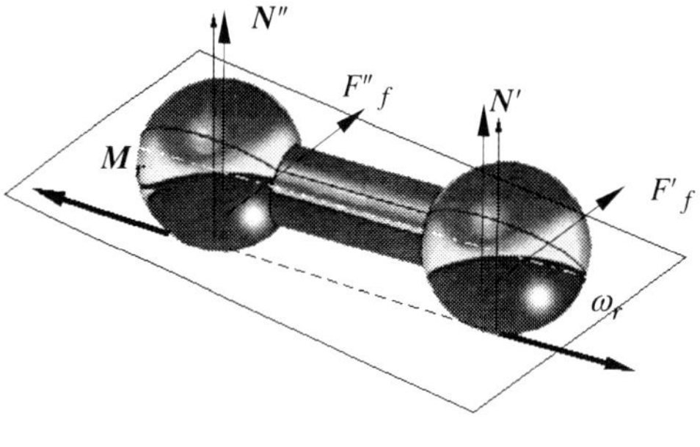

Figure 1. The rolling body, the relative angular velocity and the moment of rolling friction.

An inextensible wire is winded on the cylindrical part and passed over a pulley of negligible mass. A body is suspended by the end of the wire and moves on vertical direction as in figure 2. For reasons to be explained afterward, in figure 2 there were considered both the case when the body moves upward ("u" index) and the case when the body moves downwards ("d" index) along the line of the steepest slope of the plane. The motion of the body is a plane-parallel one. The unknowns of the problem are the reactions from the contact point (the normal component N, the friction force F_f, the moment of rolling friction M_r), the axial tension in the wire T and the characteristics of the motion (the angular acceleration ε and the acceleration of the center of mass a).

The hypothesis of pure rolling is expressed using the relation:

$$a_{u,d} = R\varepsilon_{u,d} \tag{3}$$

Three scalar equations can be written for each of the situations presented in figure 2: two equations for the motion of the center of mass and an equation corresponding to the moment of momentum theorem. Additionally, for the motion of the suspended body a scalar equation can be written. As conclusion, for any of the cases from figure 2, five scalar equations are available and the number of the unknowns is six, namely: $T, a, \varepsilon, N, F_f, M_r$. Therefore, a supplementary equation is required to ensure the compatibility of the problem. This equation is a constitutive equation describing the dependency of the moment of rolling friction on the normal force. Subsequently, the model that considers the rolling friction torque proportional to the normal force is proposed [7]:

$$M_r = s_r N \tag{4}$$

To be remarked that for both cases specified in figure 2, the normal force N and the moment of rolling friction M_r, have the same values, respectively, and so the indexes are abandoned.

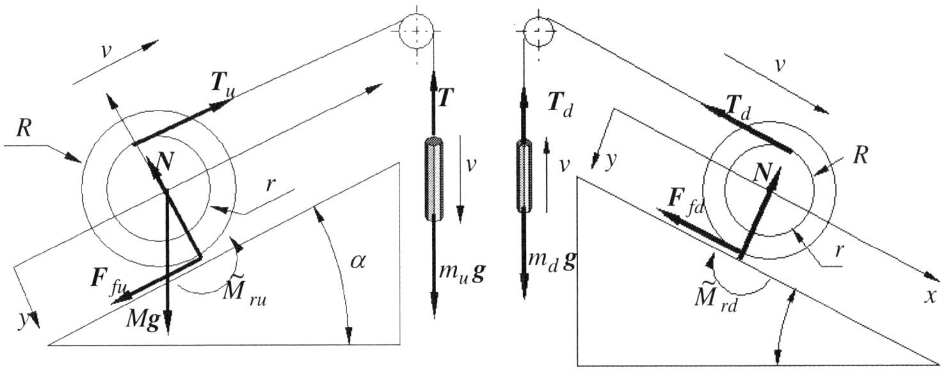

a)The upward motion of the body b)The downward motion of the body

Figure 2. The cases considered for the motion of the rolling body on the inclined plane

3. The equations of motion. Obtaining and discussions

For the ascending motion on the inclined plane, the equations of motion have the form:

$$\begin{aligned} Ma_u &= T_u - F_{fu} - Mg\sin\alpha \\ 0 &= N - Mg\cos\alpha \\ J_z a_u / R &= T_u r + F_{fu} R - M_{ru} \end{aligned} \tag{5}$$

and the equation of motion for the suspended body is:

$$m_u a_u = m_u g - T_u \tag{6}$$

When the body moves downhill, the equations of motion have the following form:

$$\begin{aligned} Ma_d &= Mg\sin\alpha - T_d - F_{fd} \\ 0 &= N - Mg\cos\alpha \\ J_z \frac{a_d}{R} &= -T_d r + F_c R - M_{rd} \end{aligned} \tag{7}$$

and the equation of motion for the suspended body takes the form:

$$m_d a_d = T_d - m_d g \tag{8}$$

The acceleration of the body in uphill motion is obtained from the equations (5) and (6):

$$a_u = \frac{\frac{m_u}{M}\left(1+\frac{r}{R}\right) - \sin\alpha}{1+\frac{J_z}{MR^2}+\left(1+\frac{r}{R}\right)\frac{m_u}{M}} g - \frac{\cos\alpha}{1+\frac{J_z}{MR^2}+\left(1+\frac{r}{R}\right)\frac{m_u}{M}} \frac{s_r}{R} g \tag{9}$$

and using the equations (7) and (8), the acceleration for the downward motion results as:

$$a_d = \frac{\sin\alpha - \frac{m_d}{M}\left(1+\frac{r}{R}\right)}{1+\frac{J_z}{MR^2}+\left(1+\frac{r}{R}\right)\frac{m_d}{M}} g - \frac{\cos\alpha}{1+\frac{J_z}{MR^2}+\left(1+\frac{r}{R}\right)\frac{m_d}{M}} \frac{s_r}{R} g \tag{10}$$

The equations (9) and (10) may be written under the following from:

$$\begin{cases} \sin\alpha + \frac{s_r}{R}\cos\alpha - \left(1+\frac{r}{R}\right)\frac{m_u}{M} + \frac{a_u}{g}\left[1+\frac{J_z}{MR^2}+\left(1+\frac{r}{R}\right)\frac{m_u}{M}\right] = 0 \\ \sin\alpha - \frac{s_r}{R}\cos\alpha - \left(1+\frac{r}{R}\right)\frac{m_d}{M} - \frac{a_d}{g}\left[1+\frac{J_z}{MR^2}+\left(1+\frac{r}{R}\right)\frac{m_d}{M}\right] = 0 \end{cases} \tag{11}$$

Assuming that the accelerations a_u and a_d are known, the system of equations (11) allows for finding the unknowns α and s_r. At this moment, the demarcation of the two situations, for upward and downward motion, can be explained. The problem requires finding the two parameters, α and s_r. For two parameters are necessary two tests performed under different conditions. To exemplify, it is considered that these would be finding the accelerations a_{u1} a_{u2}, for the uphill motions for two values $m_{u1,2}$. It results that the system (11) has two equations of the form of the first relation in the system (11). In this case, the parameters s_r and α could not be found since in both equations of the obtained system there is no change in the left member where the unknowns are contained, ($\sin\alpha + s_r \cos\alpha / R$). As consequence, two test must be performed, one for the upwards moving body and the other for the downwards moving body and so, the moment of rolling friction changes its sign.

From the system of equations (11) the angle α is first obtained:

$$\alpha = a\sin\frac{\frac{m_u+m_d}{M}\left(1+\frac{r}{R}\right) + \frac{a_d - a_u}{g}\left(1+\frac{J_z}{MR^2}\right) + \frac{m_d a_d + m_u a_u}{Mg}\left(1+\frac{r}{R}\right)}{2} \tag{12}$$

and afterwards, using the found expression, the relation of the coefficient of rolling friction is deduced:

$$s_r = \frac{\frac{m_u-m_d}{M}\left(1+\frac{r}{R}\right) - \frac{a_d + a_u}{g}\left(1+\frac{J_z}{MR^2}\right) - \frac{m_d a_d + m_u a_u}{Mg}\left(1+\frac{r}{R}\right)}{2\cos\alpha} \tag{13}$$

4. Conclusions

The paper proposes a method for evaluation of the coefficient of rolling friction based on the inclined plane technique. A revolution body is considered, consisting in two identical bearing balls and a cylinder, with the radius smaller than the radius of the balls, attached between the balls and thus shaping an axi-symmetric body. A wire is coiled over the cylinder and a mass it added to its end. The body is placed on an inclined plane with the revolution axis normal to the steepest slope of the plane. The wire passes over a pulley positioned at the highest point of the plane. The attached mass is chosen so that the body, starting from the base, moves uphill. Assuming dry friction conditions, the motion of the center of mass is uniformly accelerated with constant acceleration. The test is repeated with another mass chosen for the downward motion, with constant acceleration. The accelerations for the two cases are found by means of the time required to pass a stipulated distance.

The theoretical expressions for the uphill and downhill acceleration are obtained and the system of two equations has as unknowns the coefficient of rolling friction and the tilting angle of the plane. To be mentioned that, in the case of finding the coefficient of rolling friction, the task to evaluate the angle of the inclined plane is not simple.

Acknowledgment. This work was supported by a grant of the Romanian Ministry of Research and Innovation, CCCDI – UEFISCDI, project number PN-III-P1-1.2-PCCDI-2017-0404 / 31PCCD/2018, within PNCDI III.

References

[1] Johnson K L 1985 *Contact Mechanics* (Cambridge University Press)
[2] Townsend D P, Dietrich M W, and Zaretsky E W 1969 *Speed effects on ball spinning torque* (Lewis Research Center Cleveland, NASA Ohio Washington DC)
[3] Dzhilavdari I Z, Riznookaya N N 2008 Studies of the Dynamics of Free Microoscillations of a Pendulum Supported by Two Balls *J. Frict Wear* 29 (1) pp 1-6
[4] Olaru D N, Stamate C and Prisacaru G 2009 Rolling Friction in a Micro Tribosystem *Tribology Letters* **35** pp 205–210
[5] Muscă I 2003 About rolling friction evaluation in a ring-ball contact *Proceedings of International Conference ROTRIB'03* pp 284-287
[6] Alaci S, Ciornei F C, Ciogole A and Ciornei M C 2017 Estimation of coefficient of rolling friction by the evolvent pendulum method, *IOP Conf. Series: Materials Science and Engineering* **200** – 012005
[7] Ardema D M 2005 *Newton-Euler Dynamics* (Springer)
[8] Cherepanov G P 2014 Theory of rolling: solution of the Coulomb problem *Journal of Applied Mechanics and Technical Physics* **55**(1) pp. 182–189
[9] Ciornei F-C, Alaci S, Ciogole V-I and Ciornei M-C 2017 Valuation of coefficient of rolling friction by the inclined plane method, *IOP Conf. Series: Materials Science and Engineering* **200** - 012006
[10] Stolarski T A and Tobe S 2001 *Rolling friction* Tribology in Practice Series (Wiley)

IOP Conf. Series: Materials Science and Engineering **514** (2019) 012005 doi:10.1088/1757-899X/514/1/012005

Method for simultaneous estimation of rolling and spinning friction in a higher pair

S Alaci[1], R D Pentiuc[2], F C Ciornei[1], I C Romanu[1], S T Siretean[1] and M C Ciornei[3]

[1] Department of Mechanics and Technologies, Stefan cel Mare University of Suceava, Romania
[2]Electrotechnics Department, Stefan cel Mare University of Suceava, Romania
[3]Department 2 Physiology I, Carol Davila University of Medicine and Pharmacy, Bucuresti, Romania

E-mail: alaci@fim.usv.ro; florina@fim.usv.ro

Abstract. A spatial higher pair is characterized by a complex motion with the direct consequence: the occurrence in the contact point of a friction torsor having all possible components (sliding, spinning and rolling friction). There are cases, like thrust ball bearings, where the components of the friction torsor are interrelated, and actual separation of these is not possible. The present paper proposes an approach for simultaneous evaluation of the coefficient of rolling friction and of the coefficient of spinning friction using a cycloidal pendulum. As principle, the experimental device consists in a steel ball set on top of two cylindrical steel rods, parallel, mounted horizontally. When a vertical rod is attached to the ball, a cycloidal pendulum is obtained. The damping of oscillatory motion is due to the two friction torques: spinning and rolling. y relevant articles in literature searches, great care should be taken in constructing both. The motion of the pendulum is analyzed for two values of the gap between the horizontal rods. A system of two equations with the coefficients of rolling and spinning friction as unknowns is obtained. Based on the hypothesis that the dependencies for the torques versus the normal force are known, these two coefficients can be obtained.

1. Introduction

In technological application there are often met cases when the contact between two elements is theoretically made in a point and therefore the relative motion between the two elements has the most complex possible form [1]. In this situation, between the contacting parts, relative sliding characterized by the sliding velocity, contained in the common tangent plane and relative rotation, characterized by relative rotation velocity ω occur. Depending on the relative position of the relative angular velocity with respect to the normal in the contact point, two limit cases are defined: spinning motion, when the relative angular velocity is parallel to the normal and rolling motion, when the rotation velocity is contained in the common tangent plane [2]. Between the two limit cases, all combinations when there are simultaneously met both spinning and rolling motion may take place. To the two types of angular velocities, torques (parallel and of opposite sign) will oppose; these friction torques take the corresponding name: spinning moment and rolling moment. The two moments are each one responsible of the tendency of decreasing the corresponding angular velocity. It is obvious that the correct evaluation of the magnitude of the two moments is one of the essential requirements in optimal design of any mobile mechanical structure. The main difficulty in reaching this task consists in the fact that the values of these moments are extremely reduced, and so the errors introduced by the measuring instruments can be greater than the parameters to be measured [3-4]. From here it results the necessity of conceiving devices as simple as possible and with the mobile bodies having technological shapes. One of the main class of methods employed in the study of the rolling friction are the pendula. The main characteristic of these devices is the fact that the effect of rolling friction has a summative effect upon their motion, and thus, after an adequate time, the effect of rolling friction becomes perceptible [5-7]. As expected, the conclusion that it is appropriate to use a pendulum for the estimation of combined effect of spinning friction and rolling friction is reached.

2. Proposed pendulum for the study of simultaneous effect of spinning and rolling friction

The pendulum proposed here is made from a steel ball of R radius to which a rod of circular cross section is attached and the whole mass of the pendulum is M , the moment of inertia with respect to an axis normal to the axis of the rod passing through the centre of mass G is J_Z and the centre of mass G is positioned on the axis of symmetry of the pendulum at a distance ξ from the center of the ball. the pendulum is brought into contact with two cylindrical steel beams, parallel to each other and positioned in an horizontal plane. The two contacts ball-beams take place in the points $C_{1,2}$.

Two coordinate systems are chosen in order to obtain the equation of motion: an immobile one $Oxyz$ and a mobile one $Ox'y'z'$ attached to the pendulum, oriented as shown in figure 1. The position of the pendulum is stipulated by two parameters: the ordinate of the center of mass of the ball and the angle of rotation of the pendulum, φ. Assuming that there is no sliding between the surfaces of the ball and of the rods in the two contact points, the horizontal straight line passing through the two points is the instantaneous rotation axis and consequently the pendulum has pure rolling motion with respect to the two rods. In each of the $C^{1,2}$ contact points the normal direction and the tangent plane are well defined. By projecting the angular velocity ω on the normal and on the tangent plane, the angular spinning velocities $\omega_{sp}^{1,2}$ and angular rolling velocities $\omega_r^{1,2}$ are obtained, respectively.

A friction torsor occurs in each of the two contact points, consisting of: the normal reaction $\boldsymbol{N}^{1,2}$ directed along the normal, the spinning friction torques $\boldsymbol{M}_{sp}^{1,2}$, oriented oppositely to the spinning velocities $\omega_{sp}^{1,2}$ together to the rolling friction torques $\boldsymbol{M}_r^{1,2}$ parallel to the angular rolling velocities $\omega_r^{1,2}$. Beside these moments, a friction force $\boldsymbol{T}^{1,2}$ parallel to the direction of motion of the center of mass occurs in each of the two contact points, given by the relation:

$$T^{1,2} < \mu N^{1,2} \tag{1}$$

which is the condition of pure rolling. By accepting the pure rolling condition in the two contact points, from the two parameters y and φ, used in stipulating the position of the pendulum only one remains independent, as the following relation exists between them:

$$y^{1,2} = -r_c \varphi \tag{2}$$

In the relation (2) r_c is the radius of the pitch circle:

$$r_c = R \cos \beta \tag{3}$$

where the angle β is a constructive characteristic of the experimental device given by:

$$\beta = asin[d/(R+r)] \tag{4}$$

and defined as the angle made by the plane containing the center of the ball and the axis of the rod and the vertical direction.

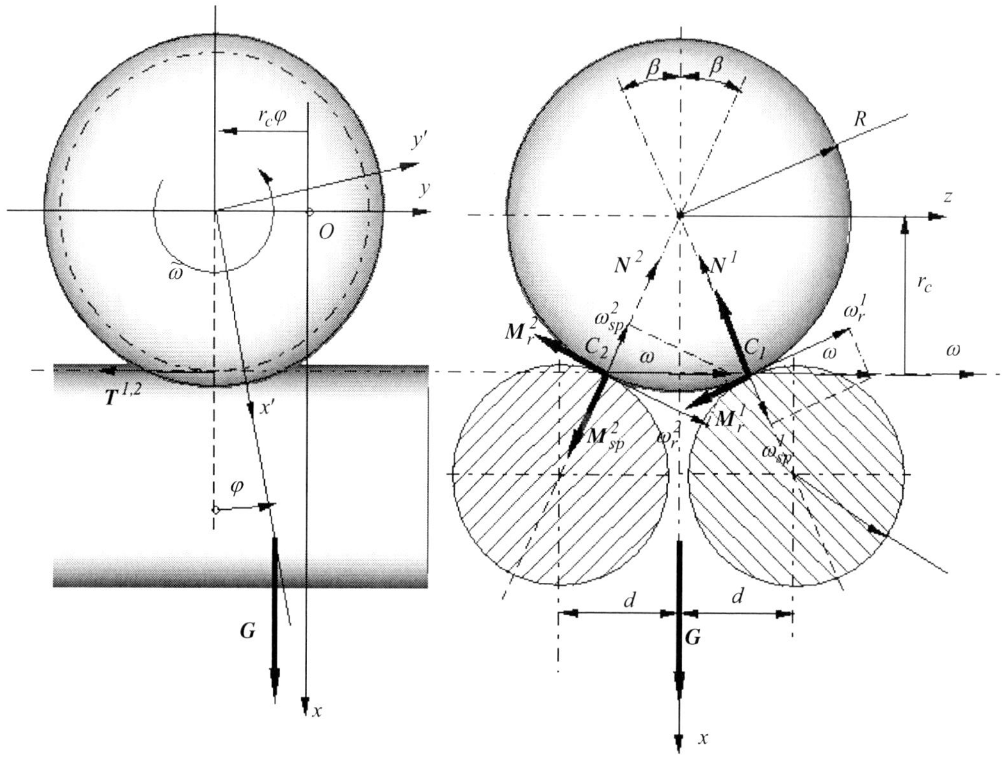

Figure 1. Principle of the device and the characteristic parameters.

From figure 1 the coordinates of the center of mass of the pendulum are easily obtained:

$$\begin{cases} x_G = \xi \cos\varphi \\ y_G = -r_c\varphi + \xi \sin\varphi \end{cases}, \qquad (5)$$

Representing these coordinates as function of the position of the center of mass with respect to the axis of rotation, as shown in figure 2, a normal (blue), curtate (red) or prolate (black) cycloid may result.

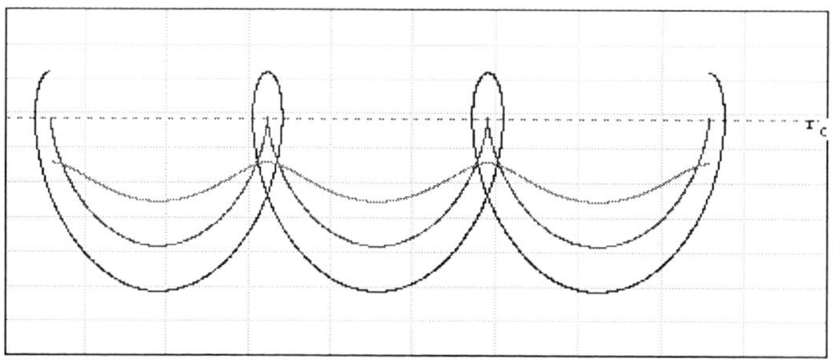

Figure 2. The trajectory of the center of mass of the pendulum.

3. Obtaining the equation of motion. Discussions

The components of the acceleration of the center of mass of the pendulum are found by differentiating with respect to time, twice, the relations (5):

$$\begin{cases} a_{Gx} = -\xi\dot{\varphi}^2 \cos\varphi - \xi\ddot{\varphi}\sin\varphi \\ a_{Gy} = (\xi\cos\varphi - r_c)\ddot{\varphi} - \xi\dot{\varphi}^2 \sin\varphi \end{cases} \tag{6}$$

The position vectors of the contact points are:

$$\boldsymbol{r}_{C_{1,2}} = \begin{bmatrix} R\cos\beta & -\varphi r_c & \pm R\sin\beta \end{bmatrix}^T \tag{7}$$

These relations are used to find the position vectors of the contact points with respect to the centers of mass, required for applying the moment of momentum theorem.

$$\overline{GC}^{1,2} = \boldsymbol{r}_C^{1,2} - \boldsymbol{r}_G = \begin{bmatrix} R\cos\beta - \xi\cos\varphi & \pm\xi\cos\varphi & R\sin\beta \end{bmatrix}^T \tag{8}$$

The components of the rolling friction torque:

$$\boldsymbol{M}_r^{1,2} = \begin{bmatrix} \pm M_r \sin\beta & 0 & -M_r\cos\beta \end{bmatrix}^T \tag{9}$$

and for the spinning friction torque:

$$\boldsymbol{M}_{sp}^{1,2} = \begin{bmatrix} \mp M_{sp}\cos\varphi & 0 & -M_{sp}\sin\beta \end{bmatrix}^T \tag{10}$$

The theorem of motion of the center of mass is required to solve the problem and it is applied as:

$$M\boldsymbol{a}_G = \boldsymbol{G} + \boldsymbol{T}^1 + \boldsymbol{T}^2 + \boldsymbol{N}^1 + \boldsymbol{N}^2, \tag{11}$$

together with the moment of momentum theorem, written with respect to the center of mass:

$$\begin{bmatrix} 0 & 0 & J_z\ddot{\varphi} \end{bmatrix}^T = \overline{GC}^1 \times (\boldsymbol{N}^1 + \boldsymbol{T}^1) + \overline{GC}^2 \times (\boldsymbol{N}^2 + \boldsymbol{T}^2) + (\boldsymbol{M}_r^1 + \boldsymbol{M}_{sp}^1 + \boldsymbol{M}_r^2 + \boldsymbol{M}_2^2). \tag{12}$$

Due to symmetry, all homologous components of the friction torsors have equal moduli. The equations (11) and (12) provide three scalar equations. The unknowns of the problem are: the φ angle that characterizes the motion and N, T, M_{sp} and M_r. Two more equations are necessary for a compatible problem, represented by the constitutive equations which describe the relation between the magnitudes of the spinning and rolling friction moments and normal force. Next, it is accepted that all friction torques are proportional to the magnitude of the normal force N, described by [8]:

$$M_{sp1,2} = s_{sp}N \tag{13}$$

$$M_{sr1,2} = s_r N \tag{14}$$

where s_{sp} and s_r are the coefficients of spinning and rolling friction respectively, both with dimension of length. The equations (11-14) may now be solved and the differential equation is found:

$$\ddot{\varphi} = -\frac{\left(\frac{s_{sp}}{\xi}\sin\beta + \frac{s_r}{\xi}\cos\beta\right)sgn(\dot{\varphi})\cos\varphi + \frac{R}{\xi}\sin\varphi\cos^2\beta}{\left[\left(\frac{s_{sp}}{\xi}\sin\beta + \frac{s_r}{\xi}\cos\beta\right)sgn(\dot{\varphi})\sin\varphi + \left(\frac{r_c}{\xi} - \cos\varphi\right)\frac{R}{\xi}\cos^2\beta + \left(1 + \frac{J_z}{M\xi^2} - \frac{r_c}{\xi}\cos\varphi\right)\cos\beta\right]\xi}\left(\frac{g}{\xi} + \dot{\varphi}^2\right) \tag{15}$$

The equation (15) is a nonlinear differential equation of second order; the solutions of which cannot be analytically found [9], hence the necessity of applying a numerical procedure of integration. A more thorough analysis of the equation (15) proves that the two friction coefficients do not occur as independent parameters, but under the form of an expression. Therefore, it is expected to use the notation:

$$s_g = (s_{sp} \sin\beta + s_r \cos\beta) \tag{16}$$

representing a global friction coefficient with dimension of length. The relation (16) shows that for small values of β angle the most important fraction belongs to the rolling friction coefficient while for greater values of β angle the parts reverse and the spinning friction is predominant. Using this expression, the equation (15) becomes:

$$\ddot{\varphi} = -\frac{\frac{R}{\xi}\sin\varphi\cos^2\beta + \frac{s_g}{\xi}sgn(\dot{\varphi})\cos\varphi}{\left[\left(\frac{r_c}{\xi} - \cos\varphi\right)\frac{R}{\xi}\cos^2\beta + \left(1 + \frac{J_z}{M\xi^2} - \frac{r_c}{\xi}\cos\varphi\right)\cos\beta + \frac{s_g}{\xi}sgn(\dot{\varphi})\sin\varphi\right]\xi}\left(\frac{g}{\xi} + \dot{\varphi}^2\right) \tag{17}$$

The numerical algorithm Runge-Kutta IV [10] was applied for the integration of differential equation (17) for a set of values of the involved parameters. The variation of the angular amplitude of the pendulum for free oscillations, with initial amplitude φ_0, is presented in figure 3. The plot shows linear decrease with time of the amplitude of the pendulum - a characteristic of damping caused by dry friction [11-12]. Once the motion of the pendulum is known by means of the dependency $\varphi = \varphi(t)$, the variations with time of the magnitude of the friction force T and of the normal reaction can be found and are expressed by the relations:

$$T = M[(r_c - \xi\cos\varphi)\ddot{\varphi} + \dot{\varphi}^2\xi\sin\varphi + g\sin\varphi]/2 \tag{18}$$

$$N = M(\xi\ddot{\varphi}\sin\varphi + \xi\dot{\varphi}^2\cos\varphi + g\cos\varphi)/(2\cos\beta) \tag{19}$$

With these dependencies known, the hypothesis of pure rolling in the contacts of the pendulum can now be verified. To this end, the dependency of the ratio T/N (representing the *static friction coefficient*) is presented in figure 4.

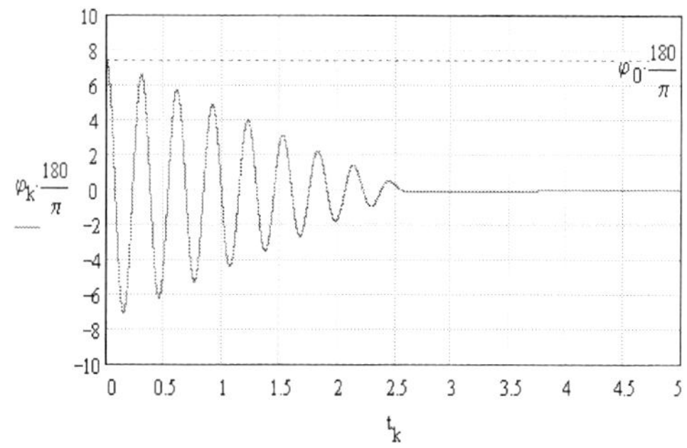

Figure 3. Variation of the angular amplitude of the pendulum.

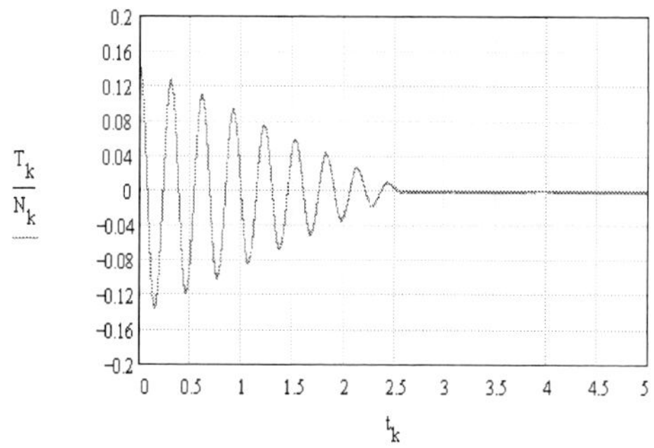

Figure 4. Verifying the pure rolling existence.

4. Conclusions

The paper proposes a method for concomitant finding of the coefficient of rolling friction and the coefficient of spinning friction under the assumption that both torques, of rolling friction and of spinning friction, are proportional to the value of the normal force. The method consists in using a cycloidal pendulum particularly designed. Specifically, the pendulum is made of a rod attached to a bearing ball. At its turn, the ball is placed on top of two cylindrical parallel beams, placed in an horizontal plane. The instantaneous axis of rotation of the pendulum is the horizontal line passing through the contact points between the ball and the beams. The normal directions in the contact point are not directed along the horizontal or vertical directions and hence the conclusion that in the contact points both rolling friction and spinning friction torques occur. The equation of motion of the pendulum is found accepting the hypothesis that pure rolling exists in the contact points. The equation obtained is a nonlinear of second order differential one and a numerical procedure must be applied for solving it. The equation is integrated for a set of parameters characteristic to the dynamical system and a linear attenuation of the amplitude of the pendulum is observed, fact that is an attribute of dry friction. The amplitude damping obviously depends on the values of the two friction coefficients asked for. The principle of the method assumes finding the velocity of angular amplitude decrease of the pendulum for two distances between the axes of the beams; thus, a system of two linear equations with the two coefficients of friction as unknowns is obtained. The method may be extended for the case when another type of dependency between the friction torques and the normal reaction is accepted.

Acknowledgment. This work was supported by a grant of the Romanian Ministry of Research and Innovation, CCCDI – UEFISCDI, project number PN-III-P1-1.2-PCCDI-2017-0404 / 31PCCDI/2018 within PNCDI III.

References

[1] Johnson K L 1985 *Contact Mechanics* (Cambridge University Press)
[2] Voinea R, Voiculescu D and Ceauşu V 1989 *Introducere în mecanica solidului cu aplicaţii în inginerie* (Editura Academiei Republicii Socialiste România Bucureşti)
[3] Olaru D N, Stamate C and Prisacaru G 2009 Rolling Friction in a Micro Tribosystem *Tribology Letters* **35** pp 205–210
[4] Townsend D P, Dietrich M V and Zaretsky E V 1969 *Speed Effects on Ball Spinning Torque* (Lewis Research Center Cleveland Ohio NASA Washington D. C.)
[5] Alaci S, Ciornei F C, Ciogole A and Ciornei M C 2017 Estimation of coefficient of rolling friction by the evolvent pendulum method, *IOP Conf. Series: MSE* **200** – 012005
[6] Ciornei M C, Alaci S, Ciornei F C and Romanu I C 2017 A method for the determination of the coefficient of rolling friction using cycloidal pendulum *IOP Conf. Series: MSE* **227** – 012027
[7] Alaci S, Ciornei F C, Bujoreanu C Ciornei M C and Acsinte I L 2018 Finding the coefficient of rolling friction using a pericycloidal pendulum, *Proceedings of ACME 2018 Int. Conference, to be published*
[8] Mangeron D, Irimiciuc N 1978 *Mecanica rigdelor cu aplicatii in inginerie. Volumul I. Mecanica rigidului* (Ed. Tehnica Bucuresti)
[9] Jordan D W and Smith P 2007 *Nonlinear differential equation. Problems and Solutions A Sourcebook for Scientists and Engineers* Oxford Texts in Applied and Engineering Mathematics (Book 11) (Oxford University Press)
[10] Hamming R 1987 *Numerical Methods for Scientists and Engineers* (Dover Publications)
[11] Torzo G and Pearnzoni P 2009 The real pendulum: theory, simulation, experiment *Lat. Am. J. Phys. Educ.* **3**(2) pp 221-228
[12] Zonetti L F C, Camargo S S, Sartori J, de Sousa D F and Nunes L A O 1999 A demonstration of dry and viscous damping of an oscillating pendulum, *Eur. J. Phys.* **20** pp 85–88

IOP Conf. Series: Materials Science and Engineering **514** (2019) 012006 doi:10.1088/1757-899X/514/1/012006

Subsurface analysis of 6082-T6 Aluminium alloy's mechanical properties subjected to small particles erosion

A A Erdoğan[1], I C Ţarcă[2], A Rus[2] and R C Ţarcă[3]

[1]Mechanical Engineering Department, Kocaeli University, Kocaeli, TR
[2]Mechanical Engineering and Automotive Department, University of Oradea, Oradea, RO
[3]Mechatronics Department, University of Oradea, Oradea, RO

E-mail: nelut@uoradea.ro

Abstract. 6082-T6 aluminium alloy is a ductile metal. When subjected to small particles erosion, sub-surface deformations are expected due to the particle impact. Plasticity is one of the properties which might reveal the rate of the deformations. As a result of plastic deformation, mechanical properties of the samples are expected to change with distance from the surface. Typically, the depth of the plastically deformed region could be between 5 and 200 μm. In order to understand the plasticity of the materials, nanoindentation researches were conducted. Aluminium alloy samples were previously subjected to hard material particles jets (Al_2O_3) under different values of incident angle, between 15 and 90 degrees. After a 3D scan of the resulted surfaces, the samples were cut in such a manner that the middle section remained visible and prepared for nanoindentation analysis. The paper presents the experiments conducted as well as some of the achieved results.

1. Introduction

Aluminium alloys are largely used in the automotive and aerospace industries, due to the combination between their low specific mass and mechanical properties. Having also good formability and machinability (such as rolling, extrusion, stamping, casting, etc) they are also suitable for industrial application.

2. Materials and methods

2.1. Material

Material subjected to tests was AA6082-T6 aluminium alloy laminated sheets, with a nominal thickness of 3mm. The dimensions of the samples provided by the Mechanical Department of the Kocaeli University were 50x50x3mm. According to (1), aluminium alloy 6082 is a medium strength alloy with excellent corrosion resistance, having the highest strength of the 6000 series alloys; it is known as a structural alloy. In plate form, 6082 is the alloy most commonly used for machining. The addition of a large amount of manganese controls the grain structure which in turn results in a stronger alloy. Table 1 shows the chemical composition of the AA6082-T6 aluminium alloy, while Table 2 offers some of its physical properties.

Table 1. Chemical composition of the AA6082-T6 aluminium alloy [1].
Spec: BS EN 573-3:2009

Chemical Element	% Present
Manganese (Mn)	0.40 - 1.00
Iron (Fe)	0.0 - 0.50
Magnesium (Mg)	0.60 - 1.20
Silicon (Si)	0.70 - 1.30
Copper (Cu)	0.0 - 0.10
Zinc (Zn)	0.0 - 0.20
Titanium (Ti)	0.0 - 0.10
Chromium (Cr)	0.0 - 0.25
Other (Each)	0.0 - 0.05
Others (Total)	0.0 - 0.15
Aluminium (Al)	Balance

Table 2. Physical properties of the AA6082-T6 aluminium alloy [1].

Physical Property	Value
Density	2.70 g/cm³
Melting Point	555 °C
Thermal Expansion	24 x10^-6 /K
Modulus of Elasticity	70 GPa
Thermal Conductivity	180 W/m.K
Hardness Brinell	~95 HB

2.2. *Method*

Experiments which generated the specimens and also results and conclusions are described in [2] and [3]. For the tests presented in the current paper three types of specimens have been considered: specimen coded 06A which corresponds to 120 mesh erodent particles at 1.5 bar pressure at an impact angle of 90 degrees, specimen coded 07A, which corresponds to 120 mesh erodent particles at 3.0 bar pressure at an impact angle of 90 degrees and specimen coded 11A which corresponds to 120 mesh erodent particles at 1.5 bar pressure at an impact angle of 15 degrees.

A nanoindentation of the specimens' subsurface along their middle line has been set, having the main aim the analysis of the Modulus of Elasticity and the Hardness of the materials in the layers right below the surface due to Al_2O_3 particles impact. To achieve the needed surface, the specimens had to be cut axially, along their middle line, so that after processing steps, the analysed surface should be as close as possible to the middle line.

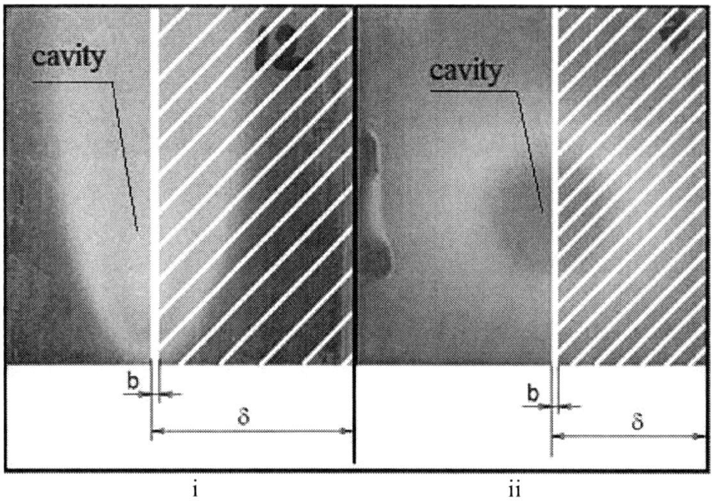

Figure 1. Specimen cutting: b – cutting tool width, δ – withdrawn part. i – specimen 11A, ii – specimen 07A

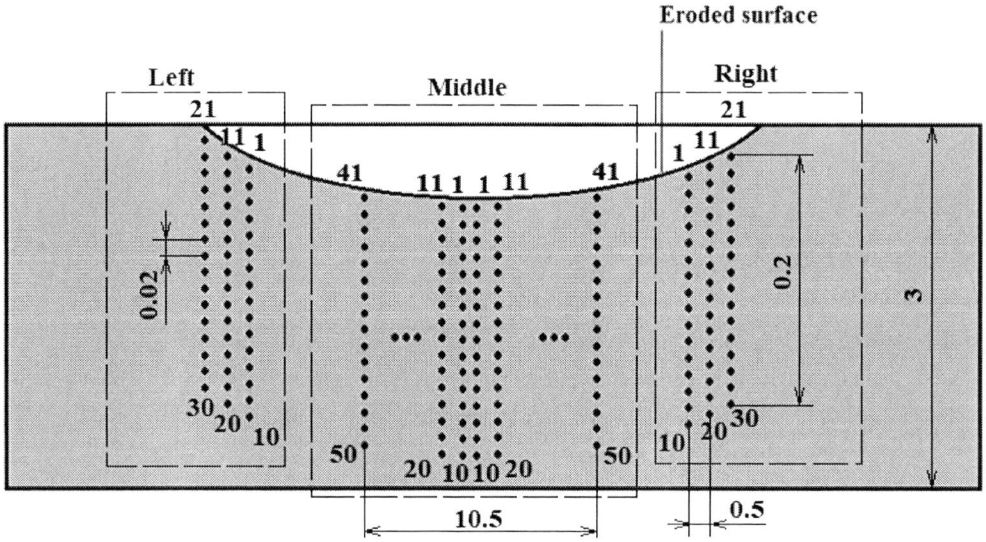

Figure 2. Geometry of nanoindentation.

Figure 2 shows the geometry of nanoindentation. Starting as close as possible from the cavity surface, there are 10 indentation points towards the base of the specimen, with a pitch of 0.02mm. To avoid an extremely long process of measurements, the surface was split in three regions, as presented in figure 2. Left and right zones are placed near the start/end of the eroded surface, containing, each of them, 50 measurement points, structured in five columns, each of 10 rows. Distance between columns is 5mm. The middle zone consists of 100 to 200 measurement points (10 columns to 20 columns), organized similar to left and right zones, symmetrical to the maximum depth of the cavity.

The nanoindentation tip characteristics are presented in table 3 and test inputs in table 4.

Table 3. Tip information.

Name	Value	Units
Tip Name	TB22118 BASIC	
Area Coefficient 1	2.539610175e+001	
Area Coefficient 2	3.876214090e+002	

Table 4. Test inputs.

Name	Value	Units
Allowable Drift Rate	0.050	nm/s
Depth Limit	1000.000	nm
Peak Hold Time	10.000	s
Percent To Unload	90.000	%
Perform Drift Test Segment	1	

3. Results and discussions

For measurements a Keysight Nano Indenter G200 has been used, the experiments being conducted in accordance with ISO 14577-1:2015.

Figure 3 shows a typical imprint after tests, and Table 5 shows results of the tests (mean values).

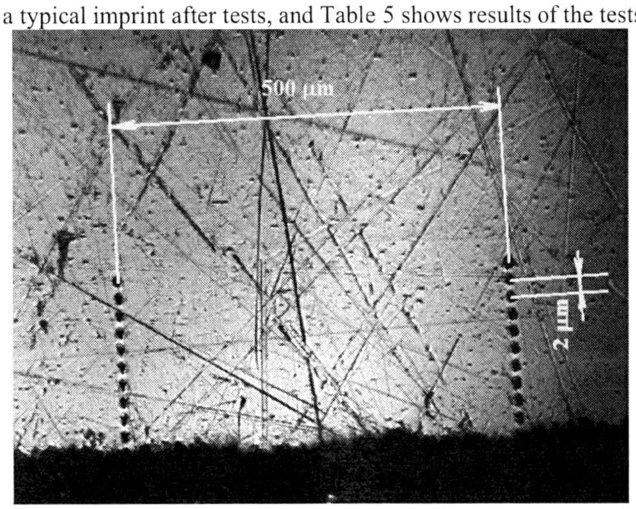

Figure 3. Geometry of the imprints during tests, for two adjacent columns.

Notation made in table 5 are as follows:

- 6A, 7A, 11A are the codes of the specimens
- LE 1-5 is used to address values for the left edge (named Left in figure 2), columns 1 to 5. As a general aspect regarding notation, column numbers increase from the center to the edges.
- LM 1-10 represents the measurements on the left side of the middle region.
- RM 1-10 represents the columns to the right of the middle region
- RE 1-5 is used to address columns near the right edge of the cavity.
- Ref is a reference measurement made in the vicinity of the specimen's center.

Analyzing values offered by tests, the following conclusions can be made:

- Material hardness shows small changes with the specimens, from 1.202 at 11A-LM to 1.512 for specimen 7A, left edge.
- Significant changes are noticed for the Modulus of Elasticity, both from one specimen to another, but also from one place to another at the same specimen. Also, for specimens 6A and 7A, subjected to particle jets perpendicular to their surfaces, a significant increase of the modulus has been noticed. This increase in value goes up to 88GPa related to 70GPa as stated by producers, which means an increase with 25% in properties. Specimen 11A shows almost no changes in Modulus of Elasticity values related to manufacturer specifications.

Table 5. Mean values for Modulus of Elasticity and Hardness (GPa)

Specimen	Location	Modulus (mean)	Hardness (mean)
6A	LE 1-5	80,676	1,373
6A	LM 1	87,239	1,451
6A	LM 2	85,663	1,411
6A	LM 3	85,866	1,45
6A	LM 4	80,541	1,302
6A	LM 5-7	85,218	1,399
6A	LM 8-10	83,715	1,422
6A	RM 1	84,747	1,396
6A	Ref	85,627	1,218
7A	LE 1-5	88,003	1,512
7A	LM 1-10	81,996	1,286
7A	RM 1-10	84,395	1,347
7A	RE 1-5	81,421	1,263
11A	LE 1	61,584	1,302
11A	LE 2	59,301	1,318
11A	LM 1-2	74,929	1,202
11A	LM 3-4	74,156	1,211
11A	LM 5-6	72,862	1,245
11A	LM 7-8	71,526	1,239
11A	LM 9-10	69,526	1,238
11A	RM 1-2	72,268	1,27
11A	RM 3-4	75,184	1,322
11A	RM 5-6	76,34	1,36
11A	RM 7-8	79,1	1,358
11A	RM 9-10	79,928	1,387
11A	Ref	79,109	1,166

- During tests it was noticed that Modulus of Elasticity is decreasing from the edge of the cavity to the interior of the specimen. Figure 4 shows the variation of the mean values for Modulus of Elasticity in respect to the distance from the cavity edge to the center of the specimen. Higher values are noticed near the edge.

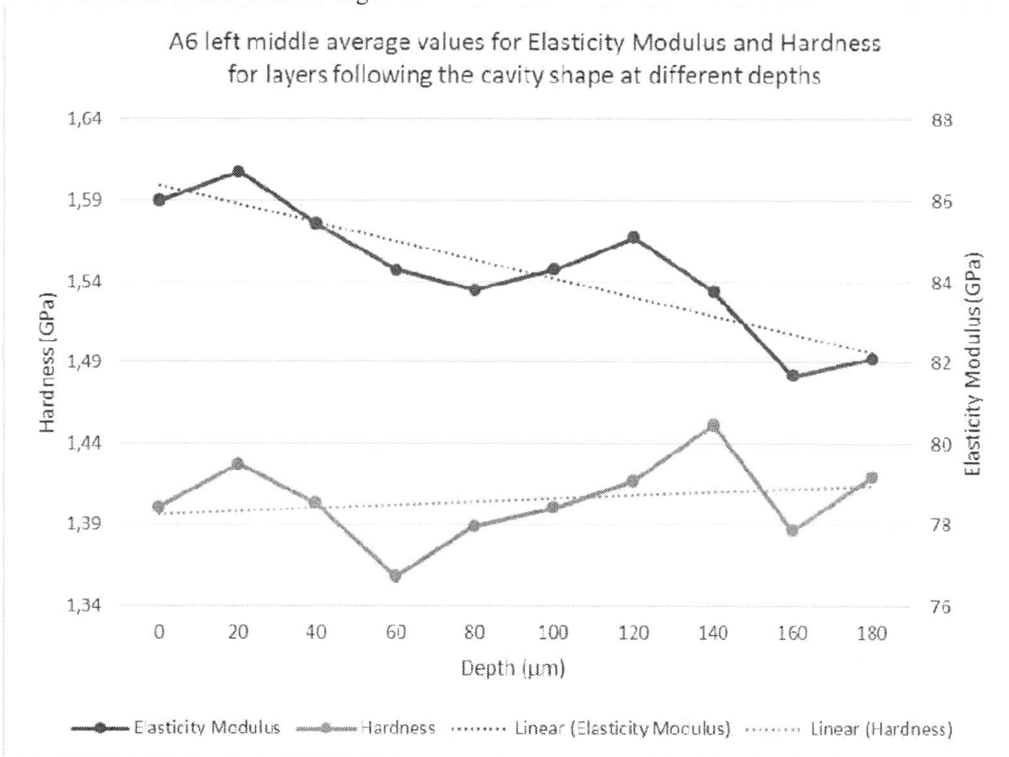

Figure 4. Modulus of Elasticity and Hardness variation with the distance to the cavity edge.

As a final conclusion, it can be stated that important changes in material properties occur during impact with hard materials. In [2] and [3] experiments regarding the dependency between particle dimensions and particle speed to the shape of cavities have been conducted. Current research noticed that changings in material properties are also influenced by the way in which materials are exposed to wear particles. Further researches have to be oriented toward two aspects:

- Investigation of the way in which changes in material properties are related to the changes in shape of the material grains;
- Investigation related to the ratio between wear and plastic deformation of the material in the process of generating the cavity. If the cavity is generated mainly as an effect of grain deformation, thus increasing material properties, maybe a sandblasting similar procedure using Al_2O_3 after aging might be an advantage in obtaining materials with better properties.

References

[1] Aluminium Alloy - Commercial Alloy - 6082 - T6 Extrusions. *AALCO.* [Online] [Cited: 02 10, 2018.] http://www.aalco.co.uk/datasheets/Aluminium-Alloy-6082-T6-Extrusions_338.ashx

[2] Erdoğan A. A., Feyzullahoğlu E, Fidan S, and Sınmazçelik T, 2016 Effect of Erodent Particle Size on Erosive Wear Behavior of AA6082-T6 Aluminum Alloy *1st International Conference on Engineering Technology and Applied Sciences Afyon Kocatepe University, Turkey 21-22 April 2016,* **92** pp 387-392

[3] Erdoğan A. A., Feyzullahoğlu E, Fidan S, and Sınmazçelik T, 2016 Effect of Erodent Particle Velocity on Erosive Wear Behavior of AA6082-T6 Aluminum Alloy *1st International Conference on Engineering Technology and Applied Sciences Afyon Kocatepe University, Turkey 21-22 April 2016,* pp 393-399

[4] ISO 14577-1:2015 Metallic materials — Instrumented indentation test for hardness and materials parameters

[5] Keysight Nano Indenter G200 Precise mechanical testing for micro-to-nano range of loads and displacements. Data Sheet, Keysight Technologies, 2013-2014, Published in USA, August 3, 2014, 5990-4172EN, https://www.keysight.com/en/pc-1674481/atomic-force-microscopes-fe-sem-nanoindenters-utm?cc=US&lc=eng

Acknowledgments

Special thanks to the colleagues from Mechanical Department, University of Kocaeli, Turkey, for providing the eroded specimens and to the staff of the Nanoindentation Laboratory of the University of Oradea for their intense activity. Authors also express their gratitude to the specialists of the FAIST Mekatronik SRL Oradea for their help in preparing the probes.

IOP Conf. Series: Materials Science and Engineering **514** (2019) 012007 doi:10.1088/1757-899X/514/1/012007

Experimental Researches concerning the Rheology of Cutting Fluids

I Radulescu[1], A V Radulescu[1], S Ramalchanov[2], S Yankov[2] and V S Rupetsov[2]

[1]University POLITEHNICA Bucharest, Romania
[2]"Paisii Hilendarski" University of Plovdiv, Technical College of Smolyan, Smolyan, Bulgaria

E-mail: irina.radulescu@upb.ro

Abstract. Cutting fluids are used to reduce the negative effects of the heat and friction on both tool and work piece. The cutting fluids produce three positive effects in the process: heat removal elimination, lubrication on the chip–tool interface and chip removal. This paper proposes to study the rheological properties of four different type of cutting fluids, as function of their chemical and physical characteristics. The tests were performed using a cone and plate Brookfield viscometer, by determining the rheological parameters and the variation of the apparent viscosity with the temperature. This approach has been used to quantify the performance of the tested products and to compare products in the marketplace.

1. Introduction

Improving machining processes includes high processing speeds and automated mechanical systems necessary for feeding and loading, correlated with operating flows. It also includes the use of various metallic materials for cutting tools, for different types of operations. All of these elements imply an important improving of cutting fluids for a wide range of machining processes and processing conditions [1], [2], [3]. Cutting fluids currently contain a wide variety of specialized chemical agents, designed to obtain the lubrication effect under different processing conditions, providing the stability of the physical, chemical and rheological characteristics, and also the antiwear protection [4].

The main functions of the cutting fluids are to remove the amount of heat released during the cutting process, to reduce the friction between the tool and the work piece, to increase the work life of the cutting tools by reducing the wear and to improve the quality of the surface for the work piece.

Depending on the specifics of the processing operations and the processed materials, cutting fluids are used in the following typological diversification [5], [6], [7]: water-immiscible oils, emulsifiable fluids such as oil/water, aqueous solutions of organic or inorganic compounds, suspensions or pastes of solid lubricants. The important rheological characteristics required of a cutting fluid for good operational performance are focused on sustaining a sufficiently large stress to maintain chips in suspension, particularly when fluid circulation is stopped. In the same time, cutting fluids need to have low enough viscosity in order to assure an efficient pumping. In steady shear flow, the cutting fluids have to be shear-thinning fluids, which means that the effective viscosity decreases by the increasing of the shear rate [8], [9].

The present paper proposes to study the rheological properties of four different type of cutting fluids, as function of their chemical and physical characteristics. The tests are performed using a cone and plate Brookfield viscometer, by determining the rheological parameters and the variation of the apparent viscosity with the temperature.

IOP Conf. Series: Materials Science and Engineering **514** (2019) 012007 doi:10.1088/1757-899X/514/1/012007

2. Experimental methodology

The cutting fluids used for the experimental tests are the following:

- INSA Rezinol C-22, which is a non-emulsifiable oil – cooling lubricant, used for lubrication and cooling of cutting tools in deep drilling of metals and alloys, in automatic lathe processing, highly alloyed steels, stainless steels and heat-resistant steels [10];
- INSA Emulsol, which is an emulsifiable lubricant – coolant, containing special emulsifiers, anti-corrosion inhibitors and bio-resistant components prepared using special technologies and refined mineral oil. It finds application in metalworking for high quality machining of cast iron, steel and other metals in operations - turning, milling, grinding, reaming, cutting [11];
- Würth Cut & Cool, Cutting Oil, which is used for drilling, thread cutting, thread moulding, lathing, countersinking, rubbing and sawing. It is ideally suited for stainless steel, but also for high-alloy steels, constructional steels, non-ferrous and precious metals. It can be also used as a preservative agent on semi-finished/finished parts and machine parts, and can be used as gun oil [12];
- Rhenus RL 3 DW is a thin corrosion preventive fluid with water displacing properties solved in hydrocarbons free of aromatics, which is free of heavy metals. It is universally used for grinding and machining operations of cast iron, steels and aluminium alloys [13]

The main physicochemical properties of these cutting fluids are presented in Table 1.

Table 1. Main physicochemical properties of cutting fluids [10], [11], [12], [13]

Properties	INSA Rezinol C-22	INSA Emulsol	Würth Cut & Cool, Cutting Oil	Rhenus RL 3 DW
Color	Light brown	Light Brown	Tan	Brown
Physical state	Liquid	Liquid	Liquid	Liquid
Odour	Characteristic	Characteristic	Characteristic	Characteristic
Kinematic viscosity at 40°C,mm^2 /s	22	24	-	100
Dynamic viscosity at 40°C, mPa.s	-	-	3.5	-
Flash point /C.O.C./, °C	210	200	-	-
Pour point , °C	-15	-15	-	-
Density at 20 °C , g/ml	0.860	0.855	-	0.78

Rheological tests were performed using a Brookfield viscometer CAP 2000+, with a cone and plate geometry. Three type of tests have been used:

- Determination of the soaking time, which represents the time necessary for the measurement to become stable and for the fluid to become homogeneous. Three „Time to stop" tests (variation of the apparent viscosity versus time) were performed for each type of fluid, with different rotation speeds: 100, 200 and 500 rpm.
- Determination of the rheological model, using a "velocity imposed gradient" test (variation of the shear stress versus shear rate), with the rotation speed range between 100 rpm and 1000 rpm, using cone geometry number 8.
- Determination of the variation of apparent viscosity with temperature, using a "velocity imposed gradient" test, with the rotation speed 500 rpm and temperature range between 20^0C and 75^0C.

IOP Conf. Series: Materials Science and Engineering **514** (2019) 012007 doi:10.1088/1757-899X/514/1/012007

The experimental results corresponding to the determination of the rheological parameters were numerically treated assuming the validity of power law rheological model:

$$\tau = m \left(\frac{\delta u}{\delta y}\right)^{n} \qquad (1),$$

where: τ – shear stress
m – consistency index
n – flow index
$\frac{\partial u}{\partial y}$ – shear rate

Regarding the variation of the apparent viscosity versus temperature, the Reynolds model was used:

$$\mu = \mu_{50} e^{m(t-50)} \qquad (2),$$

where: μ – apparent viscosity
μ_{50} – viscosity at 50 ^{0}C
m – temperature parameter
t – temperature

3. Results

The experimental results corresponding to the determination of the soaking time, for all four fluids, are presented in Figures 1, 2, 3 and 4.

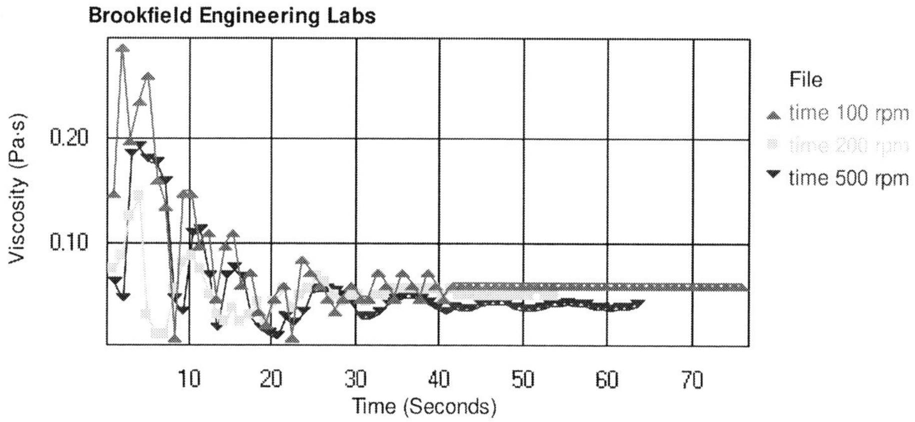

Figure 1. Variation of the apparent viscosity versus time for INSA Rezinol C-22

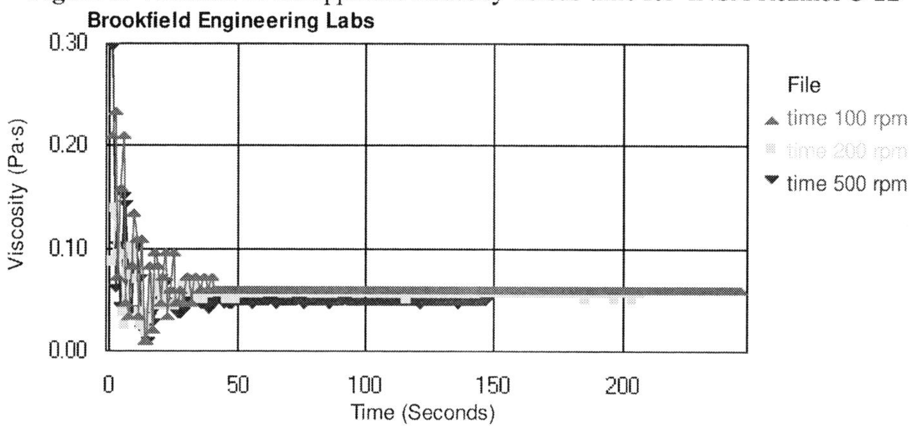

Figure 2. Variation of the apparent viscosity versus time for INSA Emulsol

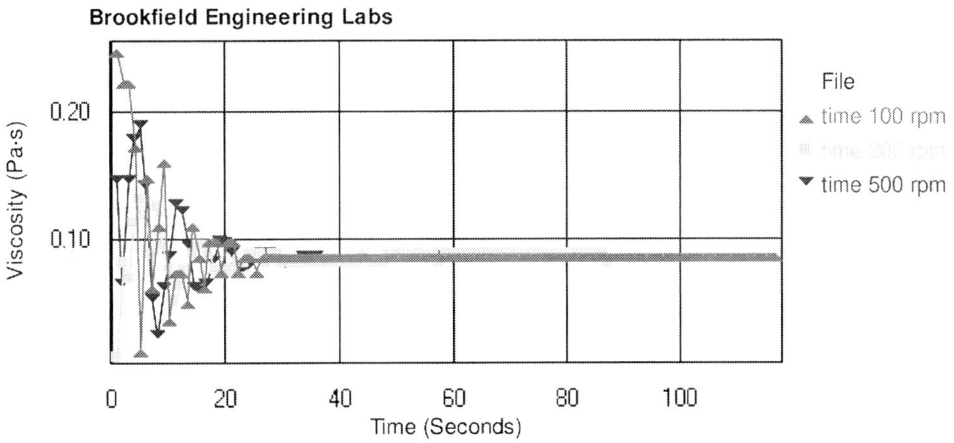

Figure 3. Variation of the apparent viscosity versus time for Würth Cut & Cool, Cutting Oil

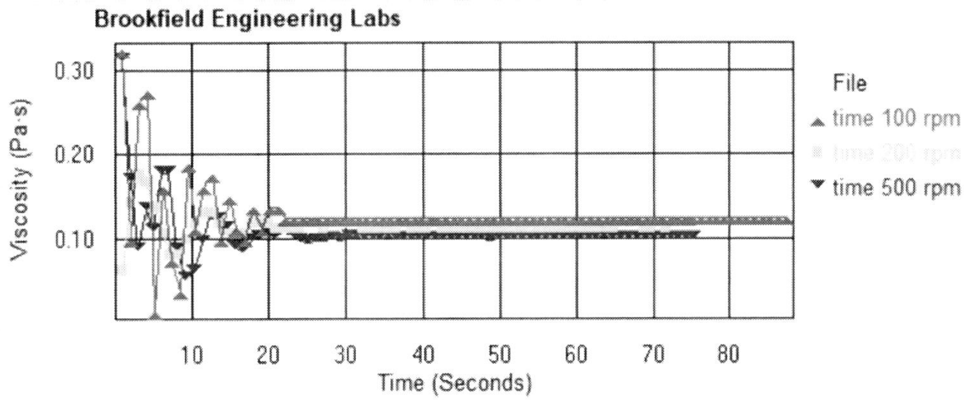

Figure 4. Variation of the apparent viscosity versus time for Rhenus RL 3 DW

Table 2 presents the effective values of the soaking time for each one of four fluids. These values are necessary for the second type of test – determination of the rheological model, because they serve as input parameters for test designing.

Table 2. Effective values of the soaking time

	INSA Rezinol C-22	INSA Emulsol	Würth Cut & Cool, Cutting Oil	Rhenus RL 3 DW
Soaking time, sec.	40	50	30	30

The experimental test regarding the determination of the rheological parameters consists of a load from the 200 s^{-1} to 2000 s^{-1} shear rate gradient, followed by an unload in order to highlight the thixotropy of the lubricant - "shear memory". The test is repeated three times for each cutting fluid, at a temperature of 20 ^{0}C, and the duration of homogenization (soaking time) of the sample at a certain shear rate was given by the values from Table 2. The comparative rheograms (Figure 5) are obtained by plotting shear stress as a function of the shear rate, as an average of three points, using the software Capcalc 32 specific for the viscometer.

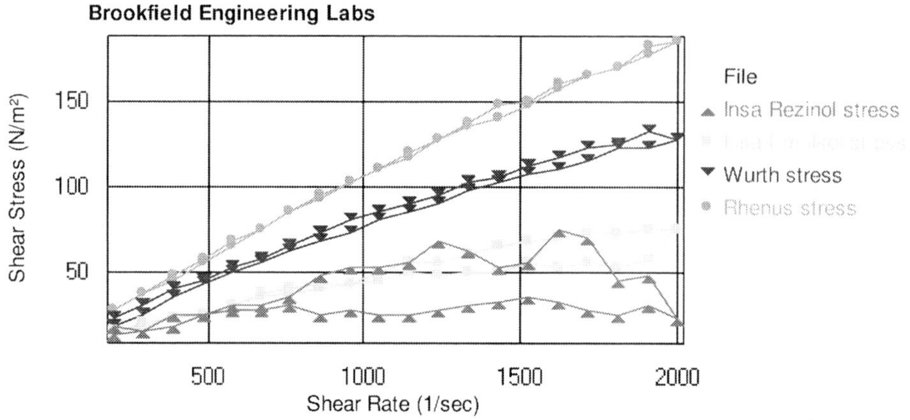

Figure 5. Comparative rheograms for all four cutting fluids, at 20 ^{0}C

The results regarding the variation of the apparent viscosity with temperature, using a "velocity imposed gradient" test, are presented in Figure 6.

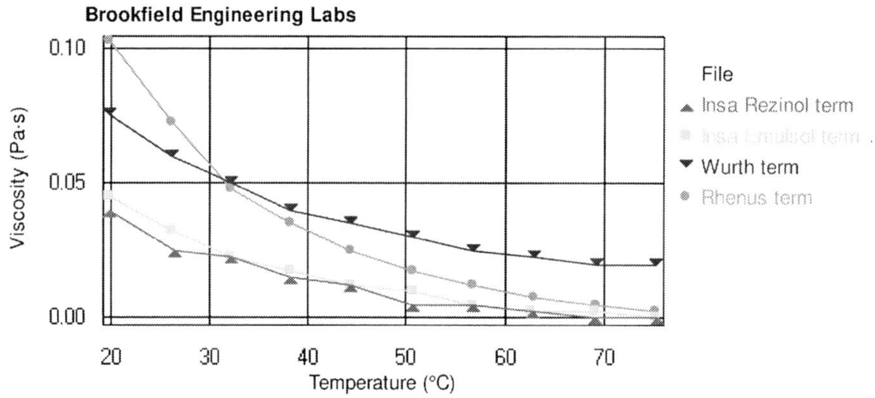

Figure 6. Variation of the apparent viscosity versus temperature for all four cutting fluids, at a constant shear rate gradient of 1000 s^{-1}

4. Discussions

The experimental results concerning the determination of the rheological parameters, obtained in the hypothesis of validity of power law model (eq. 1), are presented in Table 3. These results are also synthesized in Figures 7 and 8.

Table 3. Rheological parameters according to power law model, for all four cutting fluids

Cutting fluid	Consistency index, Pa·s^n	Flow index	Correlation coefficient, %
INSA Rezinol C-22	1.393	0.461	72.1
INSA Emulsol	0.344	0.703	92.4
Würth Cut & Cool, Cutting Oil	0.278	0.815	96.3
Rhenus RL 3 DW	0.313	0.842	98.7

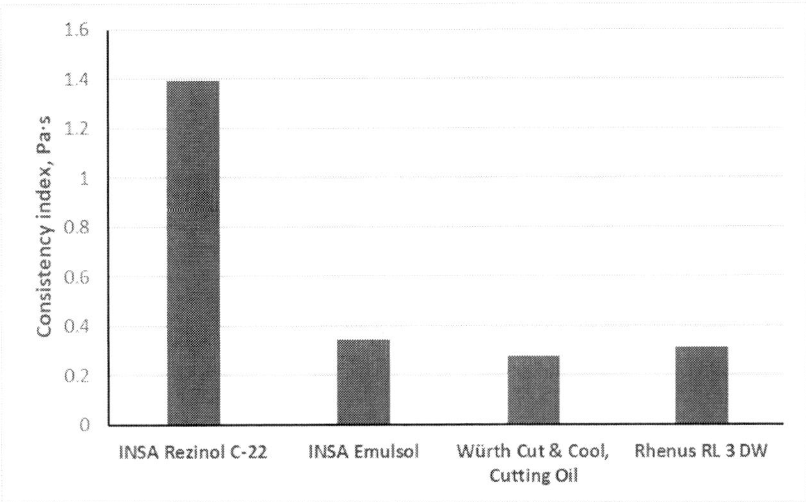

Figure 7. Variation of the consistency index, for all cutting fluids

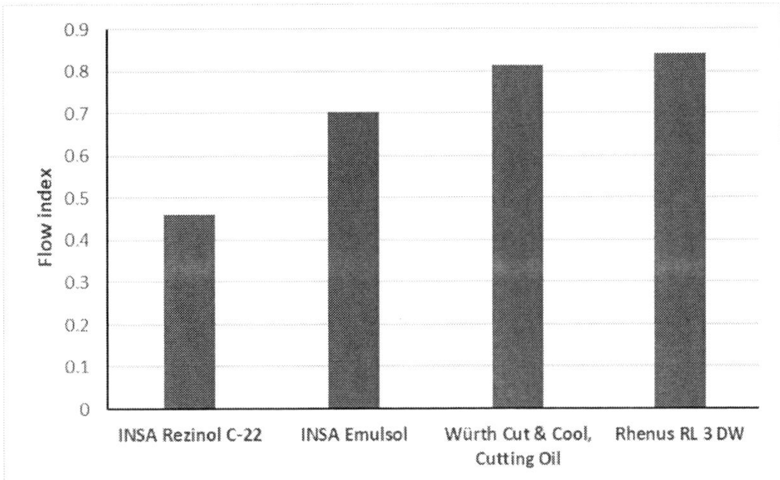

Figure 8. Variation of the flow index, for all cutting fluids

Analysing the rheological results (Figure 5), it can observe that INSA Rezinol C-22 and INSA Emulsol cutting fluids have a pronounced thixotropic behaviour, which is due to the inhomogeneity of the internal structure. For the other two cutting fluids, Würth Cut & Cool, Cutting Oil and Rhenus RL 3 DW, the thixotropic behaviour is almost absent, which denotes a very stable and homogeneous structure.

Regarding the rheological model, from Table 3 results that power law model is very appropriate for INSA Emulsol, Würth Cut & Cool, Cutting Oil and Rhenus RL 3 DW cutting fluids, because for these three fluids the correlation coefficient is very high, over 92%. For INSA Rezinol C-22 cutting fluid the correlation coefficient is very reduced, due to the pronounced thixotropy of the fluid, so the power law model cannot be considered.

From the point of view of the consistency index, it can be observed that there is approximate no difference between cutting fluids INSA Emulsol, Würth Cut & Cool, Cutting Oil and Rhenus RL 3 DW. Regarding the flow index, this parameter has different values according to the type of cutting fluid, in the range of 0.45 … 0.85.

The experimental results concerning the variation of the apparent viscosity versus temperature, according to the Reynolds model (eq. 2), are presented in Table 4 and Figures 9 and 10.

Table 4. Rheological parameters for Reynolds model

Cutting fluid	μ_{50}, Pa·s	m, $^0C^{-1}$	Correlation coefficient, %
INSA Rezinol C-22	0.005	-0.0715	89.2
INSA Emulsol	0.012	-0.0571	97.4
Würth Cut & Cool, Cutting Oil	0.031	-0.0253	95.0
Rhenus RL 3 DW	0.018	-0.0634	98.6

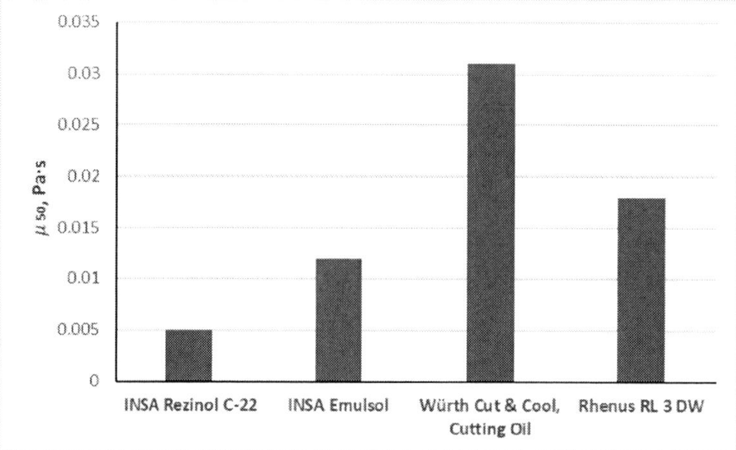

Figure 9. Variation of the apparent viscosity at 50 ^0C, for all cutting fluids

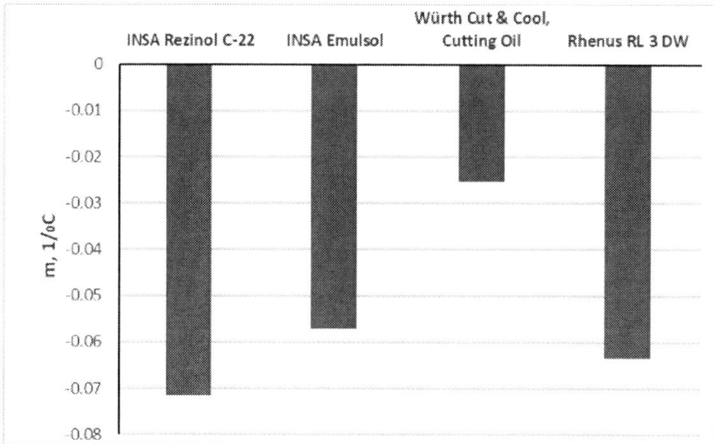

Figure 10. Variation of the temperature parameter, for all cutting fluids

Analysing the thermal results (Figure 6 and Table 4), it can observe that Würth Cut & Cool, Cutting Oil and Rhenus RL 3 DW cutting fluids are more stable on the whole range of temperatures than INSA Rezinol C-22 and INSA Emulsol cutting fluids. The apparent viscosity of Rhenus RL 3 DW is greater than for Würth Cut & Cool, Cutting Oil for temperatures between 20 ^0C and 32 ^0C. If the temperature is greater than 32 ^0C, Würth Cut & Cool, Cutting Oil has a greater apparent viscosity than Rhenus RL 3 DW.

Another interesting observation is the fact that the variation of the apparent viscosity for Würth Cut & Cool, Cutting Oil is less affected by the increasing of the temperature than for Rhenus RL 3 DW. This means that the viscosity index of Würth Cut & Cool, Cutting Oil is greater than that of Rhenus RL 3 DW, which is an important quality of a cutting fluid.

5. Conclusions

In this paper was performed the study of the rheological properties on a range of temperatures for four different cutting fluids: INSA Rezinol C-22, INSA Emulsol, Würth Cut & Cool, Cutting Oil and Rhenus RL 3 DW. It results that INSA Rezinol C-22 and INSA Emulsol cutting fluids show very important thixotropic behavior, which means that they have reduced stability during functioning and they are less homogeneous compared to Würth Cut & Cool, Cutting Oil and Rhenus RL 3 DW. Also, the apparent viscosity of INSA Rezinol C-22 and INSA Emulsol cutting fluids is very small, so from rheological point of view, it is recommended to use them in common cutting processes, without intense cutting regimes.

Würth Cut & Cool, Cutting Oil and Rhenus RL 3 DW cutting fluids are much more stable and homogeneous and have almost no hysteresis or thixotropy. They are recommended, from rheological point of view, to be used for intensive cutting processes, with high surface quality requirements. If the local temperature during the cutting process is low (20 ^{0}C ... 40 ^{0}C), Rhenus RL 3 DW cutting fluid is recommended, because it has a greater viscosity than Würth Cut & Cool, Cutting Oil. In the case when the local temperature during the cutting process has important variation, from the ambient temperature to high values, Würth Cut & Cool, Cutting Oil cutting fluid is recommended, because its apparent viscosity decrease less with temperature and it is more stable than that of the other studied fluids.

6. References

[1] S Kalpakjian, K S Vijai Sekar and S R Schmid 2014 *Manufacturing engineering and technology* Pearson

[2] W Grzesik 2008 *Advanced machining processes of metallic materials: theory, modelling and applications* Elsevier

[3] N Gopikrishna, M S Chander and M Srikiran 2016 Determining the Influence of Cutting Fluid on Surface Roughness during Machining of EN24 and EN8 steel by using CNC Milling Machine *Journal for Research| Volume 1*(**11**)

[4] E Kuram, B Ozcelik, E Demirbas and E Sik 2010 Effects of the cutting fluid types and cutting parameters on surface roughness and thrust force *Proceedings of the World Congress on Engineering* **2** pp 978-988

[5] S Boran and A Tamas 2014 Rheological behavior of castor oil mixed with different pyromellitic esters *Journal of the Serbian Chemical Society* ***79*(2)** pp 241-251

[6] D R Brewer, J M Franco and L A Garcia-Zapateiro 2016 Rheological properties of oil-in-water emulsions prepared with oil and protein isolates from sesame (Sesamum Indicum) *Food Science and Technology* ***36*(1)** pp 64-69

[7] E D S Santos, A P D P Camargo, E A D Faria, F A F D Oliveira Junior, S M Alves and E L D Barros Neto 2017 The Lubricity Analysis of Cutting Fluid Emulsions *Materials Research* ***20*** pp 644-650

[8] N J Alderman, D Ram Babu, T L Hughes and G C Maitland 1988 The rheological properties of water-based drilling fluids *Proceedings of the Xth International Congress on Rheology, Sydney* **1** pp 140-142

[9] D Dichev, H Koev, D Diakov, N Panchev, R Miteva and H Nikolova 2017 Automated System for Calibratind Instruments Measuring Parameters of Moving Objects *Proceedings of 59th International Symposium ELMAR* pp 219-224

[10] *** http://insa-motoroils.bg/INSA_RESINOL_C22_18L

[11] *** http://insa-motoroils.bg/other-products/EMULSOL/INSA_EMULSOL_10_L

[12] *** http://pim.wurth.ca/Technical/SDS_893.050004.PDF

[13] *** http://www.pingyiao.com/en/rhenus/rh8/r.rhenus%20RL%203%20DW.htm

IOP Conf. Series: Materials Science and Engineering **514** (2019) 012008 doi:10.1088/1757-899X/514/1/012008

Evaluation of the corrosion wear speed of different equipment in the water injection treatment plant

C Dudu[1], R G Ripeanu[1], A C Drumeanu[1], A Dinita[1] and A M Lospa[1]

[1]Mechanical Engineering Department, Petroleum-Gas University of Ploiesti, Romania

E-mail: adinita@upg-ploiesti.ro

Abstract. The paper presents the results of the theoretical and experimental researches carried out for the evaluation of the corrosive rate of different equipment from water injection treatment plant used for water flooding. The degradation are occurring mainly at pipelines and connecting elements of the flotation skid, in the form of advanced corrosion wear because the water injection contains an inconsistent composition mixture of dissolved gases, hydrocarbons, solid particles. The manifestation of corrosion wear at the skid connections indicates that the mechanical-physical-chemical treatments used to reduce aggressiveness are not sufficiently effective. The tests were carried out on the corrosion specimens taken from the critical areas of the skid connections and using the different formation water taken from the same critical points. Corrosion rate was obtained through experimental tests made in accordance with ASTM G5-15 and ASTM G1-03. The results materialized in the form of an influence of corrosion inhibitor and temperature on the corrosion rate for different formation water samples.

1. Introduction

Oil from reservoirs that have come to the surface through various exploitation methods contains water, gas and suspended solids. The fluid mixture coming from the wells to the park through the mixing pipes is guided through a manifold to different separators to separate the phases. Under the operating conditions in Romania, the deposit water represents on average about 60 ... 70% of the total amount of fluid extracted from the wells.

The collection consists of transporting the heterogeneous mixture of crude oil, reservoir water, associated gases and solid impurities from the production wells to treatment plants. Collecting and centrally separating well production has the advantage of requiring a smaller number of equipment's (separators, pumps, reservoirs, etc.), which increases economic efficiency. At the same time, these facilities serve to calibrate the production of each production wells.

Formation water resulting from separation and treatment is sent to specially prepared tanks for treatment and then injected into specially prepared wells. Formation waters separated from oil and gas is used as injection water (technological water) in secondary recovery processes or to be eliminated by injection (waste water) in specially designed wells.

Regardless of the destination, injection water must meet the following qualities, [1]:

- a high degree of purity obtained thought the lowest possible mechanical and oil content;
- lower degree of aggressiveness (corrosion);
- the highest stability;
- compatibility with the fluids and minerals in the reservoir formation;
- low oxygen content and bacteria.

Factors to be taken into account when choosing an injection water treatment system are related to the distances to sources and injection points and also to the possibilities of location on the ground so that gravity can be used in the process of pumping water [2].

The treatment and injection systems (open, closed or semi-closed systems) are selected according to the characteristics of the injection water.

A complete water treatment plant (Figure 1), comprising a chemical conditioning kit, a SPINSEP dissolved gas flotation unit and a MONOSEP walnut filter.

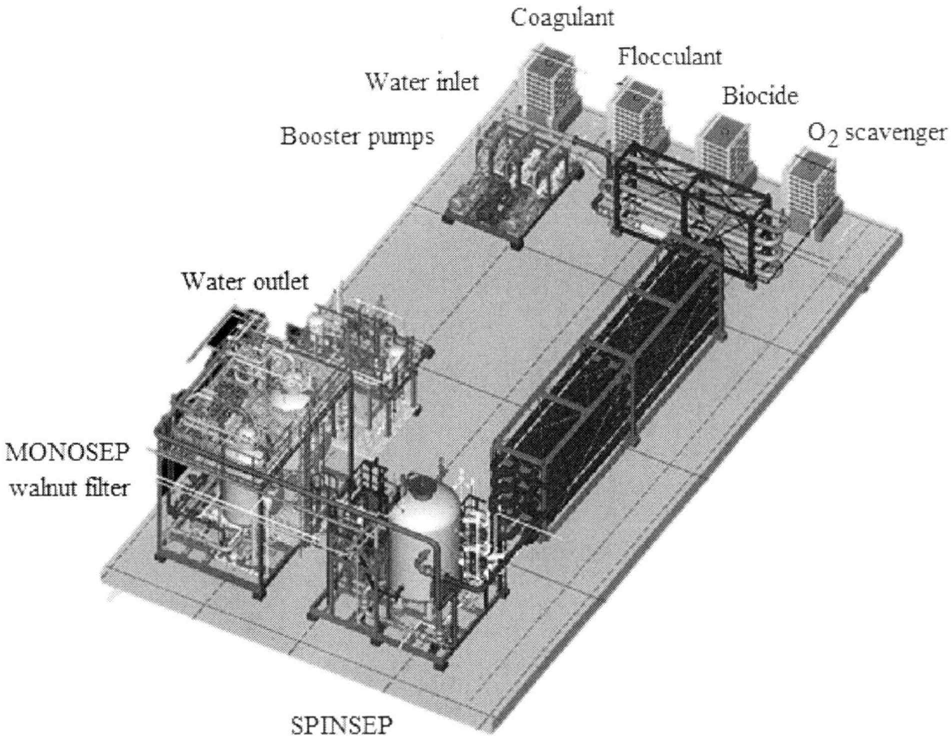

Figure 1. Water treatment plant design. [1]

The water is taken from a tank with a pump set and sent to the chemical conditioning units. Here two chemical agents are added and conditioned to modify the water content of the particles, facilitating separation. Both the SPINSEP flotation system and the MONOSEP filter have a vertical design for a minimum surface area. Formation water is pumped from storage tanks to the treatment plant, the chemical conditioning subsystem, with three centrifugal water transfer pumps. The water is transferred through the flocculant and coagulant injection system into the secondary separation system, SPINSEP. In addition to the flocculant and coagulant, seawater is also treated with an oxygen inhibitor to remove dissolved oxygen in water and a biocide to remove bacteria.

The main degradations occurred at the equipment in the water treatment plants were recorded mainly in the pipes and the connecting elements of the SPINSEP gas flotation skid, in the form of their advanced corrosion. Locations of these degradations were founded at bends and elbows where the direction of the pipes is changed to 90° to ensure a turbulent tangential flow of the fluid at the entrance to the flotation vessel, at section changes of the pipes, to the connecting elements (flanges, threads) provided at the pipes of the gas flotation skid, welding joints.

2. Experiments tests

2.1. Corrosion tests
Experimental tests were performed in order to establish the influence of corrosion inhibitor and temperature on corrosion rate in different stages of treatment in an injection plant. Electrochemical parameters were obtained using Voltalab 10 system and Voltamaster software, tests being performed according to ASTM G5-15, [4] and ASTM G1-03(2017), [5] ed.1 standards.

2.2. Samples preparation
It is known that stress move the material electrochemical potential to became more anodic, and also fluid speed, turbulence, changing directions and sections increase corrosion wear [6,7,8] so at injection water circuit in injection plant elbows were critical zones. So, samples were extracted from extrados 6 in. elbows (P285NH - chemical composition is presented in Table 1 according to EN 10222-4 and compare with experimental test made with Oxford Foundary-Master Pro) without affecting the material structure by waterjet cutting machine. Disks with Ø 16 X 3 mm were machined by grinding at low intensive parameters using cooling fluid.

Table 1. Chemical composition % of steel P285NH (1.0477).

	C	Si	Mn	Ni	P	S	Cr	Mo	V	N
EN 10222-4	max 0.18	max 0.4	0.6 - 1.4	max 0.3	max 0.025	max 0.015	max 0.3	max 0.08	max 0.05	max 0.02
Experimental	0.189	0.23	0.908	0.006	0.007	0.003	0.02	0.001	0.003	-

The surface exposed to corrosion test medium were polished with abrasive paper of 500 Mesh gradually. In Figure 2 is presented the corrosion cell and potentiostat. Corrosion cell works with a saturated calomel reference electrode ($Pt/Hg/Hg_2Cl_2$), graphite counter electrodes and specimen holder expose 1 cm^2 of the specimen to the 1000cm^3 test solution.

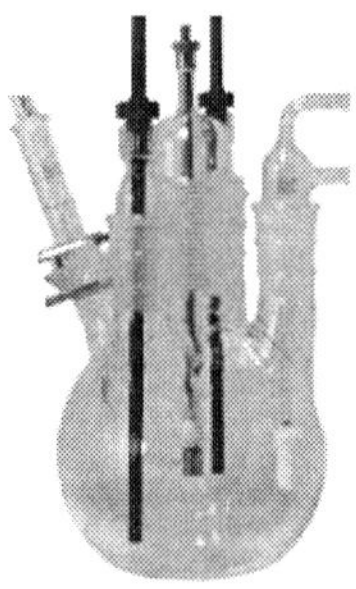

(a)

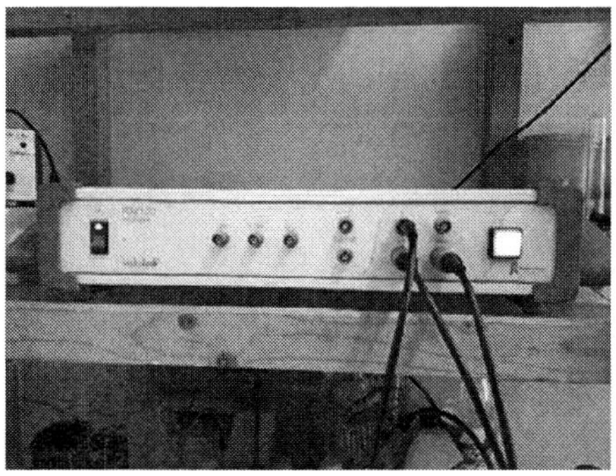

(b)

Figure 2. Corrosion cell and potentiostat.
(a) Corrosion cell; (b) Potentiostat.

2.3. Corrosion rate calculus
There are several electrochemical techniques that can be used to evaluate the behavior of materials in aggressive medium such as [9,10,11,12]: potentiodynamic anodic, cathodic or both polarization measurements, galvanic corrosion measurements, potentiostatic measurements, linear polarization, pitting scans, Evans plots measurements etc. Evans plots technique quickly yields corrosion rate information. The linear portion of the anodic or cathodic polarization logarithm current vs. potential plot is extrapolated to intersect the corrosion potential line (Tafel plots). This permits rapid, high accuracy measurement of extremely low corrosion rates. For this reason, to determine electrochemical

parameters we used this technique.

According to the mixed potential theory [6,7], any electrochemical reaction can be divided into two or more oxidation and reduction reactions and can be no accumulation of electrical charge during the reaction. In a corroding system, corrosion of the metal and reduction of some species in solution is taking place at same rate and the net measurable current, i_{meas} is zero. Electrochemically, corrosion rate measurement is based on the determination of the oxidation current, i_{ox} at the corrosion potential, E_{corr}. This oxidation current is called the corrosion current, i_{corr}.

$$i_{mas} = i_{red} - i_{ox} = 0 \quad (1)$$

When a metal or an alloy came in contact with an aggressive medium the material potential in open circuit is the corrosion potential E_{cor}. Many of modern corrosion techniques are based on theoretical of shapes of the polarization curbs Stern and Geary.

If a potential is applied on material from an external source and reaction rate is controlled, we have, [8]:

$$i_{red} = i_{ex} - \frac{\eta}{\beta_1} \text{ and } i_{ox} = i_{ex} - \frac{\eta}{\beta_2} \quad (2)$$

where: η is overpotential, equal with the difference between the material external potential applied, E_{apl} and the corrosion potential, E_{corr};
β_1 şi β_2 – constants.

By logarithm of above relations, we obtain the J. Tafel equations:

$$\eta = -\beta_C \log \frac{i_{red}}{i_{ex}}$$
$$\eta = -\beta_A \log \frac{i_{ox}}{i_{ex}} \quad (3)$$

where: $\beta_C = 2{,}3\beta_1$ and $\beta_A = 2{,}3\beta_2$ are Tafel constants, cathodic respective anodic.

Once established corrosion potential E_{corr} in volts and corrosion current density i_{corr} in $\mu A/cm^2$ the corrosion rate could be calculated.

According to Faraday law we have:

$$Q = \frac{nFW}{M} \quad (4)$$

where: Q – charge, Coulomb;
n – number of electrons involved in electrochemical reactions;
F – Faraday constant, 96487 Coulomb;
W – weight loss of the material, g;
M – atomic mass of the material, g.

We could obtain corrosion rate, usually expressed in length units/time (mm/year), CR:

$$CR = i_{corr}\left(E.W.\right) / \rho F \quad (5)$$

where: $E.W. = M/n$ – equivalent weight, g;
n – number of electrons involved in electrochemical reactions;
ρ – density of the material in g/cm^3.

2.4. Test mediums

Test fluids were injection waters collected in different stages of treatment at injection plant. In Table 2 are presented the injection water composition collected from outlet header of collecting park which are transferred to injection plant.

Table 2. Injection water at outlet collecting park header composition.

Element	Free CO_2	Dissolved CO_2	Sulfurs	Barium	Calcium	Dissolved Fe	Total Fe	Potassium
Value, mg/l	88	28	0.04	103	3573	34	35	170
Element	Magnesium	Natrium	Strontium	Bromides	Chlorides	Iodides	Bicarbonates	Lithium
Value, mg/l	1779	17550	127	143	39440	13	195	2.4

The injection water from collecting park enter in the injection treatment plant and was labelled as Sample 1. The water samples were labelled according to the collecting point at different stages in the injection treatment plant as presented in Table 3. The pH of water samples was established with pH-meter type Radiometer PHM - 201 Bc with combined pH electrode and an electrode for temperature.

Table 3. Injection waters samples.

Labelling	Sample 1	Sample 2	Sample 3	Sample 4	Sample 5	Sample 5
Collecting stages position	Outlet collecting park	Decantation tank outlet	Buster pump inlet	SPINSEP inlet	SPINSEP outlet	Outlet from plant
pH value at 25^0C	6.46	6.78	6.88	6.94	6.98	7.06

In order to evaluate the temperature and corrosion inhibitor influence on corrosion rate were performed tests at 25^0C, 30^0C and 35^0C and at different corrosion inhibitor concentrations. The corrosion inhibitor must be correlated with material and medium. Was used a corrosion inhibitor type CRW 85579 which contains etandiol, ethers, dodecenylsuccinic acid, quaternary ammonium compounds, benzyl (hydrogenated tallow alkyl), dimethyl and chlorides. Tests were performed at 0ppm, 30ppm, 60ppm, 100ppm and 480ppm.

3. Results

3.1. Corrosion test

In Figure 3 it is shown the Voltamaster capture with Evans plots for tested elbow sample material (P285NH) in Sample 3 injection water. The plots (fig. 3 and 4) were obtained from the Voltamaster software, applying the relation presented in paragraph 2.3, based on the increase rate of corrosion potential with an 1 mV/sec and the corrosion current witch occurs.

Electrochemical parameters obtained for extrados elbow material samples tested in injection water samples at 25^0C are presented in Table 4.

The Voltamaster software can calculates the following input and output parameters:

- *Input*
  - Calomel reference electrode (the potential measured according to used electrode);
  - Sample exposed to medium surface, 1 cm^2;
  - Material sample, P285NH (iron element witch is corroded – Fe with atomic mass of 55.85 g, valence of 2 and density 7.85 g/cm^3);
  - Analyses potential domain – 1000 mV to + 1000 mV;
  - Smoothing, calcul zone and segment were choosed in order to obtain for multiple determination coefficient, *Coef* =1.
- *Output*

- *E(i=0)* – corrosion potential, mV;
- *Rp* – polarization resistance, ohm.cm^2;
- i_{corr} – corrosion current density, μA/cm^2;
- B_c, B_a – Tafel constants, cathodic respective anodic;
- *Coef* – multiple determination coefficient;
- *Corrosion* - corrosion rate, mm/year.

Table 4. Electrochemical parameters for extrados elbow material in injection water samples.

Water samples	Electrochemical parameters		
	Current density, μA/cm^2	Corrosion potential, mV	Corrosion rate, mm/year
Sample 1	57.0656	-693.2	0.6632
Sample 2	50.8816	-565.0	0.5913
Sample 3	50.8673	-641.7	0.5911
Sample 4	44.6664	-661.8	0.5191
Sample 5	41.8391	-687.2	0.4862
Sample 6	37.7738	-667.2	0.4389

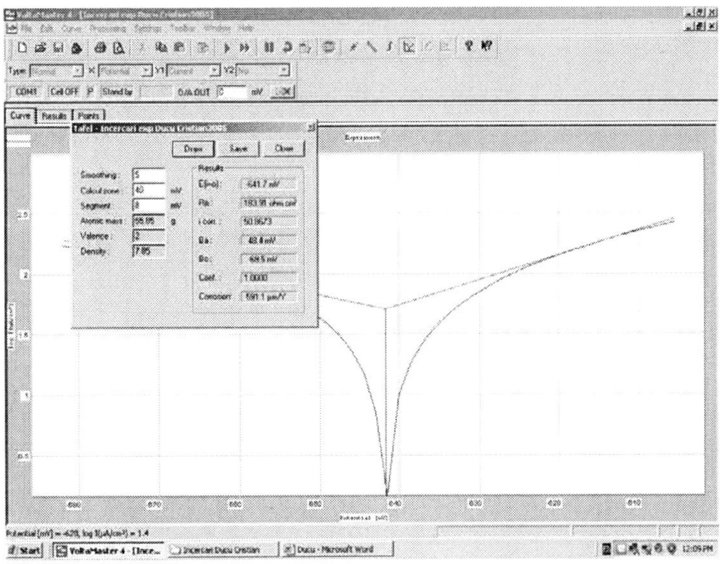

Figure 3. Electrochemical parameters and Evans plots for Sample 3 injection water at 25^{0}C.

Analyzing the results showed in table 4 we could conclude that the corrosion rate obtained have important values even in treatment injection water plant these values are diminished the final water which will be injected in wells has still a great value.

Formation water composition presented in table 1, show that it is aggressive but decantation, corrosion inhibitor used, and filtration must conduct to smaller values of corrosion rate. This higher water aggressivity produce erosion-corrosion wear at elbows which have a durability of 6 month.

Increasing the temperature, the electrochemical reactions are accelerated, and corrosion rate became greater. In Figure 4 it is shown the electrochemical parameters and Evans plot at 30^{0}C.

Electrochemical parameters obtained at 25^{0}C, 30^{0}C and 35^{0}C are presented in Table 5. This testing temperatures for injection waters are usually obtained in spring and summer.

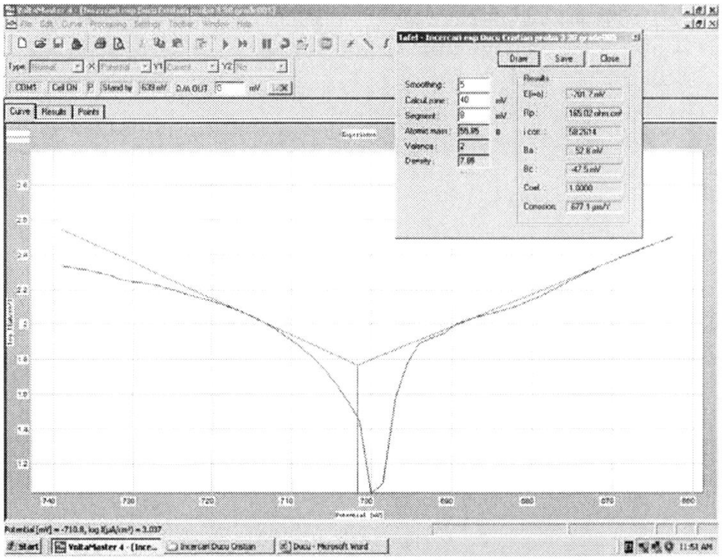

Figure 4. Electrochemical parameters and Evans plots for Sample 3 injection water at 30^{0}C.

The corrosion inhibitor optimum dilution is very important to inhibitor effectiveness. The corrosion rate values presented in table 5 suggest that inhibitor concentration was not proper used in correlation with injection water composition and temperature.

Table 5. Electrochemical parameters for extrados elbow material in injection water sample 3 at different temperatures.

Water sample 3 temperature, ^{0}C	Electrochemical parameters		
	Current density, μA/cm^2	Corrosion potential, mV	Corrosion rate, mm/year
25	50.8673	-641.7	0.5911
30	58.2614	-701.7	0.6771
35	72.4090	-656.5	0.8415

From table 5 we observe an important increase of corrosion rate with temperature which confirm the small elbow durability.

In Table 6 are shown the results of electrochemical parameters obtained for different corrosion inhibitor type CRW 85579 concentrations.

Table 6. Electrochemical parameters for extrados elbow material in injection water with different CRW corrosion inhibitor concentration.

Corrosion inhibitor concentration in injection water, ppm	Electrochemical parameters		
	Current density, μA/cm^2	Corrosion potential, mV	Corrosion rate, mm/year
0	57.0656	-693.2	0.6632
30	24.8311	-625.6	0.2885
60	6.5744	-669.2	0.0764
100	36.9839	-654.1	0.4298
480	54.5723	-634.9	0.6342

We observe from values presented in table 6 that optimum corrosion inhibitor concentration is 60 ppm (0.06 ml/l). With the increase of corrosion inhibitor concentration, the corrosion rate is greater. The corrosion rate obtained at CRW 480 ppm in injection water suggest that at injection plant were used such values of CRW concentration. The optimum CRW concentration obtained is for formation water composition presented in table 2 and at 25^0C. At different temperatures and compositions of waters the optimum value will be different.

4. Conclusions

Experimental program carried out in this paper consisted of the corrosion test on specimens (Ø 16 X 3 mm disks) extracted from elbows used in pipes water circuit in injection plant in 6 water samples, taken from water injection treatment plant, labelled *Sample 1 … Sample 6* (Outlet collecting park, SPINSEP inlet, SPINSEP outlet, Outlet from plant), in order to establish the injection waters aggressiveness in different stages of treatment in the plant and to obtain the optimum corrosion inhibitor concentration.

Experimental analysis of pH values for all 6 formation water samples leads to the conclusion that water samples have a slightly acidic character reaching neutral pH at water outlet from injection treatment plant (*Sample 6* pH = 7.06).

Corrosion tests carried out at 25^0C, 30^0C and 35^0C on specimens extracted from an elbow lead to the general conclusion that the corrosion rate is high, decreasing from the water inlet to outlet of the treatment plant. The corrosion rate increases significant with the temperature. We notice that in summer the water temperature easily reaches 30^0C or 35^0C.

The concentration of corrosion inhibitors introduced into the formation water is very important, the optimal value of corrosion rate (60 ppm at 25^0C) is obtained. An important observation is that with increasing concentration of inhibitors, from a certain value of this concentration, an increase in the corrosion rate is obtained.

References

[1] Ripeanu RG 2005 *Tribocoroziunea pompelor de extractive* (Ed. Universitatii din Ploiesti)
[2] Thomas S, Enhanced Oil Recovery – An Overview, *Oil and Gas Science and Technology – Rev. IFP*, 63 (2008), No.1, pp. 9-19
[3] ***AUSWAHL - REFERENZEN. Andritz Papiermaschinen 1997 Detailplanungsprojekte
[4] ***ASTM G5-14, Standard Reference Test Method for Making Potentiodynamic Anodic Polarization Measurements, ASTM International, West Conshohocken, PA, 2014, www.astm.org
[5] ***ASTM G1-03 2017, Standard Practice for Preparing, Cleaning, and Evaluating Corrosion Test Specimens, ASTM International, West Conshohocken, PA, 2017, www.astm.org
[6] Tudor I, Ripeanu RG 2002 *Ingineria coroziunii Vol.I and II* (Ed. Universitatii din Ploiesti)
[7] Roberge PR 2000 *Handbook of corrosion engineering* (Mc.Graw-Hill, New York)
[8] Papavinasam S 2014 *Corrosion Control in the Oil and Gas Industry*, (Gulf Professional Publishing)
[9] Ripeanu RG, Antonescu L 2005 The influence of welding joints above corrosion behavior at methane gas pipes, *Proceedings of the 2nd International Conference on Manufacturing Engineering „ICMEN", October 4-7, Greece,* pp. 457-463
[10] * * * Electrochemical Techniques for Corrosion 1978 (NACE Publication, Houston, U.S.A. 77027)
[11] Mansfield F 1973 Simultaneous Determination of Instantaneous Corrosion Rates and Tafel Slopes from Polarization Resistance Measurements, *J. Electrochem. Soc.*, pp. 515-521
[12] * * * Application note 140, (1978), Princeton Applied Research Corporation, Princeton, AN 140-10M-5

IOP Conf. Series: Materials Science and Engineering **514** (2019) 012009 doi:10.1088/1757-899X/514/1/012009

CFD Evaluation of sand erosion wear rate in pipe bends used in technological installations

A M Lospa[1], C Dudu[1], R G Ripeanu[1] and A Dinita[1]

[1]Mechanical Engineering Department, Petroleum and Gas University of Ploiesti, Romania

E-mail: adrian.lospa@gmail.com

Abstract. In this paper are presented the results of the CFD analysis evaluations performed by the authors to determine the erosion rate in pipe bends used in technological installations that circulate fluids together with a small percentage of solid particles under the form of sand. The pipe bends that were analysed have different curvatures and diameters and the same wall thickness. The CFD analysis is focused on determining the areas where the erosion occurs together with the erosion rate values. The main area where erosion occurs is at the pipe bend extrados. The differences that have been noticed are influenced by the pipe bend type, as the maximum erosion rate increases as the bend curvature increases. The research will continue with designing an experimental test rig for erosion testing to compare the CFD and experimental program results.

1. Introduction

Erosion has been long recognized as a potential source of problems in oil and gas production systems, being a complex process affected by multiple factors present in the operational conditions. Potential mechanisms that could cause significant erosion damage are, [1]:

- Particulate erosion;
- Liquid droplet erosion;
- Erosion-corrosion;
- Cavitation.

Particulate erosion is defined as the loss of material or loss of material integrity as a result of solid particle impact on the material surface. A minimum tolerance for erosion needs to be considered when designing oil and gas production systems. This can be done by setting an acceptable erosion rate, based on which the remaining target service life of the system and the system maintenance costs can be computed, [2].

According to Venkatesh, [3], the most vulnerable parts of production systems are the components in which flow direction changes suddenly, high flow velocities caused by high volumetric flow rates are present or high flow velocities caused by flow restrictions are present.

Some of the most important factors that need to be taken into consideration when determining the rate of particulate erosion are, [1, 4]:

- Fluid velocity – Material wear is a function of the fluid velocity by the means of a velocity exponent, n: $Wear = f(velocity)^n$;
- The mass of sand and the sand flowrate;
- The average sand particle size, its shape, sharpness and hardness;
- The fluid flow regime, fluid viscosity and density;

- The system design and configuration, the flow path profile and shape;
- The materials used for the piping system;
- The angle of impingement of the sand particles.

Extensive research on erosion has been carried out, based on which theoretical and empirical erosion models were developed, see Table 1.

Table 1. Theoretical and empirical erosion models.

Erosion Model	Mathematical expression	Description
Finnie and McFadden - Theoretical [5]	$V = \frac{c \cdot M \cdot U^2}{4P\left(1+m \cdot r^2 / I\right)}\left[\cos^2(\alpha) - \left(\frac{x_0'}{U}\right)^2\right]$	V is the volume removed from surface, M the mass of eroding particle, m the mass of individual particle, I the moment of inertia of the particle, r the average particle radius, the impact angle, U the particle impact velocity, P the horizontal component of flow stress, c the fraction of particles cutting in an idealized manner, x'_0 the horizontal velocity of the particle after cutting
Neilson and Gilchrist – Theoretical [6]	$\alpha \le \alpha_0 : W = \frac{M}{2}\left(\frac{V^2\cos^2(\alpha)\sin(n\alpha)}{\varphi} + \frac{V\sin(\alpha) - K^2}{\varepsilon}\right)$ $\alpha \ge \alpha_0 : W = \frac{M}{2}\left(\frac{V^2\cos^2(\alpha)}{\varphi} + \frac{(V\sin(\alpha) - K)^2}{\varepsilon}\right)$ $E = \frac{W}{M}$	M is the mass of erosive particles, W the eroded mass produced by M at impact angle of and velocity of V, K the velocity component normal to the surface below which no erosion takes place in certain hard materials
ANSYS Fluent Standard – Theoretical [7]	$R_{erosion} = \sum_{p=1}^{N_{particles}} \frac{m_p \cdot C(dp) \cdot f(\alpha) \cdot v^{b(v)}}{A_{face}}$	$C(dp)$ is the particle diameter function, the impact angle of the particle, $f(\ )$ the impact angle function, the relative particle velocity, $b(\ )$ the relative particle velocity function, A_{face} the area of the cell face at the wall. Default values: C = 1.8E-09, f = 1, b = 0.
McLaury and Ahlert - Empirical [8]	$ER = A \cdot F_s \cdot V^n \cdot f(\alpha)$ $\alpha \le a : f(\alpha) = a \cdot \alpha^2 + b \cdot \alpha$ $\alpha > a : f(\alpha) = x \cdot \cos^2(\alpha) \cdot \sin(\alpha) + y \cdot \sin^2(\alpha) + z$	a, b, c, w, x, y, z are impact angle function constants A = 1559HB - 0.59E-09, HB is the Brinell Hardness
Tulsa – Empirical [9]	$ER = \frac{A \cdot F_s \cdot F_p \cdot V^n \cdot m}{D_p^2 \cdot B^b}$	A is a constant, F_s the sand sharpness correction factor, F_p the wall material penetration factor, B the material Brinell hardness, n the velocity exponent, D_p the pipe diameter, b the hardness exponent
Salama and Venkatesh – Empirical [10]	$ER = \frac{S_k \cdot V^2 \cdot m}{d_{pipe}^2}$	S_k is a geometry dependent constant, S_k = 0.038 for short radius elbows and S_k = 0.019 for ells and tees d_{pipe} is the diameter of the pipe, m the sand flow rate, V the particle impact velocity
Oka et al. – Empirical [11]	$E(\alpha) = g(\alpha) \cdot E_{90}$ $g(\alpha) = \left(\sin(\theta)\right)^{n_1} \cdot \left(1 + H \cdot v\left(1 - \left(\sin(\theta)\right)\right)\right)^{n_2}$ $E_{90} = K \cdot (H \cdot v)^{k_1} \cdot \left(\frac{V_p}{V}\right)^{k_2} \cdot \left(\frac{D_p}{D}\right)^{k_3}$ $n_1 = s_1 \cdot (H \cdot v)^{q_1}; n_2 = s_2 \cdot (H \cdot v)^{q_2}; k_2 = 2.3 \cdot (H \cdot v)^{0.038}$	K, k_1, k_2, k_3, s_1, s_2, q_1, q_2, n_1, n_2 are constants and exponent values for particles material and targeted material

According to [12], the flow velocity should be the main factor used for design calculations. A threshold for the pipeline flow velocity, V_e [feet/sec], called the "erosional velocity", should be set, indicating the velocity above which erosion may occur.

$$V_e = \frac{C}{\sqrt{\rho}} \tag{1}$$

where: C represents an empirical constant; ρ is the fluid mixture density [lbs/ft^3].

Considerations related to the values for C, the empirical constant [12]:

- Conservative approach for solid-free fluids: C = 100 for continuous service and C = 125 for intermittent service;
- For solid-free fluids, where corrosion is not anticipated or is controlled: C = 150... 200;
- Values up to C = 250 have been successfully used for intermittent services.

According to Salama and Venkatesh, [10], erosion in solids-free fluids occurs only at very high velocities. To avoid this, a value of C = 300 must be used.

Based on [2], particle erosion in pipe bends can be estimated with the following expression:

$$E_{L,y} = \frac{K \cdot F(\alpha) \cdot U_p^n}{\rho_t \cdot A_t} \cdot G \cdot C_1 \cdot GF \cdot m_p \cdot C_{unit} \tag{2}$$

where: $E_{L,y}$ represents the annual surface thickness loss [mm/year]; K - the material erosion constant [(m/s)$^{-n}$]; U_p - the particle impact velocity [m/s]; n - the velocity exponent (n = 2.6 for ductile steels); ρ_t - the density of target material [kg/m^3]; A_t - the area exposed to erosion [m^2]; G - the particle diameter correction function [-]; C_1 - a model/geometry factor [-], GF - the geometry correction factor [-], m_p - the mass rate of sand [kg/s]; C_{unit} - the unit conversion factor (from m/s to mm/year) [-]; α represents the particle impact angle [°] and $F(\alpha)$ characterizes the ductility of the pipe bend material and has the following expression (the maximum erosion is experienced for impact angles of 20°... 50°) [-]:

$$F(\alpha) = 0.6 \cdot \left[\sin(\alpha) + 7.2 \cdot \left(\sin(\alpha) - \sin^2(\alpha)\right)\right]^{0.6} \cdot \left(1 - e^{-20\alpha}\right) \tag{3}$$

According to [2], the preferred option for complex piping configurations or for situations where the erosion location is of utmost importance, the preferred option is to run CFD erosion simulations.

2. CFD Analysis

Computational fluid dynamics (CFD) is a branch of fluid mechanics that uses numerical analysis and data structures to solve and analyse problems that involve fluid flows, [13]. A CFD analysis translates in solving the Reynolds averaged Navier-Stokes equations for a specified fluid domain, [2].

2.1. Geometric model

In order to evaluate the erosion impact on pipe bends, several configurations have been designed, taking into account different diameters and curvatures.

Table 2. Theoretical and empirical erosion models.

Bend Type	30° Bend	45° Bend	90° Elbow
Drawing	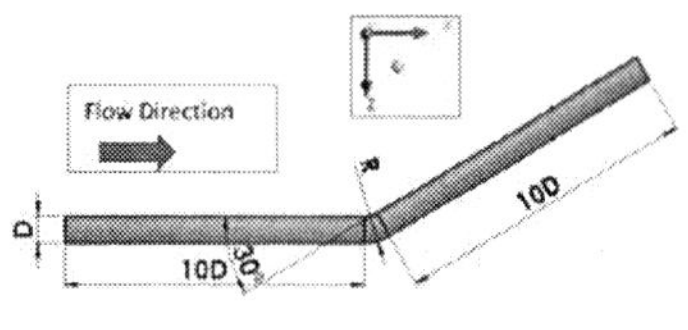	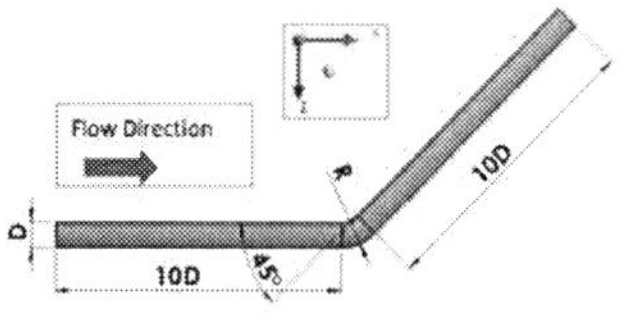	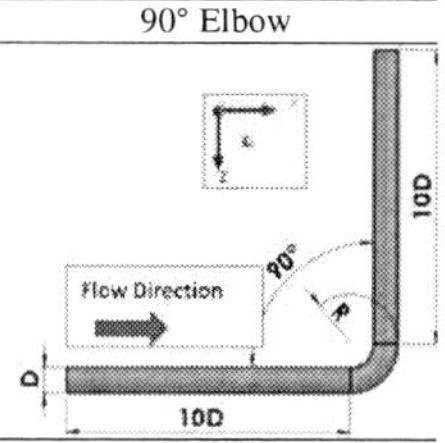

Diameter (D_e), mm	76.1	88.9	114.3	76.1	88.9	114.3	76.1	88.9	114.3
Wall thickness, mm	4.5	4.5	4.5	4.5	4.5	4.5	4.5	4.5	4.5
Bend radius, mm	95	114	152	95	114	152	95	114	152

The 9 bend configurations plus the specific dimensions can be seen in Table 2. Fluid flow enters through a straight section of 10*D* long, followed by a 30°, 45° or 90° bend section and continues through another 10*D* straight section.

Figure 1 shows the mesh details of the 30°, 45° and 90° bends for one of the diameters, D_e = 114.3 mm. Details about the number of cells and nodes of the mesh for all the bend configurations can be observed in Table 3.

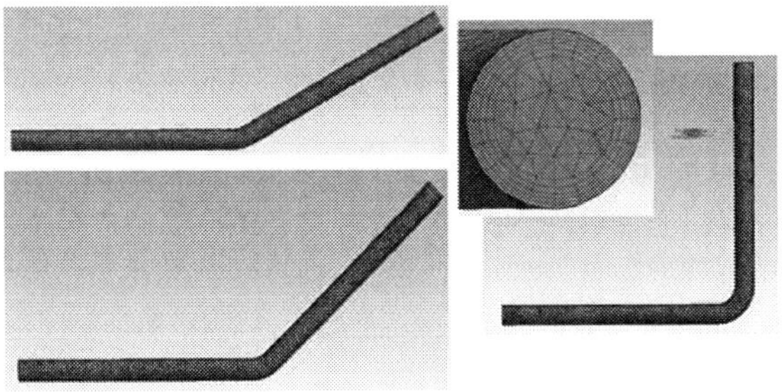

Figure 1. Mesh details for 30°, 45° and 90° pipe bends.

Table 3. Pipe bends mesh nodes and elements.

Bend Type	30° Bend		45° Bend		90° Elbow	
Mesh details	Nodes	Elements	Nodes	Elements	Nodes	Elements
D_e=76.1mm	22140	57168	22489	58199	22717	58191
D_e=88.9mm	21520	55609	22165	57196	22880	58958
D_e=114.3mm	21323	54970	21562	55701	22417	57872

2.2. *Flow model*

The problem is treated as the steady flow of water through the 9 bend configurations on the XZ horizontal plane, with gravitational effects included, g = -9.8 m/s^2 on Y axis). The specified temperature is 20°C, water density ρ = 998.2 kg/m^3 and viscosity η = 0.001003 kg·m^{-1}·s^{-1}. The flow velocity at the inlet on X direction is U = 5 m/s.

A standard k-ε two equations turbulence model based on turbulence kinetic energy (k) and its dissipation rate (ε) transport equations is used to model the turbulence. The model is valid only for fully turbulent flows. Scalable wall functions are used to model the wall-bounded turbulent flows. The turbulence model is modified to enable the viscosity-affected region to be resolved with a mesh all the way to the wall, including the viscous sublayer, [7].

2.3. *Simulation Conditions*

The following parameters have been adopted for the boundary conditions:

- Inlet velocity - U = 5 m/s;
- Solid Wall - wall roughness R_a = 6.3 μm; Reflection coefficients are set as per [14]: 4 polynomial coefficients for the normal and tangent direction; Impact angle function as per [14]: 5 points piecewise linear; Diameter function *C(dp)* = 1.8E-9 as per [7]; Velocity exponent n = 2.6 as per [14];
- Outlet - Pressure Outlet.

2.4. *Discrete phase modelling*

The Euler-Lagrange approach is used to model the discrete phase. The fluid phase is treated as a continuum by solving the Navier-Stokes equations, while the dispersed phase is solved by tracking a large number of particles through the calculated flow field. The dispersed phase can exchange momentum, mass, and energy with the fluid phase, [7].

Sand is used as discrete phase, with a density of ρ = 2500 kg/m^3, flow rate = 0.1 kg/s and a particle diameter of 1000 μm.

A Non-spherical Drag Law is used to model the non-spherical sand particles, [7]:

$$C_D = \frac{24}{\mathrm{Re}_D}\cdot\left(1+b_1\cdot\mathrm{Re}_D^{\,b_2}\right)+\frac{b_3\cdot\mathrm{Re}_D}{b_4+\mathrm{Re}_D} \tag{4}$$

where: C_D represents the drag coefficient and Re_D the Reynolds number.

The values for the b_1, b_2, b_3 and b_4 coefficients are calculated as follows:

$$b_1 = \exp\left(2.3288-6.4581\phi+2.4486\phi^2\right) \tag{5}$$

$$b_2 = 0.0964+0.5565\phi \tag{6}$$

$$b_3 = \exp\left(4.905-13.8944\phi+18.4222\phi^2-10.2599\phi^3\right) \tag{7}$$

$$b_4 = \exp\left(1.4681+12.2584\phi-20.7322\phi^2+15.8855\phi^3\right) \tag{8}$$

$$\phi = \frac{s}{S} \tag{9}$$

where: Φ represents the shape factor, s - the surface area of a sphere having the same volume as the particle and S - the actual surface area of the particle.

A value of = 0.9 for the sand particle shape factor was adopted to run the CFD analysis.

3. Results and conclusions

The maximum erosion rates in mm/year as per the CFD computation for steel pipe bends with density ρ = 7850 kg/m^3 are shown in Table 4.

Table 4. Maximum erosion rates shown as mm/year.

Bend Type	30°	45°	90°
D_e = 76.1mm	0.767 mm/year	1.25 mm/year	3.35 mm/year
D_e = 88.9mm	0.540 mm/year	2.01 mm/year	2.58 mm/year
D_e = 114.3mm	0.674 mm/year	1.08 mm/year	1.19 mm/year

The sand particle trajectories, together with the velocity magnitude, for the 30°, 45° and 90° bends, with D_e = 114.3 mm are presented in Table 5.

Table 5. Sand particles trajectory and velocity.

Bend Type	30° Bend	45° Bend	90° Elbow
D_e=114.3 mm	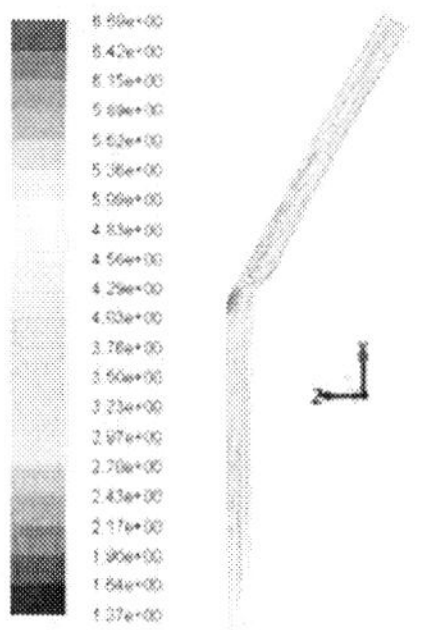	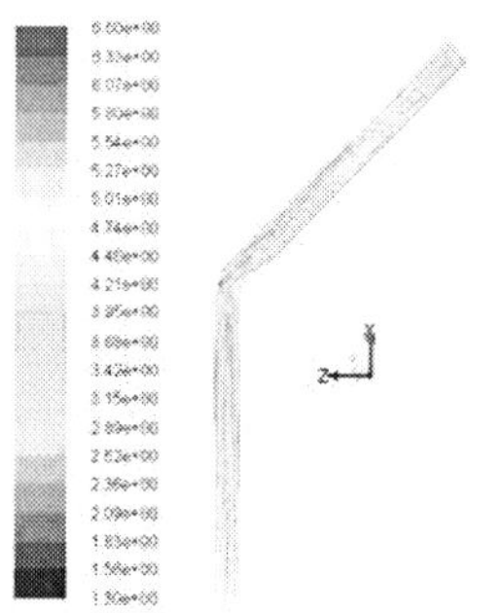	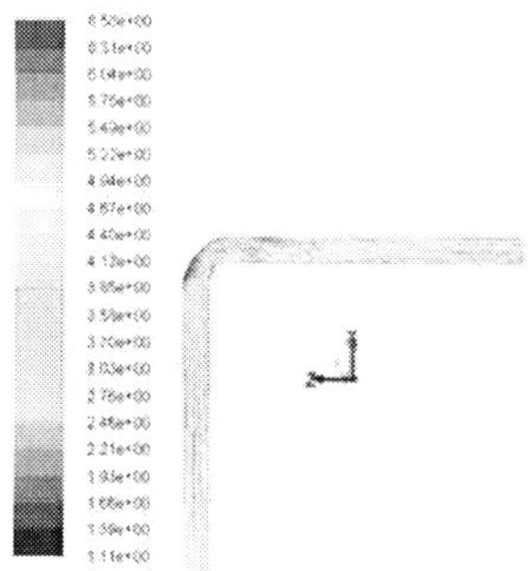

In Table 6, all the CFD simulation results for the 9 bend configurations are showcased.

IOP Conf. Series: Materials Science and Engineering **514** (2019) 012009 doi:10.1088/1757-899X/514/1/012009

Table 6. CFD Simulation Results – Erosion Rate.

<table>
<tr><th>Bend Type</th><th>30°</th><th>45°</th><th>90°</th></tr>
<tr><td>D_e = 76.1mm</td>
<td>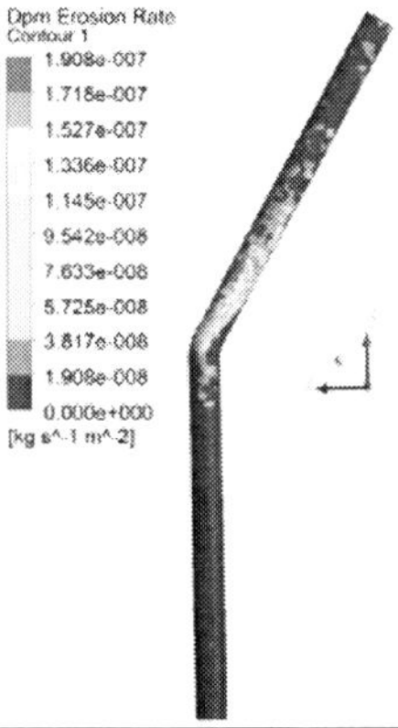</td>
<td>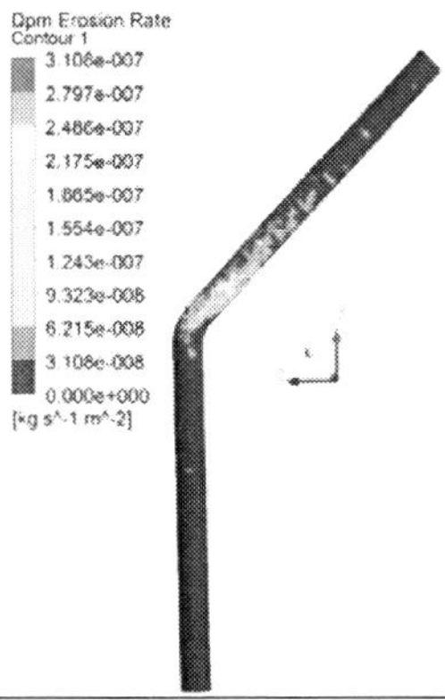</td>
<td>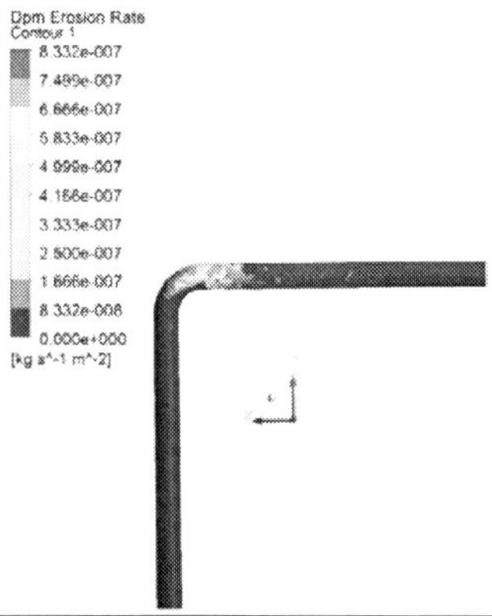</td></tr>
<tr><td>D_e = 88.9mm</td>
<td>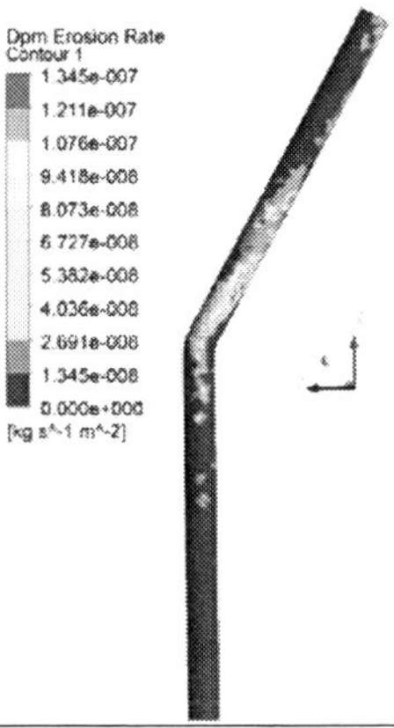</td>
<td>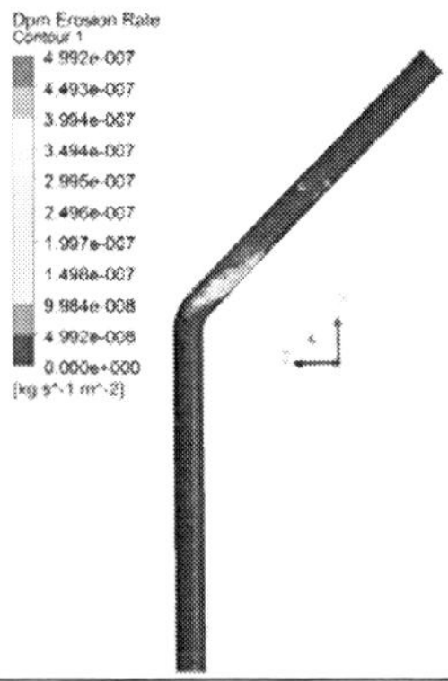</td>
<td>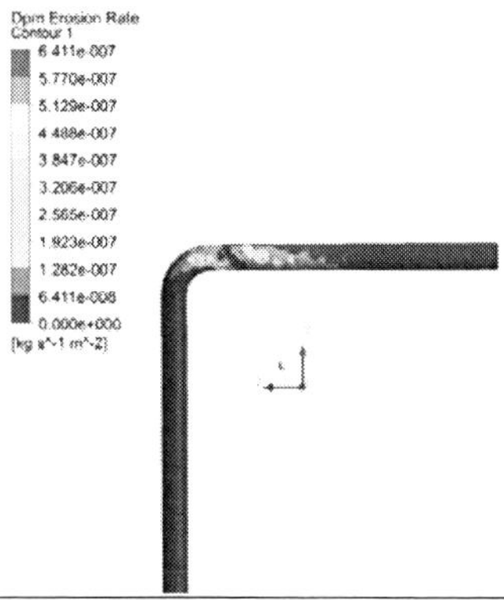</td></tr>
<tr><td>D_e = 114.3mm</td>
<td>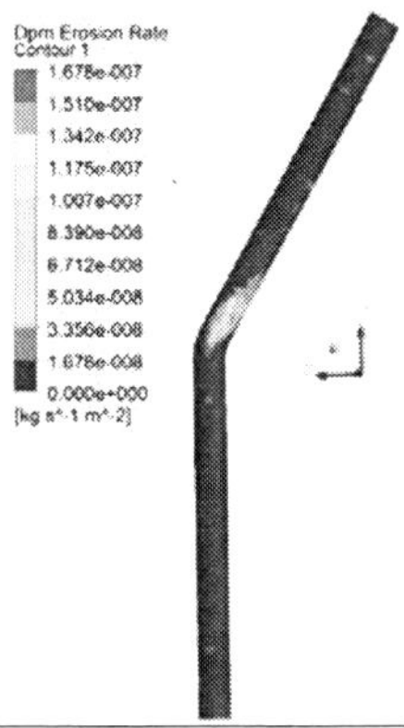</td>
<td>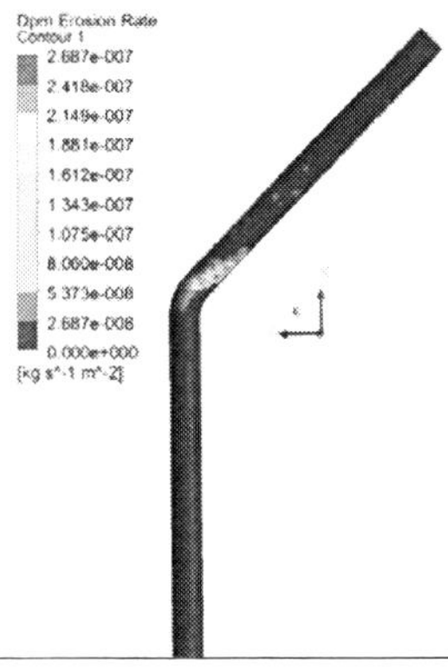</td>
<td>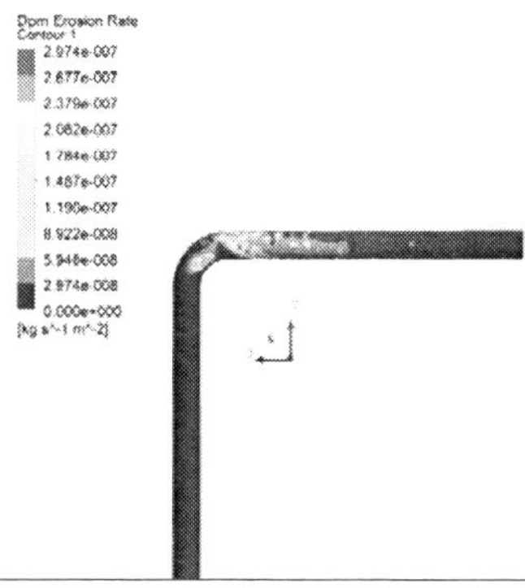</td></tr>
</table>

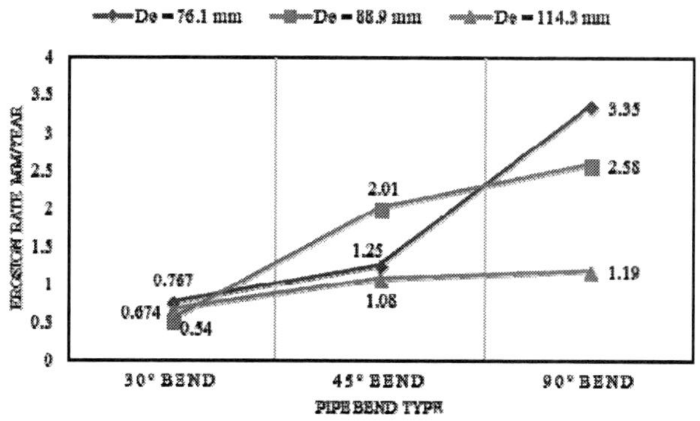

Figure 2. Maximum erosion rates based on bend types and diameters.

Simulation results are in good agreement with the experimental results of Bourgoyne [17] for 90° elbows conveying water and sand particles with an average diameter of 350 μm (see Table 7). Bourgoyne studied failure in diverter systems due to erosion produced by sand particles entrained in the flow stream. An experimental diverter system with various fittings was built to test and to compare erosion severity for different geometries (elbows and tees with a nominal diameter of 50.8 mm). The difference between simulation and experimental results is due to: different elbow diameters (76.1/88.9/114.3 mm vs. 50.8 mm), different sand flow rates (0.1 kg/s vs. 1.44/1.72/6 kg/s) and different sand particle diameters (1000 μm vs. 350 μm).

Table 7. Bourgoyne experimental data on erosion rate [17].

Geometry	r/D	U (m/s)	Sand flow rate (kg/s)	ER (kg/m^{2}s)	ER (mm/year)
Seamless elbow	1.5	9.45	1.44	3.42 x 10^{-9}	0.014 mm/year
Cast elbow	3.25	14.6	1.72	2.41 x 10^{-8}	0.1 mm/year
Cast elbow	3	11.5	6	3.49 x 10^{-9}	0.014 mm/year

Analysing the results from tables 4, 5 and 6, the following conclusion can be drawn:

- Maximum erosion rates are obtained for 90° bends, graphically shown in Figure 2;
- For 90° bends, the maximum erosion rate decreases as the pipe diameter increases;
- From a bend curvature perspective, the calculation is showing that the erosion rate increases as the bend curvature increases;
- After hitting the pipe bend wall, the sand particles are getting a swirl. The particles that are hitting the bend intrados are getting accelerated, the highest velocity increase being calculated for the 30° bend types;
- The working fluid type should be taken into consideration, for corrosive fluids the calculations need to account for erosion-corrosion;
- As next steps, the influence of the particle shape factor and the particle material type will be looked upon by the means of CFD analysis;

Experimental data availability for elbow erosion due to sand particles present in a liquid flow stream is quite rare based on the literature review performed by the authors. The research will continue with designing and building of an experimental test rig for erosion testing to generate the experimental data to compare with the CFD simulation results.

References

[1] Barton N A 2003 Erosion in elbows in hydrocarbon production systems: Review document prepared by TÜV NEL Limited for the Health and Safety Executive
[2] Det Norske Vertitas Recommended practice DNVGL-RP-O501 Managing sand production and erosion 2015
[3] Venkatesh E S 1986 Erosion damage in oil and gas wells *Proceedings Rocky Mountain Meeting of SPE, Billings MT May 19-21 1986* pp 489-497
[4] Peters J 2014 NEL UK Erosion Management Using Erosion Testing Combined with Erosion *6th European Sand Management Forum Aberdeen 26-27 March 2014*
[5] Finnie I and McFadden D 1978 On the velocity dependence of the erosion of ductile metals by solid particles at low angles of incidence *Wear 48(1)* pp 181-190
[6] Neilson J and Gilchrist A 1968 Erosion by a stream of solid particles *Wear 11(2)* pp 111-122
[7] ANSYS Fluent 15.0.7 Theory Guide https://www.sharcnet.ca/Software/Ansys/15.0.7/
[8] McLaury B S 1996 Predicting Solid Particle Erosion Resulting from Turbulent Fluctuations in Oilfield Geometries *University of Tulsa PhD Thesis*
[9] Ahlert K R 1994 Effects of particle impingement angle and surface wetting on solid particle erosion of AISI 1018 Steel *University of Tulsa MSc Thesis*
[10] Salama M M and Venkatesh E S 1983 Evaluation of Erosional Velocity Limitations in Offshore Gas Wells *15th Annual Offshore Technology Conference Houston Texas May 2 – 5 1983* pp 371 – 376
[11] Oka Y I, Okamura K and Yoshida T 2005 Practical estimation of erosion damage caused by solid particle impact: Part 1: Effects of impact parameters on a predictive equation *Wear 259(1)* pp 95-101
[12] API 14E Recommended practice for design and installation of offshore production platform piping systems 5th edition 1991 (revised 2000)
[13] Wikipedia Computational fluid dynamics https://en.wikipedia.org/wiki/Computational_fluid_dynamics
[14] Tutorial: Using FLUENT's Erosion Model to Investigate Erosion in a 90 degree Elbow Bend *Fluent Inc. November 6 2006*
[15] J. K. Edwards, B. S. McLaury, and S. A.Shirazi. "Evaluation of Alternative Pipe Bend Fittings in Erosive Service". In Proceedings of ASME FEDSM'00: ASME 2000 Fluids Engineering Division Summer Meeting, Boston, MA. June 2000
[16] Mohyaldin M E, Elkhatib N and Ismail M C 2011 Evaluation of different modelling methods used for erosion prediction in *Proceedings of the NACE Corrosion Shanghai Conference & Expo. 2011* pp 1-19
[17] Bourgoyne, A. T. (1989, January 1). Experimental Study of Erosion in Diverter Systems Due to Sand Production. Society of Petroleum Engineers. doi:10.2118/18716-MS

IOP Conf. Series: Materials Science and Engineering **514** (2019) 012010 doi:10.1088/1757-899X/514/1/012010

Some aspects regarding the roughness of the railway surface and rolling noise at locomotives

L M Babici[1], A Tudor[2]

[1]Romanian Railway Authority -RRA, 393 Calea Grivitei, 010719, Bucharest, Romania
[2] Polytechnic University of Bucharest, Mechanical and Mechatronics Faculty, Romania

E-mail: lbabici@yahoo.com, andreitudor1947@gmail.com

Abstract. Railway noise is activated by irregularities on the running surfaces of rails and wheels, these being more significant for the most part of the wavelength and the frequency range of interest. The rolling noise generation mechanism consists of two major structural components: the wheel and rail. The contribution to the noise of the track can be further separated into noise radiation on the rails and in the sleepers. To mitigate the rolling noise, these components must be analysed. To direct and manage mitigation methods within Faurei Railway Testing Center, the rail surface roughness in accordance with EN 15610 was determined in a line of alignment, this being a direct method of measuring the surface roughness of the rail, in association with the rolling noise introduction. In this paper,it is proposed a fractal analysis of the microgeometry of the rail surface, with the possibility of using it as an acoustic parameter.

1. Introduction

Noise is an unwanted sound due to its volume and can be harmful. As well as noise, vibration from railways can cause annoyance. This may be due to vibration, usually in the range 2 to 80 Hz, or due to the radiation of low frequency sound transmitted through the ground, usually in the range 30 to 250 Hz. Noise is an important and annoying side-effect of the process of rolling the wheel on the rail.. Different types of noise are generated by the wheel-rail interface and means of controlling them. They have as main source of rolling noise that occurs in a line of alignment and has a wide frequency content [-1-3].

The first step in noise control is to identify the dominant source. There are many different sources of noise from a railway, and in different situations the dominant source may vary [1,2].

- *Rolling noise* is the most important source of rail noise caused by wheel/rail vibrations induced at wheel/rail contact.
- *Curve squeal* noise is caused by the interaction between wheel and curve rails. It is strongly tonal, being associated with the vibration of the wheel in one of its resonances.
- *Bridge noise* is caused by dynamic forces acting on it from the track. Depending on the type of bridge, when a train runs over it, the emitted noise increases considerably.
- *Aerodynamic noise* is caused by a non-uniform airflow over the train compared to other railway noise sources, which are caused by the sound radiation through the vibrations of solid structures. Aerodynamic noise sources generally increase much faster with the train speeds than the mechanical sources.
- *Ground-borne vibration and noise* can be experienced in two ways. Low frequency vibration

between about 2 and 80 Hz is perceived as vibration and tends to be associated most with heavy freight trains. Higher frequency ground-borne vibration from about 30 to 250 Hz is associated more with trains in tunnels in urban areas such as metro operations but can also be significant for surface railways, particularly where noise barriers block out the direct airborne sound.

- Other sources of railway noise: Traction noise, warning signals from trains, track maintenance equipment, shunting noise.

2. The rolling noise and roughness

The rolling noise from the wheel-rail contact is the noise generated by the vehicle-track interaction. The most important form of noise generated at the wheel/rail interface is the rolling noise which is caused by the roughness of the wheel and rail running surfaces [3-4].
Significant rolling noise mechanisms attributable to wheel and rail roughness are:
(a) continuous deformation from gradually varying wheel and rail profiles,
(b) impact of wheel and rail asperities and
(c) parametric excitation associated with spatially varying rail and, possibly, wheel impedance.
Wheel and rail vibrations are induced by wheel-rail contact, in particular by irregularities on the surface of the wheel and rail (roughness and corrugation).
Severe quasi-periodic roughness, known as corrugation, may also develop in some situations. This may have peak-to-trough amplitudes of typically 50 μm at a wavelength of 50 mm; at longer wavelengths the amplitudes can be greater than this. [1]
At the wavelengths of relevance to rolling noise, roughness may be caused by manufacturing variability in the rail section or variable wear of the wheel and rail surfaces. Roughness of the wheel and rail surface is a major cause of wheel-rail noise at all speeds. Typical wavelengths of roughness relevant to rolling noise are between about 5 and 500 mm.
When crossed at typical train speeds, the corrugation of wheel and rail surfaces with wavelengths between 5 and 250 mm produces vibrations in the acoustic frequency range.

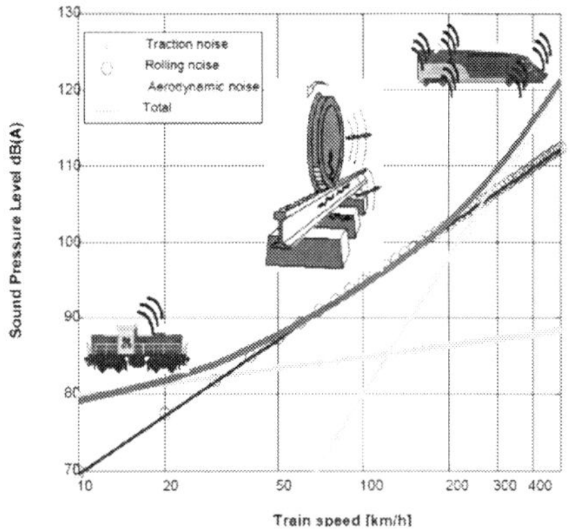

Figure 1. Reduction of railway noise pollution of the rolling noise [Thompson, 2009].

Figure 1 shows that between 30 and 200 km/h the rolling noise is the dominant source. Reduced speed can only be found in shunting yards or near stations. Speeds above 200 km/h are found only on high-speed lines. Speed between 30 and 200 km/h is also the speed at which freight trains operate (approximately 100 km/h). Many sources identify freight trains as the noisy trains.

2.1.Mechanism of rolling noise generation

The mechanism of rolling noise generation consists of two structural components: the wheel and rail. To mitigate rolling noise, these components need to be analysed. To direct and manage mitigation methods, it is desirable to be able to determine the roughness of the wheels and rail.
From the first series of works published by Remington, the rolling noise generation mechanism has

been proven to be due to vibrations of the wheel and the rail, excited by their surface heat. This is also known as roughness in the rail system (figure 2).

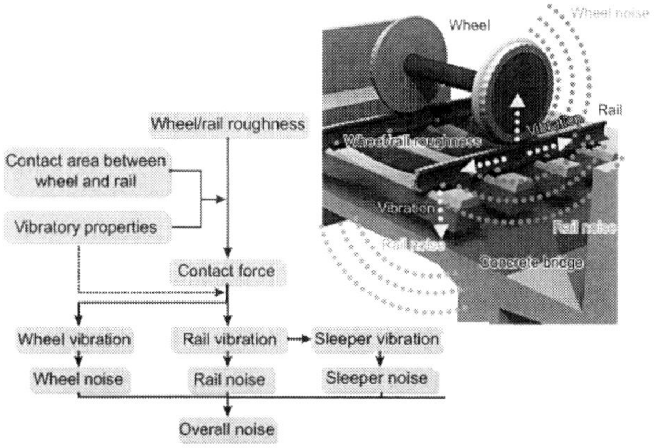

Figure 2. Noise generation mechanism [Remington, 1976].

3. Roughness standard parameters

The roughness of the contact surfaces is determined by the simultaneous action of several phenomena, among which the most important are the following: elastoplastic deformations of the material, vibrational phenomena of the wheel-rail system and friction phenomena.

The interaction of these phenomena leads to the conclusion that microgeometry, in terms of the height of the roughness, is characterized as a random size. Roughness is of great importance in surface response in relation to mechanical solicitations. The original topography is modified in friction phenomena.

Surface roughness is a component of surface texture. This is quantified by deviations in the direction of the normal vector of a real surface from its ideal shape. If these deviations are high, the surface is harsh; if deviations are small, the surface is smooth.

In general, roughness is considered to be a high frequency component on short wavelengths of the micro-irregularities of the surface to be measured. Roughness is the microgeometric measurement of the height variations of a physical surface (according to surface metrology). This measurement is different from the determination of deviations from the correct geometric shape that characterizes surface geometry or undesirable misalignments. Rigidity with high values may be an undesirable feature, as it causes increased friction and wear.

The effect of wear can be assessed by surface topography analysis.

The purpose of this study was to examine the microgeometry of the rail surface by different methods for a sequence of the thread 1 outer ring center line, such as:

- Statistical sizes (average, mean square deviation, centered moments);
- Limit curve according to EN 15610 and EN ISO 3095 standards;
- Abott-Firestone bearing area curve;
- Fractal features.

By recording the results on the height of the roughness, the main statistical characteristics are determined and the Abbott-Firstone bearing curve and fractal character are derived.

Random character of roughness height is evidenced by experimental tests and different types of post-processing. Roughness on the wheel and rail surface gives rise to the vertical vibrations of the wheel and rail systems, depending on their dynamic properties. The main roughness wavelengths that are relevant to rolling noise are between approximately 5 and 500 mm (Thompson, 2009). [1]

Determination of rail and wheel roughness according to Regulation no. 1304/2014 of the EU Commission; EN ISO 3095: 2013 [5].

The rolling noise of railway vehicles depends on the roughness of rails and wheels. Therefore, with regard to the acoustics of railway vehicles, it is essential to have a detailed knowledge of the unevenness of the running surface.

Rail and wheel roughness measurement shows the system requirements for high precision measurement because irregular size of a micrometer or less be determined reliably.

For the roughness of the rail, the tests on the Faurei Railway Testing Centre were carried out. This is managed by Romanian Railway Authority and it is used for dynamic testing of rolling stock.

Tests carried out at the Centre for Railway Faurei Testing (Figure 3) as well as those from the specialized laboratories Romanian Railway Notified Body (RRNB) are performed by technical staff that meets the highest standards of training and experience.

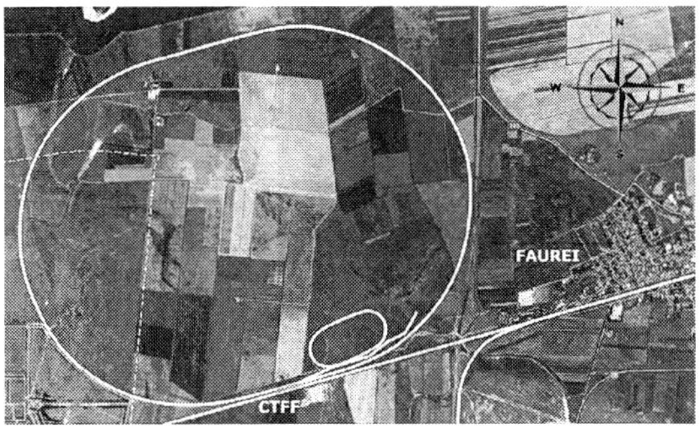

Figure 3. Railway testing center graphic presentation.

To determine the roughness of the rail, a segment in a straight line of a ballasted track with concrete sleepers, with no visible joints or defects, was chosen on the line segment between 0 + 100 m and 0 + 200 m, without sound impact caused by line welding or any faulty gripping of the sleepers. The surface roughness of the rail was determined in accordance with EN 15610 standard, which shows a direct method of measuring the surface roughness of the rail associated with rolling noise.

The test area was chosen so that the rail does not present localized geometric marks (rail defects, skid marks, etc.), which can lead to rolling noise amplification.

The track area on which the test was carried out, in accordance with ISO 3095 was conducted.

Thus, the level of the soil surface from the upper part of the rail was within the [0 - 2] m limits.

No access was allowed in the area and there were no sound absorbing or reflective surfaces.

A test surface length of 30 m was considered. As for the width of the reference surface, since it was greater than 30 mm, the acoustic roughness was measured on the center line of the reference surface as well as on two additional lines located at left and right of the center line at a distance of 10 mm from this, for both rails, over the reference surface length, both on the right and on the left side of the reference surface.

3.1. Roughness Measurement

Tests for rail roughness were performed with m | rail trolley (Figure 4). The m | rail trolley is a high-precision measuring system to continuously record the rail roughness. The m | rail trolley is a track acoustic rail roughness measurement system using an interdependent acceleration sensor with a mobile holder that is moved manually along the rail (carriage);

The system can measure amplitude deficiencies below a micrometer at wavelengths from a few millimeters to a few meters.

In Figure 5a, a stretch of the roughness profile (y1) evaluated on the rail in the area 0+150m-0+152.5m for thread 1 outer ring center on the Faurei Railway Testing Center is shown, on a 2.5 m length and with a 1 mm mesh (2500 points).

The acoustic rail roughness of the test section shall be assessed according to EN 15610:2009. For type test measurements, the default upper limit provided in Figure 5b shall apply, taking into account.

IOP Conf. Series: Materials Science and Engineering **514** (2019) 012010 doi:10.1088/1757-899X/514/1/012010

For train speeds up to 190 km/h, the wavelength bandwith shall be at least 0.003 m to 0.10 m. For higher speeds at least 0.003 m to 0.25 m.

Figure 4. m|rail trolley.

Acoustic roughness r(x)is defined that variation in the height of the rail running surface associated with rolling noise excitation expressed as a function of distance x along the rail.
The acoustic roughness expressed in dB is defined as:

$$L_{r(x)} = 10\log\left[\left(\frac{r(x)}{1\mu m}\right)^2\right] \tag{1}$$

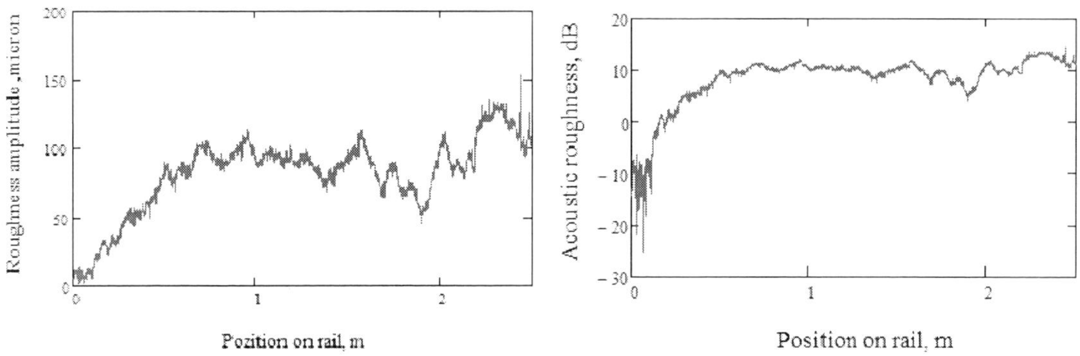

a) b)

Figure 5. Evaluated rail roughness meter (a) and acoustic (b) from the area 0+150m-0+152.5m for thread 1 outer ring center.

Subsequently, in order to analyze the state of the rail roughness, we selected and analyzed the $r(x)=y(x_i)$ experimental sequence for N points measured with the equipment m | rail trolley.
Experimental results $y(x_i)$ are analyzed according to standardized criteria specific to rolling wheel/rail (EN 15610: 2009, ISO 3095: 2013), according to general statistical criteria and by criteria used in fractal analysis of microgeometry of contact surfaces. The results are analyzed with an original program developed in MATCHAD software. In this sense, according to EN 15610: 2009, the data set is corrected by spikes and pits [5, 6]. Thus, the aberrant peak is the coordinate point (x_s, y_s) for which the second derivate conditions are simultaneously satisfied and derived first:

$$\frac{d^2}{dx^2} y \le S_{cr} \frac{\mu m}{m^2}, \text{ where } S_{cr} = -10^7 \tag{2}$$

and product of derivative dy/dx, for $x = x_i$ and $x = x_{i+1,}$ is negative or zero.
Derivatives are determined numerically based on the coordinates (x_{i+1}, y_{i+1}) and (x_i, y_i) of the neighbouring points. If these conditions are fulfilled, the spike is considered aberrant and ordered (yi) is corrected by linearization with the order of the neighbouring points $y_i = (y_{i-1} + y_{i+1}) / 2$ (figure 6, (b)- ycor (j…k) is the corrected profile).

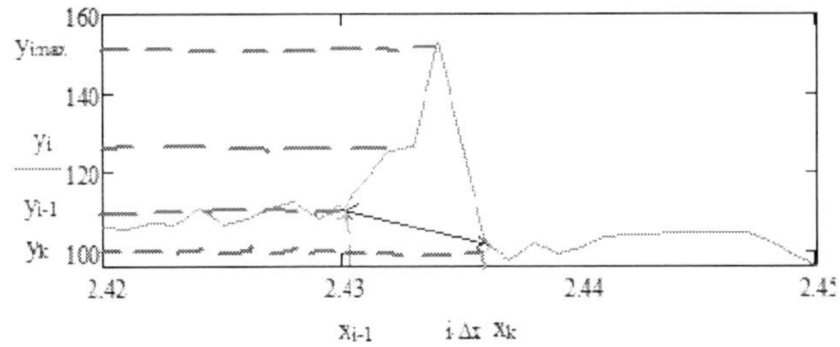

a)

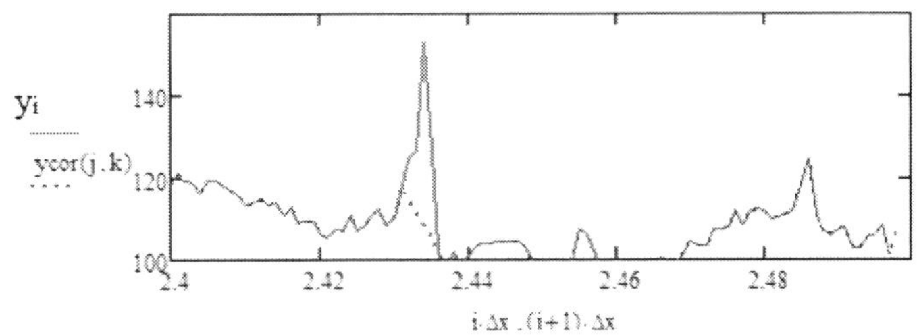

b)

Figure 6. —Spike removal a- area of the spike; b- modified spike.

The valley (hole, pit) is considered excessive (aberrant) if the intersection of a circle of radius R = 0.375 m and the coordinate center (x_i, y_i + R) with the roughness profile (y_i) will be in the point (x_k, y_k) than (x_i, y_i). The explicit equation of circle C is:

$$yc_k = y_i + R - \sqrt{R^2 - (k\Delta x - i\Delta x)^2}\,. \tag{3}$$

The correction of the profile at the abscissa point x_i is:

$$ycp_i = y_i + \max(y_k - yc_k)\,. \tag{4}$$

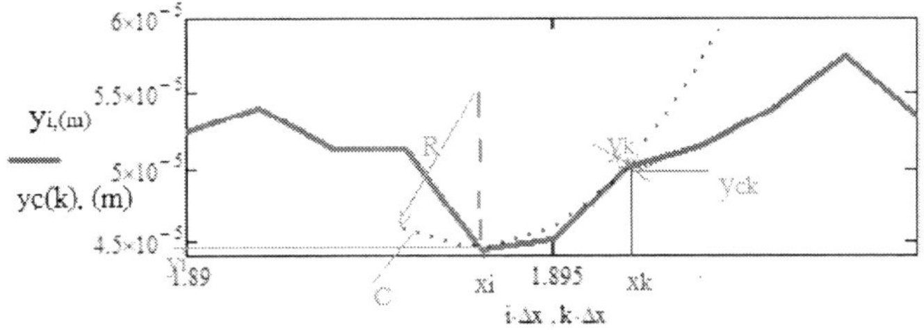

Figure 7. Curvature processing.

After correcting peaks (spikes) and aberrant valleys, the roughness profile (ycp_i) analyzes the general statistical parameters and functional parameters.

As statistical parameters useful for the qualitative and quantitative assessment of rolling noise, the following are analyzed:

- **Arithmetic mean deviation of the assessed profile (Ra):** defined on the sampling length. Ra is used as a global evaluation of the roughness amplitude on a profile; for the corrected roughness profile in example figure 5,

$$R_a := \frac{1}{N}\sum_i (ycp_i)\,; \qquad \text{Ra} = 81.18\ \mu\text{m}, \tag{5}$$

where N is the number of meshing points of the profile;

- **Root mean square deviation of the assessed profile** (Rq)**:** corresponds to the standard deviation of the height distribution, defined on the sampling length. Rq provides the same information as Ra;

$$R_q := \left(\frac{1}{N}\sum_i (ycp_i)^2\right)^{1/2}\,; \quad R_q = 85.7\ \mu\text{m}. \tag{6}$$

Ten point height: defined on the sampling length (**Rz**): this parameter is frequently used to check whether the profile has protruding peaks that might affect static or sliding contact function:

$$Rz_{ISO} := (ysl_{N-2} + ysl_{N-5} + ysl_{N-6} + ysl_{N-3} + ysl_{N-4} - |ysl_1| - |ysl_2| - |ysl_3| - |ysl_4| - |ysl_5|)\frac{1}{5}\,; \tag{7}$$

Rz = 67.98 μm,

where y_s is the roughness vector sorted in ascending order.

- **Maximum profile peak height (Rp)**: height of the highest peak from the mean line,

$$R_p := \max(ycp) - mean(ycp)\,; \quad R_p = 71.82\mu m\ ; \tag{8}$$

-Maximum profile valley depth (Rv): depth of the deepest valley from the mean line,

$$R_v := \min(ycp) - mean(ycp)\,; \ R_v = -81.17\mu m. \tag{9}$$

-Skewness of the assessed profile (Rsk): asymmetry of the height distribution, defined on the sampling length. It is the third central moment of the profile amplitude probability density function, measured over the assessment length:

$$R_{sk} := \left[\sum_i (ycp_i)^3\right] \ , R_{sk} = 1.114\mu m. \tag{10}$$

- **Kurtosis of the assessed profile (Rku):** sharpness of the height distribution, defined on the sampling length. The kurtosis coefficient is the fourth central moment of a profile amplitude

$$R_{ku} := \frac{1}{N1R_q^4}\left[\sum_i (ycp_i)^4\right]; \qquad R_{ku} = 1.288\mu m. \tag{11}$$

An important parameter of rail and wheel microgeometry used to evaluate rolling noise is acoustic roughness as a function of wavelength. For the evaluation of the roughness (y) dependence on the wavelength (λ), based on the experimental record y (x_i), the discrete Fourier Transform (CFFT) of the vector y (x_i) is used.

The corrected roughness (ycp) acoustic level ($y\lambda_c$) is:

$$y\lambda_c = 10\log\left(\frac{ycp^2}{stdev(ycp)^2}\right). \tag{12}$$

IOP Conf. Series: Materials Science and Engineering **514** (2019) 012010 doi:10.1088/1757-899X/514/1/012010

and the wavelength (λc) is the complex number of the discrete Fourier transform (CFFT) (implicit function in the Matchad program)

$$\lambda c_k := \left|YTc_k\right| \qquad \text{where } YTc := \mathrm{CFFT}(yc). \tag{13}$$

$$\Delta x = 1 * 10^{-3} m, \quad k = 1, 2 .... N - 2. \tag{14}$$

Thus, in Figure 8 shows the acoustic level of the spikes and pits (yλc and sorted values of yλc, y_{sort}) of the rail running surface (Figure 5) as a function of the wavelength.

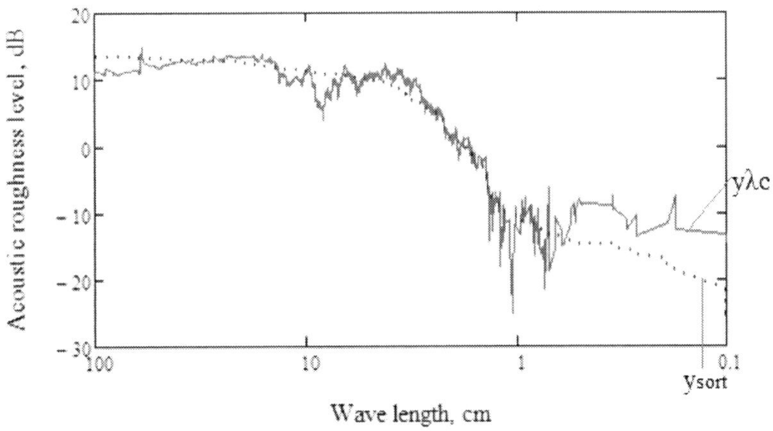

Figure 8. Acoustic roughness levels vs wave length.

This roughness in the surface profile has significant wavelengths between about 10 and 300 mm and amplitudes between about 0.1 μm and 30 μm greater for severe corrugation.
The validity of the linear relationship assumed between roughness and noise depends on the quality and completeness of roughness measurements.
This is illustrated by comparing different rail roughness systems. Particular attention should also be paid to processing the roughness data before a representative excitation spectrum can be derived.

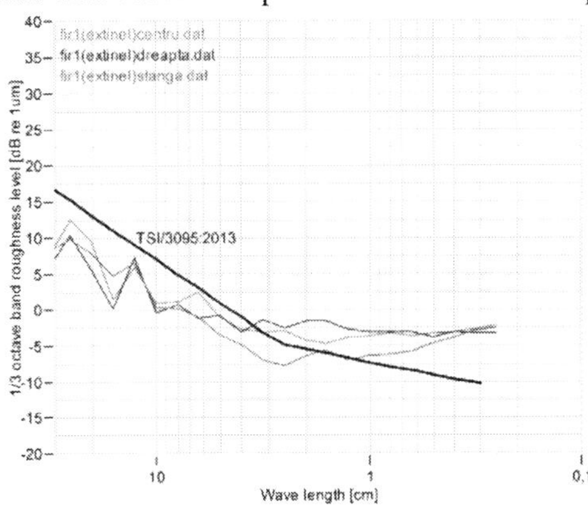

Figure 9. Comparison of acoustic roughness spectra thread 1 outer ring centre.

Future research will investigate and perform experimental determinations from a range of sources to confirm the linear relationship between roughness and noise, at least for roughness amplitudes that are not too severe. Consideration will also be given to the validity of the assumption made in the models that the roughness of the wheels and rails can be added to an equal basis to give the total roughness spectrum.

Acoustic roughness was initially performed using equipment software m | rail trolley. It has been chosen that the acoustic roughness on the outer ring 1 string center; it is determined on the three parallel lines: left, center and right (figure 9).

4. Bearing Area Curve (BAC)

Bearing Area Curve Abbott-Firstone represents the intersection of the profilogram with parallel planes and the cumulation at the same level of the lengths of the intersection of the profile with the respective plane.

BAC of the roughness profile on a certain direction deduced by Abbott-Firstone highlights the microgeometry's behaviour during the processing of external loads.

BAC can be determined by intersecting the roughness profile with n equidistant planes relative to a reference plane (generally the roughness base plane) as shown in Figure 10 [7, 8, 9].

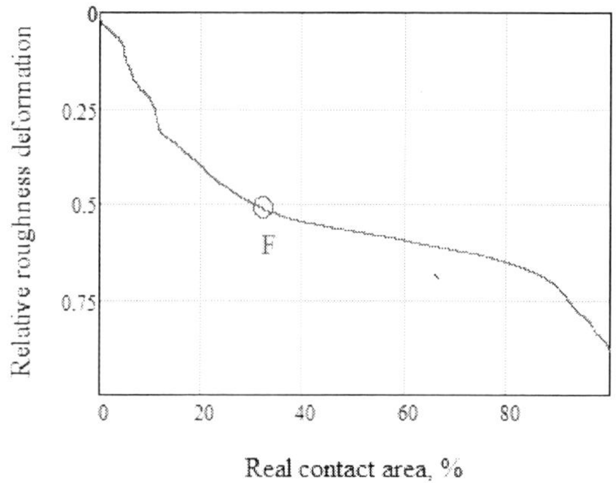

Figure 10. -Bearing Area Curve (BAC).

The roughness profile from figure 5 corresponds BAC from figure 10. It should also be noted that the area above the inflection point F, i.e. the area above the plane of F point, would represent the roughness that will elastically deform; below this line would follow the plastic deformation regime.

The bearing area curve can be used as a local stiffness parameter and as the source of excitation for torsional and flexural vibrations of the wheel.

5. Characterization geometry of surface topography by fractal

Fractal geometry as a tool for the characterization of surface topography has gained much attention in recent years [10-13]. This is due in part to the observations that fractal geometry can reflect the natural and intrinsic properties of random phenomena and that it can overcome several disadvantages of conventional statistics and random process methods of surface analysis.

Several methods have been developed to characterize the dimension of a fractal set, such as the compass

dimension, box dimension, mass dimension and area-perimeter dimension [9-10]. If a homogeneous and isotropic rough surface has fractal dimension D, and its profile in an arbitrary direction has fractal dimension D, then a simple relation exists between them; namely, $D_s = 1{+}D$ [9].

We use the structure function method to calculate fractal parameters dimension (D) and topothesy length (L_t). It is considered a signal $y1(x_i)$ whose statistical properties for a delay are to be investigated.

The increment of signal $y1(x{+}t) - y1(x)$ is assumed to have a gaussian distribution with zero mean and the variance is the sums of the squares of the delayed amplitude differences. This function is structure function in continuum form:

$$S(t) = E\left[\left(y_1(x) - y_1(x+t)\right)^2\right] = c\left(|t|\right)^{4-2D} \tag{15}$$

Where, $E\left[\left(y_1(x) - y_1(x+t)\right)^2\right]$ denotes an expectation, C is a constant, D is the fractal dimension of the y1(x).

The chord joining *y1* values separated by a distance *t* has a finite mean-square slope. If there is a displacement $t = L_t$ such that the string has an r.m.s. slope of unity statistically:

$$S(L_t) = L_t^2 \tag{16}$$

where L_t is a characteristic parameter of the fractal function called the topothesy.

Thus, the structure function for fractal roughness has an expression

$$S(t) = L_t^{2D-1} \cdot (|t|)^{4-2D} \tag{17}$$

The fractal parameters (D, L_t) can be determined by graphical form (log- log) of the structure function. The signal y1 (x_i) is discretized in N equal intervals, $\Delta x = x_p / N$, where x_p is the reference length of the rail profile as standard (2.5m). Thus, an "i" point of the signal has the coordinates $x_i = \Delta x.\ i$ and $y_i = y1_i$.

This point differs from another located at a distance $x_k = \Delta x.\ k$ and with order $y_k = y_{i+k}$. The structure function in discrete form, S1(N,k), of a signal y1(x_i) is:

$$S1(N,K) = \frac{1}{N-k} \sum_{i=1}^{N-k} (y1_{i+k} - y1_i)^2 \tag{18}$$

This function is independent of the mean plane; thus, any profile structure function, irrespective of its mean line, is a section of the surface structure function.

For example, in Figure 11 shows the structure function of the roughness profile corrected for the rail with the roughness of Fig. 5.

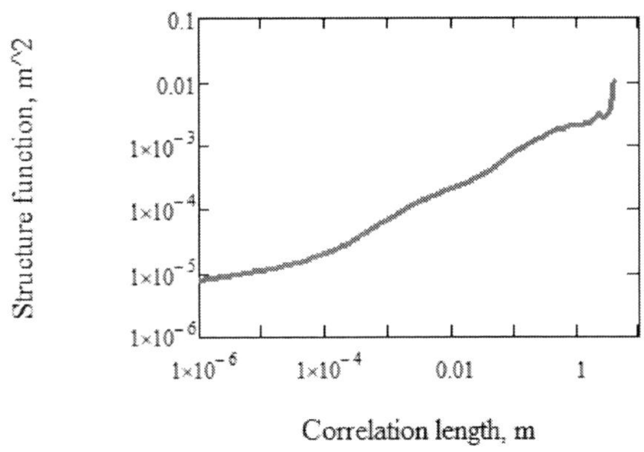

Figure 11. Structure function of roughness rail.

If the curve representing the structure function in the logarithmic coordinate, is a right with a non-integer slope, then the roughness can be considered as a fractal character [7-14].

On the basis of the analysis of the curve, $\left(\log\left(S(N,k)\right) - \log\left(x(k)\right)\right)$, it is inferred by the smallest square method that the rightmost approximates the function of the structure having slope $\beta = 1.476$. Thus, the fractal parameters are evaluated in Matchad program:

$$D = \frac{(4-\beta)}{2} = 1.262 \text{ and} \tag{19}$$

$$L_t = 3\Delta X = 0.003 \text{ m} \tag{20}$$

For fractal character roughness (parameters D and L_t), we can associate a continuous and undifferentiated Weierstrass-Mandelbrot (W-M) function y (x) [7, 12, 14, 16]:

$$y(x, D, \gamma) = G^{D-1} \sum_{n=n_1}^{\infty} \frac{\cos(2\pi\gamma^n x)}{\gamma^{(2-D)n}}, \qquad 1 < D < 2 \text{ and } \gamma > 1 \tag{21}$$

where G is a scaling constant and the frequency modes γ^n correspond to the reciprocal of the wavelength (λ_n) of roughness as $\gamma^n = 1/\lambda_n$.
The scaling constant G can be defined and the hypotheses that increment of signal y1(x+t) – y1(x) is assumed to have a gaussian distribution with zero mean:

$$\text{where } C4(D) = \left[\frac{\Gamma(2D-3)\sin\left[\pi\frac{(2D-3)}{2}\right]}{2-D}\right]^{\frac{1}{2D-1}}, \qquad 1 < D < 2 \text{ and } \gamma > 1 \tag{22}$$

where $\Gamma(x)$ is Gamma function at the x argument.

In the case of the corrected roughness rail from Figure 5, the scaling constant G = 1.377 mm.

$$G(D) = \frac{L_t}{C4(D)} \tag{23}$$

The W-M associated function can be used to characterize the height of the roughness at any scale regardless of the radius of the probe for experimental determination.
The state of stresses and deformations between the rail and wheel roughness as well as the wear behavior can be assessed using the W-M function [7, 9, 11].

Figures 12 a, b exemplify the fractal evaluation of the rail roughness by the Weierstrass-Mandelbrot (W-M) function and the acoustic roughness level for small part of rail (0.7--- 0.72 m).

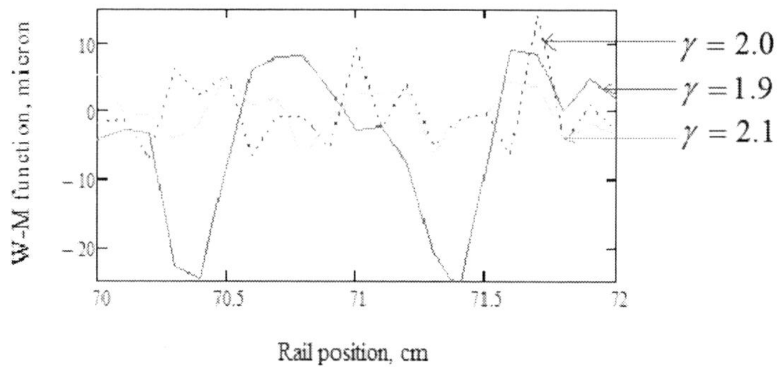

a)

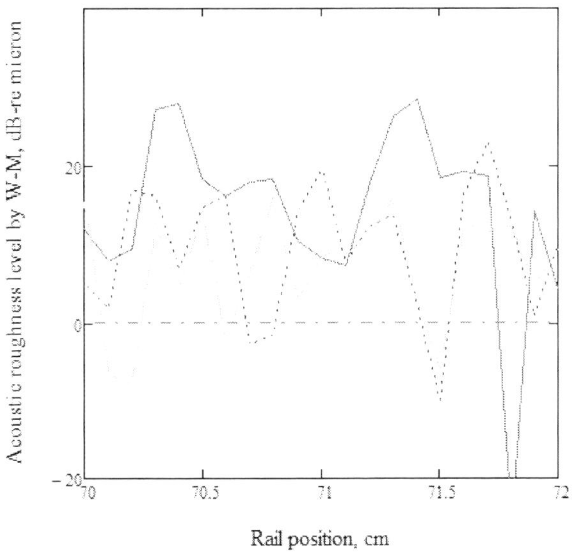

b)

Figure 12. Fractal roughness of rail, W-M function (a) and acoustic roughness level W-M (b).

In a future paper we will compare the acoustic roughness and the fractal roughness in terms of the rolling noise of a driving wheel to the railroad locomotives.

6. Conclusions

Microgeometric deviations (roughness), corrugations and circularity of wheel are the major cause of rolling noise in locomotives. The current parameters characterizing microgeometry are acoustic roughness (dB) and wavelength (cm) and compare to an acoustic-curve-wavelength curve standard limit.

The microgeometry of the rails at the Faurei ring of testing corresponds to the EN 15610-2011 standard for wavelengths longer than 0.8 cm and it has the fractal character.

To validate the results, it is necessary to remove or correct peaks and pits with the criteria of standard EN 15160-2011.

Matchad's computational program allows statistical analysis, correction of aberrant spikes and pits, deduction of the bearing area curve, and fractal parameters of the microgeometry of railway tracks.

Standard experimental attempts on locomotive and wagon noise will be correlated with fractal parameters and bearing curves of roughness in the next paper.

Acknowledgments

The autors are grateful for support from Faurei Railway Testing Center and from Romanian Railway Authority.

References

[1] Thompson D, 2009 Rolling noise and vibration. Mechanism noise and means of control, Elsevier Ltd.

[2] Remington P.J., 1976 Wheel/Rail noise- Part IV: Rolling noise, *Journal of Sound and Vibration*, 46(3), pp.419-436.

[3] Galaitsis G. and Bender K., 1976 Wheel/Rail noise- Part IV Measurement of wheel and rail roughness, *Journal of Sound and Vibration*, 46(3), pp.437-451.

[4] Ruiten van C.J.H., 1988 A new method for measurement of wheel/rail roughness, *Journal of Sound and Vibration*, 120(2), pp.287-295

[5] **** EN ISO 3095:2013" Acoustics. Railway applications. Measurement of noise emitted by railbound vehicle"
[6] **** EN 15610:2011" Railway application- Noise emission- Rail roughness measurement related to rolling noise generation"
[7] Tudor, A, 1990 Contactul real al suprafețelor de frecare, *Editura Academiei Române*, Bucureşti.
[8] Tudor, A, 2002 Frecarea si uzarea materialelor, *Editura Bren*, Bucureşti.
[9] Zhou G.Y., Leu M.C.,1993 Fractal geometry model for wear prediction, *Wear* 170, pp 1-14.
[10] Hasegawa M., Liu J., Okuda K., Nunobiki M., 1996 Calculation of fractal dimensions of machined surface profiles, *Wear* 192, pp 40-45.
[11] Sahoo P., Roy Chowdhury S.K., 2002 A fractal analysis of adhesive wear at the contact between rough solids, *Wear* 253, pp 924-934.
[12] Bigerelle M., Mathia T., Bouvier S., 2012 The multi-scale roughness analyses and modeling of abrasion with the grit size effect on ground surfaces, *Wear*, pp 286-287, 124-135.
[13] Ciavarella M., Delfine V., Demelio V., 2006, A new 2D asperity model with interaction for studying the contact of multiscale rough random profiles, *Wear,* pp 556-567.
[14] Majumdar A., B. Bhushan, 1991 Fractal model of elastic-plastic contact between rough surfaces, *ASME J. Tribol.* 113 pp 1-11.
[15] Mandelbrot B.B, 1985 Self-affine fractals and fractal dimension, Phys. Scr. pp 257-260.
[16] Gagnepain J.J. and C. Roques-Carmes, 1986 Fractal approach to two-dimensional and three-dimensional surface roughness, *Wear*, 109, pp 119-126.
[17] Ling F.F., 1990 Fractals, engineering surfaces and tribology, *Wear* 136, pp 141-156.
[18] Majumdar A. and Bhushan B, 1990 Role of fractal geometry in roughness characterization and contact mechanics of surfaces, *ASME J. Tribal.* 112 pp 205-216.
[19] Sayles, R.S. and T.R. Thomas, 1977 The spatial representation of surface roughness by means of the structure function: A practical alternative to correlation. *Wear*, 42 (1977), pp 263-276.

IOP Conf. Series: Materials Science and Engineering **514** (2019) 012011 doi:10.1088/1757-899X/514/1/012011

Simulation of Charpy test for different impact velocities

A E Musteaţă[1], C Pirvu[2], L Deleanu[1] and C Georgescu[1]

[1]Department of Mechanical Engineering, „Dunarea de Jos" University, Galati, Romania
[2]National Institute for Aerospace Research „Elie Carafoli" INCAS, Bucharest, Romania

E-mail: lorena.deleanu@ugal.ro

Abstract. This paper presents how a finite element model of Charpy sample fails when it is hit with an impactor. Studying the step-by-step evolution of failure of polymeric sample allows for establishing stages of the process characterising the Charpy impact, pointing out processes and phenomena that are difficult to be noticed during the actual impact or by investigating the broken pieces. The simulation was run for different impact velocities (1...3 m/s) and the sample material was considered a bilinear orthotropic hardening model. After analysing the images of the simulations, the following conclusions may be formulated. Simulation helps identifying the stages of the material failure. Using the bilinear hardening model for the polymer failure at low impact velocity (1...3 m/s) gave reasonable simulations. The time for failure process is longer at lower speed and the aspect of the failed surface depends on the impact velocity.

1. Introduction

The Charpy test was developed by Russell and Georges Charpy, at the beginning of the 20th century. It is one of the most common impact testing methods due to the simple shape of samples and the simplicity of the mechanical rig and thus, the results are easy to obtain and compare [1], [2]. The test rig consists of a weighted pendulum, which is dropped from a specified height to impact the sample. The energy consumed to deform or break the sample is simply calculated by the difference in the height of the pendulum, before and after the sample impact.

Procedures for Charpy testing are included in standards and the most preferred are ASTM E23, ASTM A370, ISO 148 or EN 10045-1. While the test has been most frequently performed on metallic materials, there are several standards applying for polymers, including ASTM D6110 and ISO 179.

Ductile crack initiation and propagation in Charpy specimens change the stress distribution ahead of the notch root. The stress level ahead of this notch root is significantly increased after ductile crack initiation. The quasi-static and dynamic formulations yield similar stress-strain distributions, confirming the results from the energy balance. Adiabatic heating, accounted for both the quasi-static and dynamic formulations, slightly favours crack initiation. The Gurson-Tvergaard-Needleman model allows a good failure prediction with strain rate and temperature independent damage parameters [3].

2. Model

The method of finite elements could help the engineer to avoid some unsolved issues in the analytical models. Many times, the transition from brittle to ductile fracture was described by experimental data that were, sometimes, misinterpreted [4]. The results of a Charpy test, especially the fracture energy on temperature, depend on structural parameters of the material. Polymers have complex structure and, thus, the influence of their structure on this curve is very peculiar, including aspects of composition, morphology and molecular structure. But these parameters are also inter-conditioned, so the problem becomes more complicated.

A Charpy sample is typically a bar with a notch machined into one face (figure 1 presents the draw of the sample used in this simulation). The notch is placed on the opposite face to that bearing the impact and helps concentrate the stress and fracture initiation in the middle zone of the sample. Testing can be performed on a large temperature range. For this model, the impactor has 0.504 kg, this value being characteristic for Charpy test machine for polymers and the impact surface is plane. Its geometry is very close to that proposed in standards.

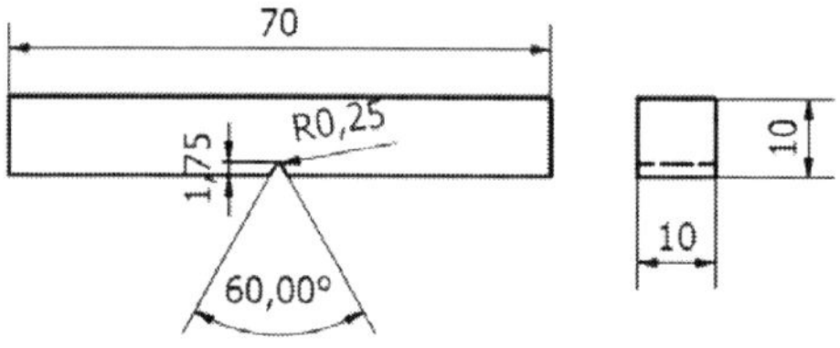

Figure 1. Dimension of the sample (for this simulation).

The authors used the code AutoDyn and the explicit Lagrangian dynamic solver. The materials being considered bilinear orthotropic with hardening, characterized by a yield strength and a tangential modulus. The failure criterion was considered the equivalent plastic strain thus, the failure occurs when this value is reached. The body erosion is controlled by the program and imposed by the same factor as for the other bodies in the model [5], based on the geometrical deformation limit, one for each material.

Shokrieh and Joneidi [6] designed the mesh of the Charpy specimen with three distinct regions: the region parallel with the contact surface is most refined in order to guarantee a good contact with the striker, the region around the notch is refined as the stress is concentrated at the notch tip and the rest of the material volume.

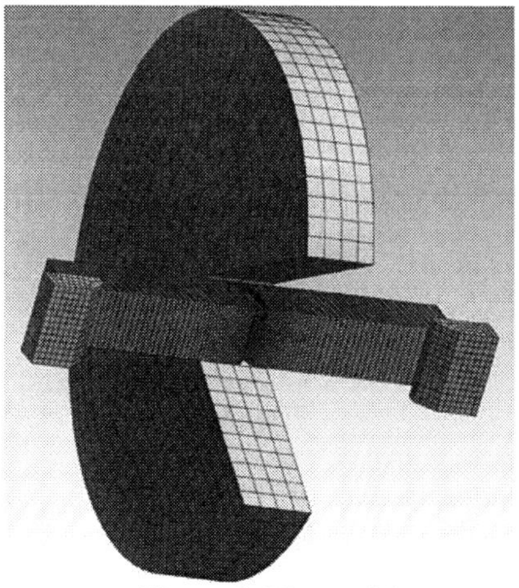

Figure 2. The model.

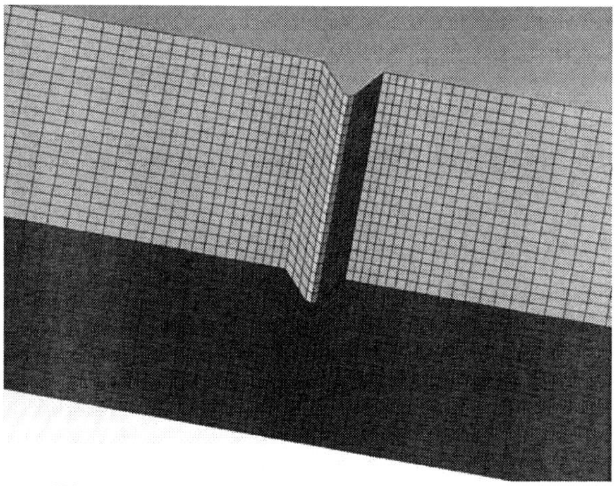

Figure 3. Detail of mesh in the notch zone.

The model has four bodies (figure 2), the polymeric sample and three bodies made of stainless steel (the impactor and two supports with round corners).

Rossol et al. [7] proposed the following initial and boundary conditions: striker velocity imposed, rate-dependent material behavior, quasi-static or fully dynamic (including inertia terms) solution procedure and adiabatic (not allowing any transfer of the heat generated by plastic deformation, since the characteristic heat diffusion time is considered to be about one order of magnitude longer than the duration of an impact test).

The authors took into considerations the followings: adiabatic conditions at room temperature (22°C), imposed impact velocity. Also, they introduced friction between the impactor and the sample and between the rigid supports and the sample (COF = 0.25, a typical value for sliding contact polymer-metal). Even some recent papers [8], [9] neglect friction in their simulations. But other authors consider the friction between surfaces in contact. Ghaith and Khan [10] introduced in their model two types of frictional contact interactions; sample-support and impactor-sample.

The contact conditions are: the contact between the Charpy sample and its supports and also the contact between the polymeric sample and the impactor are considered friction contacts, characterised by a constant friction coefficient of 0.25. Boundary conditions refer to the supports with fixed surface ends, that is the nodes on the surfaces laying on the machine frame have no movement and no rotation. Initial conditions consider that the impactor velocity just before reaching the sample surface could vary, this paper presenting simulation results for impact velocity (v_0) of 1 m/s, 2 m/s and 3 m/s, characteristic for Charpy tests: 3.7 m/s [11], 5...9 m/s for steel modelling [12]. The mesh sensitivity depends on the element size that should be as smaller as possible and a finer mesh could be requested on the notch zone. For models with bodies with very different levels of dimensions (here, the sample and its bottom-round notch), the mesh presents some issues. An improved mesh will ask for a more performant computer. Minimum edge length is 0.5236 mm and the growth rate are set at 1.2. Element size is 0.4...0.7 mm (figure 3) for the notch zone (the radius of the notch being r = 0.25 mm). The maximum energy error is set for 0.9. The end time is set at 10^{-3} s.

Taking into account the narrow range of the impact velocity (1...3 m/s), this simulation was done for constitutive bilinear hardening polymeric material, similar to a polyethylene grade (table 1). An improved model may use constitutive models depending on strain rate and temperature, as that proposed by Johnson and Cook [13].

Table 1. Material characterization.

Property	**Polymer**	**Stainless steel**
Density, kg/m^3	950	7750
Young modulus, MPa	1100	193000
Poisson ratio	0.42	0.31
Bulk modulus, MPa	2291.7	169300
Shear modulus, MPa	387.32	73664
Yield strength, MPa	25	207 (compression and tensile)
Tangent modulus, MPa	350	
Maximum equivalent plastic strain	0.05	

3. Results

Figures 4, 5 and 6 present several moments of the failure process, pointing out that, even if the velocity range is not large (1...3 m/s), the behavior of the sample is quite different.

The difference in failure process may be point out by comparing the graphs in figures 9, 10, 11. At v_0 = 1 m/s, the maximum von Mises stress is reached after a longer time from the impact, $\Delta t = 1.75\times10^{-3}$ s. At v = 2 m/s, the maximum value is recorded at $\Delta t = 1.24\times10^{-3}$ s and at v_0 = 3 m/s, this time interval is $\Delta t = 1.0\times10^{-3}$ s. Sample impacted at v = 1 m/s is not yet totally broken (the sample is not yet split in two pieces). Thus, 10^{-5} s is not enough to simulate the failure at this speed and the high value of equivalent stress is a prove. It seems that the separation advances smoothly at slow velocity and in steps at higher ones (see the several stress peaks for v_0 = 2 m/s and v_0 = 3 m/s and the peaks are sharper for the highest velocity.

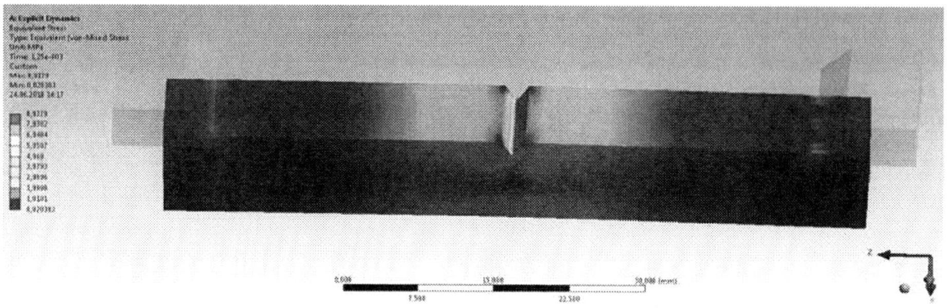

a) t = 1.25×10^{-3} s

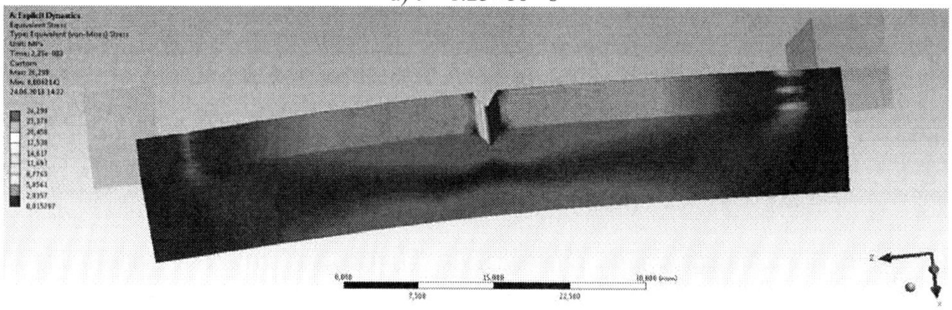

b) t = 2.25×10^{-3} s

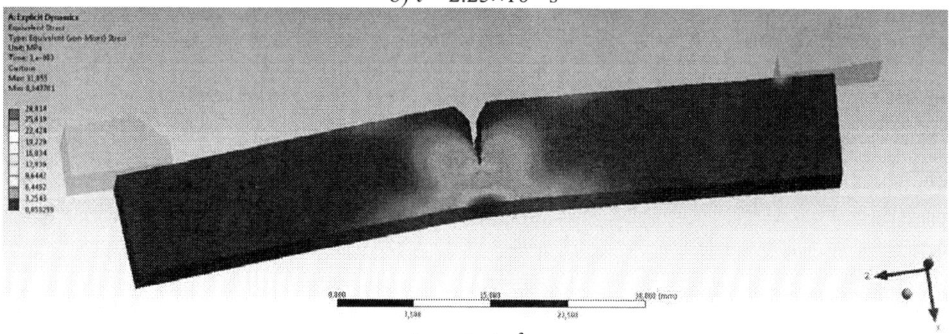

c) t = 3×10^{-3} s

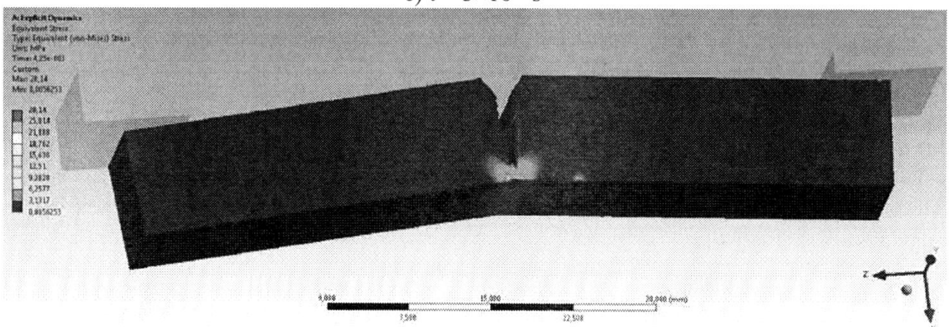

d) t = 4.25×10^{-3} s

Figure 4. Moments of failure for the impactor velocity of v_0 = 1 m/s (impact start at t = 1.0×10^{-3} s).

IOP Conf. Series: Materials Science and Engineering **514** (2019) 012011 doi:10.1088/1757-899X/514/1/012011

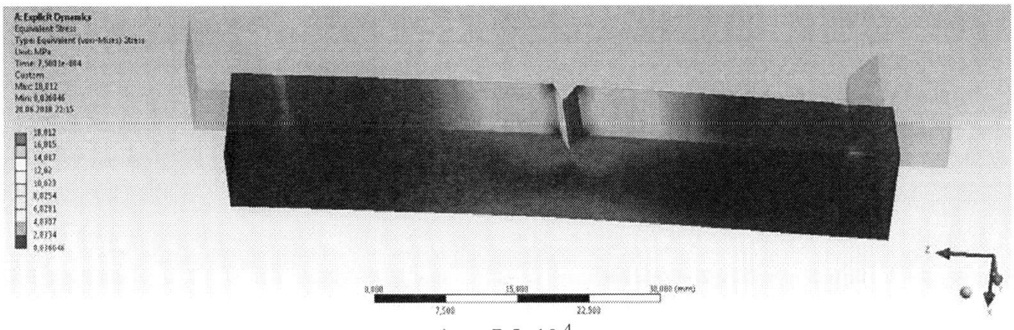

a) t = 7.5×10^{-4} s

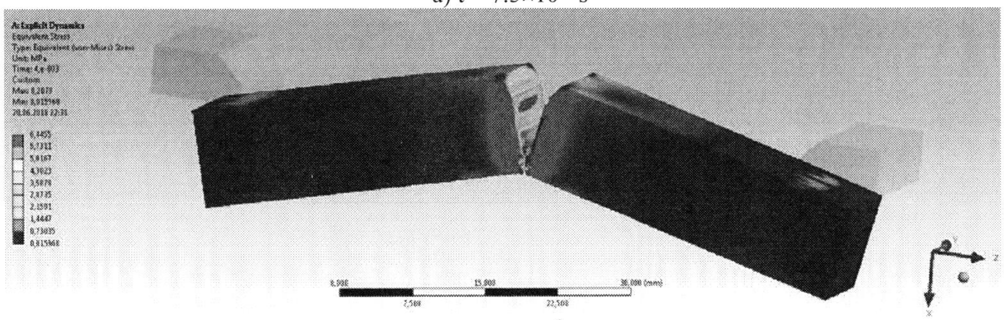

b) t = 4×10^{-3} s

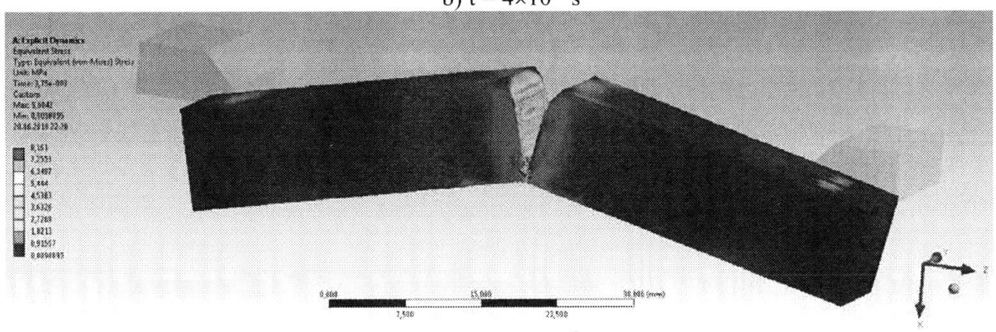

c) t = 3.75×10^{-3} s

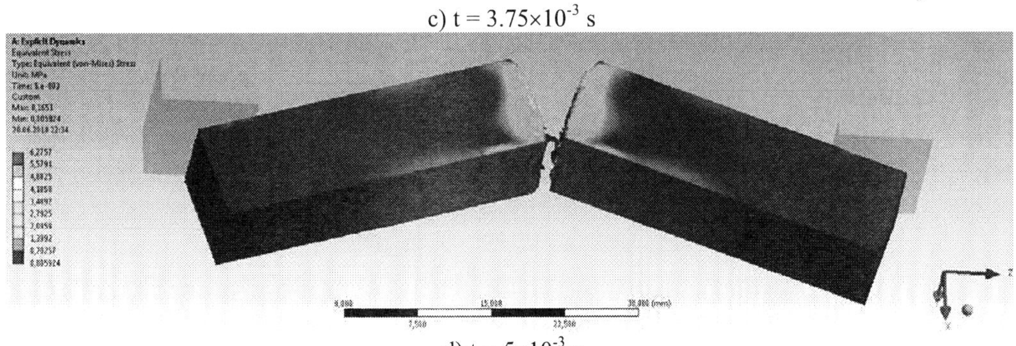

d) t = 5×10^{-3} s

Figure 5. Moments of the failure for the impactor velocity of v_0 = 2 m/s (impact start at t = 5×10^{-4} s).

IOP Conf. Series: Materials Science and Engineering **514** (2019) 012011 doi:10.1088/1757-899X/514/1/012011

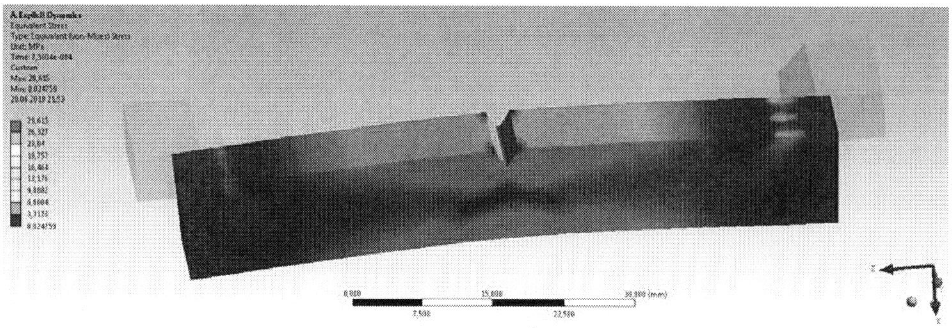

a) t = 7.5×10^{-4} s

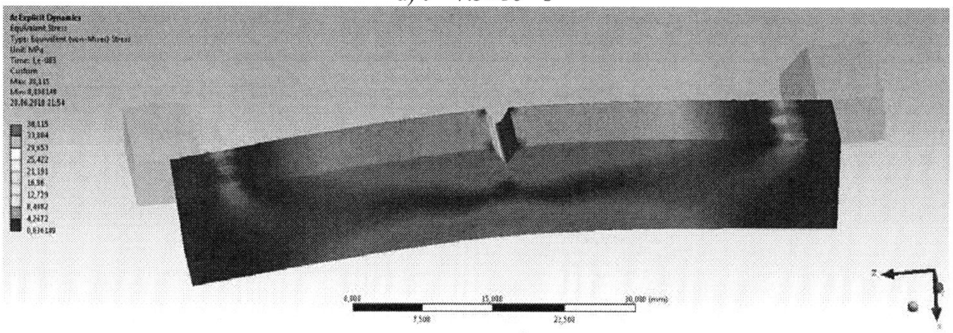

b) t = 1×10^{-3} s

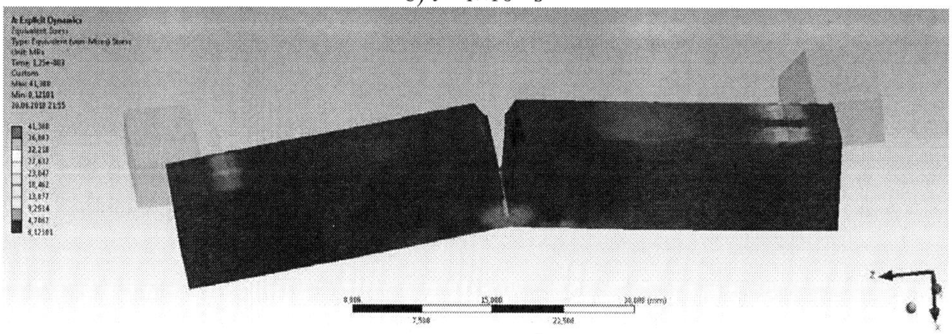

c) t = 1.25×10^{-3} s

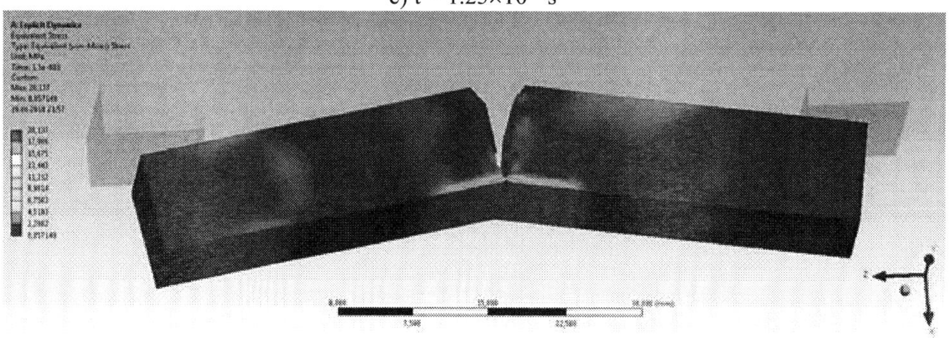

d) t = 1.5×10^{-3} s

Figure 6. Moments of the failure for the impactor velocity of v_0 = 3 m/s (impact start at t = 2.5×10^{-4} s).

IOP Conf. Series: Materials Science and Engineering **514** (2019) 012011 doi:10.1088/1757-899X/514/1/012011

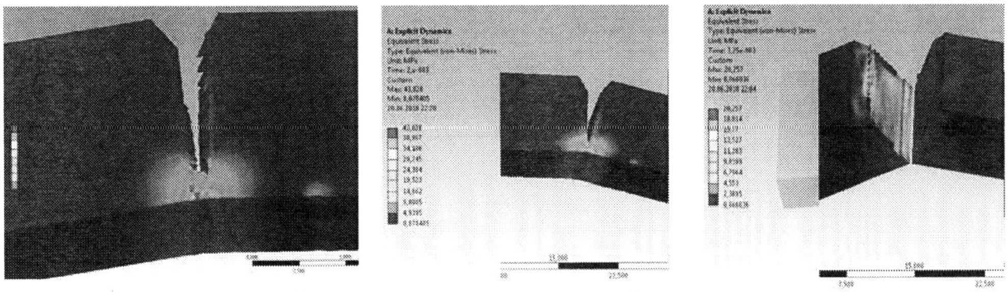

a) $v_0 = 1$ m/s, $t = 3.75\times10^{-3}$ s b) $v_0 = 2$ m/s, $t = 2\times10^{-3}$ s c) $v_0 = 3$ m/s, $t = 3.25\times10^{-3}$ s

Figure 7. Details of the sample failure for different impact initial velocity (von Mises stress distribution).

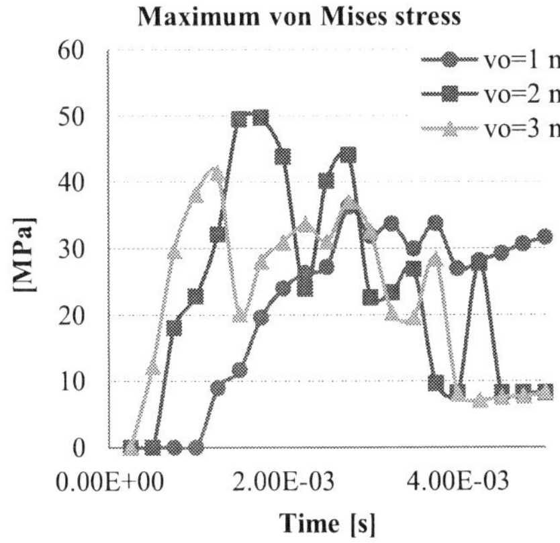

Figure 8. Maximum von Mises stress during the impact.

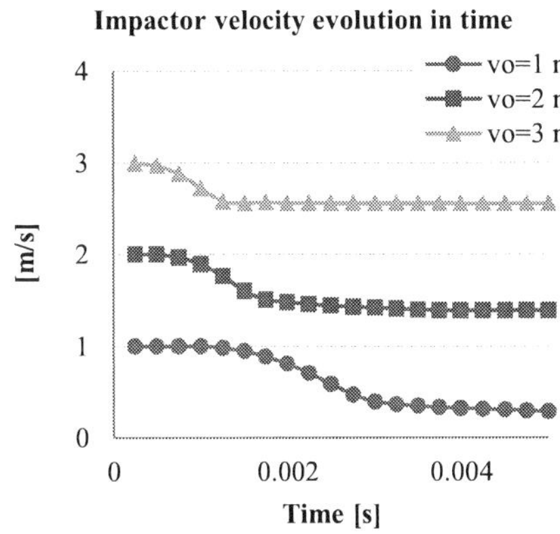

Figure 9. Impactor velocity during the impact.

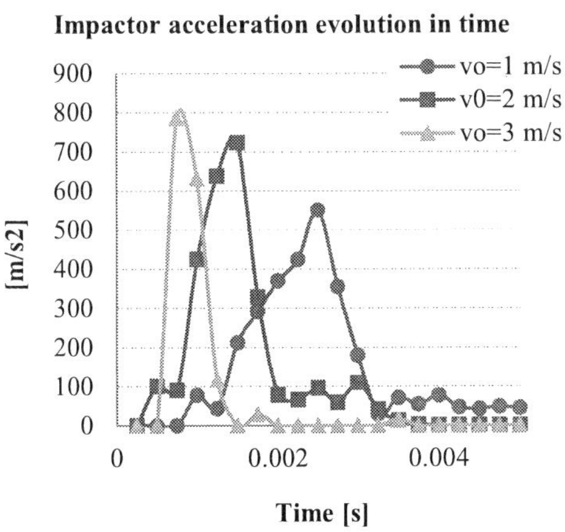

Figure 10. Impactor acceleration during the impact.

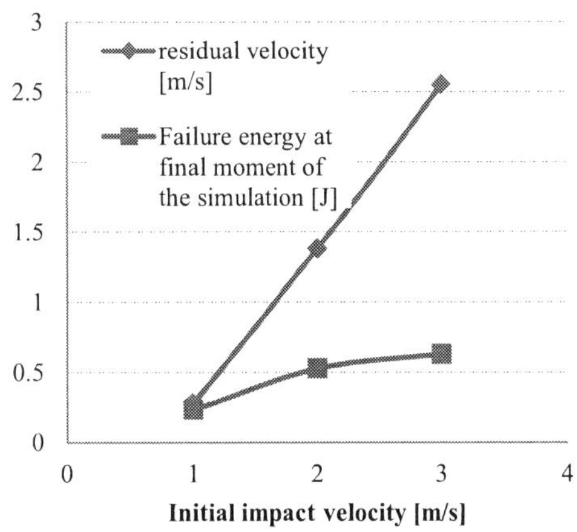

Figure 11. Residual velocity of the impactor and the failure energy at the final moment.

Acceleration has the highest value for the highest velocity, the peak is narrower than those at very low values. For v = 2 m/s, there is a small shoulder, explained by the elastic behavior of the material and than an increase on a larger time interval. Small oscillations are proving that breakage does not constantly advance, as one may see in figure 8. At lower impact velocity, the acceleration has a lower peak value, but larger shape. Very low values of acceleration (figure 10), a constant velocity (figure 9) and images in figure 7 suggest that the separation is almost finished for samples run at v = 2...3 m/s but it continues for v = 1 m/s. Same conclusion could be drawn from Figure 12. The energy spent for deformation and breakage is too small for accepting the separation of the two ends of the sample. But for the higher velocities, the energy "lost" for sample failure, in conjunction with the images in figures 9-11, differs with only 16% (taking into account the value for v = 3 m/s). Constitutive models from literature [14], [15], [16] also consider that higher strain rate makes the material more resistant.

The evolution of impactor velocity in time is similar for all velocities, but its decrease last shorter for a higher velocity.

A polymeric material tested by the authors (blend of PA6 + PP) has similar aspect (figure 12) with that obtained for the simulation with 1 m/s (figure 7.a and figure 4.d).

a) general view of the failed surface (edge of the notch zone is at right side of the image).

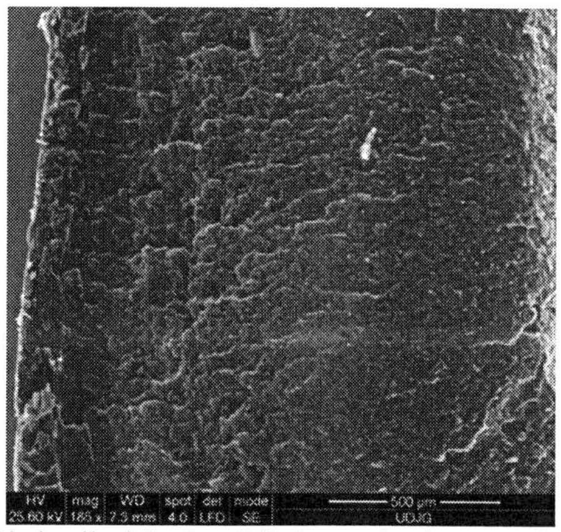

b) detail of the last failed zone (at the opposite surface to that containing the notch).

Figure 12. SEM images of a polymeric blend (PA + PP) tested by the authors at 1 m/s with an impactor of 0.5 kg.

4. Conclusions

After analysing the images of the simulations, the following conclusions may be formulated.

Simulation helps identifying the stages of the material failure.

Using the bilinear hardening model for the polymer failure at low impact velocity (1...3 m/s) gave reasonable simulations.

The time for failure process is longer at lower impact velocity and the aspect of the failed surface depends on this velocity.

5. References

[1] Holt J M 1990 *Charpy Impact Test: Factors and Variables* (Philadelphia: ASTM)
[2] Francois D and Pinneau A 2002 *From Charpy to present impact testing* (Oxford: Elsevier)
[3] Haušild P, Berdin C and Rossoll A 2005 Modelling of the Charpy impact test in the DBTT range *Materials Science Forum* **482** pp 331-334
[4] Báránya T, Czigány T and Karger-Kocsis J 2010 Application of the essential work of fracture (EWF) concept for polymers, related blends and composites: A review *Progress in Polymer Science* **35** pp 1257–1287

IOP Conf. Series: Materials Science and Engineering **514** (2019) 012011 doi:10.1088/1757-899X/514/1/012011

[5] Năstăsescu V, Ştefan A and Lupoiu C 2001 *Analiza neliniară prin metoda elementelor finite. Fundamente teoretice şi aplicaţii* (Bucureşti: Academia Tehnică Militară)

[6] Shokrieh M M and Joneidi V A 2015 Characterization and simulation of impact behavior of graphene/ polypropylene nanocomposites using a novel strain rate–dependent micromechanics model *Journal of Composite Materials* **49** pp 2317–2328

[7] Rossoll A, Tahar M, Berdin C, Piques R, Forget P, Prioul C and Marini B 1996 Local Approach of the Charpy Test at Low Temperature *Journal de Physique IV Colloque* **06** pp C6-279-C6-286

[8] Mohan Kumar K, Devaraj M R and Lakshmi Narayana H V 2012 Finite element modelling for numerical simulation of Charpy impact test on materials *Proceedings of the International Conference on Challenges and Opportunities in Mechanical Engineering, Industrial Engineering and Management Studies ICCOMIM 2012* pp 32-36

[9] Serizawa H, Zhengqi W U and Murakawa H 2001 Computational analysis of Charpy impact test using interface elements *Transactions of JWRI* **30** pp 97-102

[10] Ghaith F A and Khan F A 2013 Three dimensional nonlinear finite element modeling of charpy impact test *International Journal of Mechanical Engineering and Technology* **4** pp 377-386

[11] Visser H A, Caimmi F and Pavan A 2013 Characterising the fracture toughness of polymers at moderately high rates of loading with the use of instrumented tensile impact testing *Engineering Fracture Mechanics* **101** pp 67–79

[12] Madhusudhan D, Chand S, Ganesh S and Saibhargavi U 2018 Modeling and simulation of Charpy impact test of maraging steel 300 using Abaqus *IOP Conf. Series: Materials Science and Engineering* **330** 012013

[13] Johnson G R and Cook W H 1985 Fracture characteristics of three metals subjected to various strains, strain rates, temperatures and pressures *Engineering Fracture Mechanics* **21** pp 31-48

[14] Capaldi F M 2012 *Continuum mechanics: constitutive modeling of structural and biological materials* (New York: Cambridge University Press)

[15] Puzrin A M 2012 *Constitutive Modelling in Geomechanics* (Heidelberg: Springer)

[16] Ottosen N S and Ristinmaa M 2005 *The mechanics of constitutive modeling* (Oxford: Elsevier)

IOP Conf. Series: Materials Science and Engineering **514** (2019) 012012 doi:10.1088/1757-899X/514/1/012012

Additivation of vegetal oils for improving tribological characteristics

C Georgescu[1], L C Solea[1] and L Deleanu[1]

[1]Department of Mechanical Engineering, Faculty of Engineering, Dunarea de Jos University of Galati, Romania

E-mail: constantin.georgescu@ugal.ro

Abstract. This paper presents a review on issues on additivating vegetal oils with friction and anti-wear modifiers. A SWOT analysis on vegetal oils reveals the shortcomings of these oils and the trend in their modification by the help of additives. Based on research reports available from literature and their experimental work, the authors formulated a synthesis on how the friction and wear modifiers act in lubricants and especially in vegetal oils. They discussed the influence of some friction and anti-wear modifiers added in vegetal oils, especially in soybean oil and rapeseed oil. The conclusion is that additivation of vegetal oils is still at the beginning as the reported results are not so efficient as compared to aditivation of classical alternative - mineral oils and even less as compared to synthetic oils. But research must be continued as these vegetal oils become an actual resource for lubricants base-stock.

1. SWOT analysis of vegetal oils as lubricants

The OECD presented a report on lubricants and additives in 2014 [1], [2] stating that vegetable oils have an upward trend in their use as lubricants, especially for areas with a stronger impact on the environment. Worldwide, 40 million tons of lubricants are consumed annually, while petroleum-based lubricants are still dominant. They begin to be challenged and replaced by synthetic oils and even by vegetal oils [3]. In 2016, Romania was the third largest producer of industrial seeds in the EU (rapeseed, soybean and sunflower).

Vegetal oils themselves may act as anti-wear additives and friction modifiers, due to strong interactions with the surfaces they come into contact with, especially metallic surfaces [4]. The long molecular chains of fatty acids and the presence of polar groups in the vegetal oil structure give them the ability to adhere and maintain on surfaces in contact, even under relatively severe regimes [5].

A SWOT analysis for introducing vegetal oils as lubricants showed that a set of properties should be considered when the designer decides this lubrication solution [6], [7].

Strengths include:

- biodegradability [8],
- environment protection or/and acceptability (non-polluting or environmental friendly lubricants) [9], [10], [11],
- extraction from renewable resources (even with a reference to 100 years) or the possibility of recycling or re-use of the lubricant [3],
- high viscosity index [7],
- better flammability characteristics, auto ignition points and higher ignition temperature on hot surfaces [12], [13].

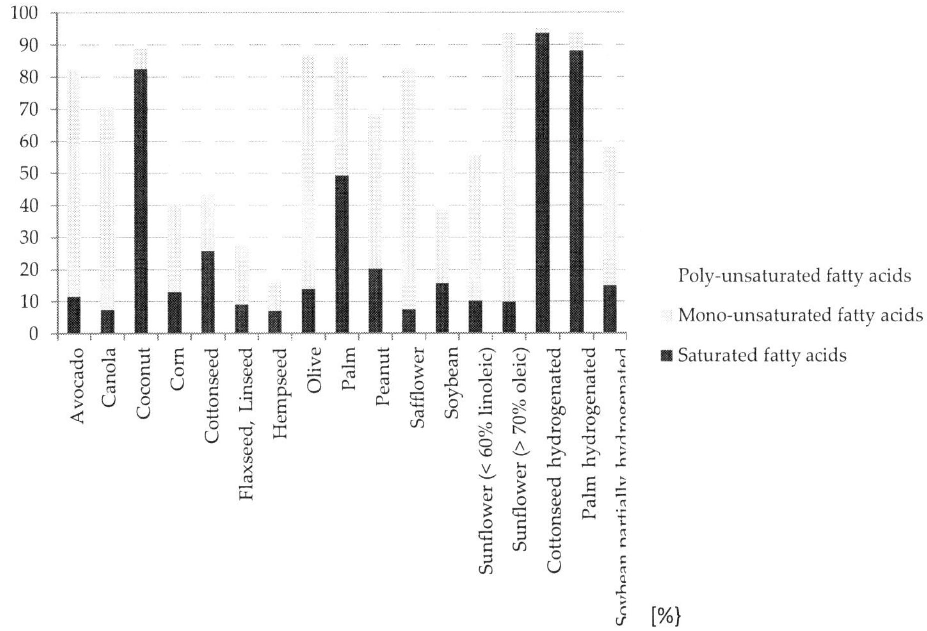

Figure 1. Diversity in fatty acid composition for vegetal oils. [14], [15]

Weak points are especially related to the chemical composition of these oils as they consist of a mix of fatty acids (figure 1), the non-saturated having a great influence in weakening the reliability of oils as lubricants:

- lower viscosity as compared to mineral and synthetic oils [16], [17], [18], [19],
- oxidation [20], [12], [21] and poor chemical stability; there is a correlation between the unsaturation degree of a vegetal oil and its oxidative stability; the oxidation temperature and the oxidative stability can be predicted using each oil fatty acid composition rather than individual fatty acid percentage [22],
- sensitivity to moisture,
- operating temperature range lower than that of mineral and synthetic oils [23]; a bio-based lubricant will cause premature and possibly catastrophic equipment failure if the functioning temperature is over 105°C (and as low as 70...80°C for some vegetal oils); in extreme high-temperature, environmentally sensitive applications, biodegradable synthetic fluids are recommended,
- many of their properties are more time-dependent than those of mineral and synthetic oils,
- low temperature properties are poorer for vegetal oils than other lubricants [24],
- lower limit to seizure on four ball machine and sharp gradient to seizure limit (figure 2) [25].

Opportunities are supported by the following facts:

- complying with more stringent environmental protection requirements will minimize health and pollution risks,
- the high risk of exhaustion of petrol resources and the necessity of replacing them as an issue to be solved for the global economy.

The new market shares for organic and biodegradable lubricants (obtained from renewable resources, especially plants) have increased for areas such as hydraulic fluids, chain lubricants, mold lubricants, two-stroke engines, turbine fluids, etc. [9].

Environmentally acceptable lubricants are classified according to the type of base oil used in their formulation. In general, lubricants consist of approximately 75 to 90 % base oil [10].

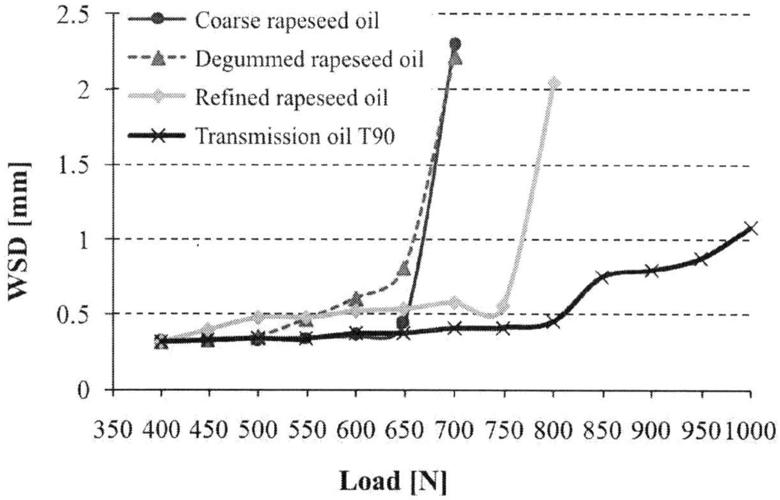

Figure 2. Evolution of WSD (wear scar diameter) with load to seizure limit for several vegetal oils [25].

Threats include:

- the need to re-design systems using bio-liquids, a solution that is possibly costlier,
- accepting lowering some system operating parameters (especially load and maintenance, but not limited to) [26],
- the still high price (but not forgetting that synthetic oils in the 1990s were almost 10 times more expensive than mineral ones and today the ratio is only 3 to 1), market and users inertia, the diversity of environmental and safety specifications and a global policy that has not yet been clearly addressed on environmental issues.

Literature reported relatively low or inconclusive results on nano lubricants [27], [28], [29], [30] mostly on transformer oil, gear oil and heat transfer oil. Limited investigations on the influence of nano additives on vegetal oils are presented [31], [32]. Even if modern equipment working under high load, speed and thermal conditions, require cooling and efficient lubrication, and for this concept, mineral and synthetic oils are still preferred, investigations on vegetal oils are needed for particular applications with environmental impact and in the perspective of oil resources extinction.

Chan et al. [33] propose criteria for the molecular structure of vegetal oils in order to have a good tribological behavior, based both on friction coefficient (stability, value and spread range) and wear measurements. The authors add several other parameters that would make the difference between poor and good quality of a lubricant: temperature in contact, durability (including oxidation), rates of different modifications, produced in time on the bodies involved in the tribosystem (solids and lubricant, environment). The bio-based oils include vegetable oils (VO), epoxidized VO, ring-opened products from epoxidized VO, estolides, and polyol esters. And the issue of additivating these oils is presented with recent advances: plant-derived compounds and polymers, particulate and layered materials, and ionic liquids.

Bio-based lubricants have also gained importance as alternatives to conventional petroleum-based lubricants in various applications, especially the automotive industry and green industries (food, wood, transport etc.). But these lubricants are still far from being practical substitutes. However, these shortcomings can be addressed by modifying the vegetable oils chemically or incorporating additives into these oils [31]. This review provides an overview of lubricants using vegetal oils as base and additives for improving their tribological behavior and useful information from authors' experimental work for supporting the introduction of such lubricants in actual applications.

2. Nano additives for lubricants

Nanofluids containing solid nanoparticles could have a higher viscosity than common working fluids. But the actual effects of volume fraction, temperature, particle size, and shape on the viscosity of nano fluids are determined through experiments [34].

Early formulated vegetal oils used performance additives similar to those for petroleum oils that are, mostly, nonpolar, whereas triglycerides are highly polar. Thus, conventional petroleum additives have solubility problems when added in vegetal oils. Frequently, a dispersant agent has to be used.

Amine phosphate compounds are used in biodegradable products, but they are not efficient in petroleum as in vegetable oils. Some anti-wear additives used in vegetal oils reduce oxidative stability when combined with certain antioxidants. These include amine phosphate compounds and molybdenum dialkyl phosophordithioate with ZDDC [22].

Gupta et al. [35] presented a review on thermophysical properties of nanofluids, but vegetal oils are mentioned only once. Based on recent literature [29], [36], additives for lubricants could be grouped as in table 1.

Multifunctional additives in vegetal oils are of great interest [33], as they may offer an improvement on several characteristics, especially viscosity and oxidative stability.

Table 1. A classification of additives for lubricant. based on [22], [36] [30], [37], [38], [39], [40]

Modifiers of chemical properties	Modifiers of physical properties	Anti-wear and friction modifiers		Extreme pressure additives
		Inorganic O	rganic	
deposit control additives anti-oxidation additives detergents anti-toxicity agents biodegradability promoters	viscosity control additives poor point depressants anti-foaming additives dispersants	**metals** **salts** **oxides** (metallic or not) **carbonic materials**	carboxylic acids and derivatives, stearic acid and esters, amides, imides, amines and their derivatives (oleylamide, phosphoric or phosphonic) acid derivatives, organic polymers (methacrylates)	ZDDP, sulfurized isobutene, fatty oils and olefins, sulfurized synthetic esters, sulfurized fatty oil + olefin

Nanoadditivation could increase the seizure load in a lubricant, as for instance, the hexagonal boron nitride (hBN) (70 nm) in SAE 15W-40 diesel engine (figure 3) [42]. A concentration of 0.3 vol% of surfactant (oleic acid) was added in order to prevent the sedimentation of the nanoparticles. Ten increments of loading were applied starting from 196 N to 1,570 N, with 10 seconds for every test.

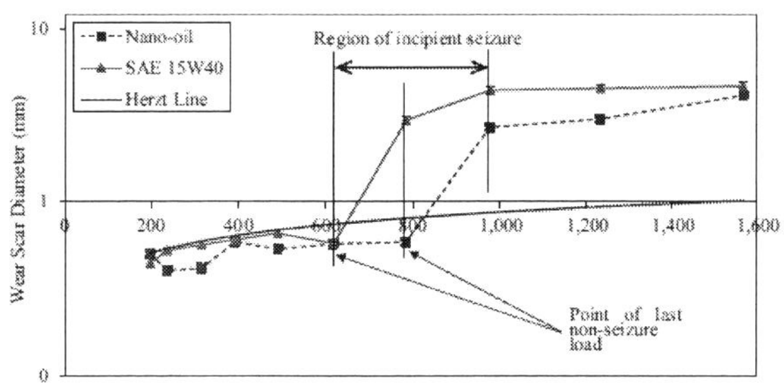

Figure 3. Wear scar diameter (WSD) and load to seizure evolution for SAE 15W-40 and additivated SAE 15W-40 with 0.5 vol. % hBN [42]

Ionic liquids are attractive lubricant additives for their nonflammability, low vapor pressures, structural diversity and reasonable tribological properties. Phosphite binds to aliphatic acid as anion and triethylbenzylammonium serves as a cation, which could be applied as lubricant additives in mineral oil and refined vegetal oils. These additives were added in Arawana 1:1:1 oil (this being the proportion of saturated, mono-unsaturated and poly-unsaturated fatty acids in a combination of vegetal oil). Results demonstrated that these ionic liquids could enhance the tribological properties of tested oil. The chain length and unsaturation of fatty acids are significant in improving the tribological properties of the base oil. [41].

3. Issues related to vegetal oil additivation

The number of additives compatible with vegetal oils, synthetic esters or polyalkylene glycols is small relative to the number of additives compatible with mineral base oils [10].

A package of additives for vegetal oils may contain: detergents, dispersants, anti-corrosion and anti-oxidation agents, anti-foam agents, viscosity modifiers and anti-wear additives, pour-point depressants. It is hard to determine the synergic influence of such a package. But additives also should be both ashless (containing Ca, Na, K, Mg and not other metals) and non-toxic [43]. Among the soaps, calcium-based soaps are considered less toxic compared to other types (e.g., Li-based), and soaps in general are considered less toxic than graphite thickeners [44].

Issues of nanoadditivation of vegetal oils are related to:
- concentration, shape and size of nano particles,
- compatibility with the oil composition,
- chemical and physical stability in time and under the exploitation regime parameters,
- nanoparticle agglomerations, sedimentation, dispersion,
- additive fall-down caused by water that could be present in any lubricant system; water also increases acid formation, seals deterioration, rust generation and wear acceleration. Most bio-fluids are more susceptible to hydrolytic breakage, the result may be acid formation, additive precipitation etc.

A dedicated chapter, "Additives for bioderived and biodegradable lubricants" [22], started by comparing petroleum oils, vegetal oils, saturated esters and polyalphaolephins and concluded that vegetal oils have to be formulated for their individual set of properties.

Reviews on additives for lubricants appear in literature [37], [30], [45], [46], but only few deals with the influence of nanoadditives in vegetal oils, the analysed results being not enough satisfactory as compared to the additivation of mineral and synthetic oils. The addition of nanoparticles to a base oil (mineral, synthetic or vegetal) is a promising approach towards improving certain characteristics, such as friction and wear resistance, thermal and chemical resistance, but literature and reports do not have yet any clear recommendation on how to formulate and to use new additivated lubricants without test laboratory, especially those that could offer data that are compared to the "classical" ones.

Small particles with the size of nanometers, such as graphite [47], boron nitride (BN) [42], natural and synthetic minerals, MoS_2 [48], WS_2 [49] and polytetrafluoroethylene (PTFE) [50], have been used both as solid lubricants and as additives in lubricants. These particles have a tendency to deposit during storage and use because of their size and their tribological properties remain poor in the presence of humidity and oxygen and, thus, limiting their applications [51]. There have been investigations on the addition of nanoparticles with the typical size in the range of 2–120 nm to lubricants as friction modifiers for efficiently reducing friction and wear. In particular, nanoparticles based on carbon compound, metal, metal oxide, metal sulphide, metal borate, metal carbonate, rare earth compound and SiO_2 have been investigated [28], [52], [53] and their tribological performances as friction modifiers are dependent on the degree of crystallinity, size, shape and concentration [30].

There are many studies on neat vegetal oils [54], [55], [13], but the additivation of these lubricant fluids is still at the beginning. A synthetic research work on this subject was done by Zulkifli [38] and the same author gives experimental results for a metal oxide (TiO_2) [56]. Glycol was used as solvent. The nanoparticles reduced friction coefficient (COF) up to 15% at high load and improved WSD especially at low load (40 kgf, four ball test) by creating an additional protective layer.

There are two tendencies on vegetal oil additivation:

- to use the same additives as for the above-mentioned oils,
- to formulate new additives, based on particular characteristics of vegetal oils.

Concentration of nano additives in the base oil strongly influences tribological properties, but the optimum values could be hard to be evaluated by theoretical models and the actual influence of additive(s) is determined only by testing. As there are many tribological characteristics that could be affected by additivation and the engineer has to accept a compromise: to optimize a parameter (of high interest for a particular application), to improve others and to make acceptable the induced effects. For instance, the nanoadditives will reduce wear but they could increase the friction coefficient and the temperature in contact. Tests will tune this system approach especially when they are done with parameters in the ranges of actual application.

Alves et al. [57], [58] studied the development of vegetal lubricants with addition of oxide nanoparticles (ZnO and CuO) as additive for extreme pressure (EP), exploring their and base-oil influence on tribological behavior. They considered the additives as for extreme pressure (EP), but it is hard to draw a neat boundary between wear and friction modifiers and EP additives, based on classical literature [29], [36], especially for vegetal-based lubricants, due to their particular physical and chemical behavior in the presence of the additive and under working conditions (load, speed, temperature etc.). The results showed that, with the addition of nanoparticles to conventional lubricant, the tribological properties can be significantly improved. EP additives and reaction substances smoother and form a more compact film on the worn surfaces, which is responsible for the further reduced friction and wear. However, the addition of some nanoparticles in vegetable-base lubricants is not beneficial to wear reduction. A conclusion obtained also by Cristea for adding graphite and graphene in soybean oil [47], [79].

There are researchers that include metal oxides in the category of EP additives [59] but, due to the poor chemical activity on the substrate and to the fact that the layers are not continuous and stable in time, these additives could be treated more like friction and wear modifiers.

Even if many specialists wrote about a smoother and more compact tribolayer formed on the worn surfaces, responsible for friction and wear reduction, recent report and SEM investigations reveal that the tribolayer of wear and friction modifiers is not continuous, very often being a powdery layer of nanosheets, nanoparticles, rolled or not [60], [61] [62].

Manny additives are considered as multifunctional ones as they modify two or more characteristics of interest (for instance, they increase oxidative stability and reduce wear). Ethyl cellulose (EC) could be considered as a multifunctional additive. Delgado [63] reported that adding EC in high oleic sunflower (HOSO) oil and castor (CO) oil make them able to reach stable and non-gel-like blends, with viscosities at 40°C in a large range (62 and 493 cSt) and viscosity indexes into API group III (VI 120). The HOSO/CO/EC blend showed a reduction in friction coefficient at low speed and generated a stable EHD-film at 100°C of around 20 nm, suggesting better boundary properties than HOSO/EC or CO/EC oils. Ethyl cellulose hindered wax crystallization process of these vegetal oil-based lubricants at 5°C. Therefore, the suitable combination of both castor and HOSO oils with EC as multifunctional additive allows for formulating eco-friendly fluids with a wide viscosity range, better viscosity-temperature dependence than many mineral or synthetic oils and excellent boundary lubrication, in applications like hydraulics, metal working, gear transmissions etc.).

The use of vegetal oils as lubricants implies a research on the set of tribological features, including those that can be highlighted on the four-ball tribotester [64], [46]: the influences caused by the nature of the lubricant, chemical and rheological changes, load and speed conditions, the influence of additive(s) nature and concentration, etc. [65], [66].

In the literature, there are reported tests on vegetal oils but the data are far from being comparable and useful for industrial scale applications. The information on the behavior of these oils on the four-ball machine is of interest because it is possible to compare the vegetal oils, whether additivated or not, with those already in use, mineral or synthetic ones [67], [47], [68], [25], [55]. Cermak [69] tested vegetable oils on the four-ball tester in accordance with the American Standard ASTM D4172 and obtained good values for friction coefficient and diameters of wear scar of 0.53 mm for Cuphea oil and 0.89 mm for Lesquerella crude oil. It is interesting to note that soybean oil had a wear scar diameter higher (0.70 mm) than other oils tested in this study (grasshopper oil - 0.629 mm, cress oil - 0.59 mm).

4. Mechanisms of nano additive for reducing friction and wear

The influence of an additive on the tribology of the contact depends on how the additive evolved in contact as size, shape, structure and chemical composition. A schematic representation of how the additives protect the surface from being worn is given in figure 2.

For instance, a spherical nano additive in contact [70] could keep its shape and acts like a damper or/and spacer between two asperities in contact. But also it could change its shape, becoming flatter when the load increases, protecting a larger surface against rubbing under severe load, the relative motion of surfaces producing a shearing process between additive and surface and not one among asperities [70].

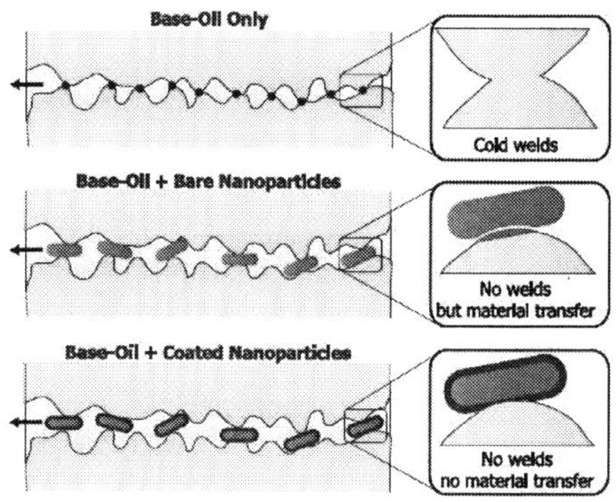

Figure 4. Types of lubrication: a) lubricant without additives: micro jonctions with cold weld, b) mixt regime with additive that reduces adhesion and abrasion wear (the additive is mechanically fixed on the surfaces), c) coated and/or more flexible nanoparticles minimizing the number or micro-welds and the debris resulted from asperities fractures. [70]

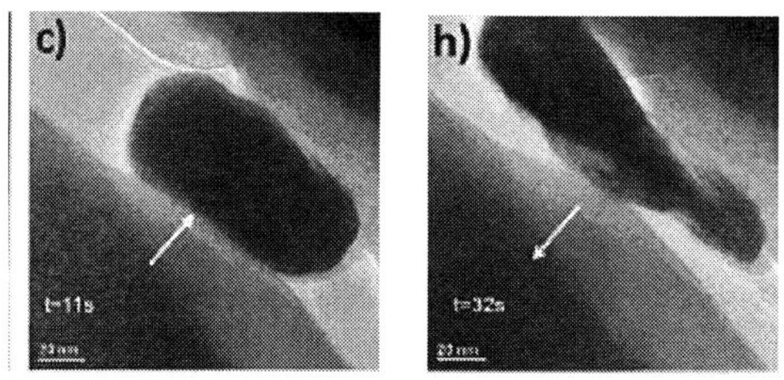

Figure 5. Photos highlighting two of the steps through which a particle of WS_2 passes into a loaded contact [71].

In 2012, a study by Lahouij et al. [71] shows how the WS_2 particle protects the direct contact between metal asperities (figure 5). The void structure functions as a shock absorber and either the structure collapsed, or the particle was fragmented, it continued to remain between the two solid bodies. The hollow core of the particle was visible and the deformation was large at the beginning of the stress, but as the load increased, the particle behaved like a variable-elasticity spring, the elastic characteristic actually increasing. Then the particle begins to tear or scissor, even to fracture fragments.

The relative size of nanoparticles and asperities is also important in improving the tribological behavior of a contact. A larger additive particle could be trapped in a valley of the texture and could redistribute pressure in a local contact. When the particles are smaller than the profile heights, they agglomerate in the valleys, smoothening the surface, making decreasing the local pressure distribution. Figures 6 and 7 are helpful for understanding how some additive act to protect the solid bodies in contact and, thus, reducing wear [62], [70], [78].

Lee J et al. [72], Lee K et al. [73] and Hwang et al. [74] considered that the addition of nanoadditives in lubricants enhanced the lubrication characteristics as compared to microparticles, but for vegetal oils, the nanoadditivation is still contradictory.

Spheroidal nanoparticles in contact take up some of the load and play the role of very small rollers or ball bearings. Fibrous particles in oils (carbon nano fibers and nanotubes [75], [76], for example) have higher friction coefficients than spherical or close 3D-sized nanoparticles, probably because of the way they fix on the textures in contact and the fact that spherical particles tend to roll and not to be dragged into contact. Fibrous nanoparticles agglomerate more easily than spherical ones and this is why their thickness becomes larger than the thickness of the lubricant film, resulting in an increase in roughness, an embarrassment of the lubricant circulation, especially when entering into contact.

Wu et al. [62] proposed a model that takes into account the additive concentration (figure 6). Although the model was created for TiO_2 in water as lubricant, it can be used to explain the behavior of lubricants with other nanoscale particles (metal oxides, carbon materials etc.). The fluid lubrication mechanism with nano additives has been also described in [77], [78], [79].

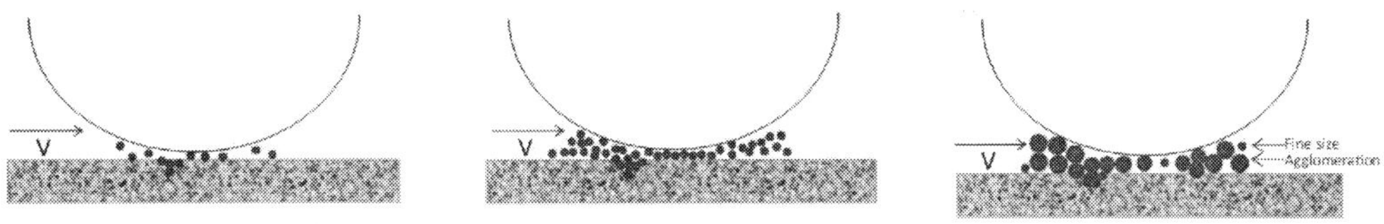

Figure 6. Effect of particle size [62]

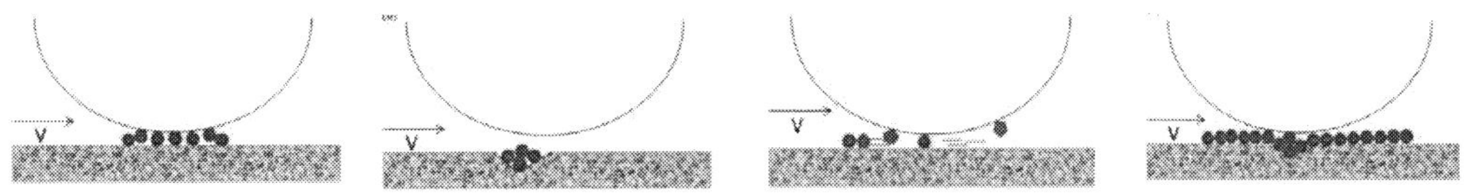

a) rolling b) mending c) polishing d) protective

Figure 7. Lubrication mechanisms when using nano particles in lubricants [78]

One or more of the mechanisms for reducing friction and anti-wear mechanisms of nanoparticles in lubricants has been also reported as (figure 7):

- micro-rolling, [62],
- smoothing/leveling [80], [81],
- polishing [73],
- forming protective film [57], [82], [83].

The first two mechanisms have a direct effect on lubrication [73]. In the case of rolling, no chemical reactions occur and spherical or oval nanoparticles are willing to roll, reducing friction.

Shape, size and even size distribution of nanoadditives during functioning of the tribosystem are important factors affecting its durability and performance. The shape directly determines pressures experienced in contact. When starting, a spherical additive will support such a high pressure that it will be flattened, similar to a nanoplatelet and its contact with the solid body becomes planar: under a higher load, the additive could be laminated into a nanosheet. Thus, nanosheets reduce the risk of indenting, scratching or/and deforming the texture of bodies in contact. The problem is that nanosheets tend to roll instead of remaining laminated on the texture and to form agglomerations of nanorolls that could alter the distance between bodies in contact and to cause friction fluctuation and non-uniform exposure of the surfaces to direct contact and, thus, to generate more intense wear. This aspect was pointed out by Cristea [47], [79] for graphite and graphene in soybean oil and by Cristea et al. [60] for TiO_2 in rapeseed oil.

Despite the general idea discussed several decades ago, the nanoparticles do not form a continuous film on the rubbing surfaces [47], but they intermediate the load transfer and partially transform the sliding into rolling friction.

5. Anti-wear additives and friction modifiers

Anti-wear additives and friction modifiers are used for medium stresses and their protection (not like a film, but like a powdery third body) relies heavily on the adsorption processes and on the regeneration of this intermediate powdery layer, even though particles partially deteriorate or/and they are thrown away from contact during the exploration.

However, due to their poor chemical and physical reaction with the oil and the solid surfaces in contact, nanoparticles have only poor efficiency in oils. This issue could be mitigated by surface modification techniques, for either the particles or the substrates.

A classification of nanoadditives as friction and wear modifiers is presented in Table 2.

Table 2. Improvers of tribological behavior

Anti-wear and friction modifiers					
			Inorganic		
Carbon materials	**Metals**	**Rare earth elements**	**Metal oxides**	**Metallic salts**	**Organic**
black carbon [47], [61] fullerene [84], [85], [72] graphene [47], [86] graphite [47] diamond powders [81], [87] graphene oxide [88]	Cu [59], [89] Ag [90] Cu, Fe, Co [68], [91]	rare earth [92] $La(OH)_3$ [93] La borate [105]	TiO_2 [60], [67], [87] hBN /Al_2O_3 [94] CuO [32], [95], [96] FeO [95] Al2O3 [96], ZrO_2 [97] $ZnAl_2O_4$ [104], Al_2O_3+TiO_2 [98], SiO_2 [32], [99]	WS_2 [49], WS_2+ZDDP [49] hBN [42] PTFE +hNB [50] MoS_2 [99], CaCo3, BN [94] Zr phosphate [100] hydrosilicate [101]	tri(hydroxymethyl) propane esters [102] PTFE [103]

Friction modifiers can be classified into two distinct groups depending on the friction reduction mechanism:

- through an adsorbed film,
- by friction with the third body (the additive particles act like spacers, dampers or/and intermediate rolling elements).

The first is generally due to polar molecules having a polar functional radical (alcohols, aldehydes, ketones, esters and carboxylic acids) and a nonpolar terminal group. The polar group of the molecule adheres to the surface with long chains exposed to moving surfaces, reducing friction. They may also have polar elements that may chemically react with the surface to form a protective film. Vegetal oils and animal fats have such molecular structures and, therefore, they have good results in reducing friction.

Dubey [103] studied the effect of particles' size on the entry in the work zone of the contact and higher the load less is a clear influence and, hence, less nanoparticles are accommodated, but the conclusion for PTFE particles is that they are easily agglomerated and, under compression and shear loads, they may "weld" enough to grow in size and do not entry in contact.

Table 3 presents the diversity of testing conditions, lubricants and additives in order to underline that it is hard to make comparison among results and conclusions and to point out that using the four ball tribotester will be helpful for making a step forward in additivating the vegetal oils.

Table 3. Diversity of lubricant, friction modifiers and test conditions

Ref.	Friction and wear modifier	Oil Test	conditions	Results
[95]	ZnO and CuO, 11.71 nm and 4.35 nm, 0.5%wt	soybean and sunflower, by epoxidation reaction	**Reciprocating ri g** (20 Hz 1 mm stroke), ball-on-disk steel ball (570-750 HV, Φ6.0 mm) slides on a softer steel disk (190–210 HV) 10 N, 60 min, 50 °C	biolubricants without additives are slightly more tribologically effective than lubricants with additives
[96]	CuO (50 nm, 3.5 Mohs), Al_2O_3 (50 nm, 8-9 Mohs), (0.5, 1.0, 2.0 wt%)	GL-4 (SAE 75W-85), Poly-alphaolefin 8 (PAO8)	**reciprocating r ig,** 2 h, 200 N, 50 Hz, 1 mm stoke AISI 52100 ball (Φ10 mm, 60 HRC) against an AISI 52100 disk, at 50°C **four-ball tribotester:** ball made of AISI 52100, Φ12.7 mm, 60 HRC **seizure test**: 18 s run, 500 rpm, increasing load till 7200 N, room temperature (25 °C)	COF decreased in time, more pronounced at 1.0 and 2.0 wt% additive; CuO more effective than Al_2O_3; Al_2O_3 acted as an abrasive (COF increased at all concentrations), with re-agglomeration; CuO has good results at all concentrations, but similar to PAO8; at 2.0 wt% CuO in PAO8, the limit of seizure increases up to 273%
[89]	Cu 18.2-80.2 nm	SF15W/40	**ball-on-disk** GCr15 steel ball (61-63 HRC) and C45 carbon steel disk (hardness 2100 MPa), 15 N, 0.13 m/s, amplitude 5 mm, 20 min.	optimum concentration 0.8%wt, mass loss is half of the neat oil
[101]	mineral (China): Mg5.82Al0.02Fe0.05 Ca0.0K0.07Mn0.01(S i4.21O10.36)(OH), Cu (50 nm), 5% oleic acid as dispersant	Diesel engine oil (CD 15w/40)	**ball-on-disk** 50 N (Hertz pressure of 1.15 GPa), frequency 10...30 Hz, oscillating amplitude of 1 mm, 30 min	optimum concentration 7.5 wt% Cu
[106]	ZnO (125 nm) 1.0%, 2.0%, 3.0%, and 4.0%	commercial refined base oil (not identified)	**four ball tester** 30 min, 392 N, 1450 rpm, Φ12.7 mm, 59~61 HRC.	minimum WSD and maximum seizure load for 3%wt ZnO
[104]	$ZnAl_2O_4$ (95 nm), spherical: 0.05, 0.1, 0.5, 1 wt.% ZnO, Al_2O_3	pure lubricant oil (not identified)	**four-ball tester** F= 147 N 30 min, 75°C 1450 rpm,	optimal concentration of $ZnAl_2O_4$ is 0.1 wt%.
[90]	Ag 6–7 nm	low volatile multialkylated cyclopentanes lubricant	**oscillating teste r,** ball-on-disk, tests at 20°C and 100°C, a frequency of 25 Hz, amplitude 1.0 mm. F=100...900 N	wear resistance and load-carrying capacity raised and COF changed less with 2% Ag in oil
[47]	carbonic material (black carbon, graphite, graphene)	soybean oil	**four ball tester** 100, 200 and 300 N 1000, 1400, 1800 rpm	COF is increasing a little wear tends to decrease at 1% and heavier loads
[78]	TiO_2, 20 nm, 0.2%...8% 0.002...0.08 wt% PEI+10.0 vol% glycerol+water balance	water-based lubricants	**ball-on-disk** F=50 N 5 min+5 min 20 mm/s	0.8 wt% TiO_2 lubricant has good tribological properties (wear decreases at half) COF dry=0.45, COFwater=0.35 COFwater+TiO_2=0.27
[103]	nano PTFE 0–6% 90–100 nm ZDDP 1...3%	n 150 N API Group II oil	**four-ball tester,** ball of 12.7 mm, 1450 rpm weld test 60 s wear test F=392, 588, 784 N	optimal concentration 3% PTFE for minimal WSD

Table 3. Diversity of lubricant, friction modifiers and test conditions (continued)

Ref.	Friction and wear modifier	Oil Test	conditions	Results
[49]	PAO PAO + 1%WS2, PAO +1%ZDDP, PAO+1%WS_2+1%ZDDP	PAO	**reciprocating rig** Hertz pressure 1.12 GPa, lubricant temperature 100°C, velocity 3 mm/s, frequency 0.5 Hz, 8 h, 14400 cycles	Tribological properties of nano particles enhanced when ZDDP is present. The-anti-oxidant properties of ZDDP protectWS2 from oxidation.
[50]	PTFE (12 µm) + hNB (70 nm), 0–4% each; dispersant (1 wt%) and petroleum ether PIBSI (1%)	Group III	**four ball tester** weld test wear test 392, 588 and 784 N	higher weld load for 3% PTFE+ 1% hNB
[88]	graphene oxide three types 0.2 wt. % to 0.4 wt. %	rapeseed oil	**four ball tester** seizure test 1760 rpm 0.67 m/s, for 10 s. wear test 98 N; 196 N; 294 N; 392 N	max. non-seizure load and minimum WSD at 0.2% treated graphene oxide

The extreme pressure additives are chemically bonded to the rubbing surfaces, especially when the temperature reaches the threshold of additive reaction with the metallic surfaces. The addition of ZDDP additive in soybean oils effectively protected the rubbing surfaces, but it showed no clear trend on the friction coefficient. For epoxidized soybean oil and high-oleic soybean oil, the temperature and the additive interactions predominantly affected the wear scar diameters. The additive tends to improve stability and to have a less dependence on speed and temperature, but not a consistent reduction in wear scar diameter [110]. Similar results were obtained by Cristea [47] with nano friction modifiers (black carbon, graphene and graphite) in a degummed and refined soybean oil.

Figure 5 exemplifies the diversity of shape and size of wear improvers and friction modifiers.

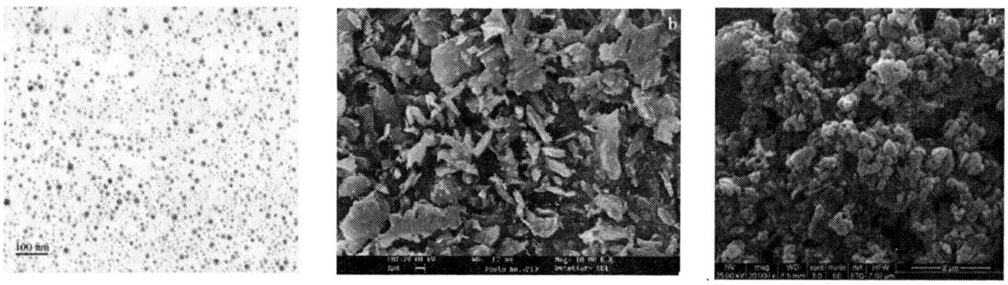

a) TEM image of Ag nanoparticles [90] b) Phyllosilicate natural mineral serpentine. 1 to 5 µm [93] c) $La(OH)_3$ nanoparticles 50 nm [93]

Figure 8. Diversity of size and shape of nano additives for lubricants.

Zareh-Desari [32] studied the influence of adding SiO_2 and CuO (0.3...1%) in soybean and rapeseed oils, containing various concentrations. Silica nanoparticles (SiO_2) have appropriate dispersivity and stability even in high volume fractions [87].

Padgurskas et al. [91] analyzed the influence of different metal nanoparticles (Fe, Cu, and Co) on a mineral oil (SAE 10) and they exhibit different behavior when added to this oil. Cu particles possess the most effective wear resistance characteristic. They reported that a mixture of nanoparticles is more effective than using them separately.

Ma [90] added Ag nanoparticles highly soluble in non-polar and weak polar organic solvents as a synthetic alkylated cyclopentane oil (having high decomposition temperature and very low vapor loss) and improved wear performance and load-carrying capacity, with low effect on friction (figure 9). During the friction process, Ag nanoparticles were deposited on the friction pair surfaces to form metal Ag boundary film, with low shearing stress and contributed to prevent the steel-to-steel contact from severe adhesion and abrasion.

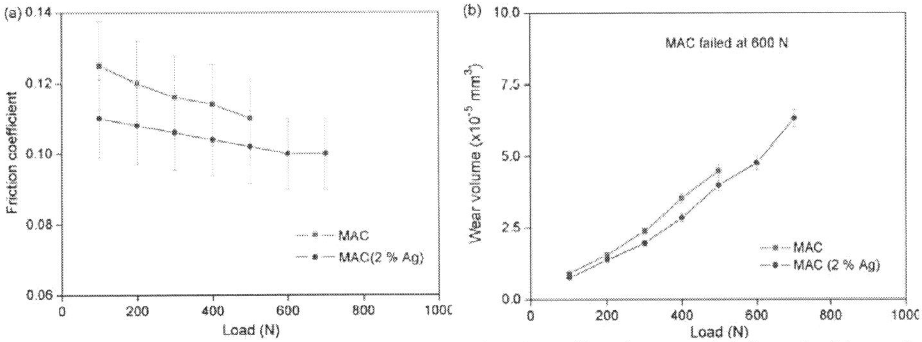

Figure 9. Friction coefficient (a) and wear volume (b) of steel/steel contact lubricated with synthetic alkylated cyclopentane oil (MAC) base oil and containing 2% Ag nano particles at 100 ◦C for 1 h [90].

Wu et al. [78] analysed TiO_2 and CuO behavior as nanoadditives in lubricant oil. The addition of two different nanoparticles in oil decreases its friction (CuO performed better than TiO_2). In addition, both exhibited uniform dispersion and distribution in the base oil. Figure 10 presents details on the wear scars on balls tested with rapeseed oil with particles of TiO_2, (1 wt% TiO_2), at v=0.69 m/s [60] and one may notice that the dispersion on the surfaces are non-uniform and some particles agglomerate, becoming of micronic size.

a) F = 200 N b) F = 200 N c) F = 300 N

Figure 10. Wear scars with particles of TiO_2, after testing the rapeseed oil with 1% TiO_2, at v = 0.69 m/s [60].

The anti-wear behavior of oxide nanoparticles depends on the lubricanting base oil. They do not show good anti-wearability when combined with epoxizided vegetal oil, like sunflower and soybean oils, because of chemical nature of these oils on film formation due to polar groups that adhere to surface. In this case, nanoparticles have a third body behavior, increasing friction. ZnO reduces friction and wear when introduced in a mineral oil. Synthetic oil has its tribological properties improved with addition of CuO [30].

Solid lubricants are added in oils with the purpose of reducing friction and wear. This group of friction modifiers includes carbon materials (fullerene, nanotubes, graphite, graphene etc.), but also molybdenum and wolfram sulphides, fluorinated polymers, such as PTFE and perfluoropolyalkylethers [30]. Solid lubricants (micro or nano) also help in situations where sliding surfaces have a more rough texture, "leveling" the profile of both surfaces. They are also recommended for reciprocal movements (in the case of the piston ring), producing a reduction in wear.

It is added to lubricants that come into contact with surfaces with which EP additives can not chemically react, such as polymers and ceramics and some of their composites [36].

The lubrication mechanisms of nanoparticles as friction modifiers include three types of friction [107]:

- rolling friction: spherical nanoparticles act as micro or nano ball roller bearings between triboelement surfaces, under light load conditions,
- sliding friction: nanoparticles serve as spacers and eliminate direct metal/metal contact between the asperities of the two triboelements, under higher load conditions,
- rubbing with the third body: exfoliating nanoparticles and their outer layers gradually transfer to the surface textures, providing easier friction under high load conditions, when the third body may be considered a mixture of oil, nanoparticles and wear particles.

The use of nanoparticles as lubricant additives is a top issue of research in the last decades [30], [70].

Jayadas et al. [64] have calculated the advantage of using additives in oils, based on the results of the shear rate and the temperature influence on the viscosity of the additivated lubricant. Wu et al. [62] reported increased load capacity for the nanoadditivated lubricants. Many studies were based on a single concentration of the additive. The effect of varying viscosity due to nanoparticle concentration is difficult to model and, therefore, tests become relevant.

6. Carbon nanoparticles as additives in lubricants

The use of nano-carbon materials is more recent - the last decade. Specialists divide nanocarbon materials into four classes, depending on carbon allotropy: zero dimension (fullerene, but also black amorphous carbon), single-dimensional or 1D (carbon nano wires and nano bars), two-dimensional or 2D (graphene [86]), three-dimensional or 3D (diamond powders).

For the time being, graphene friction patterns and data obtained at AFM are contradictory [86]. Tribotests that allow for more pertinent comparisons and proximity to actual tribosystems are few and, therefore, the influence of adding graphene as oil additive has to be tested. Graphene is considered a rising lubricant [86]. In spite of the efforts of developing research for existing and future applications, their tribological potential as a lubricant remains relatively unexplored. Very high chemical inertia, good strength and the ability to crack easily between platelets or very smooth surfaces are favorable attributes for a very good tribological behavior.

Lin et al. [108] studied the stability of micro-graphene plates in oil and tested as dispersants: sodium dodecyl benzene sulfonate, stearic acid, dodecyl trimethyl ammonium chloride, oleic acid, sorbitan monooleate and polysorbate. They concluded that, from of all these substances, stearic acid and oleic acid are the most suitable dispencers. The optimal concentration of modified graphene was 0.75 wt%, reporting an improvement in wear and contact loading capacity. The results were superior to the addition of natural graphite flakes in the same oil.

Xiao and Liu [27] compared various types of 2D nano additives. Materials with layered structures, such as graphite and MoS_2, are applied as solid lubricants. Two adjacent layers of these structures are bonded by weak van der Waals forces, leading to low shear strength, which allows adjacent layers for sliding easily against each other under shear force, providing the friction reduction, even in the presence of a fluid.

As compared to conventional organic lubricant additives like molybdenum dithiocarbamate (MoDTC), zinc dialkyldithiophosphate (ZnDTP), and zinc dialkyldithophosphate (ZDDP), carbonic nanomaterials (see exemples in figure 11) are superior in two aspects:

- they exhibit good performance in reducing friction and wear, but the authors' opinion based on recent papers [47], [79], could point out that is true for moderate regimes (figure 12),

- significantly lower toxicity than some organic and inorganic additives, making them attractive for environmentally restricted applications.

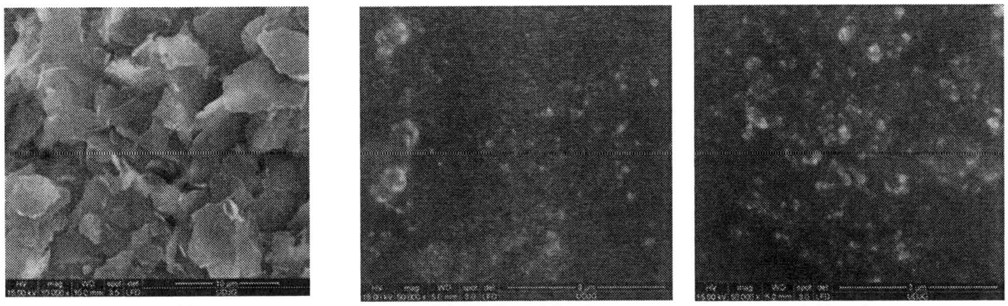

a) Graphite b) Carbon black c) Graphene

Figure 11. Different carbonic materials used in soybean oil as additives [47]

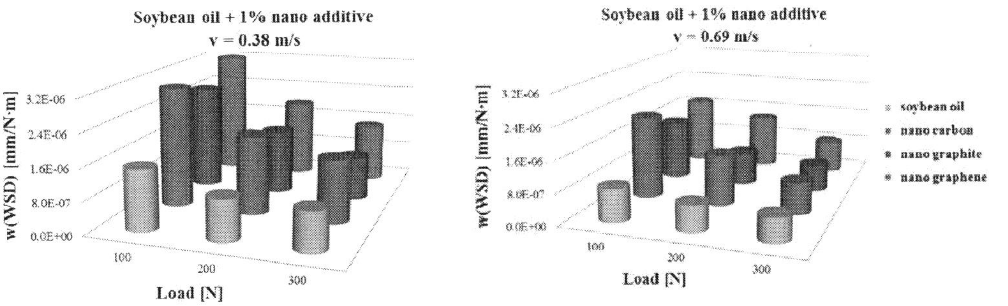

Figure 12. Maps of the rate wear of WSD, w(WSD), for soybean oil base lubricants, additivated with 1wt% nano additive (carbonic materials, tested 1 h on a four ball machine) [47]

The downward trend of w(WSD) had a higher gradient for lubricants with 1% nano additive, which would recommend further testing for more severe regimes, where additives are likely to better protect the surface of the contact. The wear rate of WSD is calculated as:

$$w(WSD) = \frac{WSD}{F \cdot L} \quad [mm / N \cdot m] \tag{1}$$

where *WSD* is the wear scar diameter average for a test [mm], *F* - the load applied on the four balls [N], *L* - the sliding distance [m].

Although micro-scale diamond particles are used for surface superfinishing, it has been found that at the nano level, these particles act on the principle of bearings, like roller balls between the two surfaces in sliding. The results on nanodiamonds are contradictory [81].

Lee [72] studied nanographite in a mineral oil for transmission (220 cSt at 20 °C) (Supergear EP220, SK, Korea), but this oil also had EP additives. The average graphite size was 55 nm. The tested concentrations were 0.1 wt% and 0.5 wt%. The alkyl aryl sulfonate was used as a dispersant. Agglomerations of particles acted as a contaminant and increased the risk of generating more abrasive wear by passing agglomerations through the contact and "falling" (continued slipping with local shock in contact).

Graphite is effective in high temperature and high load applications. Similar lubricants, such as MoS_2, will rapidly oxidize when working temperature reaches 760-1200 °C, although MoS_2 has a better lubricating capacity. Graphite is such a good solid lubricant due to the lamellar plate structure, composed of planes of hexagonal carbon atoms, with weaker van der Waals forces holding together the planes [109].

7. Conclusions

From the studied literature, there is a tendency for research on additivation of vegetal oils, especially for lubrication. Addition of nanoparticles may improve tribological properties of vegetal oils. Friction is not always reduced, but the increase in friction coefficient is acceptable as it is accompanied by a substantial reduction in wear. Most tests are done on four ball machine, pin/ball-on-disk and reciprocating tribotesters. Attention should be focused on the dispersion of nano additive and the selection of dispersant.

The reported results are still inconclusive and the applications of these oils are based more on market inertia or on the practical experience of users.

Despite the advantages of nanoparticles as oil additives, as analysed and summarized in this paper, there are also some challenges hardwired to their applications, which could potentially form future research:

- to prepare and maintain homogenous mixtures of nanostructure particles and oils. Strong van der Waals forces between the particles cause them to aggregate in solutions; therefore, various modification techniques should be investigated for the stabilization of nanoparticles in vegetal oils to produce lubricants, physically and chemically stable,
- to test under adequate conditions, closer to the considered future applications,
- to report results in a manner that allow for comparing data and translating them to the actual tribosystems.

The conclusion is that additivation of vegetal oils is still at the beginning as the reported results are not so efficient as compared to the additivation of classical alternative - the mineral oils and even less, as compared to synthetic oils. But research must be continued as these vegetal oils become an actual resource for lubricants base-stock.

8. References

[1] *** OECD series on emission scenario documents, Number 10, Emission scenario document on lubricants and lubricant additives, JT00174617, ENV/JM, 2004

[2] *** OECD, Towards the development of OECD best practices for assessing the sustainability of bio-based products, 2010

[3] Cermak S C, Biresaw G, Isbell T A, Evangelista R L, Vaughn S F, Murray R 2013 New crop oils—Properties as potential lubricants *Industrial Crops and Products* **44** pp 232–239

[4] You J, Li F, Huang Y 2010 Tribological performance of two potential environmentally friendly ashless vegetable oil additives, *China Petroleum Processing and Petrochemical Technology* **12(1)** pp 43-48

[5] Adhvaryu A, Erhan S Z, Perez J M 2004 Tribological studies of thermally and chemically modified vegetable oils for use as environmentally friendly lubricants, *Wear* **257** pp 359–367

[6] Kurth T L, Byars J A, Cermak S C, Sharma B K, Biresaw G 2007 Non-linear adsorption modeling of fatty esters and oleic estolide esters via boundary lubrication coefficient of friction measurements *Wear* **262(5–6)** pp 536–44

[7] Erhan S Z 2005 *Industrial Uses of Vegetable Oils*, Peoria, Illinois, AOCS Press

[8] Luna F M T, Rocha B S, Rola Jr E M, Albuquerque M CG, Azevedo D C S, Cavalcante Jr C L 2011 Assessment of biodegradability and oxidation stability of mineral, vegetable and synthetic oil samples, *Industrial Crops and Products* **33** pp 579–583

[9] Norrby T, 2003 Environmentally adapted lubricants – where are the opportunities?, *Industrial Lubrication and Tribology* **55(6)** pp 268-274

[10] *** *Environmentally Acceptable Lubricants*, EPA 800-R1-002 2011 United States Environmental Protection Agency Office of Wastewater Management Washington, DC 20460

[11] Rakopoulos D C, Rakopoulos C D, Giakoumis E G, Dimaratos A M, Founti M A 2011 Comparative environmental behavior of bus engine operating on blends of diesel fuel with four straight vegetable oils of Greek origin: sunflower, cottonseed, corn and olive, *Fuel* **90** pp 3439–3446

[12] Georgescu C, Cristea G C, Solea C L, Deleanu L, Sandu I G 2018 Flammability of Some Vegetal Oils on Hot Surface, *Revista de chimie*, **69(3)** pp 668-673
[13] Solea L C 2013 *Contribuţii la studiul comportării reologice şi tribologice a unor lubrifianţi biodegradabili pe bază de uleiuri vegetale*, PhD, Universitatea "Dunărea de Jos", Galaţi
[14] Tarrago-Trani M T, Phillips K M, Lemar L E, Holden J M 2006 Review. New and existing oils and fats used in products with reduced trans-fatty acid content, Nutrient database, Release 24, *Journal of the American Dietetic Association* pp 867-880
[15] Fiebig H-J 2011 Fatty acid composition of important plant and animal fats and oils, Münster, http://www.dgfett.de/material/fszus.php
[16] Paredes X, Comunasa M J P, Pensado A S, Bazile J-P, Boned C, Fernández J 2014 High pressure viscosity characterization of four vegetable and mineralhydraulic oils *Industrial Crops and Products* **54** pp 281–290
[17] Biresaw G, Bantchev G B, Cermak S C 2011 Tribological properties of vegetable oils modified by reaction with butanethiol *Tribology Letters* 43 pp 17–32
[18] Liu Z, Sharma B K, Erhan S Z, Biswas A, Wang R, Schuman T P 2015 Oxidation and low temperature stability of polymerized soybean oil-based lubricants *Thermochimica Acta* **601**, pp 9-16
[19] Quinchia L A, Delgado M A, Valencia C, Franco J M, Gallegos C 2010 Viscosity modification of different vegetable oils with EVA copolymer for lubricant applications *Industrial Crops and Products* 32 pp 607–612
[20] Erhan S, Sharma B K, Perez J M 2006 Oxidation and low temperature stability of vegetable oil-based lubricants *Industrial Crops and Products* **24**, pp 292–299
[21] Wang M 2014 *Biolubricants and biolubrication*, PhD, KTH Royal Institute of Technology, Stockholm
[22] Miller M 2008 *Chapter 18, Additives for bioderived and biodegradable lubricants* in Rudnick L R, Erhan S Z (eds) Lubricant Additive Chemistry and Applications, 2nd edition, CRC Press, Taylor and Francis Group
[23] Fox N J, Stachowiak G W 2007 Vegetable oil-based lubricants—A review of oxidation, *Tribology International* **40** pp 1035–1046
[24] Wang R, Schuman T P 2015 Oxidation and low temperature stability of polymerized soybean oil-based lubricants *Thermochimica Acta* **601** pp 9–16
[25] Georgescu C 2015 *Use of vegetal oils for obtaining environmental-friendly lubricants* (in Romanian), research report for the project POSDRU/159/1.5/S/138963 Performanţă sustenabila in cercetarea doctorală şi post doctorală – PERFORM
[26] Nagendramma P, Kaul S 2012 Development of ecofriendly/biodegradable lubricants: An overview *Renewable and Sustainable Energy Reviews* **16**, pp. 764–774
[27] Xiao H, Liu S, 2017 2D nanomaterials as lubricant additive: A review *Materials and Design* 135 pp 319–332
[28] Bakunin V N, Suslov A Y, Kuzmina G N, Parenago O P, Topchiev A V 2004 Synthesis and application of inorganic nanoparticles as lubricant components – a review. Journal of Nanoparticle Research 6 p 273
[29] Rudnick L R, Erhan S Z 2006 *Natural oils as lubricants* in Synthetics, mineral oils, and bio-based lubricants: chemistry and technology Rudnick L R, Erhan, S Z (eds) New York, CRC/Taylor & Francis Group
[30] Tang Z, Li S 2014 A review of recent developments of friction modifiers for liquid lubricants (2007–present) *Current Opinion in Solid State and Materials Science* **18** pp 119–139
[31] Zainal N A, Zulkifli N W M, Gulzar M, Masjuki H H 2018 A review on the chemistry, production, and technological potential of biobased lubricants *Renewable and Sustainable Energy Reviews* **82** pp 80–102
[32] Zareh-Desari B, Davoodi B 2016 Assessing the lubrication performance of vegetable oil-based nano-lubricants for environmentally conscious metal forming processes *Journal of Cleaner*

Production **135** pp 1198–1209

[33] Chan C-H, Tang S W, Mohd N K, Lim W H, Yeong S K, Idris Z 2018 Tribological behavior of biolubricant base stocks and additives *Renewable and Sustainable Energy Reviews* **93** pp 145–157

[34] Bashirnezhad K, Bazri S, Safaei M R, Goodarzi M, Dahari M, Mahian O, Dalkılıça A S, Wongwises S 2016 Viscosity of nanofluids: A review of recent experimental studies *International Communications in Heat and Mass Transfer* **73** pp 114–123

[35] Gupta M, Singh V, Kumar R, Said Z 2017 A review on thermophysical properties of nanofluids and heat transfer *Renewable and Sustainable Energy Reviews* **74** pp 638–670

[36] Rudnick L R (ed) 2009 *Lubricant additives. Chemistry and applications*, Second Edition, CRC Press, Taylor & Frances Group

[37] Shahnazar S, Bagheri S, Hamid S B A 2016 Enhancing lubricant properties by nanoparticle additives *International journal of hydrogen energy* **41** pp 3153-3170

[38] Zulkifli N W M, Kalama M A, Masjukia H H, Yunus R 2013 Experimental analysis of tribological properties of biolubricant with nanoparticle additive *Procedia Engineering* **68** pp 152–157

[39] Mariano G 2009 *Selection and application of solid lubricants as friction modifiers* in Lubricant Additives, ed. Rudnick L R, p 173

[40] Dai W, Kheireddin B, Gao H, Liang H 2016 Roles of nanoparticles in oil lubrication *Tribology International* **102** pp 88–98

[41] Zheng G, Dinga T, Huang Y, Zheng L, Rena T 2018 Fatty acid based phosphite ionic liquids as multifunctional lubricant additives in mineral oil and refined vegetable oil *Tribology International* **123** pp 316–324

[42] Abdullah M I H C 2016 Effect of hexagonal boron nitride nanoparticles as an additive on the extreme pressure properties of engine oil *Industrial Lubrication and Tribology* **68(4)** pp 441–445

[43] Habereder T, Moore D, Lang M 2008 *Chapter 26, Eco Requirements for Lubricant Additives* in Lubricant Additive Chemistry and Applications, 2nd edition, Rudnick L R (ed) CRC Press, Boca Raton

[44] Gow G 2009 *Chapter 14, Lubricating Grease* In: Chemistry and Technology of Lubricants, 3rd edition, Mortimer R, Fox M Orszulik S (eds) Springer, Dordrecht

[45] Ali M K A, Xianjun H, Mai L, Bicheng C, Turkson R F, Qingping C 2016 Reducing frictional power losses and improving the scuffing resistance in automotive engines using hybrid nano materials as nano-lubricant additives, *Wear* **364-365** pp 270–281

[46] Syahrullai S, Ani F, H, Golshkouh I, 2013 Wear resistance characteristic of vegetable oil, *The 2nd International Conference on Sustainable Materials Engineering*, Penang, Malaysia, pp 44-47

[47] Cristea G C 2017 *Tribological characterization of soybean oil additivated with nano materials based on carbon (black carbon, graphite and graphene*, Phd, "Dunarea de Jos" University of Galati, Romania

[48] Charoo M S, Wani M F, Hanief M, Rather M A 2017 Tribological properties of MoS2 particles as lubricant additive on EN31 alloy steel and AISI 52100 steel ball *Materials Today: Proceedings* **4** pp 9967–9971

[49] Aldana P U, Dassenoy F, Vacher B, Le Mogne T, Thiebaut B 2016 WS_2 nanoparticles anti-wear and friction reducing properties on rough surfaces in the presence of ZDDP additive *Tribology International* **102** pp 213–221

[50] Gangwani P, Gupta M K, Bijwe J 2017 Synergism between particles of PTFE and hBN to enhance the performance of oils *Wear* **384-385** pp 169–177

[51] Spikes H A 2002 Film-forming additives-direct and indirect ways to reduce friction *Lubrication Science* **14** p 147

[52] Sunqing Q, Junxiu D, Guoxu C 1999 A review of ultrafine particles as antiwear additives and

friction modifiers in lubricating oils *Lubrication Science* **11** p 217
[53] Bakunin V N, Suslov A Y, Kuzmina G N, Parenago O P 2005 Recent achievements in the synthesis and application of inorganic nanoparticles as lubricant components *Lubrication Science* **17** p 127
[54] Habibullah M, Masjuki H H, Kalam M, Ashraful A M, Habib M A., Mobarak H M 2014 Effect of bio-lubricant on tribological characteristics of steel, *Procedia Engineering* **90** pp 740–745
[55] Georgescu, C, Solea L C, Cristea G C, Deleanu L 2015 On the lubrication capability of rapeseed oil, paper 5533, *Recent Advances in Mechanics and Materials in Design, M2D'2015*, Ponta Delgada, Azores, Portugal
[56] Zulkifli N W M, Kalam M A, Masjuki H H, Yunus R 2013 Experimental analysis of tribological properties of biolubricant with nanoparticle additive *Procedia Engineering* **68** pp 152–157
[57] Alves S M, Barros B S, Trajano M F, Ribeiro K S B, Moura E 2013 Tribological behavior of vegetable oil-based lubricants with nanoparticles of oxides in boundary lubrication conditions *Tribology International* **65** pp 28–36
[58] Alves S M, Mello V S, Faria E A, Camargo A P P 2016 Nanolubricants developed from tiny CuO nanoparticles *Tribology International* **100** pp 263–271
[59] Gulzar M, Masjuk H H, Varman M, Kalam M A, Mufti R A, Zulkifli N W M, Yunus R, Zahid R 2015 Improving the AW/EP ability of chemically modified palm oil by adding CuO and MoS_2 nanoparticles *Tribology International* **88** pp 271–279
[60] Cristea G C, Cazamir D, Dima D, Georgescu C, L Deleanu L 2018 Influence of TiO_2 as nano additive in rapeseed oil, *The 8th International Conference on Advanced Concepts in Mechanical Engineering - ACME*, 01-11
[61] Cristea G C, Dima C, Georgescu C, Dima D, Solea L, Deleanu, L 2017 Evaluating lubrication capability of soybean oil with nano carbon additive, *15th International Conference on Tribology*, Kragujevac, Serbia
[62] Wu Y Y, Tsui W C, Liu T C 2007 Experimental analysis of tribological properties of lubricating oils with nanoparticle additives *Wear* **262(7-8)** pp 819–825
[63] Delgado M A, Quinchia L A, Spikes H A, Gallegos C 2017 Suitability of ethyl cellulose as multifunctional additive for blends of vegetable oil-based lubricants *Journal of Cleaner Production* **151** pp 1-9
[64] Jayadas N H, Nair K P, Ajithkumar G 2007 Tribological evaluation of coconut oil as an environment-friendly lubricant *Tribology International* **40** pp 350–354
[65] Campanella, A, Rustoy E, Baldessari A, Baltanas, M 2010 Lubricants from chemically modified vegetable oils *Bioresource Technology* **101**, pp 245-254
[66] Mobarak H, Chowdhury M 2014 Tribological Performance of Hydrogenated Amorphous Carbon (aC:H) DLC Coating when Lubricated with Biodegradable Vegetal Canola Oil, *Tribology in Industry* 36(2) pp 163-171
[67] Ilie F, Covaliu C 2016 Tribological properties of the lubricant containing titanium dioxide nanoparticles as an additive *Lubricants* **4**, p 12
[68] Padgurskas J, Rukuiza R, Prosycevas I, Kreivaitis R 2013 Tribological properties of lubricant additives of Fe, Cu and Co nanoparticles *Tribology International* **60**, pp 224-232
[69] Cermak S C, Biresaw G, Isbell T, Evangelista R, Vaughn S, Murray R, 2012 Lubrication properties of new crop oils, May, Detroit, Michigan, *Proceedings of Society of Tribologists and Lubrication Engineers Annual Meeting & Exhibition* p 68
[70] Akbulut M 2012 Nanoparticle-based lubrication systems, *Journal of Powder Metallurgy and Mining* 1 p 101
[71] Lahouij I, Bucholz E W, Vacher B, Sinnott S B, Martin J M, Dassenoy F 2012 Lubrication mechanisms of hollow-core inorganic fullerene-like nanoparticles: coupling experimental and computational works *Nanotechnology* **23** pp 3757-01
[72] Lee J, Cho S, Hwang Y, Cho H, Lee C, Choi Y, Ku B-C, Lee H, Lee B, Kim D, Kim S H, 2009

Application of fullerene-added nano-oil for lubrication enhancement in friction surfaces *Tribology International* 42 pp 440–447
[73] Lee K, Hwang Y, Cheong S, Choi Y, Kwon L, Lee J et al. 2009 Understanding the role of nanoparticles in nano-oil lubrication *Tribology Letters* pp. 35-127
[74] Hwang, Y, Lee C, Choi Y et al. 2011 Effect of the size and morphology of particles dispersed in nano oil on friction performance between rotating discs *Journal of Mechanical Science and Technology* **25(11)** pp 2853-2857
[75] Chen C S, Chen L, Xu S, Yang Z, Li W H 2005 Modification of multi-walled carbon nanotubes with fatty acid and their tribological properties as lubricant additive, *Carbon*, **43(8)** pp 1660-1666
[76] Joly-Pottuz L, Ohmae N 2008 Carbon-based nanolubricants, in Martin J M, Ohmae N (eds), *Nanolubricants* Tribology Series pp 93-148Wiley
[77] Wo H, Hu, K, Hu X G 2004 Tribological properties of MoS_2 nanoparticles as additive in a machine oil, *Tribology*, **24** pp 33–37
[78] Wu H, Zhao J, Xia W, Cheng X, He A, Yun J H, Wang L, Huang H, Jiao S, Huang L, Zhang S, Jiang Z 2017 A study of the tribological behaviour of TiO_2 nanoadditive water-based lubricants *Tribology International* **109** pp 398–408
[79] Cristea G C, C Georgescu C, D Dima and L Deleanu 2018 Influence of graphene as additive in soybean oil, *the 8th International Conference on Advanced Concepts in Mechanical Engineering - ACME 2018*, 01-12, Iasi, Romania
[80] Liu G, Li, X, Qin B, Xing D, Guo, Y, Fan R 2004 Investigation of the mending effect and mechanism of copper nano-particles on a tribologically stressed surface *Tribology Letters* **17**, p 961
[81] Liu Y, Wang X, Pan G, Luo J 2013 A comparative study between graphene oxide and diamond nanoparticles as water-based lubricating additives *Science China Technological Sciences* **56(1)** pp 152–157
[82] Gu C, Zhu G, Li L, Tian X, Zhu G 2009 Tribological effects of oxide based nanoparticles in lubricating oils *Journal of Marine Science and Applications* **8**, pp71-76
[83] Yu H L, Xu Y, Shi P J, Xu B S, Wang X L, Liu Q, Wang H M 2008 Characterization and nano-mechanical properties of tribofilms using Cu nanoparticles as additives *Surface & Coatings Technology* **203** pp 28–34
[84] Lee J, Cho S, Hwang Y, Lee C, Kim S H 2007 Enhancement of lubrication properties of nano-oil by controlling the amount of fullerene nanoparticle additives *Tribology Letters* **28(2)** pp 203-208.
[85] Yao Y, Wang X, Guo J, Yang X, Xu B 2008 Tribological property of onion-like fullerenes as lubricant additive *Materials Letters* **62** pp 2524–2527
[86] Berman D, Erdemir A, Zinovev A V, Sumant A V 2015 Nanoscale friction properties of graphene and graphene oxide *Diamond & Related Materials* 54, pp91-96
[87] Peng D, Kang Y, Hwang R, Shyr S, Chang Y 2009 Tribological properties of diamond and SiO_2 nanoparticles added in paraffin *Tribology International* **42** pp911-917
[88] Zhang G, Xua Y, Xiang X, Zheng G, Zengc X, Lia Z, Rena T, Zhang Y 2018 Tribological performances of highly dispersed graphene oxide derivatives in vegetable oil *Tribology International* **126** pp 39–48
[89] Pan Q, Zhang X 2010 Synthesis and tribological behavior of oil-soluble Cu nanoparticles as additive in SF15W/40 lubricating oil *Rare Metal Materials and Engineering* **39(10)** pp 1711-1714
[90] Ma J, Mo Y, Bai M 2009 Effect of Ag nanoparticles additive on the tribological behavior of multialkylated cyclopentanes (MACs) *Wear* **266** pp 627–631
[91] Padgurskas J, Rukuiza R, Kreivaitis R, Asadauskas S J, Brazinskiene D 2009 Tribologic behaviour and suspension stability of iron and copper nanoparticles in rapeseed and mineral oils *Tribol e Mater Surfaces Interfaces* 3(3), pp 97-102

[92] Liu R, Wei X, Tao D, Zhao Y 2010 Study of preparation and tribological properties of rare earth nanoparticles in lubricating oil, Tribology International 43(5) 1082–1086
[93] Zhao F, Bai Z, Fu Y, Zhao D, Yan C 2012 Tribological properties of serpentine, La(OH)3 and their composite particles as lubricant additives *Wear* 288 pp 72–77
[94] Hakimi M I, Abdullah C, Abdollah M F B, Amiruddin H, Tamaldin N, Nuri N. R. M 2013 optimization of tribological performance of hBN/AL_2O_3 nanoparticles as engine oil additives *Procedia Engineering* **68** pp 313 – 319
[95] Ferreira Trajano M, Freitas Moura E I, Ribeiro K S B, Alves S M 2014 Study of oxide nanoparticles as additives for vegetable lubricants, *Materials Research.* **17(5)** pp 1124-1128
[96] Peña-Parás L, Taha-Tijerina J, Garza L, Maldonado-Cortés D, Michalczewski R, Lapray C, 2015 Effect of CuO and Al_2O_3 nanoparticle additives on the tribological behavior of fully formulated oils *Wear* **332-333** pp 1256–1261
[97] Ma S, Zheng S, Cao D, Guo H 2010 Anti-wear and friction performance of ZrO2 nanoparticles as lubricant additive *Particuology* 8 pp 468–472
[98] Ali M K A, Xianjun H, Mai L, Qingping C, Turkson R F, Bicheng C 2016 Improving the tribological characteristics of piston ring assembly in automotive engines using Al_2O_3 and TiO_2 nanomaterials as nano-lubricant additives *Tribology International* **103** pp 540–554
[99] Xie H, Jiang B, He J, Xia X, Pan F 2016 Lubrication performance of MoS_2 and SiO_2 nanoparticles as lubricant additives in magnesium alloy-steel contacts *Tribology International* 93 pp 63–70
[100] He X, H Xiao, H Choi, Agustín Díaz, B Mosby, A Clearfield, H Liang, Zirconium phosphate nanoplatelets as lubricant additives *Colloids and Surfaces A: Physicochemical Engineering Aspects*, **452** 32–38
[101] Zhang B-S, Xu B, Xu Y, Gao F, Shi P, Wu Y-X 2014 Cu nanoparticles effect on the tribological properties of hydrosilicate powders as lubricant additive for steel–steel contacts *Tribology International* 44 pp 878–886
[102] Ji X, Chen Y, Wang X, Liu W 2012 Tribological behaviors of novel tri(hydroxymethyl)propane esters containing boron and nitrogen as lubricant additives in rapeseed oil *Industrial Lubrication and Tribology* 64(6) pp 315–320
[103] Dubey M K, Bijwe J, Ramakumar S S V 2015 Nano-PTFE: New entrant as a very promising EP additive *Tribology International* 87(1) pp 121-131
[104] Song X, Zheng S, Zhang J, Li W, Chen Q, Cao B, 2012 Synthesis of monodispersed $ZnAl_2O_4$ nanoparticles and their tribology properties as lubricant additives *Mater Res Bull* **47(12)** pp 4305-4310
[105] Gu K, Chen B, Wang X, Wang J, Fang J, Wu J Huan L, 2014 Preparation, friction and wear properties of hydrophobic lanthanum borate nanorods in rapeseed oil *Transactions of Nonferrous Metals Society of China* **24** pp 3578−3584
[106] Qian J, Zhang Y, Wang L, Xing J 2011 Study on lubrication properties of modified nano ZnO in base oil *China Petroleum Processing and Petrochemical Technology* **13(3)** pp 69-73
[107] Tevet O, von-Huth P, Popovitz-Biro R, Rosentsveig R, Wagner H D, Tenne R 2011 Friction mechanism of individual multilayered nanoparticles *Proceedings of the National Academy of Sciences of United States of America* **108** p 19901
[108] Lin J, Wang L, Chen G 2011 Modification of graphene platelets and their tribological properties as a lubricant additive *Tribology Letters* **41(1)** pp 209-215
[109] Pierson H O 1993 *Handbook of carbon, graphite, diamond and fullerenes properties, Processing and applications*, Noyes Publications, New Jersey
[110] Cheenkachorn K 2013 A study of wear properties of different soybean oils *The Mediterranean Green Energy Forum 2013, MGEF-13, Energy Procedia* **42** pp633-639

IOP Conf. Series: Materials Science and Engineering **514** (2019) 012013 doi:10.1088/1757-899X/514/1/012013

Friction characteristics of protein-glucose-thermoset composites

I Graur[1,2], V Bria[1], M S Drăghici[1] and C Munteniță[1]

[1]"Dunărea de Jos" University of Galati, Research and Development Centre for Thermoset Matrix Composites, Galați, Romania
[2]Diagnose and Measurement Group, Galati, Romania

E-mail: iulia.graur@ugal.ro

Abstract. Polymers are materials with a great spread and represent the center of the life comfort development. They have become since their discovery, indispensable to a modern prosperous society. In order to obtain optimal performance of the composite material, a broad variety of organic modification agents were used. The purpose of this study represents the development of modern composite materials modified with glucose particles and protein substances (gelatin and wheat gluten). The glucose-modified composite materials are used in the medical field.Gelatin is a protein derived from collagen, being bio-compatible and non-toxic. Wheat gluten is a vegetable protein used as a modifying agent to form composite materials that can be transformed into a bioplastic material. The following types of epoxy resins were used for this study: Epyphen RE 4020, Epoxy RESIN C, Epoxy RESIN HT. These composite materials are designed to withstand some mechanical, physicochemical, thermal properties and most valuable, tribological properties. Tribological tests aim to determine the friction coefficient. These tests are necessary to understand the tribological properties of composite materials and the field in which they can be used.

1. Introduction

The complex structure and unique properties of the composite materials typically depend on the nature of the materials constituting the matrix [1] and on the specific characteristics and complex geometry of the dispersion or reinforcement phase [2]. That is why it is necessary to handle separately the potential problems related to the structure and properties of the composite materials with particles, as well as the matrices that come into the composition of the composite materials [3]. Composite materials, one of the most rapidly growing classes of materials, are being used increasingly for tribological applications [4]. To obtain optimal performance of the composite material, a broad variety of organic and inorganic modification agents are used [5]. Because the polymer matrix must withstand high mechanical and tribological loads, it is usually reinforced with fillers. Almost all properties can be changed by incorporation of reinforcement or filler particles into a matrix but composite polymers have also the ability to control their frictional and wear behaviours sliding contacts [6]. The modifying agents are small particles dispersed in the matrix material [7], which can be easily made and embedded in the material [8]. The use of agents helps to improve the material by increasing the abrasion resistance and lowering the coefficient of thermal expansion [9]. Proteins are thermoplastic polymers of polar and non-polar amino acids that are capable of developing numerous intermolecular bonds and bearing various linkages [10]. Gluten is a vegetable protein [11], used as a modifier for the formation of composite materials, which can be transformed into a bioplastic material by casting. It is used because of its availability, good biodegradability and viscoelastic properties. These small molecules interact with the polymer chains, therefore reducing the forces, holding the chains together. Gluten-based composite materials are used to make food packaging as it improves food storage, having the ability to act as a barrier against water, oxygen, and light, therefore reducing oxidation of food. Gelatin is a protein derived from collagen, being bio-compatible and non-toxic. It is used most often for many biomedical applications. Gelatin films were used to protect food from light. The glucose-modified composite materials are used in the medical field. In medicine, it is mainly used as aqueous solutions for infusion [12]. Depending on their concentrations, they have numerous actions and directions. In industry, glucose is most often used in the nano-powders form. In this research paper, we studied the influence of test speed and applied pressure values on the friction coefficient of different epoxy matrix composites modified with different mixture of glucose, gelatin and gluten under dry sliding conditions against hard steel on a pin-on-disc apparatus. The friction coefficient behaviour of proteins such as glucose, gelatin and gluten, dispersed in polymeric matrix are topical [13-17].

2. Materials and method

The preparationof the modified composite materials consisted of the addition of an amount of the glucose, gluten and gelatin powder modifying agent in an epoxy matrix [19-20]. The purpose of combining these materials is to obtain multifunctional properties, superior to the reference materials. Three epoxy resins were used in this study: Epiphen RE4020-DE 4020 (Bostik), Epoxy Resin C (R&G Gmbh Waldenbuch), Epoxy Resin HT-2 (R&G Gmbh Waldenbuch). For comparison of future results, we formed materials with the three unmodified resins with the abbreviated names as: E0, C0, H0. The amounts of organic agents used for the composite materials and samples coding are shown in Table 1. The resins were modified by adding different mixture percentage of glucose, gelatin and gluten. The molds consisted of polypropylene tubes with a diameter of 8 mm and a length of 200 mm. The test regimes for friction coefficient have been established as follows: a load of 15N and speed of 1m/s- R1; a load of 20N and speed of 0.75m/s-R2; a load of 25N and speed of 0.6m/s-R3. R1, R2 and R3 represents the friction coeficient regimes used for composite materials tests. Tribological tests were done 23°C for 1000 m distance. The tests were performed at three different regimes with three loads and three sliding speeds in order to respect the same product load-speed.

Table 1. Amounts of organic agents used for the composite materials and samples coding

Modifying agents, % \ Samples coding	E1,C1,H1	E2,C2,H2	E3,C3,H3
Glucose	4	3	3
Gluten	3	4	3
Gelatin	3	3	4

3. Results

It can be seen for all the tested materials, a sudden increase in the first 10 meters of sliding after which the friction stabilizes.

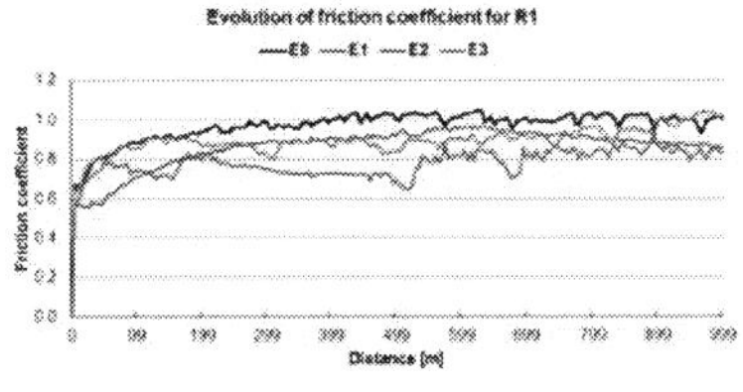

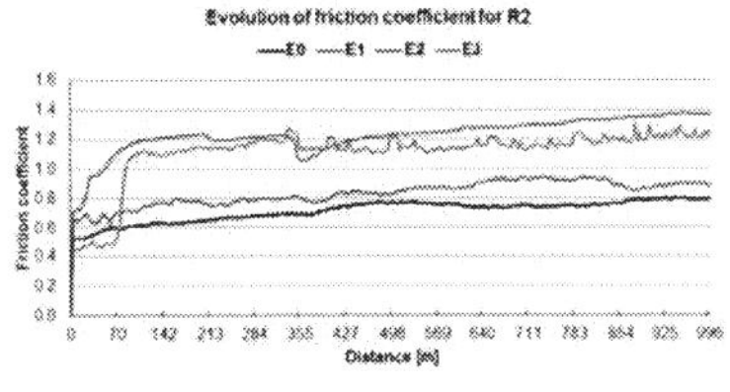

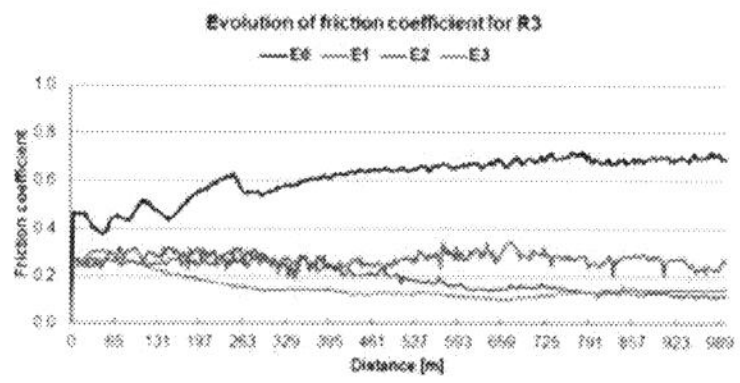

Figure 1. Evolution of friction coefficient for E epoxy system at all regimes

Figure 1 shows the evolution of friction coefficient for composite materials formed with E epoxy resin for all three test regimes. For all materials and for all three test regimes, a sudden increase of friction coefficient can be observed, after which it stabilizes. For the neat polymer (in this case E0) it can be seen that the value of friction coefficient does not change during the R2 test regime. For R1 and R3 regimes it can be seen a sudden decrease, followed by a sudden increase. The other materials exhibiting an increasing trend (in the case of the R2 test regime) and a decreasing trend (in the case of the R1 and R3 test regimes) of the friction coefficient during the sliding distance. For R1, the friction coefficient also has lower values than the neat resin E0.

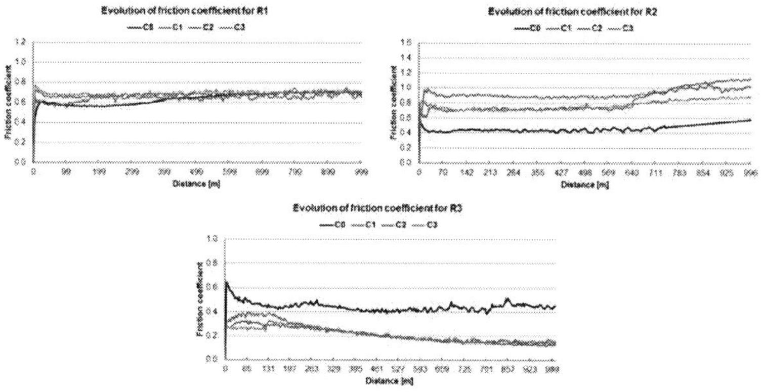

Figure 2. Evolution of friction coefficient for C epoxy system at all regimes

Figure 2 shows the evolution of friction coefficient for composite materials formed with C epoxy resin for all three test regimes. All materials behaviour for R1 regime is almost the same as the reference material C0. The friction coefficient is obviously increased for R2. Very low values of friction coefficient were obtained for R3.

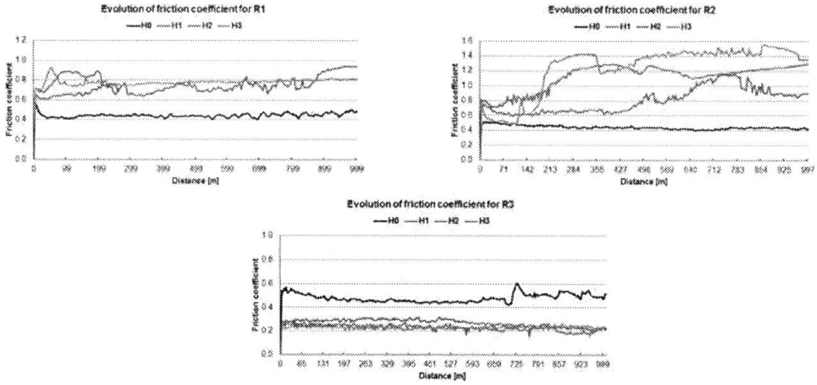

Figure 3. Evolution of friction coefficient for H epoxy system at all regimes

In Figure 3 it can be seen that for friction coefficient for composite materials formed with C epoxy resin, random friction coefficients develop throughout the test. For the test regimes R1 and R2, we note that the materials have almost the same friction behavior regardless of the applied regime. However, in R3 mode, all materials obtained from the epoxy resin H possess the lowest values compared to the neat material H0. The friction behaviour of composite materials were investigated using a TRM 1000 tribometer from Wazau on pin-on-disk geometry, with the pin made of analyzed material against steel disk (Fig.4).

Figure 4. Tribometer TRM 1000 from Wazau

4. Conclusions

The tribological analysis carried out on the three epoxy resins modified with proteins (gelatin and gluten) and glucose showed the following: E epoxy system: the friction coefficient is dramatically increased for R2, compared with the neat epoxy, by the addition of any percentage of proteins mixture and very low values were obtained for R3. For R1, the friction coefficient also has lower values than the neat resin E0.C epoxy system: for R1, C matrix composites modified with different mixture of glucose, gelatin and gluten, showed a friction stability compared to neat epoxy resin C0. The friction coefficient is obviously increased for R2. Very low values of friction coefficient were obtained for R3. H epoxy system: on R3, the smallest values of the friction coefficient are recorded.For R1 and R2, the friction coefficient is dramatically increased.Epoxy matrix composites modified with different mixture of glucose, gelatin and gluten showed a decrease in the friction coefficient varion for a load of 25N and speed of 0.6m/s-R3.

References

[1] Bilyeu B, Brostow W and Menard K P 2001 Epoxy thermosets and their applications. III. Kinetic equations and models *Journal of materials education* **23** pp 189-204.

[2] Brostow W, Deborde J L, Jaclewicz M and Olszynski P 2003 Tribology with emphasis on polymers: friction, scratch resistance and wear *Journal of Materials Education* **25** pp 119-132.

[3] Basavarajappa S, Arun K V and Davim J P 2009 Effect of filler materials on dry sliding wear behavior of polymer matrix composites–a Taguchi approach *Journal of minerals and materials characterization and engineering* **8** p 379.

[4] Menapace C, Leonardi M, Perricone G, Bortolotti M, Straffelini G and Gialanella S 2017 Pin-on-disc study of brake friction materials with ball-milled nanostructured components *Materials & Design* **115** pp 287-298.

[5] Chang L, Zhang Z, Breidt C and Friedrich K 2005 Tribological properties of epoxy nanocomposites: I. Enhancement of the wear resistance by nano-TiO2 particles *Wear* **258** pp 141-148.

[6] Suresha B, Chandramohan G, Prakash J N , Balusamy V and Sankaranarayanasamy K 2006 The

role of fillers on friction and slide wear characteristics in glass-epoxy composite systems *Journal of Minerals and Materials Characterization and Engineering* **5** p 87.
[7] Xu H, Feng Z, Chen J and Zhou H 2007 Tribological behavior of the polyamide composite coating filled with different fillers under dry sliding *Journal of Applied Polymer Science* **104** pp 2554-2560.
[8] Samyn P., Kalacska G, Keresztes R, Zsidai L and De Baets P 2007 Design of a tribotester for evaluation of polymer components under static and dynamic sliding conditions *Proceedings of the Institution of Mechanical Engineers, Part J: Journal of Engineering Tribology* **221** pp 661-674.
[9] Socalici A, Ardelean E, Pascu L and Putan V 2017 Friction and Wear Behaviour of Composite Materials Containing Organic Components *Materiale Plastice*, **54** p 269.
[10] Martínez I, Partal P., García-Morales M, Guerrero A and Gallegos C 2013 Development of protein-based bioplastics with antimicrobial activity by thermo-mechanical processing *Journal of Food Engineering* **117** pp 247-254.
[11] Gällstedt M, Mattozzi A, Johansson E and Hedenqvist M S 2004 Transport and tensile properties of compression-molded wheat gluten films *Biomacromolecules* **5** pp 2020-2028.
[12] Kunanopparat T, Menut P, Morel M H and Guilbert S 2008 Reinforcement of plasticized wheat gluten with natural fibers: from mechanical improvement to deplasticizing effect *Composites part A: Applied science and manufacturing* **39** pp 777-785.
[13] Chang L, Zhang Z, Ye L and Friedrich K 2007 Tribological properties of high temperature resistant polymer composites with fine particles *Tribology International* **40** pp 1170-1178.
[14] Ratcliffe R 1992 *Friction and Wear of Engineering Materials* (Londres:Edward Arnold).
[15] Czichos H, Klaffke D, Santner E and Woydt M 1995 Advances in tribology: the materials point of view *Wear* **190** pp 155-161.
[16] Georgescu C, Botan M and Deleanu Lorena 2014 Influence of Adding Materials in PBT on Tribological Behaviour *Materiale Plastice* **51** pp 351-354.
[17] Skopp A and Woydt M 1995 Ceramic and ceramic composite materials with improved friction and wear properties *Tribology transactions* **38** pp 233-242.
[18] Ünlü B S and Atik E 2009 Tribological properties of journal bearings manufactured from particle reinforced Al composites *Materials & Design* **30** pp 1381-1385.
[19] Mihu G, Mihalache I, Bodor M, Mircea O and Graur Iulia 2016 Tribological Characterization of Modified Epoxy Systems *Materiale Plastice* **53** pp 298-303.
[20] Munteniţă C, Eni C, Graur Iulia, Ungureanu Claudia and Bodor M 2017 A study of friction wear behaviour of nano-ferite modified epoxy resins *IOP Conference Series: Materials Science and Engineering* **174**, p 12025.

Acknowledgments
The authors would like to acknowledgmrnts the financial contribution of the Project 12 P01 024 21 (C11)/31. 08. 2012 (code SMIS 50414).

IOP Conf. Series: Materials Science and Engineering **514** (2019) 012014 doi:10.1088/1757-899X/514/1/012014

Geometry influence on silent chain – sprocket friction

L Jurj[1], R Velicu[1] and M T Lates[1]

[1]Product Design, Mechatronics and Environment Department, Transilvania University of Braşov, Romania

E-mail: rvelicu@unitbv.ro

Abstract. The silent chains in theory should not present friction at the level of the plate entering in contact with the sprocket, based on the literature. Experimental results have shown that there is a certain amount of friction, which can be inspected even visually. This scientific paper presents a geometric simulation using the exact geometry of the plate and the sprocket, to calculate the amount of mechanical work lost by friction at the entering of the plate in contact with the sprocket teeth. This modelling considers the different deviations of the pitch of the chain, resulted from nominal tolerances, elastic elongation and wear.

1. Introduction

Different studies were performed on the topic of friction reduction in internal combustion engine components. One of the field where we are interested in is the valve timing system. Within this area of the internal combustion engine, different belt and chain drives were studied. Related to chain drives, in [1] different methods were used to reduce the friction in chain transmissions with different chain types, like surface coatings, different material compositions and improved manufacturing. Different mathematical models and testing procedures were elaborated by [2], these methods helped to estimate the losses due to friction between the elements of the chain, and between the chain and the components it is in contact with.

Within the PhD thesis of [3] experimental and theoretic studies were conducted to optimize the guiding system from the point of view of reducing friction losses in chain transmissions. The contact between the chain and the guide is the subject of a model for simulation with results on the type of lubrication between the elements, and optimisation of the system.

In [4], the pitch deviation, friction and centrifugal effects were studied over the load distribution in a bush chain transmission. It is concluded that the load on the driving sprocket teeth is higher than on the driven sprocket teeth in presence of friction. The pitch deviation due to wear determines a supplementary movement of the bush on the surface of the sprocket teeth, inducing friction and wear between the elements.

Studies on bush chains concerning the contact from the joints of the chain are presented in [5]. Contact points and contact angles between bush and sprocket's tooth were determined for different pitches and the different values of pitch deviation.

The only model for friction losses in silent chains transmissions, that we found studying chain references, is the basic one considering only the pin-plate friction, neglecting possible friction between plates and sprocket's teeth.

Another aspect is that the silent chains are not standardized. Based on measuring dimensions of existing silent chains, the effects of the friction force on the contact force between the silent chain and the sprocket are presented in [6]. Contact forces between silent chain and sprocket are inputs for any model of silent chain friction losses.

This paper proposes a model for friction losses based on pitch deviations (pitch of chain is bigger than pitch on the sprocket). The influence of geometrical parameters on chain-sprocket friction losses is presented together with a comparison with the known friction losses due to pin-plate friction.

2. Theoretical model

In theory, the silent chain should get in contact with the sprocket without any friction appearing. On experimentally tested chains with their corresponding sprockets even wear marks can be found at the level of the silent chain plates and the sprocket teeth. This can be explained by sliding between these elements which can be caused by deviations in pitch on the chain. These deviations could come from manufacturing inaccuracies, or nominal deviations; from deviations caused by tensioning of the chain and from elongation of the chain due to wear. Due to wear at the level of the chain's cylindrical joints the clearances get bigger, up to 0.1…0.3% of chain pitch in case of bush chains. Because of the lack of information about pitch deviations of the silent chain, in this scientific study it will be considered an interval of 0.1…0.6% of chain pitch per link. These pitch deviations of the chain, disturb the plates positioning between the sprocket's teeth gaps, in a way that the contact point can have a certain travel at the sprocket teeth flank, while the plate rotates with the corresponding pitch angle of the sprocket.

The theoretical model is based on a few simplifying hypotheses, presented on figure 1a:

- It is considered that a single point on the chain plate is in contact with the sprocket's teeth all the time, while in reality this point is moving along the plate's outer flank. In this way, the distance travelled with sliding friction will be (with our calculation model) slightly reduced.
- The sprocket's teeth has a straight profile, while in reality, it has an curved profile, meaning that the distance travelled by the contact point with friction will be, by our calculation model, slightly increased.

$$p_{real} = p + a + a_i + a_w = p + a_p \tag{1}$$

where: p is the pitch of the chain; a is the nominal deviation of the chain; a_i is the deviation due to tensioning and a_w represents the deviation of the pitch due to wear. The total pitch deviation a_p will be considered in the range:

$$a_p = (0.1\ldots0.6)\% \text{ of } p. \tag{2}$$

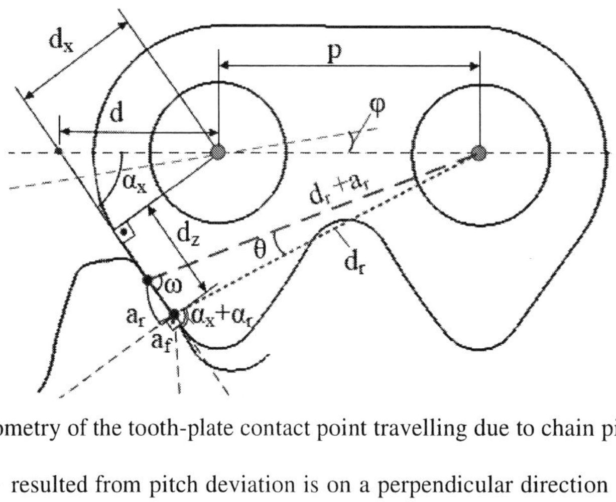

Figure. 1. Geometry of the tooth-plate contact point travelling due to chain pitch elongation.

The deviation a_f resulted from pitch deviation is on a perpendicular direction on the silent chain's outer flange and can be calculated with the following formula:

$$a_f = a_p \sin\alpha_x, \tag{3}$$

while the deviation along the line linking the contact point with the centre point of the pin, a_r can be calculated using:

$$a_r = a_f \frac{1}{\cos[90-(\alpha_x+\alpha_r)]} \tag{4}$$

Based on [7], the position of the contact point is defined by the geometrical parameters d_x and d_z, shown in Figure 1.

The first step in the theoretical model towards finding the distance over which the contact point lose mechanical work is defining the first the position of the contact point in relation with the pin centre point:

$$d_r = \left(D_x + D_y\right)^{1/2} \tag{5}$$

where:

$$D_x = p + d_x \sin\alpha_x - d_z \cos\alpha_x \tag{6}$$

$$D_y = d_x \cos\alpha_x + d_z \sin\alpha_x \tag{7}$$

Using the generalized Sinus theorem the angle between the pitch line, and the line linking the contact point with the centre of the pin can be written as:

$$\alpha_r = a\sin\left[\frac{\left(d_z + d\cos\alpha_x\right)\sin\alpha_x}{d_r}\right] \tag{8}$$

The angle between the tangent line over the silent chain plate's outer flange through the contact point and the line linking the position of the contact point after the pitch deviation can be written using the generalized Sinus Theorem in the corresponding triangle as follows:

$$\omega = a\sin\left[\frac{d_r \sin[180-(\alpha_x+\alpha_r)]}{d_r + a_r}\right] \tag{9}$$

The angle needed to calculate the length of arc described by the contact point travel:

$$\theta_1 = \alpha_x + \alpha_r - \omega \tag{10}$$

The length of the contact point displacement on the sprocket can be computed by:

$$l_{f\_sprocket} = \frac{\pi(d_r + a_r)\theta_1}{180}, \tag{11}$$

while the length travelled by the contact point calculated in function of p is:

$$l_{f\_sprocket/p} = \frac{l_{f\_sprocket}}{p}, \tag{12}$$

and the same length expressed in percentage:

$$l_{f\_sprocket/p\%} = l_{f\_sprocket/p} \cdot 100. \tag{13}$$

The mechanical work that is consumed by friction in the contact point travelling along the sprocket teeth, during a pitch angle rotation at each point of chain entering or exiting a sprocket, has the following formula:

$$L_{ap} = \mu F_{z1} l_{f\_sprocket} \tag{14}$$

where μ is the friction coefficient between plate and sprocket, F_{z1} represents the force that is coming as a reaction force between the sprocket and the silent chain teeth [5] and $l_{f\_sprocket}$ is the distance the contact point slides, during the pitch angle rotation, along the sprocket's teeth flank.

For comparison, there will be also calculated the friction torque in the cylindrical joint of the pin-plate, for each point of chain entering or exiting a sprocket [8]:

$$T_{f\_pin} = \mu d_{pin} F_1 \tag{15}$$

where μ is the friction coefficient between pin and plate (considered the same with the friction coefficient between plate and sprocket), d_{pin} - pin diameter; F_l – force in the chain arm.

The friction torque resulted from the friction between the sprocket and the plate, also for each point of chain entering or exiting a sprocket, is calculated with:

$$T_{f\_sprocket} = \mu F_x \frac{l_{f\_sprocket}}{\theta} \tag{16}$$

where is the pitch angle.

3. Results

For the theoretical model of the contact point displacement due to pitch deviation, and for the work consumed along the path described by the contact point, there were used different parameters of the silent chain LD8 with pitch of 8 mm, and a sprocket with 23 teeth. The input parameters of are shown in Table 1.

Table 1. Input parameters for the theoretical model.

p	8	mm
z	23	-
d_x/p	0.35 … 0.6	-
d_z/p	0.2 …0.5	-
x	50 … 60	deg
a_p	(0.1…0.6)p	mm
F_l	1000	N
F_x	500	N
μ	0.1	-

Working with the relative values of d_x/p and d_z/p [5] the resulting diagrams can be used to study different situations regardless of the value of the chain pitch.

The position of the contact point is defined by the distance d_z and, for involute sprockets [7], this distance depends on the number of teeth of the sprocket and the chosen profile displacement.

Figure 2 and Figure 3 are showing the effects of the contact point position (d_z/p) over the friction length and the friction torque that appears in the considered silent chain transmission.

Figure 2 presents the dependency of the friction length depending on the pitch of the chain in percentage over the pitch deviation presented in percentage. On Figure 3, values of the friction torques from the chain plate and the sprocket and the friction torque that appears between the pin and the chain plate are presented in function of the pitch deviation.

It is clear that the plate-sprocket friction losses increase with increased deviation of the chain pitch.

Figure 4 highlights the effects of the chain plate dimension d_x on the sliding friction length and Figure 5 shows the effects of the angle α_x on the sliding friction length.

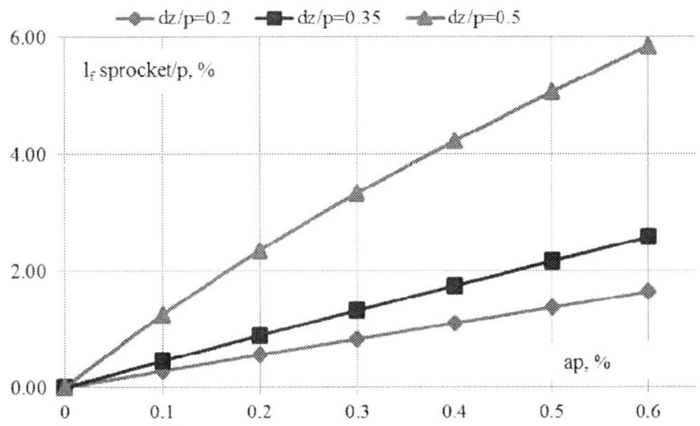

Figure 2. Sliding friction length in function of the corresponding pitch deviation with the effect of the contact point position.

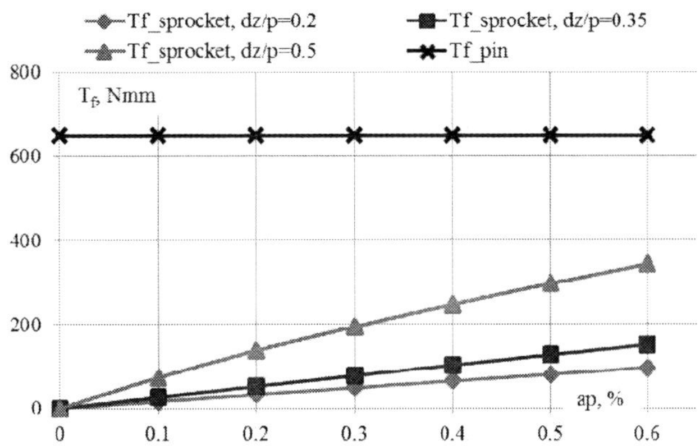

Figure 3. Friction torque at the level of the sprocket and the pin of the chain transmission with the effect of the contact point position.

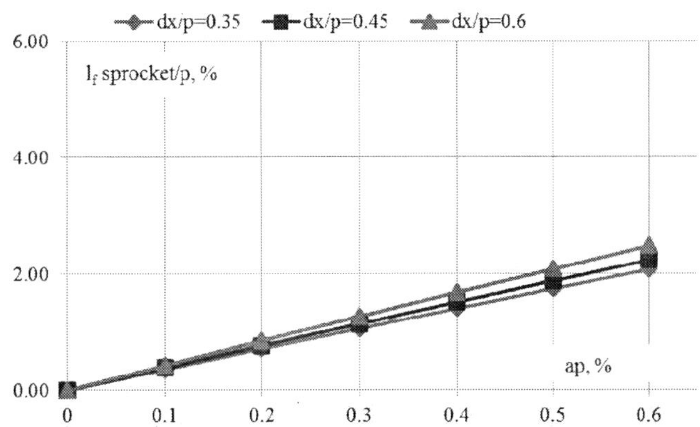

Figure 4. Friction length in function of the corresponding pitch deviation with the effect of the plate dimension.

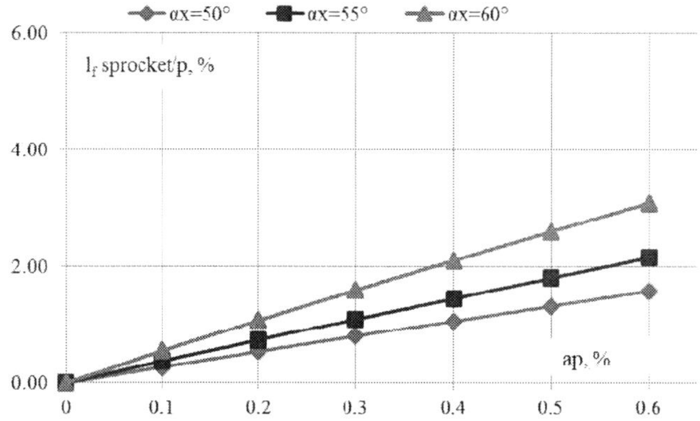

Figure 5. Sliding friction length in function of the corresponding pitch deviation with the effect of chain plate flank angle.

4. Conclusions

Based on the theoretical model of the silent chain – sprocket contact point displacement it can be seen that the contact point travel presents different values with the variation of the contact point position, the chain plate dimensions and the chain's outer flange angle. The theoretical model has shown that the position of the contact point has the greatest effect over the friction length. It also has shown that the friction torque has highest values with the increase of the parameter d_z/p.

Presenting the values of the friction torque in the cylindrical joint, on the same chart, helps the better understanding of the friction losses that can appear in a silent chain transmission. This comparison shows that the friction in the pin-plate joint, can be as much as twice the value of the friction torque from the chain-sprocket contact.

5. References

[1] Fink T and Holger B 2011 Friction reduction potentials in chain drives *MTZ* **72**

[2] Vasilievich E A, Eduardovich K K and Nikolaevich B V 2013 Inertial Method of Chain Transmission Quality Control, *Polythematic Online Scientific Journal of Kuban State Agrarian University* pP 681-696.

[3] Papuc R 2015 Theoretical and experimental research for tribological optimization of the clearance compensating system of timing chains *PhD Thesis Transilvania University of Brasov*

[4] Mohamed R N, Kurt M M 1989 The effects of the pitch difference on the load distribution of a roller chain drive, *Mech. Mach. Theory* **24 No 5** pp 351-362

[5] Todi Eftimie A 2015 Theoretical and experimental study of friction losses in bush chain transmissions *PhD Thesis Transilvania University of Brasov*

[6] Jurj L, Saulescu R and Velicu R 2017 Sprocket – silent chain force distribution with the influence of friction *Applied Mechanics and Materials* **823** *Current Solutions in Mechanical Engineering*

[7] Jurj L, Velicu R and Radu Saulescu 2018 Geometry of silent chain – involute sprocket *Annals of the Oradea University, Fascicle of Management and Technological Engineering*

[8] Jurj L, Velicu R 2018 On the benchmarking of friction in timing chains *Annals of the Oradea University, Fascicle of Management and Technological Engineering*

IOP Conf. Series: Materials Science and Engineering **514** (2019) 012015 doi:10.1088/1757-899X/514/1/012015

Micro-scratching tests of a lead-free solder alloy SAC 305 used in electronic industry

A M Petrescu[1], N A Stoica[1], A Tudor[1], A V Rădulescu[1], M Stoica[1] and G Chişiu[1]

[1]Department of Machine Elements and Tribology, University Politehnica of Bucharest, Bucharest

E-mail: am.petrescu@upb.ro

Abstract. The electronic connections need to satisfy three functionalities and those are the electrical, thermal and mechanical functionalities. The PCB (Printed Circuit Board) has the role of providing a mechanical support for an assembly of various electrical and electronical components and to connect them to obtain a functional end-product. The goal of the present paper was to identify the mechanical and tribological properties of the lead-free soldering alloy SAC 305 deposit on the copper-PCB assembly used in electronic technology. The variation of the friction coefficient for all the normal forces applied on the test samples was analyzed. The understanding of the behavior of the lead-free soldering alloy for the micro scratch test is essential for the improvement of the hole electronic system integrity.

1.Introduction

A printed circuit board (PCB) serves to mechanically support and electrically connect a set of electrical and electronic components. [1]

The mechanical function provides the mechanical integrity and the resistance of the electronic assembly. The role of the soldering alloys is to interconnect the electrical and electronical components usually by surface mount technology (SMT). They also serve as a mechanical support, holding the component in position on the PCB. [2, 3]

The environmental aspects have determined the replacement of the lead based soldering alloys with lead-free solder alloys in the electronic industry. A very important aspect in the use of the lead-free solders is to ensure at least the same quality and reliability for the electrical connection as the as the classical lead-based soldering alloys. [4]

The objective of the paper was to observe the mechanical and tribological characteristics of the soldering alloy SAC 305 deposited on the FR4 rigid support covered with a copper layer. The SAC 305 is a lead-free soldering alloy and has the following composition: 96.5% Sn, 3.0% Ag, 0.5% Cu.

The FR4 rigid support is made of a woven glass fabric with epoxy resin system. This objective is accomplished through the scratch testing method.

We can see from the profilogram that the material deforms in the proximity of the path of the indenter, where borders of materials are created. This behavior leads to the idea that the material is ductile.

The scratch test is a common testing procedure that provides fast results for the critical loads in relation to the adhesion properties of coatings. It is related to the measurement of the amount of plastic deformation caused by the indenter penetration. The scratch testing procedure involves the use of an indenter or a scratching blade in the vertical plane, generating the loading force, and in the horizontal plane, generating the friction force. This will create the scratching pattern, which indicates the failure type. [5]

2.Experiment

The objective of the experiment was to determine the mechanical and tribological properties of the SAC 305 soldering alloy deposited on the copper substrate which covers the FR4 rigid support.

The micro-scratching tests have been conducted on a Scratch-Testers Millennium 100 equipment, which performs scratches on the samples and provides as results the value of the normal force, the friction force and the friction coefficient.

The profilogram of the surface after the tests was determined with a Leica 3D Dual Core equipment, which allows the identification of the penetration depth of the indenter.

During the experiments a single type of sample has been tested, which is composed of a SAC 305 soldering alloy deposited on a FR4 rigid support with a copper substrate. The sample can be seen in Figure 1.

A Rockwell indenter with a diamond tip with a radius of 50 μm was used to apply the normal loading force, which was kept constant during the tests.

A series of 6 scratches have been made with a length of 1.1 mm. The normal loading force was increased from 10 N for the first scratch to 20 N, 30 N, 40 N, 50 N and respectively 60 N for the last scratch.

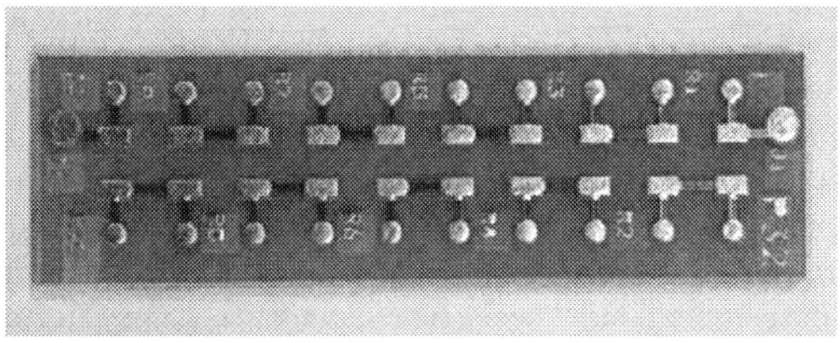

Figure 1. The sample used for the experiments

3.Results and discussion

For the 10 N normal force, the indenter moves with a speed of 0.167 mm/s over a distance of 1.1 mm for 6.587 s. The profilogram of the scratch groove can be seen in figure 2.

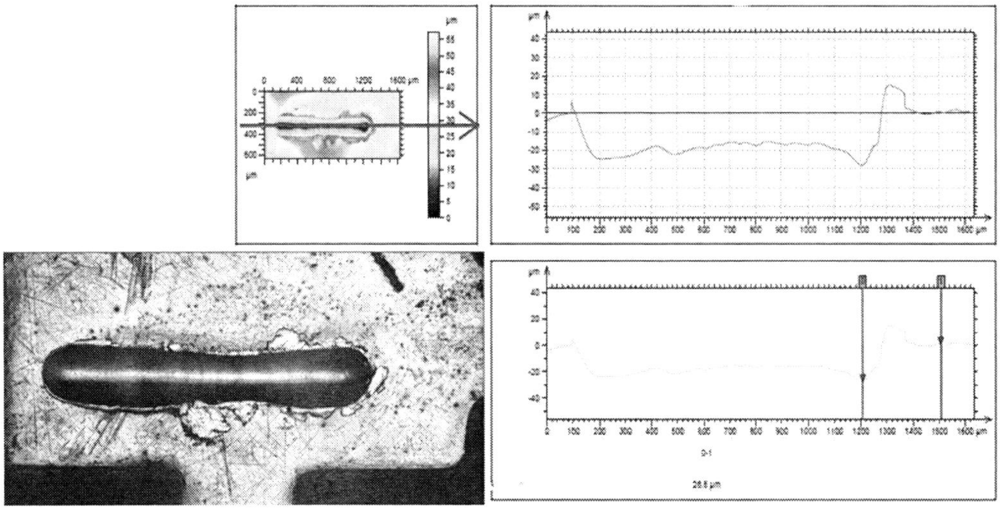

Figure 2. The profilogram for the scratch groove at normal force 10 N

When the normal force was changed to 20 N, 30 N , 40 N, 50 N and 60 N, the indenter moved with a speed of 0.167 mm/s over a distance of 1.1 mm for 6.587 s. The profilograms of the scratch grooves can be observed in figure 3 (20 N), figure 4 (30 N), figure 5 (40 N), figure 6 (50 N) and figure 7 (60 N).

In the profilograms we can obsereve that for ductile materials the borders formed on each side of the scratch are not symmetrical. This behaviour was observed in other similar studies.

The tests were made at room temperature 25°C.

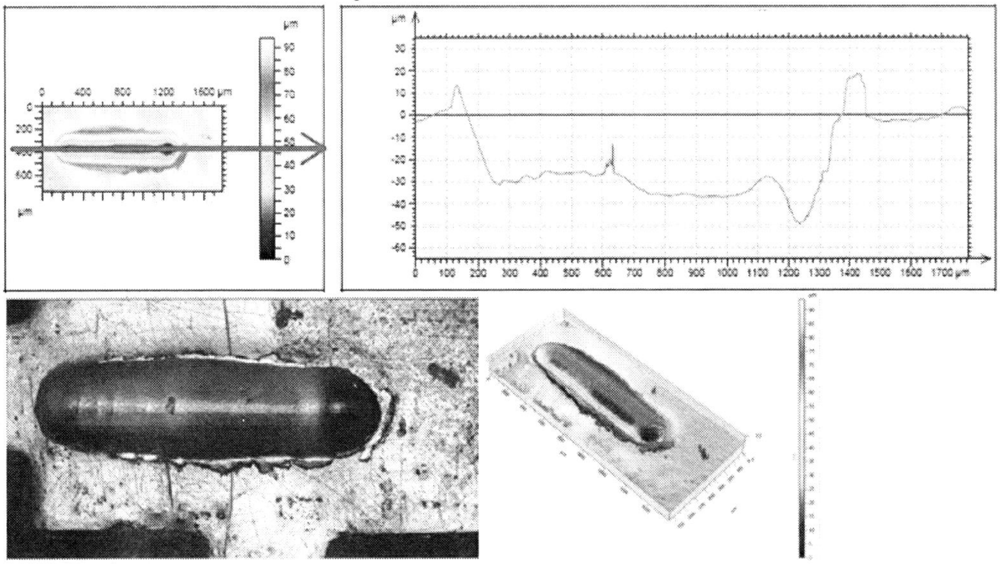

Figure 3. The profilogram for the scratch groove at normal force 20 N

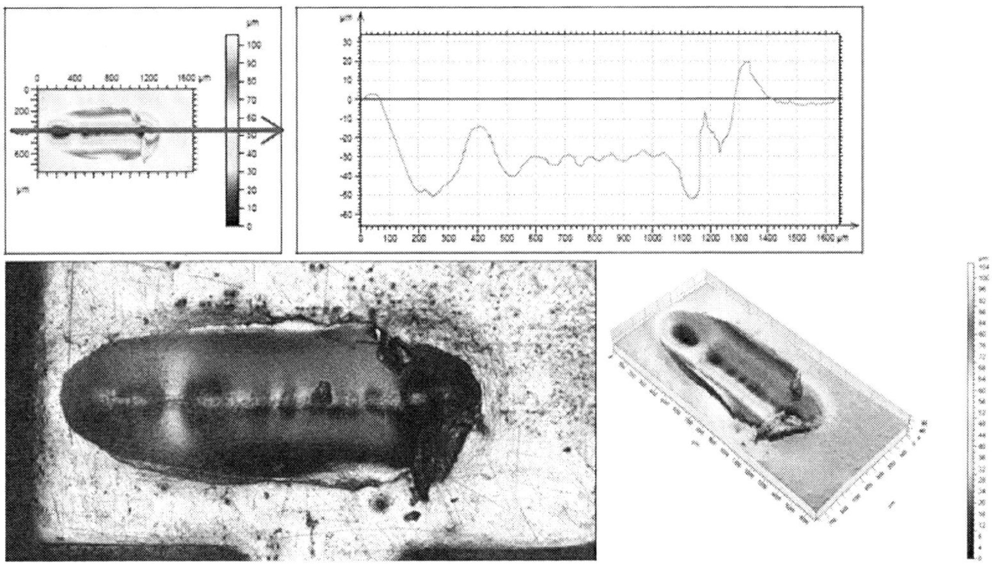

Figure 4. The profilogram for the scratch groove at normal force 30 N

1.

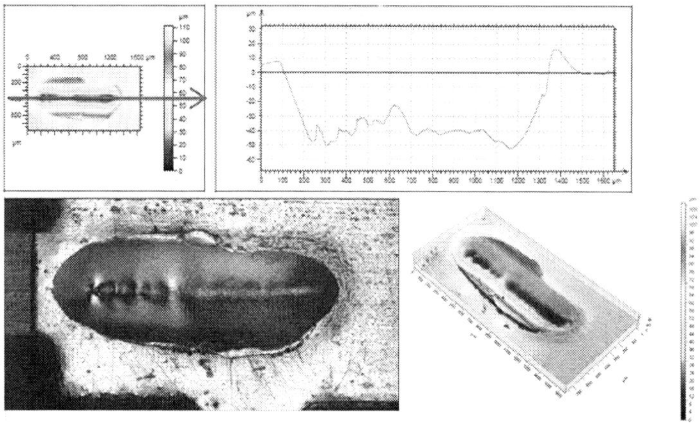

Figure 5. The profilogram for the scratch groove at normal force 40 N

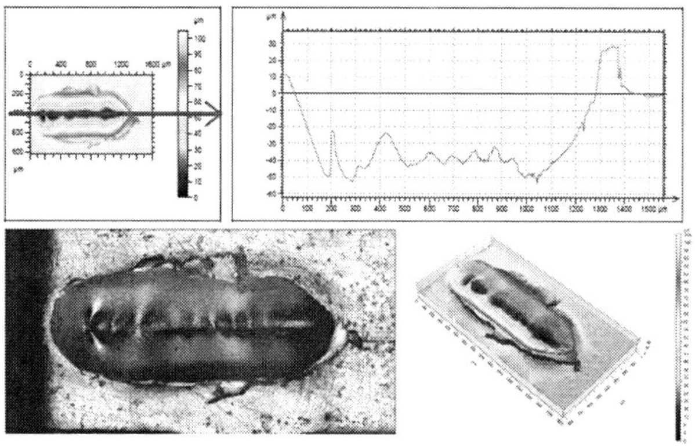

Figure 6. The profilogram for the scratch groove at normal force 50 N

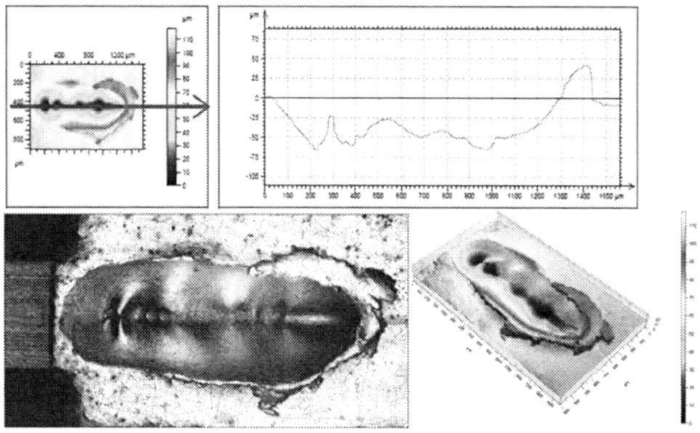

Figure 7. The profilogram for the scratch groove at normal force 60 N

The variation of the friction force for all the six normal forces applied function of the scratch length is shown in figure 8.a, while figure 8.b presents the variation of the mean friction coefficients depending on the all six normal forces applied.

IOP Conf. Series: Materials Science and Engineering **514** (2019) 012015 doi:10.1088/1757-899X/514/1/012015

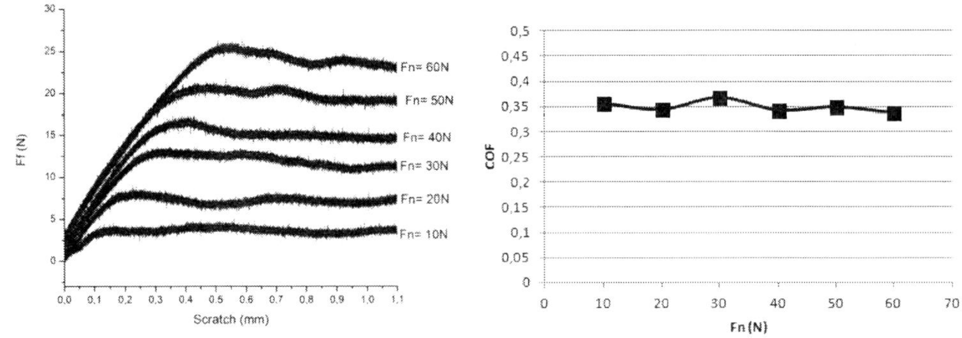

Figure 8.a The variation of the friction force function of the scratch length b. the variation of the mean friction coefficient depending on the normal forces applied

After the optical inspection and after the analysis of the profilograms, it can be concluded that the indenter did not pass through the soldering alloy SAC 305 layer, so it does not penetrate the copper layer.

The scratch groove depth for all the six normal loading forces can be observed in table 1.

Table 1. Depth of the scratch groove

Normal Force [N]	Scratch groove depth [m]
10	28.6
20	50.2
30	52.7
40	51.3
50	52.8
60	67.2

4.Conclusions

The indenter does not penetrate through the hole soldering alloy layer to reach the copper substrate deposited on the FR4 rigid support, regardless of the applied normal loading force used in the current tests.

By inspecting the profilograms it can be observed that after the micro-scratching process, the soldering alloy suffers plastic deformation. A border of material can be seen on each side of the scratch groove made by the indenter on the surface of the soldering alloy, which leads to the conclusion that the material has a ductile behaviour. The removed material is loosened by the micro-scratching process.

Some regions of the detached coating have been observed along both sides of the groove produced by the elastic recovery behind the indenter and plastic deformation in substrate.

The average friction coefficient is not influenced by the normal force applied and it is relatively constant with values between 0.34 and 0.37.

From the mechanical point of view, it is preferred for the material used for the electronic connection to have a ductile behaviour in order to take some of the elastic deformations created by the mechanical strains. A brittle connection should be avoided because it has a lower reliability.

The goal of the paper was to determine the profile of surfaces subjected to micro-scratch tests, in order to identify if the SAC305 soldering alloy is ductile or brittle, this goal being attained.

The results obtained from these tests can help in future studies for the assessment of the way a scratch or an imperfection of the contact surface can influence the electrical characteristics of the electronical systems.

We can also observe from the profilograms that both sides of the scratch are asymmetrical.

The removed material increases the volume in the process of dumping.

Electronic packaging integrity depends on the mechanical integrity of the sordered interconnects.

5.References

[1] Dumitru G I, Tudor A, Chisiu G, Plotog I, Mechanical characteristics of electronic printed circuit obtained by the vapor phase soldering process, Applied Mechanics and Materials **658**, 2014, p. 71-76.

[2] Shi X Q, Zhou W, Pang H L J, Wang Z P, Effect of Temperature and Strain Rate on Mechanical Properties of 63Sn/37Pb Solder Alloy, ASME Journal of Electronic Packaging **121** (3), 1999, p. 179-185.

[3] Hwang T K, Lee S B, Nah J W, Paik K W, Numerical analysis of the formation of coined solder bumps, Advances in Electronic Materials and Packaging, 2001, p. 388-392.

[4] Gitis N, Xiao J, Vinogradov M, Multi-Sensor Testing of Thin and Thick Coatings, Proceedings of 26th Annual Meeting of The Adhesion Society, Myrtle Beach, South Carolina, USA, 2003, p.133-135.

[5] Plotog I, Cucu T, Mihailescu B, Varzaru G, Svasta P, Vapor Phase Soldering Technology Applications for PCB with different final terminals, Conference Procedings 9th edition International Symposium on electronics and Telecommunications, Timisoara, Romania, 2010, p. 425-428.

Acknowledgments

"This work was funded by the Austrian COMET-Program (Project K2 XTribology, Grant No. 849109) and has been carried out within the Excellence Centre of Tribology. The authors would like to thank professor Franek Friedrich for his support and would also want to thank to Tetyana Khmelevska".

IOP Conf. Series: Materials Science and Engineering **514** (2019) 012016 doi:10.1088/1757-899X/514/1/012016

Twist static friction and creep between UHMWPE and bovine skin for human exoprosthesis

K A Subhi[1], A Tudor[2], A Călin[2], E Hussein[1] and Chisiu G[2]

[1]Technical college/Al-Mussaib, AL-FURAT AL-AWSAT Technical University,Iraq.
[2]Department of Machine Elements and Tribology, University POLITEHNICA of Bucharest, Romania.

Email: acalin05@gmail.com

Abstract. The twist static friction and the creep behavior of bovine skin in contact with Ultra High Molecular Weight Polyethylene (UHMWPE) indenters were studied by employing both theoretical and experimental methods. Loads (1 and 2 N) were applied to bovine skin using different indenter shapes. Moreover, the creep area and the coefficient of static friction were also discussed and analysed for different time durations. The experimental and theoretical results showed that the Voight-Kelvin rheological model could be used to represent the mechanical response of the cattle skin. As a conclusion, the creep parameter and the coefficient of static friction of bovine skin increased in time. For the circular indenter, the creep area was higher than in the case of the other shapes.
The study provides a base for more investigations on cattle skin behavior in order to be utilized on a rigid support (UHMWPE) as a new solution for "exoprosthesis support".

1.Introduction

The mechanical and physical properties of the ex vivo skin have been the main topic of numerous studies in the past twenty years. Some of these studies have also analysed the properties of in vivo skin. Several studies [1-3] investigated the mechanical properties of animal skin and compared them to properties of human skin. The studies reported that the skin properties of some animals have similar mechanical properties to human skin. One of the most important mechanical effects of the skin in contact with artificial materials is the change of elasticity and viscosity in time (viscoelasticity) [4].

It is known that the friction force in the case of contact between skin and different artificial solids has two components: a mechanical component and an adhesion component. The mechanical component of friction is defined by the coefficient of friction [5].

The adhesion component depends on the geometry of the contact and the surface energy of the skin at the interface region with the other material of the tribological system.

Soldatenkov [6] has carried out a number of investigations on the deformation component of the friction force for standard viscoelasticity. Based on the energy approach, the author proposed a new formula to estimate the viscoelasticity of the materials (standard viscoelastic – the Voight-Kelvin mechanical model) in order to calculate the force of friction for a flat indenter and a rounded corner of an indenter.

With the same objective, Cheng et al. [7] studied the viscoelastic behaviour of a standard material by indentation, where a three-element viscoelastic material (Voight-Kelvin material) has been selected under the boundary conditions of flat-punch indentation of a viscoelastic half-space.

The method for determining the viscoelastic characteristics of a Voight-Kelvin material can also be used to determine the properties of human or bovine skin. However, the skin viscoelastic properties depend on many factors such as the applied normal force, the contact area, the materials in contact and the coefficient of friction between the two mating surfaces.

In this direction, in the tribology laboratory of the Machine Elements and Tribology (OMT) department of the University POLITEHNICA of Bucharest (UPB), the behavior of the skin in contact with Ultra High Molecular Weight Polyethylene (UHMWPE) has been tested as a biocompatible anti-friction material for the joint prosthesis.

In order to provide theoretical and experimental basis for building this proposed solution, the present study includes an examination of the static twist friction and of the creep phenomenon at load step (partially). The shape of the support was assessed based on area and perimeter using three different shapes with the same cross-sectional area (circular, square, and triangular).

The present paper proposes a theoretical and experimental analysis of the cattle skin creep viscoelastic behavior and the twist behavior in contact with UHMWPE. The mechanical component of the friction is highly dependent on hysteresis losses and the distribution of contact pressure. The static twist friction torque is determined as a function of the constant loading time, the shape of punch and the same average contact pressure.

2. Experimental procedure

The classic four-ball machine (Figure 1a) is modified for this experimental study. The original device (Figure 1b) was modified with the special device (Figure 1c). The bovine skin is fixed in this device and is in contact with the rotating indenter (punch). The only possible movement between skin and indenter is twist (spin). Three different indenter shapes (circle, square and triangle), all made of UHMWPE and having the same area equal to 16 mm^2 were used to apply the load on the skin.

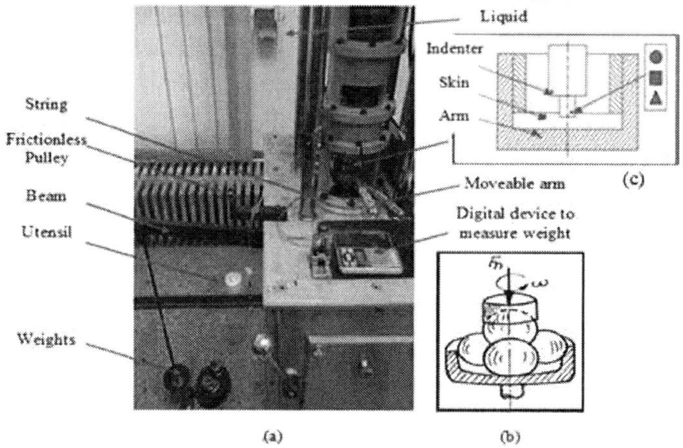

Figure 1. The four-ball tester: (a) Image of the test device during the experiment depicting the contact with the skin; (b) Original testing device of the device; (c) New device for skin.

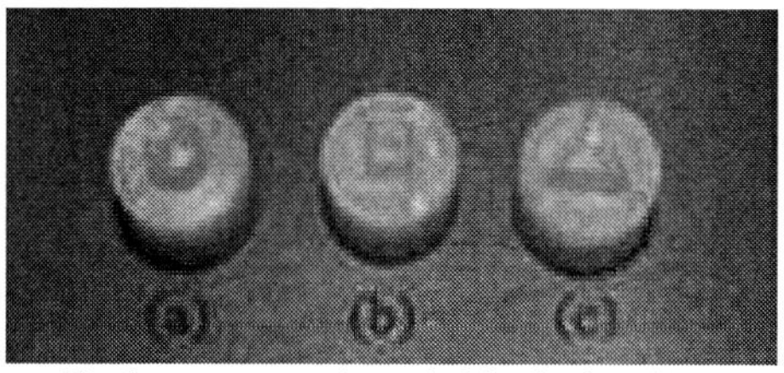

Figure 2. Types of indenters employed:(a) circle; (b) square; (c) triangle.

The ex-vivo skin was harvested from a one-year-old male calf. 25x25x4 mm pieces were prepared for testing by shaving the hair and washing the resulting samples by using only fresh water in order not to chemically degrade the biological tissue.

2.1.Testing procedure

A piece of bovine skin was placed in the device in the specified place as shown in Fig 5c. The circular indenter (radius Ro) was fixed and the normal loads (1 and 2 N) were applied. In order to determine the conventional coefficient of static friction, the following procedures were applied: the torque-measuring lever is loaded with variable weights of small amount (drops of water) until the lever begins to move continuously. The torque generated by the weight of the water in the collector immediately before the twist motion begins is considered a static friction torque. The conventional friction coefficient is defined as the ratio between the friction force and normal force. The friction force is determined based on the static friction torque and the contact radius. For the circular surface, the radius of contact is evaluated by the reduced radius of the circle, and for the surfaces of the square-shaped and the equilateral triangle-shaped, the contact radius is considered as the average of the radii of the inner circle and outer circle for square or triangle.

In order to make the arm begin or try to begin moving, water, W_{Liquid} is added resulting in a force F_{Static}. The device with skin was left in place for one minute in order for the real contact area to be determined. The test was repeated three times with the same operating conditions to reduce the variations. At the same operating conditions, the test was repeated for different times T (2, 3 and 4 minutes). Then the applied load was changed to 2 N and all the steps were repeated. The test results were collected by the computer. The indenter was changed to the square-shaped indenter (length of the square, Lso) and then to the triangle-shaped indenter (length of triangle side, Lto).

The twist friction coefficient (μ_t) is calculated based on the "r_{sf} specific friction radius"

$$M_t = \frac{M_{ft}}{F_n \, r_{sf}}, \qquad (1)$$

where: M_{ft} is the friction torque, experimentally measured on the modified 4 ball machine and F_n is the normal contact force introduced.

3. Results and discussion

The results included in this part of the study were recorded in room temperature and humidity conditions ranging from 24 to 26°C and 30 to 34%, respectively. The operating times and loads applied were variable, as well as the indenter shapes (square, circular, and triangular) for UHMWPE. The results were acquired from the average of three stable and continuously measured values. The static twist friction coefficient is calculated by equation (1) with the static friction torque measured. These coefficients are presented in the Figures 3-6. The experimental results are used to obtain the static friction coefficient in Mathcad software package. These values are written in the metrical form. The explications of these matrices columns and rows: $Ext1_{i,1}$ = time t; $Exc1_{i,2}$ = twist torque M_{tex} for normal load 1 N; $Exc2_{i,2}=M_{tex}$ for normal load 2N;

Thus, assuming the constant contact pressure on the circular surface (p_{am}), the coefficient of friction (μ_{tc}) has the following evolution in time ($t = Ext1_{i,1}$, s).

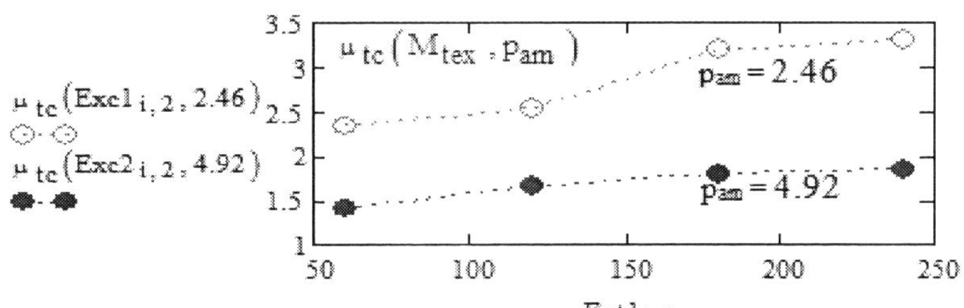

Figure 3. Static friction coefficient for pivoting for circular indenter at constant pressure

According to the theoretical model regarding the elastoplastic distribution of the contact pressure (5), the static friction coefficient of rotation is determined by the expression.

$$\mu_{exc} = \frac{M_{tex}}{\frac{2\pi}{3} R_0^3 M_{ta} P_{am} E_r} \quad (2)$$

where: M_{ta} is the theoretical twist torque.

Thus, Figure 4 shows the variation of the static friction coefficient (μ_{exc}) over time ($Excl_{i,1}$ in seconds) for two dimensionless shear strengths (τ_{ac}=0.8, τ_{ac}=0.6), two average contact pressures (p_m= 2.48 MPa, p_m= 4.96 MPa) and the twist torque measured in device for the circular shape ($Excl_{i,2}$=M_{tex} for normal load 1 N and $Exc2_{i,2}$=M_{tex} for normal load 2 N). The friction coefficient decreases when the contact pressure increases.

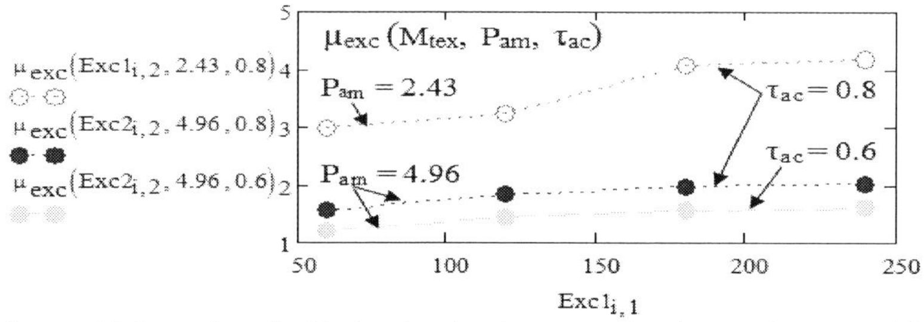

Figure 4. The coefficient of static friction in pivoting on a circular section at variable pressure

The evolution of static twist friction coefficient can be explained by the saturation contact phenomena in time and the creep of skin at the constant pressure. The decrease of friction coefficient when the contact pressure increases can be explained by the binomial friction force component.

The coefficient of friction for the triangular shape (μ_{tt}) and the square shape are shown in Figure 5 and Figure 6.

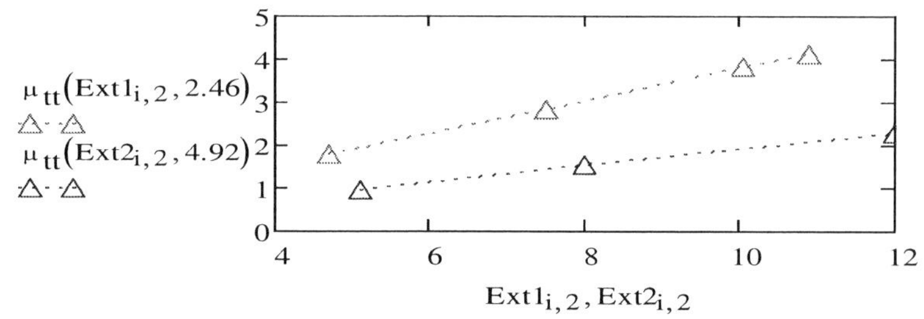

Figure 5. Static friction coefficient for pivoting on triangular shape at constant pressure

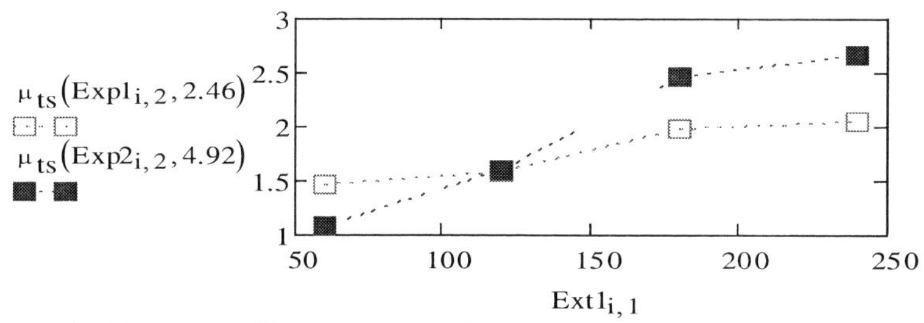

Figure 6. Static friction coefficient for pivoting on a square shape at constant pressure

The viscoelastic behavior of the bovine skin is highlighted by the analysis of the shape of the deformed zone and the variation of the penetration in time to the constant load. The effect of the creep parameter is used to analyse the changes of shapes (characteristic dimension- radius or length) of skin.

The creep parameter (φ) is the specific dimension of deformed shape (radius for the circular shape, length for the square and for the equilateral triangle) in time divided by the nominal initial dimension: $\varphi_c = R_t / R_0 : \varphi_s = L_{st} / L_{so}; \varphi_t = L_{tt} / L_{to}$. The variations of creep parameter at the step load depending on time are shown in Figure 7. The results were obtained for the circle, square and triangle indenter shapes at an applied load equal to 1 and 2 N for different durations (1, 2, 3 and 4 minutes).

The variation in time of the constant load is a consequence of the viscoelastic behavior of the skin. The viscoelastic properties of skin were determined based on the Cheng model [7]. The following values for the rheology parameters of the cow skin were obtained by an original program developed in Mathcad software and the experimental results: E_1 = 1.623 MPa; E_2 = 0.276 MPa; φ = 15.05 MPaxs. Thus, Figure 8 illustrates the dimensionless penetration of the circular-shaped indenter for a viscoelastic material, specific for cattle skin for one load evaluated by average contact pressure values.

The continuous curve is the theoretical curve of the Voight-Kelvin viscoelastic model (Figure 8), and the dotted curve is the experimental results.

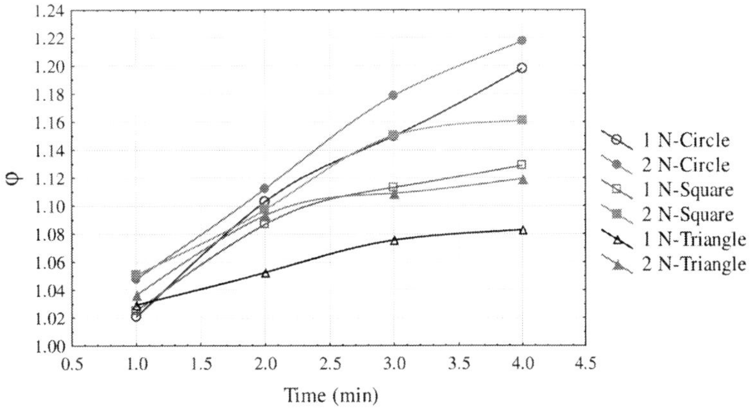

Figure 7. Variations of creep parameter (φ) respect with step load (1 N and 2 N) and shape of punch (circle, square, and triangle) for initial UHMWPE- skin contact

From the analysis of the theoretical and experimental curves of Figure 8, it can be seen that the Voight-Kelvin rheological model can be considered as suitable for bovine skin. It is appreciated that higher differences, about -12%, at the initiation of contact are due to measurement inaccuracy at zero theoretical time.

The differences between experimental and theoretical results for the analysis time interval (0-4) minutes) are between -12% and 4.3%.

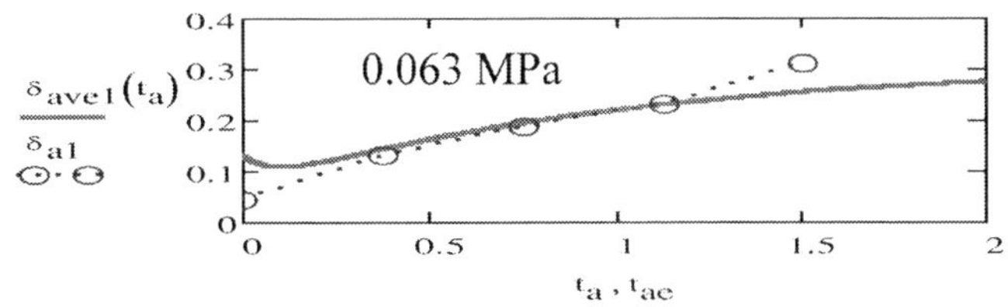

Figure 8. Theoretical penetration (continuous curve) and experimental (dotted curve) at constant average pressure 0.063 MPa

Based on this conclusion, to use the Voight-Kelvin model for tribological and damping of the bovine skin as part of the support for foot and hand exoprosthesis is proposed.

The effect of viscoelasticity of cattle skin can be seen by the evolution of the imprint in time [8]. Thus, Figure 9 shows the evolution of the imprint on the skin.

4. Conclusions

In this theoretical and experimental study, the viscoelastic behavior of bovine skin and the twist friction of rigid material (UHMW polyethylene) were highlighted.

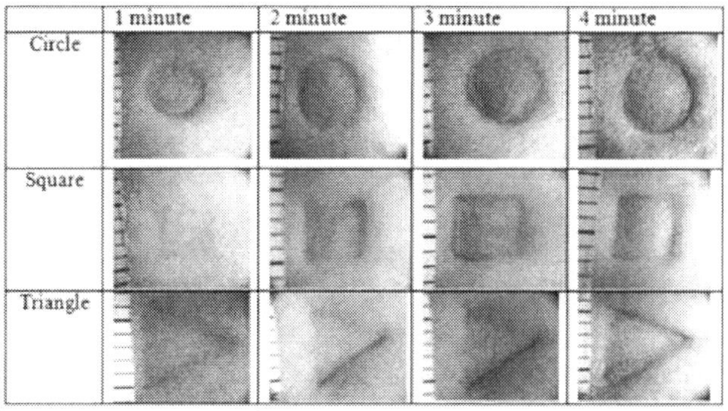

Figure 9. Imprints on the cattle skin

For establishing the viscoelastic behavior of the skin, the creep tests are used utilizing constant normal loads (1 and 2 N) that were applied using three different shaped indenters (circular, square and triangular) having the same contact area. This is an indication of the alterations in the imprints at the surface of the skin after the normal force has been removed.

The conclusions below are the outcome of the results' analysis:

- Cattle skin creep parameter increases with the increase normal load;
- The shape of contact has a high effect on the creep; for the three shapes considered (circular, square, triangular), the highest creep occurred in the case of the circle-shaped indenter and the lowest value was recorded in case of the triangle-shape indenter; the material of the punch (UHMWPE) has no effect in this case;
- There are three factors that have an effect on the conventional coefficient of static twist friction: contact shape, time and pressure;
- The static twist friction coefficient increases with time.

5. References

[1] C Pailler-Mattei and H Zahouani 2006 Analysis of adhesive behavior of human skin in vivo by an indentation test. *Tribology International* **39** pp12-21.

[2] M Zhang and A Mak 1999 In vivo friction properties of human skin. *Prosthetics and Orthotics International* **23** pp 135-141

[3] M Morales Hurtado, M Peppelman, X Zeng, P E J van Erp, E Van Der Heide 2016 Tribological behavior of skin equivalents and ex-vivo human skin against the material components of artificial turf in sliding contact *Tribology International* **102** pp103–113.

[4] Viscoelasticity https://en.wikipedia.org/wiki/Viscoelasticity.

[5] K A Subhi, A Tudor, E K Hussein and H S Wahad 2017 The adhesion and hysteresis effect in friction skin with artificial materials, *13th International Conference on Tribology, ROTRIB 16, IOP Conference Series: Materials Science and Engineering*, **174** pp 012018.

[6] I A Soldatenkov 2008 Calculation of the Deformation Component of the Force of Friction for a Standard Elastoviscous Base, *Journal of Friction and Wear* **29** no.1 pp 7–14.

[7] L Cheng, X Xia, W Yu, L E Scriven 2000 Flat-Punch Indentation of Viscoelastic Material, *journal of Polymer Science: Part B: Polymer Physics* **38** pp 10–22.

[8] K A Subhi, A Tudor, E K Hussein and H S Wahad, G Chisiu 2018 Ex-Vivo Cow Skin Viscoelastic Effect for Tribological Aspects in Endoprosthesis, *IOP Conference Series: Materials Science and Engineering* **295** no.1 pp 012018.

IOP Conf. Series: Materials Science and Engineering **514** (2019) 012017 doi:10.1088/1757-899X/514/1/012017

Optimal design through the utility function parameterization method on electromechanical actuators

R I Cotețiu[1], A Moldovan[2], I M Alexandrescu[1] and V Năsui[1]

[1] Faculty of Engineering, Technical University of Cluj Napoca,
[2] SC. Universal Alloy Corporation Europe SRL, Dumbrăvița. Romania

E-mail: vasile.nasui@cunbm.utcluj.ro

Abstract. This paper presents an analytical and computer aided procedure for the multicriteria design optimization of the electromechanical actuator. The constraints are defined to find the basic constructive parameters which defined have to by simultaneous optimized through objectively severe functions. For the defined to find the multi-criteria optimization model of the electromechanical actuator a computer program based in interactive dialogue in developed in an algorithm. The results of the program are presented considering a multicriteria optimization algorithm of the new electromechanical actuator.

1. Introduction

The electromechanical actuator has a very significant place among the transmissions, which are used in many branches of industry, in the field of machine tools, and robots have an essential importance these of programmed position. The design of roto-translation transmissions requires conditions regarding the efficiency, volume factor of safety, is presented in figure 1 [1].

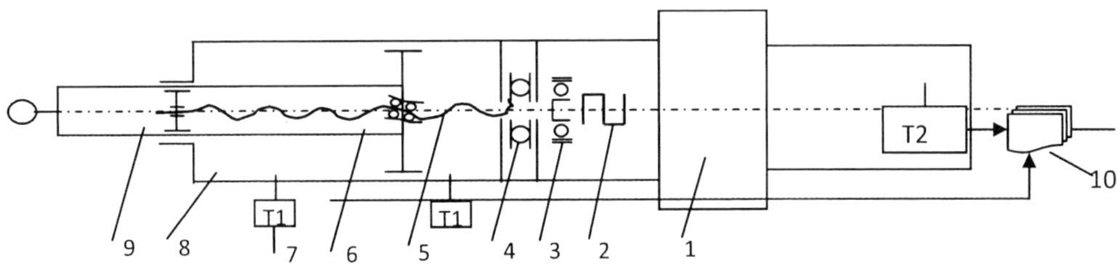

Figure 1. The mechanic system of linear electromechanical actuator: 1- engine, 2-coupling, 3,4-bearing, 5 -screws, 6- nut, 7- transducer, 8-body, 9-rod, 10-controller

The applicative and experimental research has envisaged the practical checking and complementation of theoretical models adopted for efficiency, as well as the convergence with reality of the procedures of approaching the problems observed. The creation and development of performing mechanical systems of the informatized product type, that represent the main method to increase the potential of energy conservation, has required the introduction of new optimization criteria and radical improvement solutions of dynamic quality. The general interest, in the case of the transmission of movement on the mechanical systemic to obtain efficiency as high as possible next to the ideal one.

IOP Conf. Series: Materials Science and Engineering **514** (2019) 012017 doi:10.1088/1757-899X/514/1/012017

These for the transmission of the movement of rotation the highest efficiency, a series of restrictions are imposed such as the weight, cinematic or the degrees of loading the technological ones referring to the quality of execution [2].

At the electromechanical actuator appear spatial restrictions referring to conditions of transmission of the movement (auto breaking, speed) and also technological or the execution etc. The effect of the constraints can be put in evidence by the situations in which the optimization can be done only from the point view of only one criterion, example of the minimum weigh, minimum volume or maximum efficiency resulting in constructions with axial very big and radial dimensions. The constrains in the maximization of the efficiency, taking into consideration mainly the functional conditions are of a special importance in stating the problem of optimization and in achieving the functionality of systems. In the papers on are going to present some main restrictions which affect the maximization of the mechatronic efficiency in the case of rotations system and of those of the linear transmission.

2. Theoretical comments

A method of solving this, taking into account the practical situation when we know the use and the destination of the linear electromechanical actuator would be the particular case of the functional method the multicriterial problem is reduced to a problem of defined by several objective functions. We gets the following definition of the problem to be optimized:

- minimizing: $U(f_1, f_2, ...f_k)$;
- under to: $g_j(x) < 0; j = 1, 2....m$;

Knowing the degree of importance of criterion, we give each objective function a value $0 < q < 1$, with the purpose to make a hierarchy. We can determine the best solution of the multicriterial problem by minimizing the function expressed in the following way:

$$U(x) = \sum_{j=1}^{k} \frac{q_i}{f_i\left(x^{ideal}\right)} f_i(x) \quad x \in \Omega; \tag{1}$$

where:

x^{ideal} represents the best point of each objective function;

q - coefficient that represent the weights given to deferent objective functions on the functions of degree of importance.

The sum of these weights should be: $\Sigma q_i = 1$

In these conditions, starting from a certain point from the inside of the field Ω of the allowable we can try the improvement of the value on the criterion without training a serious deterioration of each of them. The reasoning is available only in the given situation of a good knowledge of the use and the performances imposed by according to the world technical level as well as the development tendencies. In the case of the linear electromechanical actuator we raise the problem of the optimization of the construction under an acting moment and this should the following criteria:

- minimum mass F_1;
- minimum exterior diameters F_2;
- minimum distortion at torsion F_3.

By applying the example known in the specialty literature called Pareto's problem, the final expression of the optimization problem is:

- minimizing $F_1 = f_m(x) = M = M(d, l, z_s, \beta_n, p, r_0)$ [mm^3];
- minimizing $F_2 = f_9(x) = D_{ext.} = D(m, b, z_3, \propto)$ [mm];
- minimizing $F_3 = f_M(x) = M_{fr.} = M_f(F. p, \eta)$ [daN·m];

under:

- conditions of resistance at torsion;
- geometrical conditions.

The solution to the problem depends on a better mathematical expression value of each criterion

3. The formulation of problem of optimization

The optimization model, presented in figure 2, is characterized by the existence of mutually conflicting criteria and these optimum solution in relation to the one objective functions. The optimum can be described as follows, knowing the extremal of the objective function we can obtain the optimization solution for each criterion separately and the desired solution is one in solving multicriteria optimization problems it is almost impossible to define explicitly in advance the peripheries regarding the selection of the best design. In such circumstances the most suitable procedure is along a method with an interactive encroaching, in which the decision maker takes gradual decisions. The interactive method is based upon a dialogue between the designer and the computer in the process of finishing the best compromising solution [7]. The optimized parameters obtained after the global optimization are taking into account when we establish the technical solution of the correlated mechanic systems and of this destination. Because of the contradictory action of some parameters, we'll give priority to a certain criterion but only after knowing the actual working conditions next comes the systemic optimization. He facts presented a bow have a qualitative character it is not permitted the quantitative of the degree of importance and the subjective selection of the degree of importance for each criterion in general and the selection of the criteria. These results the necessity of using one of the developed techniques of analysis in the value engineering called the imposed decision. This method allows by comparison between optimization criteria, to establish the degree of importance for all the optimization criteria taken in info analysis.

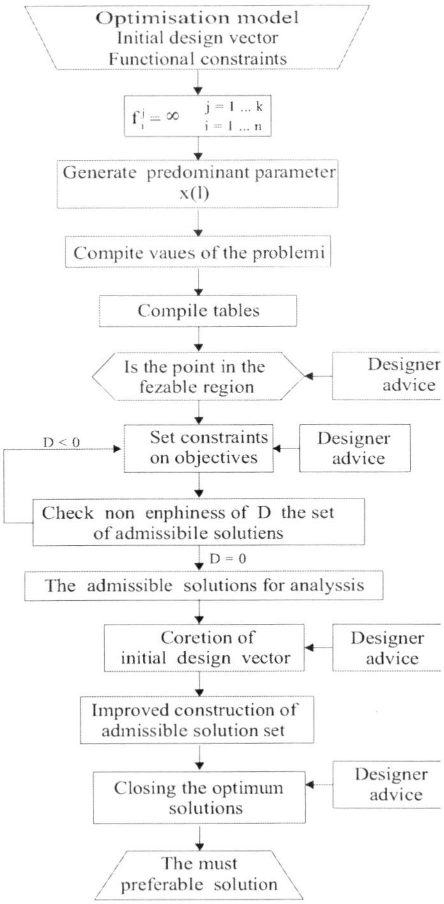

Figure 2. The iterative algorithm for optimisation

4. The operation model for optimization

The practical development of full optimization formula is hard to achier for a great number of objective functions. For a kinematics structure of screw servo driver mechanism with balls the formulation of an optimization model supposes first of all, the defining of the objective functions simultaneously with the internal and extremely constraints as within a complex mechanical system. For the kinematics relation in figure 1, the optimization solution throat simultaneous leasing or increasing of objective function: dynamic capacity, - axial rigidity - the outer speed of the screw, -the position precision: - life duration,- volume of material used for screw, the dissipation a power through scrubbing, - the information capacity: The criteria corresponding for the desired performances have the expression according to the functional criterion for the best system as extreme of the function [3].

In order to develop an optimization patterns according to very strict engineering requirement it is necessary the introduction of a number of performance criterion and the formulation of some appropriate objective functions. The mathematical pattern for the problem of the multicriteria optimization consists in nonlinear programming methods of function as in the following relation:

There is a vector:

IOP Conf. Series: Materials Science and Engineering **514** (2019) 012017 doi:10.1088/1757-899X/514/1/012017

$$x = [x_1, x_2, x_3, \quad , x_n]^q, \tag{2}$$

which has to satisfy the functional constraints under of inequality:

$$g_i(x) < 0, \quad i = 1, 2 \quad m; \tag{3}$$

of equality

$$h_i(x) = 0, \quad i = m+1, \quad p.$$

and the vector function to be optimized

$$f\{x\} = [f_1\{x\}\ldots\ldots f_k\{x\}]^q \tag{4}$$

where:

$$x = [x_1, x_2, \quad , x_n]^q$$

is a vector in n dimension in the space Euclidian R^n, $m, n, p, k \in$ N.

In many engineering optimization problems there often exist several no commensurable criteria, which must be considered. This solution is formulated as multicriteria optimization problem in which the engineer's goal is to maximize or minimize not a single but several functions at the same time [7]. In developing an optimization model, one must start with very strict engineering requirements, which a planetary gear train should fulfilling regarding the efficiency, volume, factor of safety, weight of the planetary gear train, peripheral of the planets, frictional loss power etc. [5]

The criterions' regarding the desired performance is expressed by the functions which, for the best planetary gear train should reach the extreme:

$$Extreme\ x \in D.\ f\{x\} = 0, \tag{5}$$

The following relation defines the domain:

$$D = x \in \{R^n\} \left| g(x) < 0 \wedge h(x) = 0 \right|. \tag{6}$$

Based on the objective functions given and on the functional constraints, all the relevant values of the screw - servo driver mechanism in rolling elements, so that the vector of the variable values can be written in the form of the following relations.

The variable design vector is identified with:

$$\bar{x} = [p, r_c, d_b, z_s, \beta_n]^q. \tag{7}$$

The functions of the criterion for a mechanic transmission of the screw with rolling elements can be written as it follows:

- outer diameter:

$$f_1(x) = d = \sqrt{\frac{4 \cdot \beta \cdot F}{\pi \cdot \sigma_{ac}}} \text{ [mm]}, \tag{8}$$

- axial rigidity:

$$f_2(x) = R = \Delta F / \Delta l \text{ [N/μm]}, \tag{9}$$

- volume of material used for screw

$$f_3(x) = V = V(d, l, z_s, \beta_n, p, r_0) \text{ [mm}^3\text{]}, \tag{10}$$

- the outer speed of the screw:

$$f_4(x) = n_{crt.} = 402 \cdot 10^3 \cdot a^2 \cdot d_m / l^2 \text{ [rot/min]}, \tag{11}$$

- the manufacturing precision:

$$f_5(x) = p = \Sigma \Delta p_a / l = \Sigma \Delta p / l + j_{max} \text{ [μm]}, \tag{12}$$

- the frictional moment:

$$f_6(x) = M_{fr} = F \cdot V \cdot 10^3 / 2\pi \cdot \eta \text{ [daNm]}, \tag{13}$$

- mechanical efficiency:

$$f_7(x) = \eta = \text{tg } \beta_m / \text{tg } (\beta_m + \varphi) \text{ [\%]}, \tag{14}$$

- the transmission ratios:

$$f_8(x) = i_m = \text{tg } \beta_n \text{ [/]}, \tag{15}$$

- axial rigidity:

$$f_{10}(x) = d_s = \sqrt{\frac{64 \cdot F \cdot l_f^2 \cdot c_f}{\pi^3 \cdot E}} \text{ [mm]}, \tag{16}$$

- the bending stress:

$$f_{11}(x)=\sigma_{ech}=\sqrt{\left(\frac{4\cdot F}{\pi\cdot d^2}\right)^2+\left(\frac{2\cdot M_f}{0{,}2\cdot d^3}\right)^2}\ [\text{daN/mm}^2] \tag{17}$$

- the information capacity:

$$f_{12}(x)=I_d=\log_z S^n \tag{18}$$

- life duration:

$$f_{13}(x)=L_h=\sqrt{\frac{F_m}{C}}\ [\text{ore}] \tag{19}$$

The vector to be optimized $\bar{x} = [p,\ r_c,\ d_b,\ z_s,\ \beta_n]^q$ that will satisfy the inequality constrains respectively from or respectively, equations from the

- contact stress:

$$g_1(x) = s_C = \sigma_a - \sigma_{0a} > 0\ [daN/mm^2], \tag{20}$$

where:

$$\sigma_a = (2{,}3 \div 3{,}0)\cdot 10^3\ \mu \cdot p_a,$$

$$\sigma_{0a}=0{,}388\cdot\sqrt{\frac{F_N\cdot E_{ech}^2}{R_{ech}^2}},$$

- safety factor for bending stress:

$$g_2(x) = s_F = \frac{\sigma_a}{\sigma_{ech}} - 1{,}5 > 0, \tag{21}$$

- safety factor axial rigidity:

$$g_3(x) = s_H = \frac{\sigma_a}{\sigma_{cr}t} - (0{,}5 - 0{,}8) > 0, \tag{22}$$

- safety factor for outer speed:

$$g_4(x) = s_n = \frac{n_a}{n_{crt}} - 0{,}8 > 0, \tag{23}$$

- the efficiency:

$$g_5(x) = s_e = \frac{P_a}{P_{crt}} - (0{,}80 - 0{,}99) > 0, \tag{24}$$

- life duration:

$$g_6(x) = s_L = L_{conv} - L_{necesar} > 0, \tag{25}$$

- the position precision:

$$g_7(x)=s_p=\frac{\Delta p_a}{l}-\frac{\Delta p_{\max}}{l}\le 0, \tag{26}$$

- the manufacturing:

$$g_8(x) = D_{cb\,min} \ge \frac{L}{K_a}, \tag{27}$$

where: Δp şi K_a according to the table 1

Table 1. Parameter values Δp şi K_a

Δp	0,003	0,005	0,010	0,015	0,025
K_a	25	35	50	65	90

which has to the constructive constraints as in the following

$$\frac{d_b}{D_{cb}} = 0{,}07 \div 0{,}20\,; \quad d_b = (0{,}55 \div 0{,}65)\,p, \tag{28}$$

$$\frac{r_b}{r_f} = 0{,}90 \div 0{,}98\,; \quad r_f = (0{,}515 \div 0{,}526)\,d_b.$$

5. The configuration of the experimental stand

Future development could also be optimizing from the point of view of regulating and controlling motion control parameters using fuzzy logistics based on vague crowds. In order to improve and to study the possibilities of regulating and control of the actuator movement a new research direction has been tried using complex systems of data acquiring and analysis by means fuzzy logistics [6]. Thus, a configuration has been used whose pattern is presented in figure 2. As it results from the pattern in hardware configuration, there are the following components whose main characteristics are presented in figure 3: linear actuator 1, traducer 2, and data acquisition board 3 (number of channels, maximum speed of sampling, data transfer); account unit 4 (microprocessor, memory, video screen), actuator interface and board of fuzzy logistics [5].

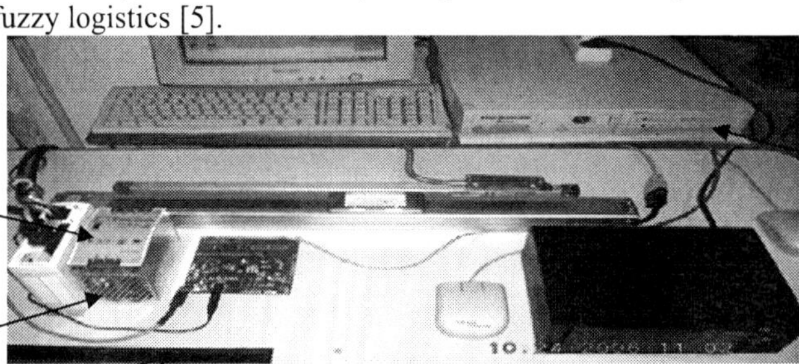

Figure 3. The stand of industrial linear

6. Conclusions

The multicriteria optimization method solves the problem of selection of optimum parameters constructive and functional of the linear electromechanical actuator. To define the multicriterial optimization model for linear mechanical electromechanical actuators, a computer program based on an interactive dialogue was developed using a new algorithm. The best comprising solution obtained through the method of the importance of the criteria of objective function gives the optimum functional parameters. The constraints are defined to find the basic constructive parameters which defined have to by simultaneous optimized, through objectively severe functions. Following the application of this methodology, we have found optimal solutions for linear electromechanical actuators using a new advanced motion control logistics. There are new developments of electromechanical linear actuators both in terms of configuration and their programming.

References

[1] Crişan N. I 2010 *Bazele designului modern,* Editura Risoprint, Cluj Napoca
[2] Jazar R. N. 2006 *Theory of Applied Robotics.Kinematics, Dynamics, and Control*, Springer, New York, Dordrecht Heidelberg London, Springer Science+Business Media LLC
[3] Marin C., Vasile G. 2011 *Tehnici de modelare si simulare in ingineria mecanica.* Editura Bibliotheca, Târgovişte.
[4] Montgomery D. C. 1996 *Design of Analysis of Experiments 4th Edition John Wiley & Sons* New–York Toronto Singapore
[5] Nãsui, V. 2006 *Actuatori liniari electromecanici.* Editura Risoprint, Cluj Napoca
[6] Popa A. 2002 *Controlul digital al sistemelor mecatronice.* Editura Orizonturi Universitare Timisoara
[7] Timiş P. G. M. 2012 *Proiectarea logică.* Editura Performantica Iaşi

IOP Conf. Series: Materials Science and Engineering **514** (2019) 012018 doi:10.1088/1757-899X/514/1/012018

On the Evolution of Contact Patch Concept

L M Dehelean[1] and N M Dehelean[1]

[1]Mechatronics Department, "Politehnica" University, Timisoara, Romania

E-mail: liana.dehelean@upt.ro

Abstract. The paper aims to highlight the evolution of the concept of physical contact between two technical bodies (machine parts). The most abstract situation occurs in antiquity when the necessary geometry concepts were created. Engineering applications from the industrial age have demanded higher performances. These have become possible through a continuous refinement of the theoretical models. The hertzian contact between two evolving parts, in general, and between the wheel and the rail, in particular, evolved and became extremely concrete and full of details that were once obscure. The contact patch between the wheel and the rail has today an asymmetric geometry correlated with the state of stress and the dynamics of the railway track. The use of digital methods for describing topologies can be a way to create a model closer to the dynamic state of wheel and rail contact. We also encounter the same contact phenomena in gear power transmission.

1. Introduction

In Materials Resistance area of activity there is an important damage phenomena that come out from a high stress concentration zona as the wheel-rail contact. Therefor the wheel-rail contact has become the main point of any research connected to vehicle improvement. The correct assessment of the contact stress values can be made possible by designing a phenomenological and computational model to follow the true real phenomenon. The wheel-rail contact governs the dynamic performance of rail vehicles through the loads it transmits and, in the same time, the machine parts are subjected to serious damage phenomena.

Starting from Hertz's wheel-rail contact model, various other models were developed, increasingly close to the real wheel-rail contact. Even the Hertz contact solution is yet used since it is of closed-form, some of its underlying assumptions may be violated quite often in wheel-rail contact. The assumption of constant relative curvature which leads to an elliptic contact patch could be much more acceptable.

Some fast non-elliptic contact models are proposed to lift this assumption while avoiding the wearying numerical procedures of calculus. These models are accompanied by a simplified approach to treat tangential tractions.

There are two alternatives for theoretic description of the rolling contact kinematic. A Lagrangian approach, in which the system of coordinates is particle-fixed. This one is mainly used for static or short-displacement applications. The other option, which seems to be more efficient in rolling contact applications, is the so-called arbitrary Lagrangian-Eulerian (ALE) approach in which the total deformation of the rolling wheel is divided into a rigid-body motion and the deformation of the material point.

The formulation of the wheel-rail contact problem is a complex task which has been the subject of several investigations that revealed different solutions for the problem. Some researchers [1-5] studied the contact forces between the wheel and the rail during the dynamic motion and, as a result of their investigations; several computer routines are now available for the calculations of the tangential forces at the contact point given the normal force and the relative velocities between the bodies.

How can we discover new sources of inspiration for a more refined models and at the same time simple for wheel-rail contact and generally between two convex bodies? This approach can help us get rid of the circle of sterile thought.

The question of contact between the circle and the right line is an ancient dispute between Plato and Protagoras. Plato, the author of the Theory of Forms, had imposed to the mathematicians of his time, the concepts of circle and right line and more the possibility of their relative position in a singular tangent point. On the other hand Protagoras had argued that the contact between circle and right line occurs in a set of distinct points. At their time Plato was the winner and so the geometry of Euclid had developed.

2. Reconsidering the ancient dispute

The mathematicians of ancient Greece operated with ideal geometric figures, these figures had metaphysical properties. For example, the right line consists of an infinity of geometric dots. These are not dimensional and therefore can be not represented graphically. Consequently, the line can be not represented graphically. But the mathematicians of ancient Greece used the graphic representations only from the need to communicate each other. The graphic figures represented merely meant to suggest the metaphysical entities they actually operated with.

The continuity of these geometric figures is difficult to define. If we use a geometrical atomistic conception, the continuity of the circle and the straight line becomes precarious. Salvation of these definitions is made by calling to infinite algebra.

The definition of continuity is precarious because we cannot avoid the dot notion. The definition of continuity without dot concept is an aporia. By accepting the continuity made by points and the number of constitutive dots to be infinite, then any cut will be made up of an infinite number of dots (1).

$$\left\{N_{AeB}\right\}=\infty;\ \left\{N_{AfB}\right\}=\infty;\ \left\{N_{\overline{CD}}\right\}=\infty. \tag{1}$$

It made a cut on the circle and one on the right line that will be tangent to the circle Figure 1. Each arc consist of an infinite number of dots. The same, the segment CD consist of an infinite number of dots. When the circle and the straight line are tangent, it forced the decrease of the cuts on the circle and the straight line (Figure 2).

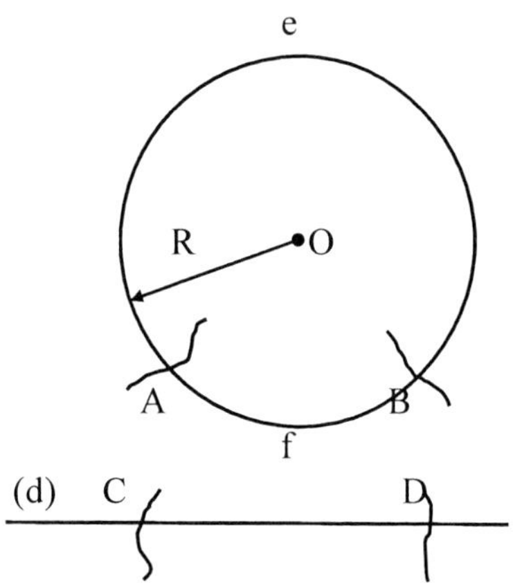

Figure 1. Cut on circle and straight line

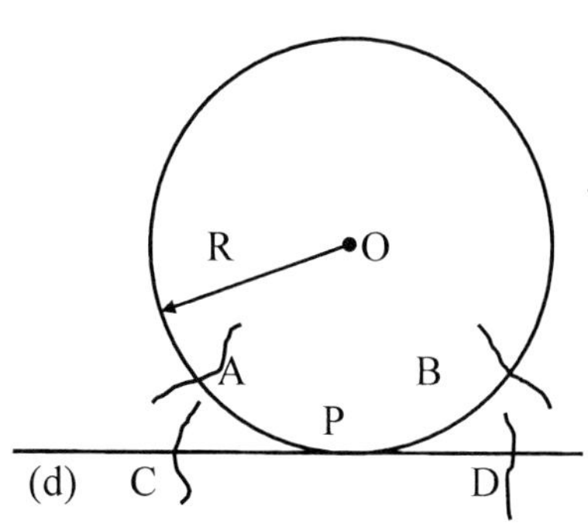

Figure 2. Cut on circle and tangent straight line

Going to limit it force that

IOP Conf. Series: Materials Science and Engineering **514** (2019) 012018 doi:10.1088/1757-899X/514/1/012018

$$A \to P; B \to P; C \to P; D \to P \tag{2}$$

$$AB \to 0\, and\, \overline{CD} \to 0 \tag{3}$$

$$N_c = \begin{cases} \infty \\ 1 \\ 0 \end{cases} \tag{4}$$

Equation (4) suggests a serious anisotropy of geometric space.
It can be admitted that the number of dots of an infinite small cut is a finite number N_f. So equation (4) become equation (5).

$$N_c = \begin{cases} \infty \\ N_f \\ 0 \end{cases} \tag{5}$$

If "p" is the geometric dot of contact between the circle and the line, according to the Euclidean geometry, 1, 2, 3, 4, ... are a set of dots adjacent to it and all of them are located at infinitely small distances, then all R_3, R_1, R_0, R_2, R_4,... rays are parallel, that means the angles φ_i are equal with 0 (Figure 3).
In this hypothesis there are a finite number of common dots of the circle and the line when they are tangent. From a macroscopic point of view, the contact between the circle and the straight is no longer at a geometric point, but in an exotic geometric entity that we can call it a point of contact. This point tends to lose its metaphysical character and become able to participate in functional models.
This assumption allows proposing P_c as a dot of contact entity. This point, P_c, will become a patch in the more concrete models.

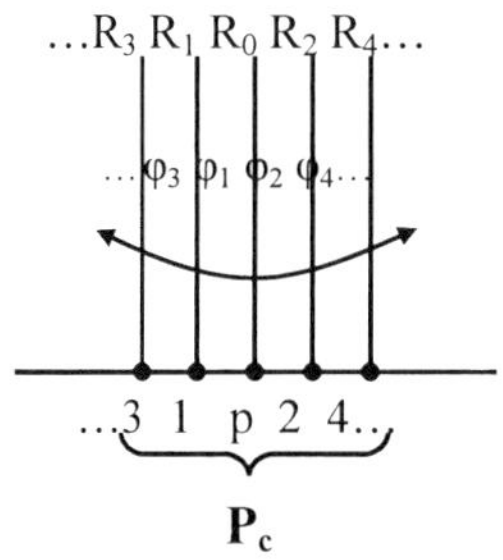

Figure 3.

3. A new wheel–rail contact model for railway dynamics

Coming from a theoretical circle-to-line contact model to a concrete rail contact model, the practical application complicates the model. The wheel-rail contact must be viewed both statically, without traction and dynamic when a creep force occurs. The contact pattern involves to estimate the distribution of contact pressure and friction coefficient.

The deterioration phenomena occurring in the wheel-rail interface involve two phenomena to blame are namely wear and rolling contact fatigue (RCF). The primary step that enables investigating such mechanisms is to clearly know the shape and size of the contact patch, and the distribution of the stresses within it. The contact point detection may be carried out by assuming the wheel and the rail as rigid or accounting for their flexibilities.

The study of rolling contact, or more accurately rolling-sliding contact, started in 1926 by the publication of Carter's paper [7] on the solution of the two-dimensional rolling contact problem.
The three-dimensional rolling contact including sliding was touched upon by Johnson [8] about thirty years later.
Johnson, in particular, attempted to model the three-dimensional rolling. He started investigating circular contact areas. In 1958, he published two papers on the rolling motion of an elastic sphere on a plane. In one of them he investigated, the effect of spin on rolling contact. He found that the spin generates a lateral tangential force as well as a moment around the vertical axis. The spin-generated lateral force is opposed by a lateral force due to lateral creepage. The case of pure creepage without spin was extended to elliptic contact area by Vermeulen and Johnson [10].

In this theory, the stick zone is a smaller ellipse with the same ratio of the axes adjacent to the leading edge as shown in Figure 4. Based on this configuration the traction vanishes only at one point in the leading edge. Pascal and Sauvage [9] realized the existence of possible secondary contact points from the observation of the sudden "jumps"(of the contact stress) in the evolution of contact parameters with respect to wheelset lateral displacement.

These jumps exist when the wheel flange comes into contact with the rail and they are indicative of the secondary contact points that come to be effective due to the elastic deformation of the profiles resulting from the contact pressure.

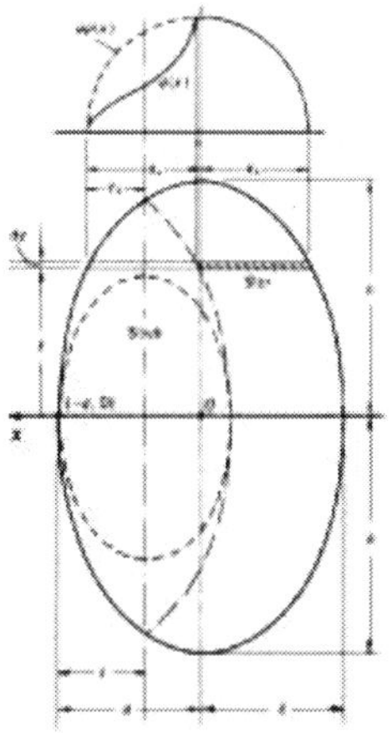

Figure 4. Stick aria along the patch wheel-rail

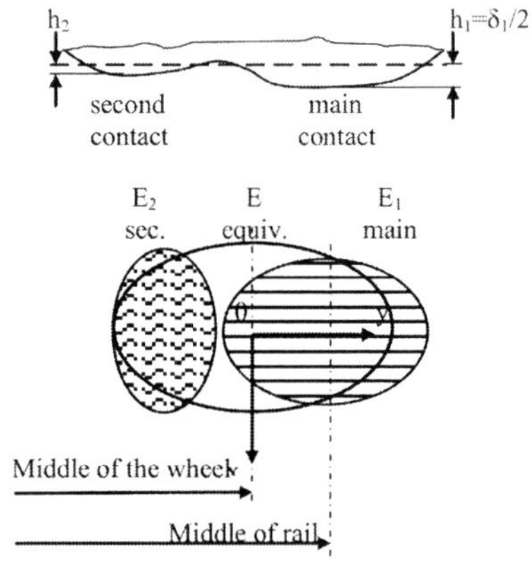

Figure 5. Pascal Model

In detecting process of the contact zones, if profiles are considered rather elastic than rigid ones, several contact zones may be possible to find. Each contact zone indicates a separate contact patch. A Hertzian ellipse is then calculated for each patch using the curvature values at that point. In order to detect the secondary contact point, the profiles in contact are deducted by a semi-elliptic function calculated using the first (main) contact ellipse width, b, and the approach, δ.

The normal forces for each ellipse are determined by taking into account the ratio between h_1 and h_2 (Figure 5) and the Hertzian relation.

The implementation of this method, known as multi-Hertzian, and the coupling with a wheelset dynamic model seemed to be very slow at the time. Therefore, Pascal decided to replace the contact ellipses by an equivalent one (as see in Figure 5).

This equivalent contact ellipse is established in a manner that results in the same creep force as the sum of the ones from two separate ellipses while the direction of the force is determined by weighted averaging.

The approach of equivalent ellipse may be adequate for dynamic simulations but it is not suitable for damage analyses such as wear calculations due to the non-physical contact patch and stress distribution.

To be able to estimate non-elliptic contact patches in a fast manner, an approximate contact method should replace the Hertz solution for the normal contact part. In fact, calculation of the elastic deformation, u_z, in Equations (6) and (7) involves numerical integration and is not possible analytically for non-Hertzian profiles (where, C denotes the area of contact).

$$d(x,y) = z(x,y) - \delta + u_z(x,y) \tag{6}$$

$$\begin{aligned} &d(x,y)=0, \quad p(x,y)>0 \qquad (x,y)\in C \\ &d(x,y)>0, \quad p(x,y)=0 \quad (x,y)(NOT \in)C \end{aligned} \tag{7}$$

Kik and Piotrowski [11] proposed an approximate method based on a concept called virtual penetration. In this concept, the deformations surface are neglected (u_z =0) and it is assumed that the bodies can rigidly penetrate into each other.

A smaller virtual penetration value d_v, than the prescribed one d_0, may result in a penetration zone which is close enough to the real contact area and can be regarded as the contact patch. The virtual penetration value can be set in different ways. As suggested by Kik and Piotrowski, it should be about half the prescribed penetration value. Figure 6 a, b, c, d, illustrates the contact patches resulting from all four virtual penetration strategies for circular wheel and rail profiles with radii 200 mm and 80 mm respectively, and wheel rolling radius 460 mm.

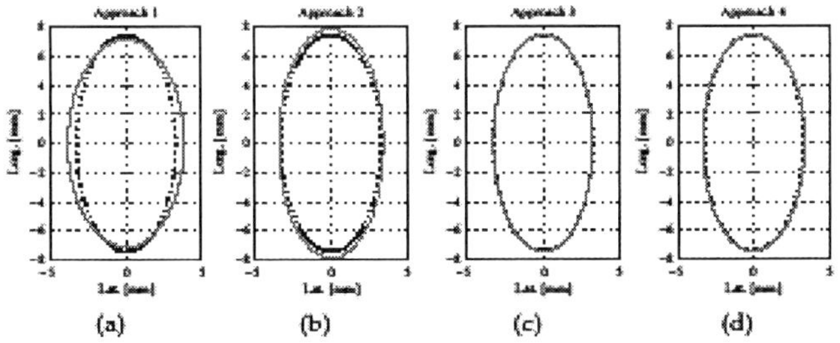

Figure 6. Different contact patches according to penetration strategies

4. Equivalent ellipse adaptation

In order to skip the calculation of no elliptic flexibility parameters and employ the elliptic ones, an imaginary equivalent ellipse may be assigned to the non-elliptic patch. Kik and Piotrowski [11] determined an equivalent ellipse for each separate contact patch by setting the ellipse area equal to the non-elliptic contact area and the ellipse semi-axis ratio equal to the length-to-width ratio of the patch.

By discretizing the contact patch into longitudinal strips, as done in the Linder method [11] and the STRIPES method [11], one may treat each strip differently.

In the Linder model, all local strip ellipses have the same lateral semi-axis. The longitudinal semi-axis is then calculated so that the strip fits into the resulting ellipse. In STRIPES, dissimilarly, the local ellipses are determined by employing relative curvatures at the centre of the strip in Hertz solution.

For engineers and researchers the real physical wheel-rail contact phenomenon is more complicated than the idealized theoretical models. In fact, several other physical phenomena, which themselves are complex in nature, are involved in practical applications: plastic flow, surface roughness and thermo-mechanical coupling and so on.

The friction itself, is an elusive phenomenon which seems to be even more difficult to fully model in the wheel-rail interface subjected to environmental changes. What makes it even more problematic to mathematically model this physical system is the lack of real-life measurements in order to validate the proposed models.

The kinematics of guidance of the wheelsets is based on the wheels and rails geometries. The movement of the wheelsets along the rails is characterized by a complex contact with relative motions on the longitudinal and lateral directions and relative rotations of the wheels with respect to the rails.

If we take the wheel material ER8, according to European standards, with nominal yield stress σ_y =54 MPa and assume μ=0.3 the calculated maximum pressure of about p = 1800MPa is the limit for which the linear elasticity assumption in the contact solution may be valid and the plastic flow in outer rail of tight curves is clearly evident. The contact pressure is reduced and therefore the size of the contact patch increases. Some attempts to include plastic flow rules in wheel-rail contact models has been done.

There are few techniques used for measuring the shape and size of the contact patch and the pressure distribution for the contact between wheel and rail profiles. Some methods utilize ultrasonic approach, while others experiment a kind of special pressure sensitive paper for this purpose. All researchers studied normal static contact, but there is no one who has done a measuring technique to measure the tangential stress (traction) distribution in wheel-rail applications.

5. Conclusions

Increasing train speeds and increased road safety will require further research into the phenomenon of rolling and the intimacy of the wheel-rail contact. New methods of experimental research will be targeted to reveal accuracy the shapes of contact patch Figure 7, in various transitory phases of the run. In order to advance in shaping the dynamic contact phenomenon of rail-wheel contact it is necessary to remodel the two bodies in contact as a single body. A differential model and a specific new differential equation will emerge.

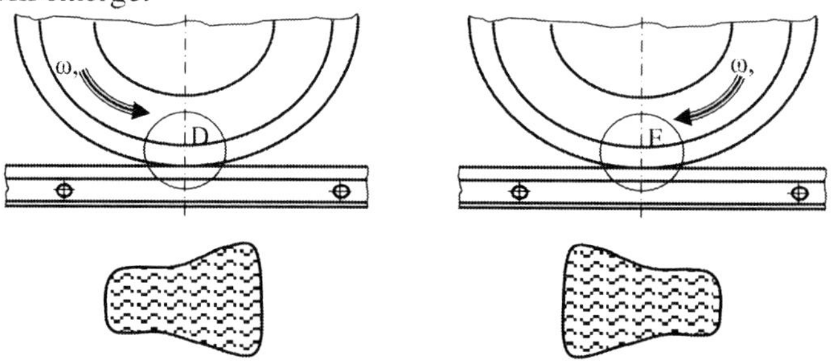

Figure 7. Shape of contact patch

Crossing over from an abstract model of circle-line tangential to a concrete wheel-rail contact model requires a strong data enriching act consist of: material structure parameters addition, elastic parameters and stress propagation functions addition. That it means it transforms the contact point into a real power transfer space that we call now the contact patch.

6. References

[1] Kalker, J. J. – Three-Dimensional Elastic Bodies in Rolling Contact, Kluwer Academic Publishers, Dordrecht, the Netherlands, (1990).
[2] Johnson, K. L. – Contact Mechanics, Cambridge University Press, United Kingdom, (1985).
[3] Kik, W., Piotrowski, J. – A Fast, Approximate Method to Calculate Normal Load at Contact Between Wheel and Rail and Creep Forces During Rolling, 2nd Mini Conference on Contact Mechanics and Wear of Rail/Wheel System, (I. Zabory, Ed.), TU Budapest, Hungary, (1996).
[4] Shen, Z. Y., Hedrick, J. K. and Elkins, J.A. – A Comparison of Alternative Creep Force Models for Rail Vehicle Dynamic Analysis, 8th IAVSD Symposium on Dynamics of Vehicles on Road and Tracks, (J. K. Hedrick, Ed.), Swets and Zeitlinger, Cambridge, Massachusetts, (1983) 591-605.
[5] Polach, O. – A Fast Wheel-Rail Forces Calculation Computer Code, Vehicle System Dynamics, Supplement 33 (1999) 728-739.
[6] João Pombo, Jorge A.C. Ambrósio, Miguel T. Silva – A new wheel–rail contact model for railway Dynamics, Vehicle System Dynamics, February 2007.
[7] Carter, F.- On the action of a locomotive driving wheel, Proc. R. Soc. London 112 (1926) 151–157
[8] Johnson, K. – Contact Mechanics, Cambridge University Press, Cambridge (1985).
[9] Pascal, J., Sauvage, G. – The available methods to calculate the wheel/rail forces in non Hertzian contact patches and rail damaging, Vehicle Syst. Dyn. 22 (1993) 263–275.
[10] Vermeulen, P., Johnson, K. – Contact of no spherical elastic bodies transmitting tangential forces, J. Appl. Mech-T ASME 31 (1964) 338–340.
[11] Kik, W., Piotrowski, J. – A fast approximate method to calculate normal load at contact between wheel and rail and creep forces during rolling, Proceedings of 2nd Mini-Conference on Contact Mechanics and Wear of Rail/Wheel Systems (1996) 338–340.

IOP Conf. Series: Materials Science and Engineering **514** (2019) 012019 doi:10.1088/1757-899X/514/1/012019

Modular omnidirectional mobile robot with four driving and steering wheels

M O Tătar[1] and R Pecie[1]

[1] Mechatronics and Machine Dynamics Department, Technical University of Cluj-Napoca, Romania

e-mail: olimpiu.tatar@mdm.utcluj.ro, pecie.robert@gmail.com

Abstract. In this paper the authors present a modular omnidirectional mobile robot with four driving and steering robot. The paper describes the experimental prototype, the kinematic analysis in two configurations (four driving and steering wheels configuration and synchro-drive configuration), the electronic components, the testing process of the robot and future research directions.

1. Introduction

The attention for mobile robots is increasing mainly due to their many application. For a robot, the ability to move increases its flexibility and ability to perform new and more complex tasks.

An important issue regarding mobile robotics is the maneuverability of a robot. The maneuverability might be improved by making use of various locomotion systems. One such locomotion system is represented by the wheels. Beside maneuverability, the design of mobile robots has to provide viable solution to the traction, stability and controllability problems. These are influenced by design choice regarding the type of wheels, their configuration and actuating solutions [1]. Lately, in this regard, the specialised literature presents several achievements [1], [2], [3], [4].

In the following pages, the paper is structured as follows: the second section presents the omnidirectional robot and in the third section is presented the kinematic analysis of the robot. The components required for the control of the robot are presented in the fourth section, the fifth section deals with the testing process, the sixth section concludes the paper and lists future development objectives.

2. Arhitecture of the robot

The robot presented next is composed out of four identical modules, each containing the driving motors, driving train and the wheels. These modules are connected to a hexagonal platform which hosts four servomotors and the electronic drivers. Synchronism is achieved by simultaneously control of all four motors, each wheel of the robot possessing its dedicated driving motor.

2.1. The prototype description

The robot platform is symmetrical, hexagon shaped with a side length of 12.5 mm, made of PMMA (Poly(methyl methacrylate), Plexiglas) and the thickness is 4 mm.

Figure 1 a presented a photo robot developed in the Robotics Laboratory of the Mechatronics and Machine Dynamics Department at the Technical University of Cluj-Napoca [1], [4]. The structural diagram of the entire robot is presented in figure 1 b. Figure 1 shows that for to steer the wheels, four servomotors (SM_1, $SM_{2,\ldots}$) need to be synchronised. For driving the robot, another four motors need to be synchronously driven (Me_1, $Me_{2,\ldots}$). The total height of the robot is of 22 mm and the weight of the robot, together with the power wiring, is of 1698 g. The robot is driven only on flat surfaces.

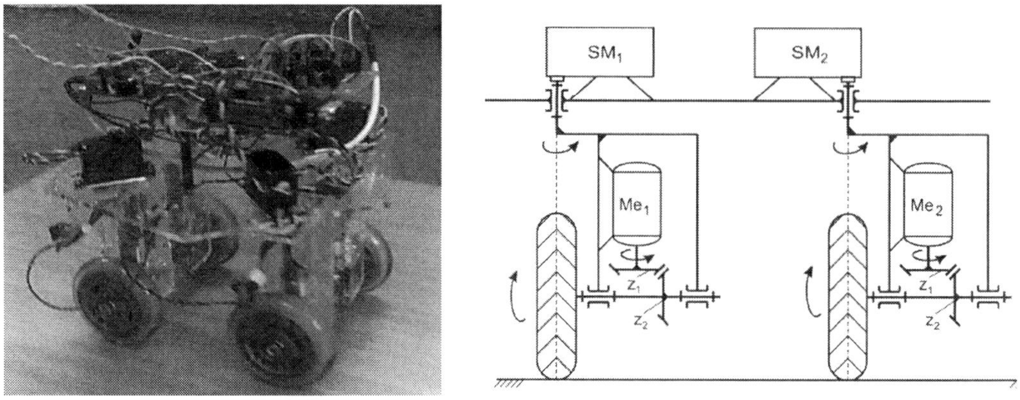

(a) (b)

Figure 1. Robot with four driving and steering wheels
a) Photo b) The structural diagram of the robot (front view) [1], [4]

2.2. *Wheeled Modules*

The wheel modules attached to the hexagonal platform are four and are also made of PMMA with a thickness of 3 mm. These modules use a DC motor with a gear reducer and conical gear with straight teeth conical geared transmission ($z_1 = z_2 = 16$ teeth) and steered by a servomotor (Fig. 2). The wheels are a diameter of 76 mm. The structural diagram and a photo of the wheel module are presented in figure 2 b, c [4].

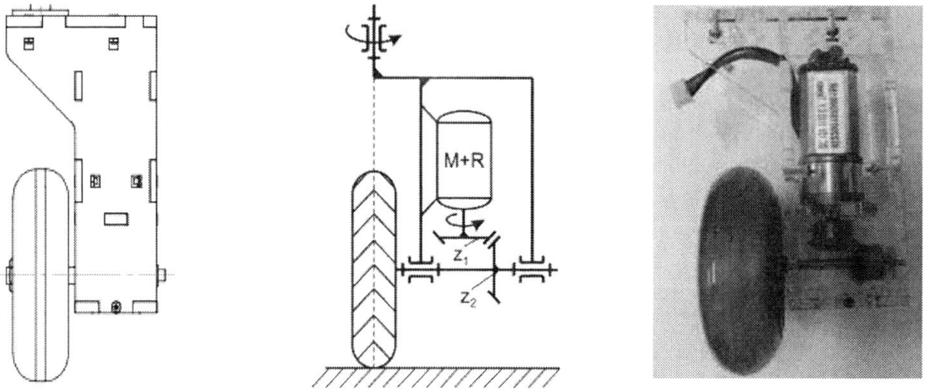

(a) (b) (c)

Figure 2. The wheel module a) 3D model a) structural diagram b) functional subassembly [4]

3. Kinematics analysis of the robot

The kinematic analysis of the robot will be presented in two instances:

- when the robot has four driving and steering wheels
- when the robot has a synchro-drive configuration

The kinematic analysis presented next is for the instance when all the wheels of the robot are both driving and steering.

The proposed kinematic model is based on the geometrical constraints of the wheels. Thus, we are defining two coordinate systems $X_IO_IY_I$ - the reference coordinate system and $X_RO_RY_R$ - the coordinate system attached to the robot's center of gravity (Fig. 3 a).

In this case the methodology proposed in [5], [6], [7] can be used.

Thus, considering the speed of the robot's platform $\dot{\xi}_I$ in the reference coordinate system $X_IO_IY_I$, it is possible to determine the speed $\dot{\xi}_R$ in the coordinate system that is attached to the robot $X_RO_RY_R$ by using the equations [5], [6]:

$$\dot{\xi}_R = R(\theta)\dot{\xi}_I \tag{1}$$

with

$$\dot{\xi}_I = \begin{bmatrix} \dot{x} & \dot{y} & \dot{\theta} \end{bmatrix}^T \tag{2}$$

$$R(\theta) = \begin{bmatrix} \cos\theta & \sin\theta & 0 \\ -\sin\theta & \cos\theta & 0 \\ 0 & 0 & 1 \end{bmatrix} \tag{3}$$

R(θ) - is the orthogonal rotation matrix expressing the orientation of the robot with respect to the base frame $X_IO_IY_I$.

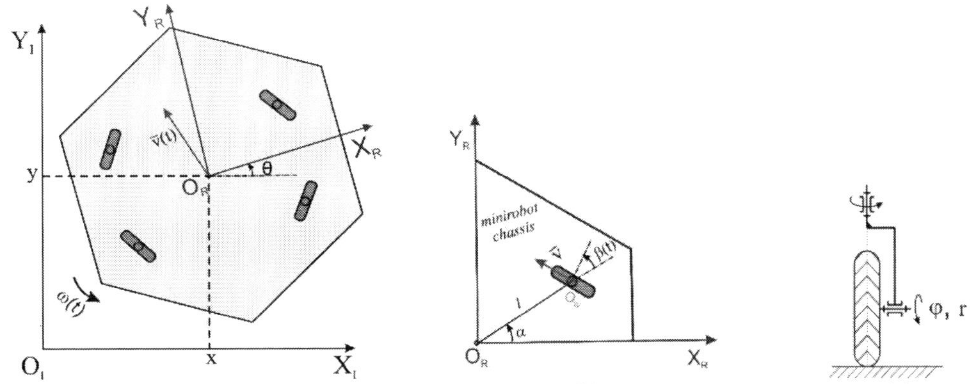

Figure 3. Setup of the coordinate systems
a) and wheel parameters for the robot b) c)

A steered standard wheel connected to the chassis and its parameters is shown in figure 3.b. The position of the wheel in the local reference frame is given in polar coordinates by the distance l and the angle α. The orientation of the plane of the wheel to with respect to l is represented by the angle β. The rotation angle of the wheel about its (horizontal) axle is denoted by φ and the radius of the wheel by r.

With $\dot{\varphi}(t)$ is noted the rotation velocity of the wheel around their horizontal axles, and $v = r\,\dot{\varphi}(t)$ the linear velocity of the wheel

The wheel can spin over time, and is steerable, so β and φ are function of time, β = β(t), φ = φ(t).

The robot kinematics constraint can be obtained from the rolling and sliding constraint of wheels caused from the no-slip condition [5], [6], [7].

The rolling constraint for the wheel is given of the equation [5], [6], [7]:

$$[\sin(\alpha+\beta) \quad -\cos(\alpha+\beta) \quad -l\cos\beta]\,R(\theta)\dot{\xi} - r\dot{\varphi} = 0 \tag{4}$$

The sliding constraint for the wheel is given of the equation [5], [6], [7]:

$$[\cos(\alpha+\beta) \quad \sin(\alpha+\beta) \quad l\sin\beta]\,R(\theta)\dot{\xi}_I = 0 \tag{5}$$

The rolling and sliding constraints, equations (4), (5) are written down for the four wheels of the robot. From equation (5) we get:

$$\begin{cases} \cos(\alpha_1+\beta_1) \quad \sin(\alpha_1+\beta_1) \quad l\sin\beta_1]\,R(\theta)\dot{\xi}_I = 0 \\ \cos(\alpha_2+\beta_2) \quad \sin(\alpha_2+\beta_2) \quad l\sin\beta_2]\,R(\theta)\dot{\xi}_I = 0 \\ \cos(\alpha_3+\beta_3) \quad \sin(\alpha_3+\beta_3) \quad l\sin\beta_3]\,R(\theta)\dot{\xi}_I = 0 \\ \cos(\alpha_4+\beta_4) \quad \sin(\alpha_4+\beta_4) \quad l\sin\beta_4]\,R(\theta)\dot{\xi}_I = 0 \end{cases} \tag{6}$$

This equation can be written as:

$$C(\beta_1,\beta_2,\beta_3,\beta_4)R(\theta)\dot{\xi}_I = 0 \tag{7}$$

where the matrix C (4x3) depends on the steering angles $\beta_i(t),\ i = 1, 2 \ldots 4$.

The mobility degree of the robot δ_m can be determined from equation (7) by knowing the number of independent constraints [5], [6], [7], [8].

$$\delta_m = 3 - rank(C)\,. \tag{8}$$

Therefore, the two following situations can be identified:

- If the axes of the four wheels of the robot do not intersect in the same point, all the kinematic constraints are independent and rank (C) = 3, thus $\delta_m = 0$. In this situation the robot's chassis is not moving.
- If the axes of the four wheels are intersecting in the same point, three or four constraints become linear dependent, thus rank (C) = 2 which results in $\delta_m = 1$.The robot can rotate around the intersection point of the four wheel axes, point that is called instantaneous center of rotation (CIR). If the axes of the four wheels are parallel (the rotation center is infinite) the robot can move in a straight line.
- The kinematic analysis detailed below applies to the case when the platform of the robot is not being turned during travel, acting like a synchro-drive robot.In this situation, for the robot to function properly, it is needed that the wheels are well aligned and their speeds need to be identical. Also, all the wheels need to turn synchronously with the same angular displacement.In figure 4 a, $X_W O_W Y_W$ is the coordinate system attached to the robot's wheels, $X_R O_R Y_R$ is the coordinate system attached to the robot's center of gravity and $X_I O_I Y_I$ is the main coordinate system. In the following kinematic equations it is assumed that the robot is travelling on a plane surface, without slipping.

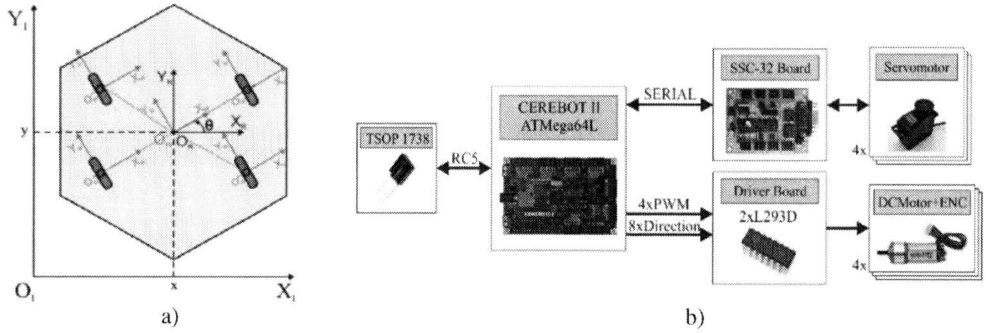

Figure 4. Setup of coordinate systems on the robot and on its wheels for the synchro-drive configuration a) and the block diagram of the electronic circuit b)

The kinematic analysis for this case is represented by the equations presented next.

3.1. Forward Kinematics

Next, (x, y) represent the coordinates of the robot's platform center of gravity related to the $X_I O_I Y_I$ coordinate system and θ represents the orientation angle of the robot or the orientation of the robot's wheels related to an axis which is parallel to $O_I X_I$. Also, v and ω are the linear and angular speeds of the robot.

The evolution of x, y and θ is given by the equations [9]:

$$\dot{p} = J(\theta)\dot{q} \tag{9}$$

$$\begin{bmatrix} \dot{x} \\ \dot{y} \\ \dot{\theta} \end{bmatrix} = \begin{bmatrix} \cos\theta & 0 \\ \sin\theta & 0 \\ 0 & 1 \end{bmatrix} \begin{bmatrix} v \\ \varpi \end{bmatrix}. \tag{10}$$

Equations (10) describe the movement of the robot on a plane surface in relation to the two inputs v and ω.

Thus, if we wish for the robot to move in a straight line, we can assign a linear speed v and setting the angular speed ω to zero. If the angular speed ω is set to a value different than zero, the robot will then travel on a curved trajectory. The travel speed of the robot v varies depending on the rotation speed of the drive motor and it can be determined using the equation:

$$v = \omega_R r = \frac{\pi n_m}{30}\frac{z_1}{z_2} r \tag{11}$$

where:
ω_R is the angular speed of the wheels
r is the radius of the wheels,
n_m is the rotation speed of the motor Me,
n_R is the rotation speed of the wheels and it is given by the equation:

$$n_R = n_m \frac{z_1}{z_2} \tag{12}$$

By using equation (10), the position and orientation of the robot can be determined [9]:

$$p = \int_0^t \dot{p}\, dt = \begin{bmatrix} x_0 \\ y_0 \\ \theta_0 \end{bmatrix} + \begin{bmatrix} \int_0^t v \cos\theta\, dt \\ \int_0^t v \sin\theta\, dt \\ \int_0^t \omega\, dt \end{bmatrix} \tag{13}$$

$$x = x_0 + \int_0^t v \cos\theta\, dt$$
$$\ldots y = y_0 + \int_0^t v \sin\theta\, dt \tag{14}$$
$$\ldots \theta = \theta_0 + \int_0^t \omega\, dt$$

where x_0, y_0, z_0 is the initial positions of the robot.
With the help of this equation, the trajectory of the mobile robot can be estimated.

Inverse Kinematics

By using equation (10), the solution $\begin{bmatrix} v \\ \omega \end{bmatrix}$ could be easlily found by determining the inverse matrix of the Jacobian J(q), if J (q) would be a square matrix. Due to the fact that J (q) is not a square matrix, $\begin{bmatrix} v \\ \omega \end{bmatrix}$ can be found by using the pseudo-inverse matrix $[J(q)]^{\dagger}$ of J (q), thus obtaining [9]:

$$\dot{q} = [J(q)]^{\dagger}\, \dot{p} = \{[J(q)^T]J(q)\}^{-1}[J(q)]^T\, \dot{p} \tag{15}$$

or

$$\begin{bmatrix} v \\ \omega \end{bmatrix} = \begin{bmatrix} \cos\theta & \sin\theta & 0 \\ 0 & 0 & 1 \end{bmatrix} \begin{bmatrix} \dot{x} \\ \dot{y} \\ \dot{\theta} \end{bmatrix} \tag{16}$$

The proposed method is simple, easy to use, can be easily implemented and is highly efficient when the robot travels onto flat surfaces.

4. The control of the robot

The robot is fitted with two microcontroller boards: Cerebot II [10] and SSC-32 [11] which communicate through a serial connection. The robot is controlled by the CEREBOT II board that is fitted with an Atmega64L microcontroller. The peripheral modules L293D and PmodHB5 are also used. The SSC-32 board, which is fitted with a AVR RISC structured ATMega8 microcontroller, generates the command impulses for the four servomotors. The block diagram of the electronic circuit is presented in figure 4b. The servomotors used are hi-tech HS-475HB and the geared drive motors are IG22 [12].

5. The testing of the robot

The robot only travels on a plane surface. During the testing process under laboratory conditions, the robot had a good performance. A few images from the testing process are shown in figure 5.

t = 0 s t = 2 s t = 6 s

Figure. 5 Images from robot testing

The robot is not energetically autonomous, being powered trough wire from an external power source. The testing was done also for a synchro-drive configuration. After testing it was concluded that the main disadvantage of this robot is the large number of motors that have to be driven and controlled. Another important disadvantage is the lack of a suspension system which recommends the robot for use only on flat, even surfaces. The authors contributed to the design of the robot's modular structure.

6. Conclusions

The paper presents the design of an omnidirectional robot with four drive and steering wheels. For the robot to travel properly, the four wheels need to be synchronously driven and steered. This synchronisation is achieved by controlling the motors and servomotors from the robot's structure.

Future research: Next, the authors want to study the dynamics of the robot, to fit the robot with sensors and to use batteries for powering the robot in order to increase its autonomy. Improving the mechanical structure of this particular robot, in order for it to be able to travel over small obstacles, proves to be a challenge.

7. References

[1] Tătar M O, Haiduc F, Szalontai A, Pop A 2016 Design and Development of the_Synchronous Mobile Robots *The 2016 IEEE International Conference on Automation, Quality and Testing, Robotics, AQTR 2016 - THETA 20* pp 197-202

[2] Hua F, Li G, Liu F, Liu Y 2016 Mechanical design of a four-wheel independent drive and steering mobile robot platform *The ICIEA 2016 - 11th IEEE Conference on Industrial Electronics and Applications* pp 235-238

[3] Grimstad L, From P J 2017 Thorvald ii-a modular and re-configurable agricultural robot. *IFAC-PapersOnLine* 50(1) pp 4588-4593

[4] Tătar M O, Cirebea C I 2017 Modular reconfigurable robots, *The 12th IFToMM International Symposium on Science of Mechanisms and Machines SYROM'2017*

[5] Campion G, Bastin G, D'Andréa-Novel B, 1996 Structural properties and classification of kinematic and dynamic models of wheeled mobile robots, *IEEE Transactions on Robotics and Automation* (1) pp 47-62

[6] Siegwart R, Nourbackhsh I, D Scaramuzza 2011 Introduction to Autonomous Mobile Robots, (Massachsetts: The MIT Press)

[7] Campion G, Chung W 2008 Cap 17 Wheeled robot in Handbook of Robotics Siciliano B, Khatib O Eds (Berlin: Springer-Verlag)

[8] Lauria M, Nadeau I, Lepage P, Morin Y, Giguère P, Gagnon F, Létourneau D, Michaud F 2006 Design and control of a four steered wheeled mobile robot. In *IECON 2006 - 32nd Annual Conference on IEEE Industrial Electronics* pp 4020-4025

[9] Zhao Y, Bement SL, 1992 Kinematics, dynamics and control of wheeled mobile robots. *IEEE International Conference on Robotics and Automation* pp 91-96

[10] http://store.digilentinc.com/

[11] http://www.lynxmotion.com

[12] http://store.digilentinc.com/

IOP Conf. Series: Materials Science and Engineering **514** (2019) 012020 doi:10.1088/1757-899X/514/1/012020

Virtual keyboard based on a brain-computer interface

O A Rusanu[1], L Cristea[1], M C Luculescu[1], P A Cotfas[2] and D T Cotfas[2]

[1] Product Design, Mechatronics and Environment Department, *Transilvania* University of Brașov, Brașov, Romania

[2] Electronics and Computers Department, *Transilvania* University of Brașov, Brașov, Romania

E-mail: oana.rusanu@unitbv.ro

Abstract. This paper proposes a novel LabVIEW based algorithm regarding the implementation of an eye-blinks strength controlled virtual keyboard aimed to help people with neuromotor disabilities. The application can be considered a brain-computer interface which is useful for the assistance of patients suffering from Locked-In syndrome or amyotrophic lateral sclerosis. A precise, repeatable and easily detectable control signal could be the eye-blink strength. The LabVIEW based virtual keyboard provides the following features: detection and counting the voluntary eye-blinks, switch command, select command, highlight action running in parallel with the previously mentioned processes, enabling cancel, delete and space commands. The working principle underlying the implementation of the proposed virtual keyboard is Divide and Conquer paradigm. Thus, it is possible the switching task across the rows, half rows and keys (characters) in order to select a row, half row, respectively key associated to the character that should be introduced in the text box.

1. Introduction

This paper proposes an application aimed for the assistance of people with disabilities. Unfortunately, the patients who suffered a brain stroke, spinal cord injuries or were diagnosed with Locked-In syndrome or amyotrophic lateral sclerosis, cope with severe speaking impairments and neuromotor diseases. Some significant examples could be the scientist Stephen Hawking and the journalist Jean Dominique Bauby. They lost their ability to communicate in a natural way (by spoken words or talking) and to interact with the outside environment via peripheral nerves and muscles, similar to the condition of healthy people.

The only available solution that has the potential to help them recover their lost independence and autonomy of doing simple and usual daily activities, without permanent support offered by a caregiver, is the development of an embedded system based on a brain-computer interface.

The brain-computer interface provides an alternative channel of communication and control. This can be achieved via advanced processing methods applied to the biopotentials recorded across the cerebral neurons using the electroencephalography technique. Thus, the mental activity detected during the execution of cognitive tasks (for instance: motor imagery based on the movement of a leg or an arm in the forward or backward direction, mental calculus solving, watching a flashing light stimulus) generates certain cerebral patterns which are firstly undergone to advanced techniques of artificial intelligence. Then, they are classified and translated into commands and signals aimed to control different biomechatronic devices: a motorized wheelchair, a robotic arm, a neuroprosthesis and the cursor position on the PC desktop.

The eye-blinking is an artefact of the electroencephalographic signal so that it can be easily detectable and it is characterized by an increase followed by a decrease in the amplitude of the acquired biopotential. The intentional or voluntary eye-blinking could be considered a precise, repeatable and easy to quantify signal so that it proves its significance in the control application.

This paper proposes a virtual keyboard controlled by eye-blinking (Figure 1). The keyboard helps people with disabilities to write or type words. Taking into account the fact that it provides an unusual and interactive method of human-machine interaction, the virtual keyboard could also raise the interest of those who are passionate about futuristic technology.

According to the scientific literature, the current variants of virtual keyboard aimed to help people with disabilities are based on the following working principles: the selection of characters caused by triggering the SSVEP (Steady State Visual Evoked Potential) [1] and P300 signals [2], eye-tracking based on image processing methods [3], electrooculography and other hybrid techniques [4].

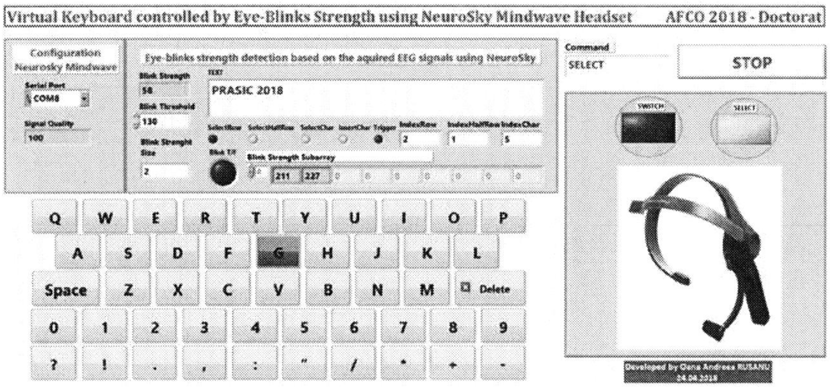

Figure 1. Front Panel (user-interface) of the LabVIEW based virtual keyboard controlled by eye-blinks strength via EEG signal acquired using NeuroSky Mindwave headset.

A rarely used method is related to providing only one control signal, respectively the amplitude or strength of the eye-blink detected via EEG signal analysis. Moreover, the application proposed in this paper is different from the previous research work [5][6][7][8] because it does not involve a given time interval in order to constrain the user to execute eye-blinking only during this time window. The currently existing applications were usually implemented using procedural programming languages, such as: C, C++, C#, Java, .net or Matlab development environment.

The software application proposed in this paper was developed in LabVIEW graphical programming environment because it provides a friendly user interface so that its progress could be improved via an interactive feedback. Another important benefit offered by LabVIEW is the efficient programming method, similar to the natural way of thinking. Therefore, it facilitates the implementation of the working principle underlying virtual keyboard design, which was based on the Divide and Conquer algorithm. It was also taken into consideration the state-machine paradigm, which is a decision-based algorithm. Clearly defined conditions should be fulfilled in order to enable the transition across the states in a proper manner.

2. Hardware system

The hardware system used for the proposed application is the NeuroSky Mindwave headset [9][10], which has an embedded sensor that should be placed on the user's forehead to the frontal cerebral lobe called FP1. The reference necessary to close the electrical circuit is a clip that should be attached to the ear's lobe. NeuroSky headset is based on the ThinkGear chip, which comprises different algorithms for the electroencephalographic signal processing in order to provide the user the possibility to read certain values corresponding to the following parameters: attention level, meditation level and the strength or the amplitude of the eye-blinking. Another feature is related to acquiring of raw EEG signal, which can be classified in different rhythms, such as: delta (0-4 Hz), theta (4-8 Hz), alpha (8-12 Hz), beta (12-30 Hz) and gamma (30-50 Hz). The sampling frequency of the NeuroSky headset is equal to 512 Hz. National Instruments provides developers with a toolkit created for NeuroSky [11][12][12], which could be used to enable the functionality of ThinkGear chip.

3. Software system

Regarding the software system proposed in this paper, both the working principle and the way of thinking underlying the implementation of eye-blinks controlled virtual keyboard are based on Divide and Conquer paradigm. This approach was also the starting point for designing similar virtual keyboards based on a brain-computer interface [5][6]. According to the previous scientific literature [7][8], LabVIEW has been rarely used to create a logical algorithm aimed for these types of applications. Therefore, it should be emphasized that the novel LabVIEW based solution proposed in this section could be used by students as valuable material to reinforce the knowledge taught by their professors.

3.1. A general overview of the available features of the virtual keyboard

The algorithm was based on the state-machine (Table 1) paradigm providing the following features:

- Counting the total number of intentional (voluntary) eye-blinks. An intentional eye-blink is detected when its strength or amplitude is characterized by a value higher than the initially set threshold parameter;
- Implementation of an instruction set aimed for enabling the SWITCH command across rows, half rows or characters (keys) when a single voluntary eye-blink was executed;
  Therefore, Divide and Conquer principle enabled the following transitions:
  i) SWITCH across the rows of keys;
  ii) SWITCH across the two parts (first and second half) of every row of keys;
  iii) SWITCH across the characters or keys placed on the two parts (first and second half) of every row of keys.
- Implementation of an instruction set aimed for enabling the SELECT command in order to choose a row, a half row or a character (key) when two voluntary eye-blinks were performed
  Therefore, Divide and Conquer principle allowed the following tasks:
  i) The selection/activation of a single row of keys;
  ii) The selection/activation of a certain part (first or second half) of every row of keys;
  iii) Inserting in the text box a specific character from either the first or the second half row.
- Implementation of an instruction set aimed for enabling the HIGHLIGHT action necessary to point out the current row, half row or key (character) followed by the typing/inserting the corresponding letter in the text box;
- Enabling CANCEL command associated with the execution of three voluntary eye-blinks. The boolean variables would be reinitialized so that the transition across rows is enabled again. Further, it is possible to select a specific row of keys.
- Enabling DELETE command triggered by recording of four voluntary eye-blinks;
- Enabling SPACE command determined by the execution of five voluntary eye-blinks.

3.2. A general overview on the LabVIEW based algorithm underlying the virtual keyboard design

In the proposed algorithm, there were defined five Boolean variables. They were initialized to the following values: SelectRow ← True; SelectHalfRow ← False; SelectChar ← False; InsertChar ← False; Trigger ← False. The meaning of these assignments is described in the Table 2.

There were also defined a series of numeric variables in order to increment the indexes associated to rows, half rows and characters (keys). There were used the below notations:

IndexRow = {1, 2, 3, 4, 5}. It can be assigned with integer values included in the interval [1, 5].

IndexHalfRow = {1, 2}. It can be assigned with either value 1 or value 2.

IndexChar = {1, 2, 3, 4, 5}. The maximum value is 3, 4 or 5. It depends on how many characters are placed on each part (first and second half) of every row of keys.

Table 1. The next running state based on the assessment of the main condition – Blink Strength > Threshold Parameter of the 'Count Eye-Blinks' algorithm.

Current State	Blink Strength > Threshold Parameter	
	TRUE	FALSE
INIT	SWITCH <One eye-blink detected> Switch Command <ArraySize ←1>	READY No Command
SWITCH	SELECT <Two eye-blinks detected> Select Command <ArraySize ←2>	READY Switch Command
SELECT	CANCEL <Three eye-blinks detected> Cancel Command <ArraySize ←3>	READY Select Command
CANCEL	DEL <Four eye-blinks detected> Delete Command <ArraySize ← 4>	READY Cancel Command
DEL	SPACE <Five eye-blinks detected> Space Command <ArraySize ← 5>	READY Delete Command

Table 2. The meaning of the opposite states (true or false) of the Boolean variables.

Variable	Value – Description of the variable state	
	True	False
SelectRow	Transition across rows is enabled.	Transition across rows is disabled.
SelectHalfRow	Transition across half rows is disabled.	Transition across half rows is enabled.
SelectChar	Transition across characters is disabled.	Transition across characters is enabled. It is also a condition necessary when someone intends to repeatedly introduce a specific character in the text box.
InsertChar	The sequence responsible for typing a character in the text box is enabled.	The sequence responsible for typing a character in the text box is disabled.
Trigger	It indicates that a character was already introduced in the text box during the last action. It is necessary in order to avoid the continuous typing of a character.	It proves that no character was previously introduced in the text box during the last action.

The transitions between rows of keys, half rows, respectively all characters included by each part of every row are indicated by highlighting those elements with blue colour.

If SelectHalfRow==True, the transition between the two parts (first and second half) of each row of keys is inactive (it is not available at the current moment of time). The further step consists of checking the SelectChar variable. Another Case Structure with two states was used in order to fulfil the instruction set for the two conditions: SelectChar == True and SelectChar == False.

If SelectChar == True, then the next step consists in checking the InsertChar variable. Both possibilities (InsertChar == True and InsertChar == False) were considered by using another Case Structure with two states. A character is introduced in the text box when both two voluntary eye-blinks were executed and the following conditions were performed: SelectRow ← False; SelectHalfRow ← True; SelectChar ← True; InsertChar ← True. The next step consists in finding the position of the character via numeric variables: IndexRow (five states), IndexHalfRow (two states), IndexChar (three / four / five states) – Table 3.

If SelectChar == False, then it is available the transition across the keys/buttons of each part (first and second half) of every row. For the case SelectChar == False, it was used the previously presented sequence of imbricated Case Structures, taking into account that currently the active key or highlighted character is coloured with blue. There are also used the Property Nodes. Variables are attached to them in order to choose the suitable colour: blue (active) or grey (inactive).

If SelectHalfRow == False, then it is available the transition across the two parts (first and second half) of each row of keys/characters (Table 3).

It is important to mention that the programming sequence aimed to highlight with blue colour the rows, half rows, each key and respectively, typing a specific character in the text box, is running in parallel with the alternative processes corresponding to recording a single voluntary eye-blink (associated with switch command) and performing two intentional eye-blinks (associated with select command for choosing the previously stated elements).

Table 3. The character introduced in the text box according to the values of Boolean variables.

IndexRow	IndexHalfRow	Visual highlighting	IndexChar				
			1	2	3	4	5
1	1		Q	W	E	R	T
	2		Y	U	I	O	P
2	1		A	S	D	F	G
	2		H	J	K	L	
3	1		Z	X	C	V	
	2		B	N	M		
4	1		0	1	2	3	4
	2		5	6	7	8	9
5	1		?	!	.	,	:
	2		"	/	*	+	-

After a character was introduced in the text box, there are two possibilities: either one eye-blink is detected or two eye-blinks are executed.

Firstly, if one eye-blink was detected, then the currently active/enabled row remains highlighted/selected. Further, if another one intentional eye-blink is recorded, then it is enabled the transition to the next row of keys. The two instructions: Trigger ← False and SelectRow ← True show that the transition across rows of keys is available again. Otherwise, if two intentional eye-blinks were performed, then it is enabled the transition to the next character. If another two voluntary eye-blinks are executed, it is possible to introduce the last active character in the text box.

Secondly, if two intentional eye-blinks were detected, then the same character remains active and it is typed in the text box one more time. Still, if another two voluntary eye-blinks are detected, then the same character is repeatedly introduced in the text box.

4. Conclusion

This paper proposed a LabVIEW application that is aimed for helping people with disabilities by providing them with an efficient method of typing via a virtual keyboard, which can be controlled using the eye-blinks. Its working principle is based on Divide and Conquer paradigm.

One eye-blink is associated to switch command and it is related to the transition (alternative highlighting) across rows, half rows or keys (characters).

Two eye-blinks are associated to select command and they are related to the alternative selection of previously mentioned elements and finally, it is possible to type a character in the text box.

Regarding the novelty of this paper, by comparing with the previous research work [6], it is important to notice that the user is not constrained by a given time interval. He only needs to execute the eye-blinks with a strength strong enough to exceed the threshold parameter that was initially set. Thanks to the attractiveness of LabVIEW environment, the graphical user-interface is friendly and it provides an intuitive feedback. Some of the useful features are related to the possibility of repeatable insertion of the same character and the advantages offered by delete and space keys.

Otherwise, the LabVIEW based virtual keyboard is associated with various applications, taking into account that it provides a simple to use and easy to learn working principle. Moreover, it can even raise the interest of healthy people.

Future research efforts will be focused on the development of an equivalent mobile virtual keyboard controlled by eye-blinks strength, which should be available on smartphones. Further, the aim of the mobile application is instant messages sending. In addition, a mechanism based on words prediction should also be implemented. Then, thorough experiments are required in order to prove or validate the usefulness of the brain-controlled virtual keyboard. Moreover, it is necessary to figure out to what extent the accuracy, precision and transfer rate of information should be improved.

5. References

[1] Gembler F, Stawicki P and Volosyak I 2017 Suitable Number of Visual Stimuli for SSVEP-Based BCI Spelling Applications *Proc. of the 2017 International Work-Conf. on Artificial Neural Networks Advances in Computational Intelligence IWANN 2017* pp 441 – 452

[2] Martínez-Cagigal, Santamaría E and Hornero R 2018 Controlling a Smartphone with Brain-Computer Interfaces: A Preliminary Study *Proceedings of the 2018 International Conference on Articulated Motion and Deformable Objects AMDO 2018* pp 34-43

[3] Garcia L, Ron-Angevin R, Loubière B, Renault L, Masson G L, Lespinet-Najib and André J M 2017 A Comparison of a Brain-Computer Interface and an Eye Tracker: Is There a More Appropriate Technology for Controlling a Virtual Keyboard in an ALS Patient? *Proceedings of the 2017 International Work-Conference on Artificial Neural Networks Advances in Computational Intelligence IWANN 2017* pp 464 – 473

[4] Saif M B and Felzer T 2016 Replacement of the Standard Computer Keyboard and Mouse by Eye Blinks *Proceedings of the 2016 International Conference on Computers Helping People with Special Needs ICCHP 2016* pp 457 – 463

[5] López A, Ferrero F, Yangüela D, Álvarez C and Postolache O 2017 Development of a Computer Writing System Based on EOG in the *Sensors* **17** pp 1 - 20

[6] Dobosz K and Stawski K 2017 Touchless Virtual Keyboard Controlled by Eye Blinking and EEG Signals *Proceedings of the 2017 International Conference on Man-Machine Interactions 5 ICMMI 2017* pp 52-61

[7] Rakesh R and Sasidharan P T 2016 Utilization of EEG Signal as Virtual Keyboard for Physically Disabled in the *International Journal of Advanced Research in Electronics and Communication Engineering (IJARECE)* **5** pp 1486 - 1491

[8] Chamayil B, Singla R and Jha R 2010 Virtual keyboard BCI using Eye blinks in EEG *Proceedings of the 2010* IEEE 6th International Conference on Wireless and Mobile Computing, Networking and Communications *2010* pp 466-470

[9] NeuroSky Store. Mindwave Mobile. Brainwave Sensing Headset, https://store.neurosky.com/pages/mindwave, available on 26.07.2018

[10] NeuroSky. EEG Biosensors., http://neurosky.com/biosensors/eeg-sensor/, available on 27.07.2018

[11] NeuroSky LabVIEW Drivers, , https://store.neurosky.com/products/labview-drivers, available on 27.07.2018

[12] NeuroSky LabVIEW Driver, https://forums.ni.com/t5/NI-Labs-Toolkits/NeuroSky-LabVIEW-Driver/ta-p/3520085, available on 27.07.2018

Virtual robot arm controlled by hand gestures via Leap Motion Sensor

O A Rusanu[1], L Cristea[1], M C Luculescu[1], P A Cotfas[2] and D T Cotfas[2]

[1] Product Design, Mechatronics and Environment Department, *Transilvania* University of Brașov, Brașov, Romania
[2] Electronics and Computers Department, *Transilvania* University of Brașov, Brașov, Romania

E-mail: oana.rusanu@unitbv.ro

Abstract. This paper proposes a LabVIEW instrument aimed to design and control a 3D virtual robot arm using hand gestures translated into commands with the help of Leap Motion Controller. In order to acquire signals and translate them into basic swipe gesture commands, a LabVIEW toolkit was used. The application is useful for providing assistance for people with disabilities who cannot move or lack strong muscle strength. They can use the virtual simulation for training purpose in order to achieve the ability to control a real robot arm. Moreover, they have a visual feedback that sustains and adjusts their movements. The 3D model is a robot arm placed on a flat surface. Translation along two perpendicular directions in the base plan is allowed and also the rotation of its joints. The virtual robot arm can provide several benefits regarding moving to a target, grasping it and bringing it to the patient.

1. Introduction

A robotic arm proves its major usefulness in assisting people with disabilities during their daily life activities, such as: eating, drinking, entertainment, household and self-care tasks. Thus, a mobile telethesis helps these patients by offering them the necessary independence, safety, comfort and self-management in different situations: grasping, raising and bringing some things of various dimensions and weights that are either placed far away or are hard to reach. A major contribution in this field was brought by the research work presented in the scientific paper [1]. It should be highlighted that [1] was the first published paper which proposed the use of the Leap Motion Device for the gestures based control of a real anthropomorphic gripper. More information on the importance of a robotic arm controlled by gestures can be found in the following resources: [2][3][4][5].

This paper presents a LabVIEW instrument which consists of a 3D virtual robotic arm. It is characterized by a natural interaction with the human operator. This means that it can be controlled by natural hand gestures. From a structural-functional point of view, this mechatronic system is considered a mobile telethesis which allows the movement in two perpendicular directions both along x and y-axes.

The virtual simulation developed in NI LabVIEW graphical programming environment can be used for the following purposes:

- testing the working principle underlying a robotic arm;
- remote control of a robotic arm via a web-based application;
- training the patients with disabilities in order to improve their accuracy, precision and effectiveness in hand gestures control applications.

2. Hardware system

The application proposed in this paper is based on using the Leap Motion Sensor (Figure 1) aimed to acquire raw signals corresponding to some hand gestures in order to translate them into certain commands to the 3D robot arm. Leap Motion Controller [6] is a low-cost and portable device allowing tracking and recording the position, orientation and speed of 3D hands and fingers. It is composed of two cameras and three infrared LEDs. These components are necessary for tracking the infrared light characterized by a wavelength of 850 nanometers meaning it is outside the visible light spectrum. Leap Motion embedded device comprises wide angle lenses which provide a large interaction space with a shape similar to an inverted pyramid [7]. The viewing range across the intersection of the binocular cameras' fields of view is estimated to 80 centimeters. This range is influenced by the propagation and intensity of LED light. Firstly, the sensor data is acquired into the local memory of the device's USB controller for performing certain resolution adjustments. Secondly, raw data is streamed via USB to be analysed by the Leap Motion tracking software. Raw data is related to grayscale stereo images of the near-infrared light spectrum, which is divided into the left and right cameras [7].

The most important part of the Leap Motion Sensor is the specialized software based on advanced algorithms applied to the raw sensor data in order to process the acquired images, compensate for background objects and ambient environmental lighting and reconstruct a 3D representation of the viewed sight. The smooth temporal coherence of the acquired data is achieved thanks to using filtering techniques [7].

MakerHub provides LabVIEW developers with an intuitive library of functions aimed to access the entire functionality of Leap Motion Controller [8]. Thus, by using the LabVIEW based toolkit, it is possible to access the position and velocity of points and hands as well as to read the following gestures: swipe, screen tap, key tap and circles. Moreover, it facilitates reading of raw frame data related to rolling, pitch, yaw, velocity and position for every point, hand or gesture in the frame.

3. Software system

3.1. Design of the 3D interactive model based on the kinematic structure of a robotic arm

This challenge was solved by taking advantage of the NI LabVIEW graphical programming environment that provides the '3D Picture Control' Functions Palette necessary to design and control certain complex 3D models. Accordingly, the virtual instrument developed in LabVIEW includes:

- the parameters used for the building of 3D geometric shapes;
- wired linking or suitable correspondence between introduced components so that they can be correctly positioned and properly oriented;
- the necessary transformations for the translation and rotation movements.

3.2. Control of the 3D interactive model based on the working principle of a robotic arm

The natural interaction between the robotic arm and human operator is based on the intuitive hand gestures identified via Leap Motion Sensor.

Regarding accessing the functionality provided by Leap Motion Sensor, it was used the free LabVIEW based Toolkit developed by MakerHub [8]. There were called the two Leap based functions: *Open Connection* and *Read One Swipe Gesture Direction* (Up, Down, Right, Left, In, Out). A subsequent stage is related to some calculus in order to map the input values received from the Leap Motion Sensor to the output values sent to the *Rotate Object* for controlling or animating the 3D robot arm.

These numerical values were set after experimenting with different examples in order to get a synchronicity between the pattern of hand gestures and the visual command executed by the robot arm. Thus, there were accomplished the following controlling sequences (Figure 3) related to swipe gestures versus position changing of the robot arm:

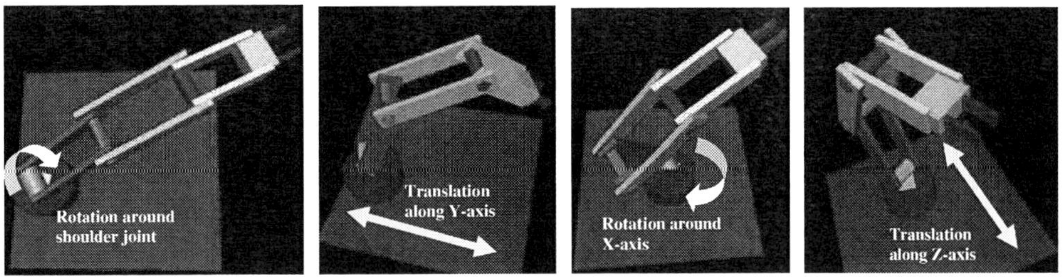

Figure 1. Different possibilities of controlling the 3D robotic arm designed in LabVIEW.

Thus, there were accomplished the following controlling sequences (Figure 3) related to swipe gestures versus position changing of the robot arm:

- Up and Down swipe gesture → move the 3D robot arm along the Z-axis (translation forwards and backwards along the virtual red coloured floor);
- Right and Left swipe gesture → move the 3D robot arm along the Y-axis (translation to the right and to the left margins of the virtual red coloured floor);
- In and Out swipe → rotate the shoulder of the 3D robot arm, respectively adjusting the angle value underlying the rotation process.

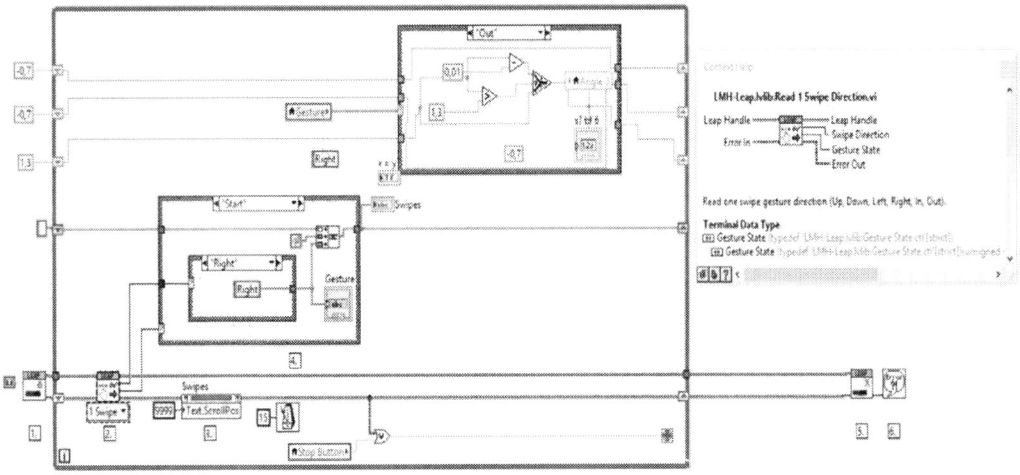

Figure 2. A LabVIEW programming sequence for the implementation of hand gestures based 3D robot arm control using Leap Motion Sensor.

According to the Figure 2, the 3D virtual robot arm is able to execute various movements so that it can easily change its position and orientation. Both design and control of the 3D structure are implemented using LabVIEW graphical programming environment.

3.3. LabVIEW based design and control of the 3D virtual robot arm

The Block Diagram is divided into the following three programming sequences (Figures 4 and 5):

- Creating the elements (geometric shapes) which are put together in order to build the 3D structure of the robot arm. There are used the following functions: Create Object, Create Cylinder, Create Box and Create Sphere. There are also invoked the following Property Nodes: Set Drawable and Add Object.
- Manual and automated control of the 3D virtual robot arm involved the following structures:

- A while loop which includes a Flat sequence structure composed by two state/frames. The first frame encompasses:
  - Some functions necessary to animating or controlling the 3D robot arm: Clear Transformations, Translate Object, Rotate Object and Scale Object;
  - A few while loops which allow the implementation of gesture-controlled 3D robot arm by using a Leap Motion sensor;
- Display the final result – the 3D scene which shows the animations of the 3D robot arm. This element is included by the second frame of the above mentioned Flat Sequence Structure.

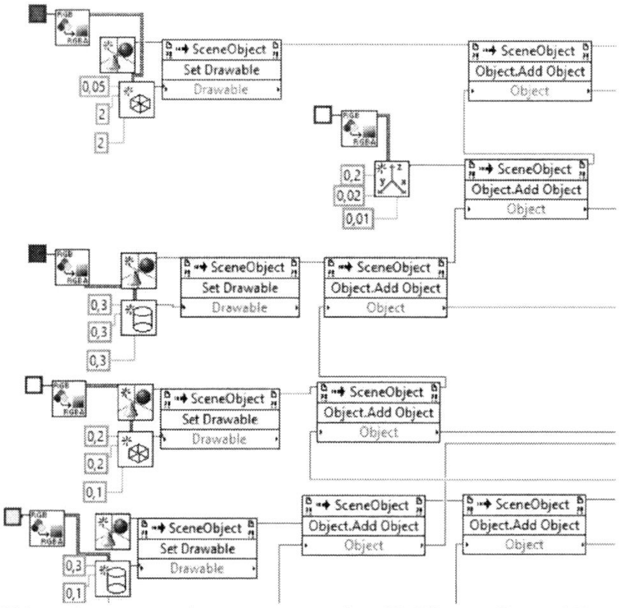

Figure 3. A LabVIEW based programming sequence using 3D Picture Control Functions and Property Nodes underlying the design of the virtual robot arm.

The programming structures and functions used in the Block Diagram are explained below.

3.3.1. Creating the elements (base, shoulder, elbow, wrist and gripper) included by the structure of the 3D robot arm. Using the *Create Object* function, it is possible to create a 3D object which will be displayed on the Front Panel. This function was necessary to create each element from the structure of the 3D robot arm. These elements could be identified by the terminal called *Name* which is a string variable. The elements have both circular and rectangular shapes.

By accessing the contextual menu of the Create Object function, the next stage consists in selecting the following features: *Create/ Method for SceneObject Class/ Set Drawable.*

Using the Drawable attribute, it was set the geometric shape (cylinder or rectangle) of the geometric elements which allowed the design of the 3D robot arm.

The *Create Cylinder* function was called in order to set the specific shape of a cylinder. There were assigned certain values corresponding to the following dimensions: detail, height and radius.

The *Create Box* function was called in order to set the specific shape of a rectangle. There were set some values according to the specific dimensions on each axis direction: X, Y and Z (height, length and width). There was also established the colour parameter for both functions: *Create Cylinder* and *Create Box*. The above mentioned values were set based on empirical experiments and continuous testing in order to get a realistic design and a robust control. The elements are also symmetrical.

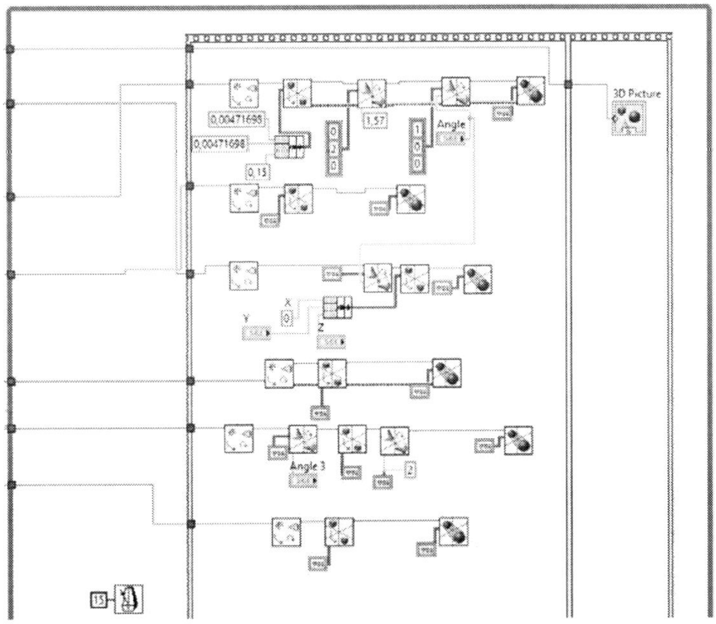

Figure 4 A LabVIEW based programming sequence using 3D Picture Control Functions enabling both manual and automated control of the virtual robot arm.

It is necessary in order to connect multiple elements to a single base element. For instance, on the virtual floor (red coloured thin rectangular area) it was placed the fundamental/base element of the robot arm (blue coloured cylinder). In a similar manner, the two green plates included by the *shoulder* joint are attached to the small green coloured cylindrical joint. The output of the Add Object Property Node was sent to the 3D Picture terminal/scene which allows the displaying of the 3D robot on the Front Panel.

An important function called *Create 3D Axis* can be also attached to the *Add Object* Property Node. An axis was defined in order to enable the rotation of the base element along with the other geometric shapes linked to it. In fact, using another Add Object Property Node, all the 3D structures are associated with that specific 3D Axis so that it is established the 3D space in the scene. The axis is characterized by the following values: length, radius and character size.

3.3.2. Manual and automated control of the 3D virtual robot arm. The manual control of the 3D robot arm is based on quickly or slowly adjusting the Knob buttons displayed on the Front Panel. The control of the 3D figure is related to running the animations shown by the 3D scene. Therefore, specific functions were used: *Clear Transformation, Translate Object* and *Scale Object*. The animations consist of the continuous movement of the elements (base, shoulder, elbow, wrist and gripper) included by the structure of the 3D robot arm. The orientation of the movement associated with these elements was achieved by calling the *Rotate Object* function. There were assigned certain values corresponding to the axis and the angle of rotation. The *Rotate Object* function plays an important role allowing the control and command of the 3D virtual figure due to setting a constant corresponding to the axis of rotation and an input numerical variable which receives specific values associated to the movement of an element (base, shoulder, elbow, etc.). These input values can be updated and modified while the simulation/application is running by adjusting the Button Knob which is attached to the *Rotate Object* function on the Block Diagram.

The input values can be introduced not only by this numerical control but also by the value sent by the Leap Motion Sensor according to the gesture executed. Therefore, the control linked to the *Rotate*

Object is substituted by another control which is used for transferring the values received as a result of performing some hand gestures.

4. Conclusion

To conclude, this paper proposed a LabVIEW based application aimed both to design the kinematic structure of a 3D virtual robot arm and to control it by using hand gestures acquired from the Leap Motion Sensor. The novelty of this virtual instrument lies in the following contributions: creating and controlling a LabVIEW based 3D model similar to a mobile telethesis and integrating the advanced functionality offered by the Leap Motion Device. Thus, it results an interactive simulation useful for achieving a training environment which provides people with motor disabilities the possibility of learning how to use hand gestures in order to control a realistic robot arm.

The Block Diagram is consisting of certain Functions and Property Nodes included by the `3D Picture` Palette necessary to firstly, create a 3D object (rectangle or cylinder) and secondly, add an object to the base in order to build the entire virtual robot arm. Further, there are also used a few functions necessary to manually or automatically control and animate the 3D model. There are used some geometrical transformations in order to translate or rotate objects across the initially set axis. The most significant feature of the LabVIEW based application proposed in this paper is related to the possibility of hand gestures based control. A programming sequence allows the process of acquiring raw data from Leap Motion Sensor and mapping them to the corresponding numerical values necessary to be transferred to the Translate Object and Rotate Object functions in order to enable the automated control of the 3D virtual robot arm. This application proves its usefulness by providing assistance to people with motor disabilities if the commands sent to the virtual model are associated with a real device. The 3D robot arm helps the disabled patients by offering them an attractive feedback during their training. Another utility is related to the remote applications of such an assistive device used by security staff during operating in dangerous media.

One of the future research directions is related to the development of a web-based instrument in order to transfer gesture-based commands across a long distance enabling the control of a real robot arm. Moreover, the research efforts are focused in the direction of extending the types of hand gestures necessary to control the virtual robot arm.

5. References

[1] Stareţu I and Moldovan C 2016 *Leap Motion Device Used to Control Real Anthropomorphic Gripper* in the International Journal of Advanced Robotic Systems (IJARS) 13 pp ...

[2] Gunawardane H, Madagedara N and Madhusanka A 2016 *Control of Robot Arm Based on Hand Gestures Using Leap Motion Sensor Technology* in the International Journal of Robotics and Mechatronics **3** pp …

[3] Sathiyanarayanan M, Azharuddin S, Kumar S and Khan G 2014 Gesture controlled robot for military purpose in the *International Journal For Technological Research In Engineering (IJTRE)* **1** pp 1300 – 1303

[4] Suresh A, Gaba D, Bhambri S and Laha D 2018 Intelligent Multi-fingered Dexterous Hand Using Virtual Reality (VR) and Robot Operating System (ROS) *Proceedings of the 2018 Robot Intelligence Technology and Applications 5. Advances in Intelligent Systems and Computing 2017* pp 459 – 474

[5] Chen M, Liu C and Du G 2018 A human–robot interface for mobile manipulator in the *Journal of Intelligent Service Robotics* **11** pp 269 - 278

[6] Leap Motion, https://www.leapmotion.com/, available on 05.05.2018

[7] How does the Leap Motion Controller work, http://blog.leapmotion.com/hardware-to-software-how-does-the-leap-motion-controller-work/, available on 05.08.2018

[8] LV MakerHub – LMC, https://www.labviewmakerhub.com/doku.php?id=libraries:leap:start, available on 05.05.2018

IOP Conf. Series: Materials Science and Engineering **514** (2019) 012022 doi:10.1088/1757-899X/514/1/012022

Modeling the mechanical and electrical characteristics of the synchronous motor

C Cristea[1] and I Stroe[1]

[1] Product Design, Mechatronics and Environment Department, Transilvania University of Braşov, Romania

E-mail: cosmin.cristea@unitbv.ro

Abstract. This paper presents the mechanical and electrical characteristics of the synchronous motor for variations of the required moment. It is presented the test configuration used to determine the mechanical and electrical characteristics of the synchronous motor. Engine torque modelling allows the shaft to be loaded with various resistive moments generated by an electromechanical brake. The presented characteristics highlight the mechanical characteristic of the synchronous motor, which does not depend on the resistant mechanical torque of the working machine during the technological process. There are presented the mechanical characteristics of the rotation speed depending on the variation of the resistant moment until the synchronism exit of the synchronous motor. The determination of these characteristics is dependent on the variation of the excitation current of the synchronous motor.

1. Introduction

For generating electricity in three-phase alternating current, the power plants use in most cases synchronous generators. Also, the synchronous machine has a wide spread both as synchronous motor and as synchronous compensator.

The study of the static and dynamic stability of the energy systems is indissolubly related to the behavior of synchronous and asynchronous machines. The impetuous increase of power installed in energy systems forced the deepening and modernization of synchronous machine theory, subsequently extended to the asynchronous machine, as well as other alternative current electrical machines. Particularly favorable has proved to be in this sense, for the analysis of the salient-pole synchronous machines functioning, „the theory of the two reactions" or the „Blondei — Park" theory. Lately, particular attention has been paid to the systematization of this theory, given the large number and variants of the equations that define the operation of the synchronous machine. [1]

The synchronous machine is characterized by the fact that the rotational speed n of the rotor is in constant ratio to the frequency of the network to which it is connected, and the inductive magnetic field is produced by a system of pairs of poles whose excitation winding is fed into continuous current. [2]

The operating mode of an electric machine is usually considered in relation to the active power. From this point of view, two regimes of the synchronous machine are distinguished: the generator regime and the motor regime. Generator and engine regimes are two aspects of the working regime. However, it is customary to emphasize these aspects because it affects certain linear dimensions of the machine elements. [3]

2. The synchronous motor

Compared to other electric motors, the synchronous motor has advantages in terms of non-load-dependent rotation speed, maintains constant speed, has a high efficiency and high power factor. [4]

A particular case of the engine regime is the compensator regime, in wich the machine takes the necessary active power from the grid to cover its own losses and provides reactive power in the grid.

The main disadvantage of the synchronous motor is that it develops electromagnetic torque only at the synchronous speed and it also has a rigid mechanical feature until the resistive torque reaches the maximum torque value at which the engine exits from synchronism, [5]. Other drawbacks are related to the oscillations that may occur at sudden variations in load on the shaft.

The mechanical characteristic of the synchronous motor is linear because its velocity does not vary with the load on its sahft, the gradient of the feature being defined by the derivative of the angular velocity relative to the moment derivation and it is null.

3. Mechanical and electrical characteristics of the synchronous motor

The choice of an electrical motor for a specific aplication, the adjustment of the speed and its behavior at different torque values, can be made if we study their mechanical and electrical characteristics.

For obtaining the mechanical and electric characterisitcs of the motor it was used the standard shown in the THE STUDY OF THE SYNCHRONOUS MOTOR, C. CRISTEA I. STROE, Bulletin of the Transilvania University of Braşov • Vol. 10 (59) No. 2 – 2017, Series I: Engineering Sciences paper. This was used to simulate an increasing value of the torque.

The synchronous motor's stator has a star configuration and is powered from a three-phase 400V line. The rotor is supplied with a voltage and a variable current through an excitation autotransformer. Their values are measured and monitored with the help of a digital multimeter. The servomotor feeds information to the controler which it displays via a dedicated software called ActiveServo. The connection of synchronous machine to the power grid is made with a switch, the voltage and excitation current are increased and variable resistive torque is applied [6].

3.1. Determining the motor's characteristics through application of a resistive torque curve

For determining these characteristics, a complete system is used which includes a servomotor that communicate through the dedicated software ActiveServo. This system is able to apply different resistive torque values at the motor's shaft. The requested reference torque values are enforced via the torque characteristic.

The servomotor is a self-cooled asynchronous motor, which allows the adjustment of the speed within a wide range of values, has a strong starting torque with protecton, is monitored from a thermal point of view and also by the controller.

The controller allows the generation of the resistive torque imposed by the torque characteristic. By connecting it to a PC the system is capable of supplying very useful and complex data which characterises the studied motor's parameters: voltages, curents, speed, torque and mechanical power. In addition it allows the application of resistive torques for multiple types of machines.

ActiveServo is the software which allows the recording and monitoring in time of the characteristics of the electrical machines. It also measures, computes and displays the varying electrical and mechanical quantities. In addition it has an integrated control system for the speed and torque. It is able to display the dependence between two quantities and also between one quantity and time. It allows the displaying of multiple characteristics, as well as exporting of graphs and measurement results [7].

4. Mechanical and electrical characteristics depending on time for imposed increasing values of the engine torque value.

In the experimental determinations performed to study the characteristics, a synchronous motor with the following characteristics was used: Pn = 0.27kW, In = 1.5A, f = 50Hz, n = 1500rot/min, U_{err} = 20Vcc; I_{err} = 4A. The excitation source generated a 20V DC voltage at a 3A excitation current.

IOP Conf. Series: Materials Science and Engineering **514** (2019) 012022 doi:10.1088/1757-899X/514/1/012022

For the synchronous engine studied, we will track the time dependence of the following dimensions: The intensity of the electric current in the rotor circuit I[A]; Engine torque M[Nm]; Speed n[rpm].

The engine is brought to the synchronism speed for 5 values of the excitation current, between 0.5 and 3A with a step of 0.5 A. By programming we have imposed the following torque characteristic:

For 100 s we plotted the torque diagram so that for 1 step = 0.1Nm to start the engine braking from 5s to 95s. At this value the engine will be braked with 1.8 Nm.

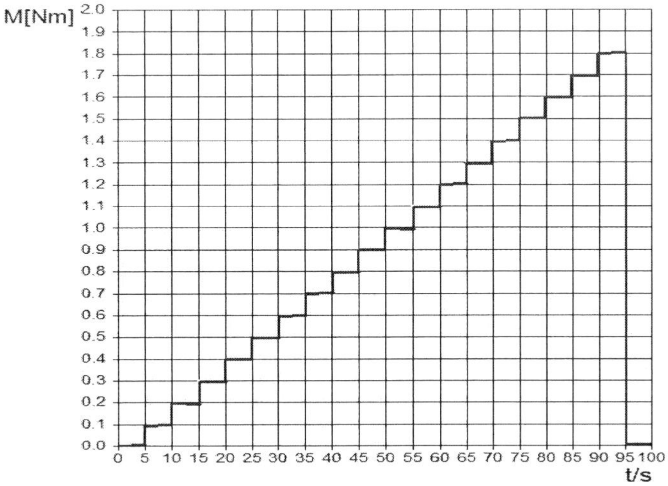

Figure 1. Torque diagram.

There are obtained the following characteristics:

For I_{err} = 0.5A The current intensity increases with the torque curve imposed up to 15s. After this value, as the speed changes, the value of the current increases unstable to 0.98A, after which at 0.82Nm the motor torque comes out of synchronism. The speed is kept constant for up to 15s at 0.3 Nm and then it is destabilized and at the value of 1356 rpm comes out of synchronism and the engine stops.

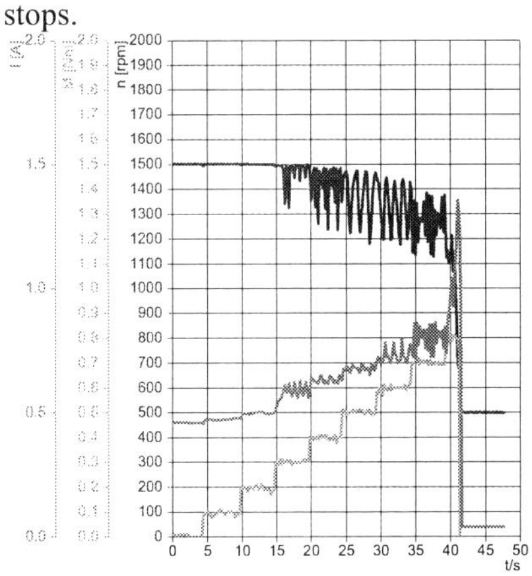

Figure 2. Characteristic representation I_{err} = 0.5A.

Table 1. Values for I_{err} = 0.5A.

t[s]	n [rpm]	M [Nm]	I [A]
0	1499	0	0.46
5	1502	0.1	0.47
10	1500	0.19	0.49
15	1495	0.3	0.52
20	1356	0.4	0.64
25	1470	0.49	0.68
30	1310	0.6	0.69
35	1135	0.71	0.71
40	1132	0.82	0.98
45	498	-0.21	0.03

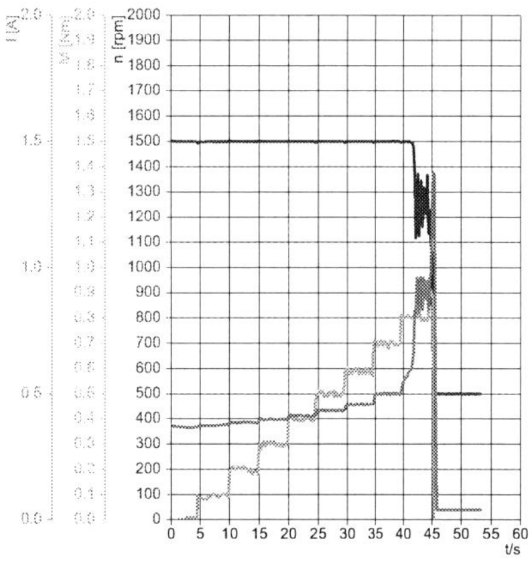

Figure 3. Characteristic representation I_{err} = 1A.

Table 2. Values for I_{err} = 1A.

t[s]	n [rpm]	M [Nm]	I [A]
0	1506	-0.01	0.37
5	1497	0.09	0.37
10	1503	0.19	0.38
15	1502	0.3	0.39
20	1500	0.41	0.40
25	1499	0.5	0.43
30	1499	0.6	0.45
35	1499	0.71	0.49
40	1498	0.8	0.54
45	1042	0.89	1.18
50	499	-0.19	0.03

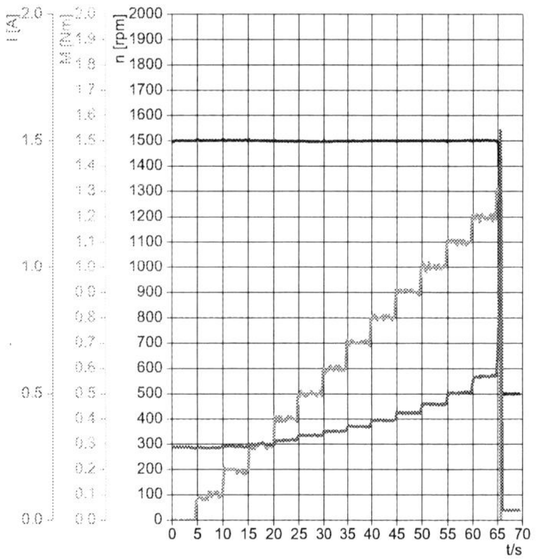

Figure 4. Characteristic representation I_{err} =1.5A.

Table 3. Values for I_{err} = 1.5A.

t[s]	n [rpm]	M [Nm]	I [A]
0	1496	0	0.28
5	1506	0.08	0.28
10	1502	0.11	0.28
15	1501	0.2	0.28
20	1498	0.31	0.29
25	1499	0.5	0.33
30	1499	0.58	0.35
35	1498	0.71	0.37
40	1500	0.79	0.39
45	1500	0.89	0.42
50	1500	1.01	0.45
55	1501	1.11	0.50
60	1500	1.21	0.55
65	1496	1.31	0.64

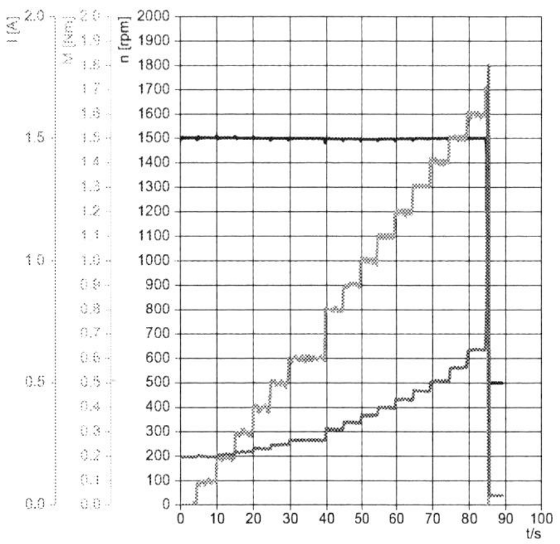

Figure 5. Characteristic representation I_{err} = 2A.

Table 4. Values for I_{err} = 2A.

t[s]	n [rpm]	M [Nm]	I [A]
0	1492	0	0.19
10	1510	0.19	0.20
20	1507	0.4	0.22
30	1501	0.59	0.26
40	1500	0.81	0.31
50	1498	1	0.36
60	1499	1.19	0.43
70	1500	1.41	0.50
80	1502	1.61	0.63
90	1004	1.81	1.82

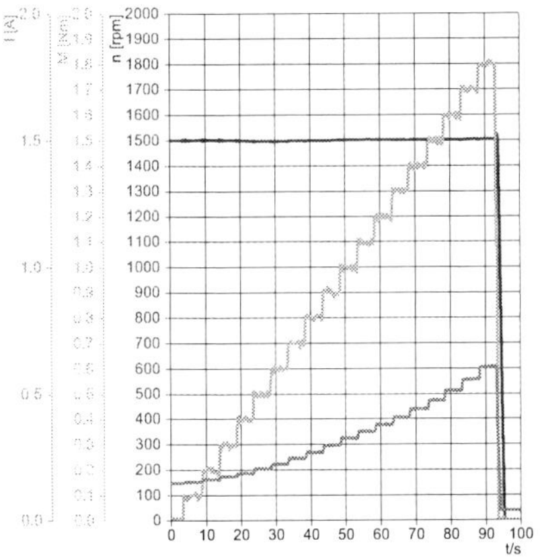

Figure 6. Characteristic representation I_{err} = 2.5A.

Table 5. Values for I_{err} = 2.5A.

t[s]	n [rpm]	M [Nm]	I [A]
0	1498	0	0.14
10	1499	0.19	0.16
20	1498	0.39	0.18
30	1496	0.61	0.22
40	1499	0.81	0.26
50	1502	1	0.32
60	1502	1.2	0.37
70	1501	1.41	0.43
80	1501	1.61	0.50
90	1503	1.8	0.60

For I_{err} = 1 A The current intensity increases with the torque curve imposed up to 40s. After this value, as the speed changes, the value of the current increases unstable to 1.18A, after which at 0.89Nm the motor torque comes aut of synchronism. The speed is kept constant for up to 40s at 0.8 Nm and then it is destabilized and at the value of 1042 rpm comes aut of synchronism and the engine stops.

For I_{err} = 1. 5A The current intensity increases with the torque curve imposed to 65s. After this value, because the speed is destabilizing, the value of the current increases unstably to 0.64A, after which at 1.31Nm motor torque comes aut of synchronism. The speed remains constant until 65s at 1.21 Nm after which it is destabilized and at 1496 rpm it comes aut of synchronism and the engine stops.

For I_{err} = 2A The current intensity increases with the torque curve imposed to 85s. After this value, because the speed is destabilizing, the value of the current increases unstably to 0.63A, after which at 1.61Nm the motor torque comes aut of synchronism. The speed remains constant for up to 85s at 1.61 Nm after which it is destabilized and at 1004 rpm comes out of synchronism and the engine stops.

For I_{err} = 2.5A The current intensity increases with the torque curve up to 92s. After this value, because the speed is destabilized, the value of the current increases unstably to 0.6A and then at 1.8 Nm the motor torque comes out of synchronism. The speed remains constant to 92s at 1.8 Nm after which it is destabilized and at 1491 rpm comes out of synchronism and the engine stops.

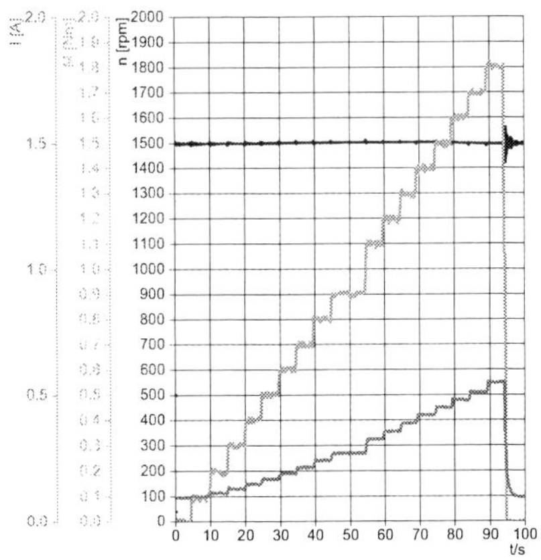

Figure 7. Characteristic representation I_{err} = 3A

Table 6. Values for I_{err} = 3A.

t[s]	n [rpm]	M [Nm]	I [A]
0	1498	0.01	0.09
10	1491	0.19	0.11
20	1493	0.39	0.14
30	1504	0.6	0.18
40	1504	0.81	0.24
50	1502	0.89	0.26
60	1502	1.21	0.35
70	1504	1.38	0.42
80	1501	1.58	0.48
90	1497	1.81	0.54

For I_{err} = 3A The current intensity increases with the torque curve imposed to 94s. After this value, because the speed is destabilizing, the value of the current increases unstably to 0.55A and then at 1.81 Nm the motor torque comes out of synchronism. The speed remains constant for up to 94s at 1.81 Nm after which it is destabilized and at 1498 rpm comes out of synchronism and the engine stops.

5. Conclusions

To improve the mechanical and electrical characteristics depending on time for required increasing values of the torque motor value it has been used an imposed torque diagram in which the engine is driven at synchronism speed for 5 values of the excitation current, between 0.5 and 3A with a step of 0.5 A, so that for 100s we plotted the torque diagram for 1 step = 0.1Nm to start the engine braking from 5s to 95s, at this value the engine being braked with 1.8 Nm. For the synchronous motor we studied, we tracked the time dependence of the following dimensions: The intensity of the electric current in the rotor circuit I[A]; Engine torque M[Nm]; Speed n[rpm]

Analyzing the current intensity value in the rotor circuit we notice that it decreases for increasing values of the excitation current. If for I_{err} = 0.5A the maximum current value at detaching is 0.98A, for the value I_{err} = 3A the motor is detached at the value of 0.54A. The same as with torque characteristics analysis, from the previous graphics we notice that although as the torque continues to increase, the motor speed remains constant at 1500 rpm until it reaches a value outside of the range of synchronism. It is noticed that at I_{err} = 0.5A the motor is disconnected after 45 seconds at 0.89 Nm, while at the maximum value studied for I_{err} = 3A the motor is disconnected at 95 seconds at 1.81 Nm. The graphics indicate that for a synchronous motor the value of the excitation current is very important for keeping the constant speed according to the torque value. The value outside of the range of synchronism is closely related to the maximum torque value but also to the excitation current value.

By adjusting the excitation current value of the synchronous motor, we can control the exit from the synchronism condition at mechanical torque shocks at the shaft. Based on this premise we can continue experimental research on the synchronous machine when it operates in generator regime.

6. References

[1] V Nedelcu, Maşini electrice Editura Didactică şi Pedagogică, Bucureşti, 1968
[2] Toma Bordea, Masini electrice - Constructie, Editura ASAB, Bucureşti, 2003, ISBN 973-86010-3-7
[3] Toma Bordea, Maşini electrice, *Teorie Editura ASAB, Bucureşti, 2002, ISBN 973-86010-0-2
[4] http://www.euedia.tuiasi.ro/uee/uee_files/Lucrarea%2014.pdf
[5] E C Teodoru, M Gogu Maşini electrice. Editura Universităţii Tehnice Gh.Asachi , Iaşi, 2000
[6] C Cristea, I Stroe, THE STUDY OF THE SYNCHRONOUS MOTOR, Bulletin of the Transilvania University of Braşov • Vol. 10 (59) No. 2 – 2017 Series I: Engineering Sciences, ISSN 2065-2119 (Print), ISSN 2065-2127 (CD-ROM)
[7] D Călueanu, I Voncilă Maşini Electrice. Editura Fundaţiei Universitare Dunărea de Jos din Galaţi, 2003.

IOP Conf. Series: Materials Science and Engineering **514** (2019) 012023 doi:10.1088/1757-899X/514/1/012023

Classification of industrial robots according to the number of degrees of mobility-structural synthesis and useful configurations

I Staretu[1,2]

[1]Product Design, Mechatronics and Environment Department, Transilvania University of Brasov, Brasov, Romania
[2]Technical Sciences Academy of Romania, Bucharest, Romania

E-mail: staretu@unitbv.ro

Abstract. Industrial serial robots can be classified according to several criteria, the most important being: the number of mobility degrees or independent movements, the number of independent kinematic axes, the shape of the workspace, the value of the useful load, the precision of touching a predetermined point, etc. Among these criteria, the number of degrees of mobility directly determines the structure of the kinematic chain corresponding to the robot and implicitly its construction, functionality and utility. In this paper, we define the three classes of industrial serial robots according to the degree of mobility: the class of industrial serial robots of minimum configuration, the class of redundant industrial serial robots and the class of hyper-redundant industrial serial robots. For each class of industrial serial robots the structural synthesis method is presented and representative examples of structures are given.

1. Introduction

Serial industrial robots currently exist in multiple variants in continuous diversification. So far, however, there has not been a classification of serial industrial robots to be useful to both designers and users. Obviously, it is first necessary to establish the main criterion or criteria for classification.

By analyzing the technical datasheets of the already manufactured industrial robots, we can note that the main classification criterion is the number of degrees of mobility, respectively the number of independent movements, of independent kinematic axes. Thus, we obtained the structural synthesis of open series cinematic chains with 3 or 6 independent movements. Therefore, the category or *class of minimal configuration serial industrial robots* has been defined [1] required and sufficient to guide (position and orient) an object in the three-dimensional space relative to an Oxyz three-orthogonal reference system. In time, one or more axes corresponding to independent movements have been added to the minimally configured serial robotic kinematic chains to increase the workspace and enhance the robot functionality. Such structures have been practically obtained individually without systematization and are now available on the industrial robotics market. If we consider the situations where the minimum configurations reach a double number of independent movements, a reasonable degree of redundancy, a second category of industrial robots is defined, namely the category or *class of redundant serial industrial robots* [2] characterized by 7 to 12 independent movements, respectively, mobility degrees or independent kinematic axes. However, structures with over 12 independent kinematic axes were imagined by the addition of a large or very large serial number of independent movements, resulting in robotic trunk-like serial structures [3] or type spine [1], obtained by adding, in series, a single cinematic coupling or a limited number of kinematic couplings. These structures with more than 12 independent movements, generating non-degenerated workspaces, form the category or *class of hyper-redundant serial industrial robots*. The paper presents for the first time the general structural synthesis of each class of industrial serial robots and gives suggestive examples of structures specific to each class.

2. Class of industrial serial robots with minimal configuration

As it has already been shown for positioning an object in three-dimensional space, three independent movements are required to obtain the three linear coordinates corresponding to the three Ox, Oy and Oz axes of an Oxyz three-orthogonal reference system. According to [1] in order to obtain the possible variants of these structures, we distinguish the repeating arrangements of *m* distinct elements grouped by *n*, namely, arrangements of two rotation or translational couples (R, T) grouped by three (XZY), X, Z, Y $\in$ {R.T}of two relative positions: perpendicular or parallel ($\perp$,||), grouped by two (θ, β), θ, β $\in$ {$\perp$, ||}, and respectively two values($\neq$ 0, = 0) assigned to three sizes: a_1, a_2 , b. We obtain $\alpha_{(R,T)}^{(XZY)} \rightarrow \alpha_2^3=2^3=8$ and $\alpha_{(\perp,\|)}^{(\theta,\beta)} \rightarrow \alpha_2^2=2^2=4$ and $\alpha_{(\neq 0,\, =0)}^{(a_1,\, a_2,\, b)} \rightarrow \alpha_2^3=2^3=8$. It also follows that two couplings R and T can be grouped in threes (XZY) in eight modes, of which four XRY groups: $\alpha_{(R,T)}^{(XY)} \rightarrow \alpha_2^2=2^2=4$ and four XTY types: $\alpha_{(R,T)}^{(XY)} \rightarrow \alpha_2^2=2^2=4$. A kinematic chain of the XRY type is determined if the relative position of the coupling R reported to the X coupling and the position of the coupling Y reported to R is known, so four kinematic *positioning chains*(PC) can be formed with a XRY-type group: $\alpha_{(\perp,\|)}^{(\theta,\beta)} \rightarrow \alpha_2^2=2^2=4$: X$\perp$R$\perp$Y; X$\perp$R||Y; X||R||Y; X||R$\perp$Y and with the four XRY groups 4 * 4 = 16 PC can be formed. Similarly, PC type XTY are obtained. Thus the four XTY variants: RTR, RTT, TTR, TTT correspond, however, to 5 variants depending on the relative position: $\perp\perp^{\perp}$, $\perp\perp\perp_{\|}$, || ||, ||$\perp$. So there are 4 * 5 = 20 XTY type PC variants. In total, we obtain 16 + 20 = 36 kinematic positioning chains, of which 20 describe non-degenerated workloads [1]. Figure 1 gives three examples of PC describing non-degenerated workloads.

The kinematic orientation chain (OC) provides the orientation of an object in space according to the three angles α, β and γ and consists of three rotation couplings with perpendicular axes: R$\perp$R$\perp$R.

Depending on the relative position of the three axes, we obtain $\alpha^k{}_{2=}\alpha^3{}_2=2^3=8\rightarrow\alpha^3{}_2*2*2=8*2*2=32$ variants taking into account the a, b, c relative positioning distances of the axes and the distinct montage variants reported to the robot arm (L, N; l, n), in which the notations mean: L-montage along the robot arm, N-montage perpendicular to the robot arm, l-montage of the gripper in the extension of the last axis of orientation, n-montage of the gripper perpendicular to the last axis of orientation) [1].

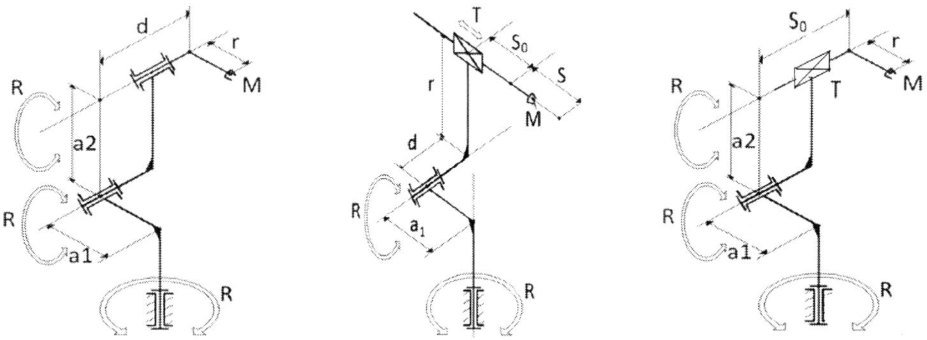

Figure 1. Positioning chains(PC) with minimal configuration.

Figure 2 gives 4 examples of OC of the 32 variants [1]. By adding to the kinematic positioning chain (PC) the kinematic orientation chain (OC), the kinematic guiding chain (GC) is obtained, which ensures the unambiguous positioning and orientation of an object in the space associated to an Oxyz tri-orthogonal reference system. If for the 20 PCs that correspond to non-degenerated workspaces, the

IOP Conf. Series: Materials Science and Engineering **514** (2019) 012023 doi:10.1088/1757-899X/514/1/012023

32 OC variants are considered, that results in 20 * 32 = 640 variants of GC that correspond to non-degenerated workspaces.

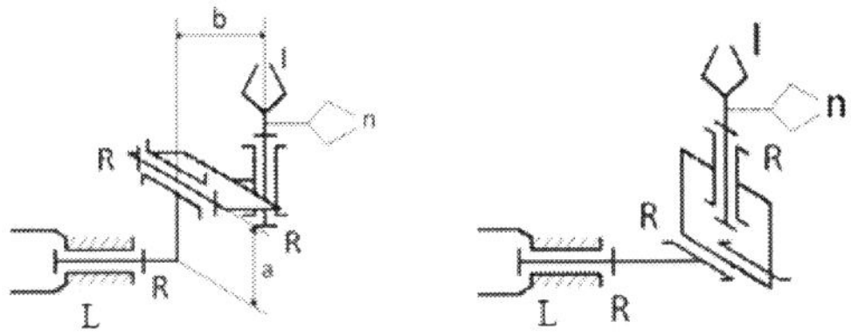

Figure 2. Kinematic orientation chain (OC).

Figure 3 gives 3 examples of such GC(orientation chain(OC: 3 rotations R⊥R⊥R) is symbolically represented).

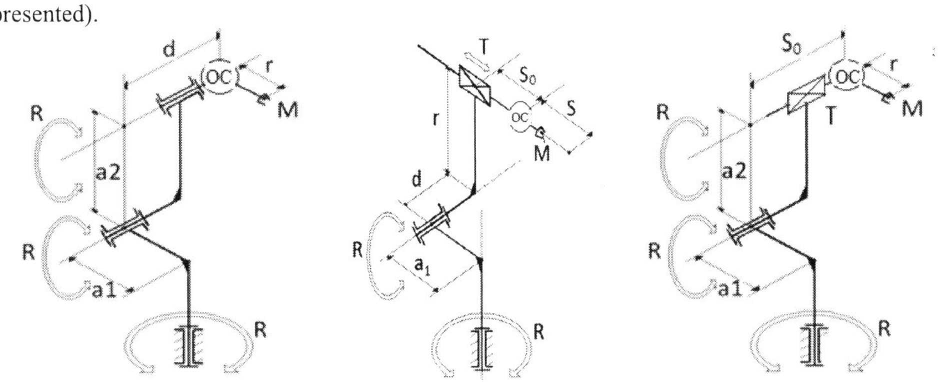

Figure 3. Example of kinematic guiding chain (GC).

3. Redundant serial industrial robots class

Any serial industrial robot that adds at least a degree of mobility, so an independent movement beyond the 6 necessary minimum mobility degrees, becomes a redundant robot because the additional degree of mobility or degrees of mobility are not necessary for univocal positioning and orientation of an object in the three-dimensional space, but only increase the working space and the functionality of the robot by increasing the capacity of accessing additional areas of these structures to structures with minimal configuration. After some time it was considered that robots with more than 6 degrees of mobility, so 6 axes, are not necessary, but 7-degree mobility, so 7, 8, even 9-axe structures were made [4-13], without a systematization of the possible variants and implicitly a method of rigorous structural synthesis. In the paper [2] a method of structural synthesis of redundant serial robots has been proposed and described, of which all structures with 7 to 12 degrees of mobility, i.e. 7 to 12 independent kinematic axes, are considered to belong. Thus, for redundant robots with 7 degrees of mobility, i.e. 7 independent kinematic axes, there are structures of the type: R⊥(PM)(OM), R||(PM)(OM) or T⊥(PM)(OM), T||(PM)(OM). By highlighting all the possible situations, there are 12 redundant structures with 7 axes: R⊥(PM)(OM), R||(PM)(OM), T⊥(PM)(OM), T||(PM)(OM), (PM)⊥R(OM), (PM)||R(OM), (PM)⊥T(OM), (PM)||T(OM), (PM)(OM)⊥R, (PM)(OM)||R, (PM)(OM)⊥T, (PM)(OM)||T. As a result, the 20 positioning structures (PM) characterized by non-degenerated workspaces correspond to 20 * 12 = 240 redundant structures with 7 degrees of mobility.

In Figure 4 there are two variants of redundant structures with 7 degrees of mobility and 7 independent kinematic axes: R ⊥(R⊥R⊥T)(OM: R⊥ R⊥ R) și T ⊥(R⊥R⊥T)(OM: R⊥ R⊥ R).

Structures with 8 axes are obtain by adding an additional axis of R or T in position ⊥or || to the structures with 7 axes obtained in the previous step, resulting in structures of the type: R⊥[R⊥(PM)(OM)], R||[R⊥(PM)(OM)], T⊥[R⊥(PM)(OM)], T||[R⊥(PM)(OM)].

IOP Conf. Series: Materials Science and Engineering **514** (2019) 012023 doi:10.1088/1757-899X/514/1/012023

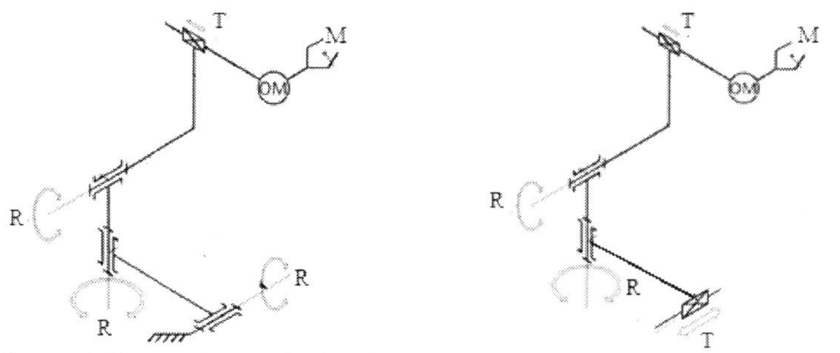

Figure 4. Two variants of redundant structures with 7 degrees of mobility.

Figure 5 gives two examples of structures with 8 axes.

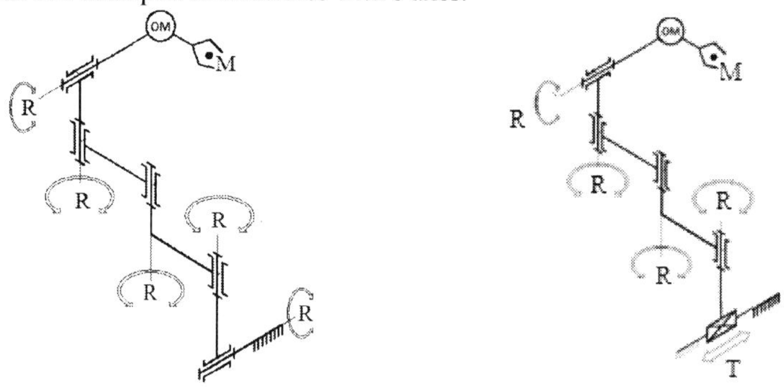

Figure 5. Structures with 8 degrees of mobility- 8 independent axes.

9-axis structures are obtained by adding an additional R or T axis in ⊥ or || position to structures with 8 axes, resulting in structures of the type: R⊥[R⊥[R⊥(PM)(OM)]], R||[R⊥[R⊥(PM)(OM)]], T ⊥[R⊥[R⊥(PM)(OM)]], T|| [R⊥[R⊥(PM)(OM)]]. Two examples of nine-axis structures are shown in Figure 6.

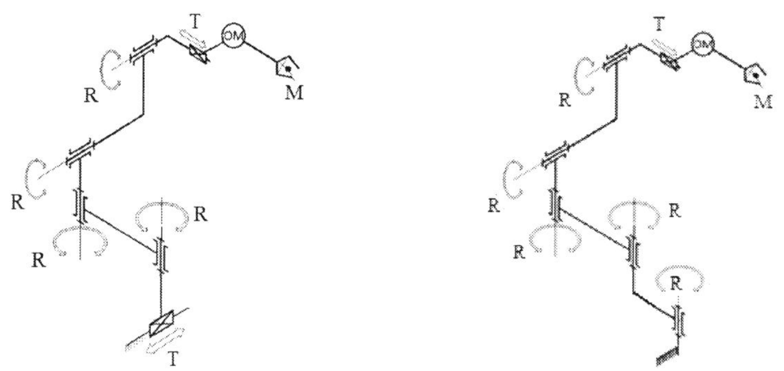

Figure 6. Structures with 9 degrees of mobility- 9 independent axes.

Structures with 10 axes are obtained by adding an additional axis of R or T in position ⊥or || to structures with 9 axes, resulting in structures of the type: R⊥[R ⊥ [R ⊥ [R⊥ (PM)(OM)]]], R||[R ⊥[R⊥[R⊥(PM)(OM)]]]], T⊥[R⊥ [R⊥ [R⊥ (PM)(OM)]]], T||[R⊥ [R⊥ [R⊥ (PM)(OM)]]].

Figure 7 gives two examples of structures with 10 axes. Structures with 11 axes are obtained by adding an additional axis of R or T in ⊥position or || to structures with 10 axes, resulting in structures of the type: R⊥[R⊥[R ⊥[R⊥[R⊥(PM)(OM)]]]], etc.

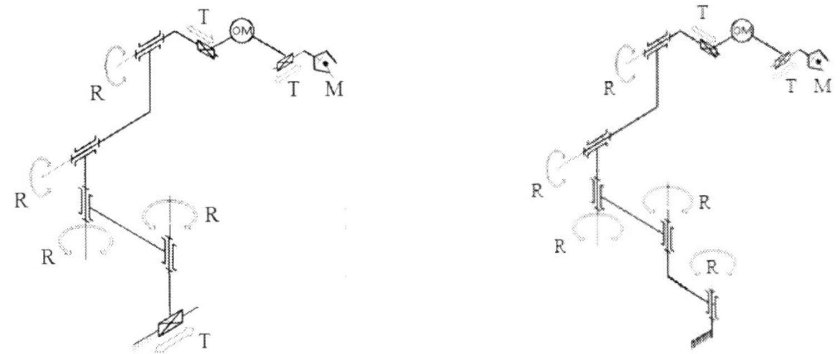

Figure 7. Structures with 10 degrees of mobility - 10 independent axes.

In Figure 8 there are two examples of structures with 11 axes.

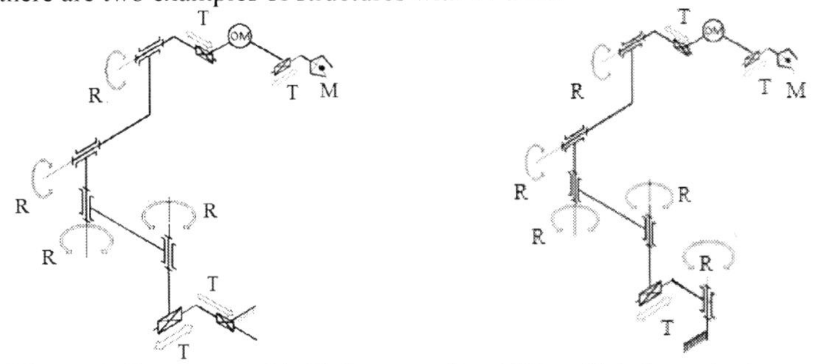

Figure 8. Structures with 11 degrees of mobility - 11 independent axes.

Structures with 12 axes are obtained by adding an additional axis of R or T in position $\perp$or || to structures with 11 axes, resulting in structures of the type: R$\perp$[R$\perp$ [R$\perp$ [R$\perp$ [R$\perp$ [R$\perp$(PM)(OM)]]]]], etc. Two examples of structures with 12 axes are given in Figure 9.

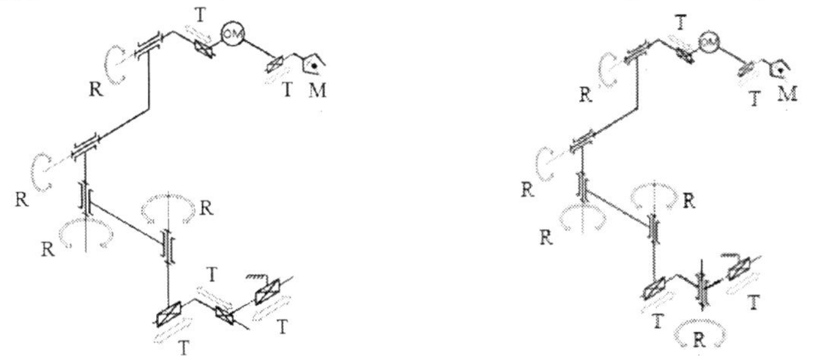

Figure 9. Structures with 12 degrees of mobility - 12 independent axes.

4. The class of serial industrial hyper-redundant robots

Industrial serial hyper-redundant are characterized by more than 12 degrees of mobility, so 13 or more independent kinematic axes. The structures of these robots are obtained by adding a degree of mobility, i.e. an independent axis corresponding to a mono-mobile rotation (R) or translation (T) coupling in a parallel (||) or perpendicular ($\perp$) relative position to structures with 12 independent kinematic axes. For hyper-redundant structures, the number of independent axes above 12 is not limited. These structures are also known as trunk-like or type spine structures [1].

Obviously, any redundant serial structure to which one or more degrees of mobility are added, so one or more kinematic independent axes of rotation or translation becomes a hyper-redundant structure. In Figure 10, two hyper-redundant structures are represented, derived from a redundant structure with 12 degrees of mobility.

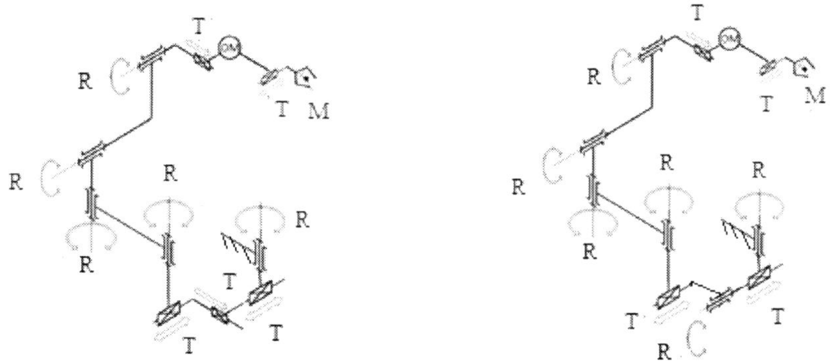

Figure 10. Structures with 13 degrees of mobility - 13 independent axes(OM: R⊥R⊥R is schematic represented).

5. Conclusions

On the basis of the facts presented in this paper, the following conclusions can be drawn:

1. After a systematization of the *minimum configuration serial industrial robot* structures with 6 degrees of mobility was completed in the 1980s, the continuation of such systematization for serial robot structures with 7 or more degrees of mobility was given up, but later such structures have emerged through disparate practical achievements.
2. The author of this paper proposed for the first time the definition of the class of *redundant serial industrial robots* with 7 to 12 degrees of mobility for which he proposed a method of structural synthesis, which allows highlighting the most useful structures possible.
3. Serial robots with 13 or more degrees of mobility form the class of *hyper-redundant robots*.

6. References

[1] Dudita Fl, Diaconescu D and Gogu Gr 1989 *Linkages mechanisms (in Romanian)*(Bucharest, Editura Tehnica)
[2] Staretu I 2015 Class of redundant serial robots with 7 up to 12 axes-definition, structural synthesis and workspace *Proceedings of the 6th International Conference Computational Mechanics and Virtual Engineering COMEC 2015(CD)* pp 1-6
[3] Ivanescu M 2004 *On the dynamic control of hyper-redundant manipulators, Advances in Automatic Control*(Dordrecht-Kluwer Academic Publishers pp 141-158)
[4] Brell-Cokcan S, Reis M, Schmiedhofer H and Braumann J 2009 Digital Design to Digital Production: Flank Milling with a 7-Axis Robot and Parametric Design *Proceedings of the 27th eCAADe Conference* pp 323-330
[5] Qi R, Lam TL, Qian H and Xu Y 2011 Arc tracking on an eight-axis robot system *Proceedings IEEE International Conference on Robotics and Biomimetics* (*ROBIO*) pp 678-683
[6] Wang J, Li Y and Zhao X 2010 Inverse Kinematics and Control of a 7-DOF Redundant Manipulator Based on the Closed-Loop Algorithm *International Journal of Advanced Robotic Systems* 7(4) pp 1-9
[7] Information on: www.foodengineeringmag.com/.../seven-ax.
[8] Information on: www.motoman.co.
[9] Information on: www.yaskawa.co.jp/en/.../robotics/01.html.
[10] Information on: www.ragroup.com.au/robots/.../sda_sia_seri....
[11] Information on: www.cloos.de/QIROX/.../roboter/.../index.p...
[12] Information on: www.densorobotics.com/news/34.
[13] Information on: www.directindustry.com › ... › Articulated robot.

IOP Conf. Series: Materials Science and Engineering **514** (2019) 012024 doi:10.1088/1757-899X/514/1/012024

The state of the modular element structure of a prehensor through virtual prototyping

I C Frincu[1] and I Stroe[1]

[1]Product Design, Mechatronics and environment Department, Transilvania University of, Brașov, Romania

E-mail: cezar.frincu@unitbv.ro

Abstract. The paper presents the study of the geometric scheme of the structure of a prehensor module. The prehensor has the role of generating the force required to manipulate an object respecting the prehensor's plan with the manipulated object. It is proposed the static concept of the prehensor according to the geometric parameters of the prehensat object. Starting from the structural scheme of the proposed module, the 3D component structure is generated using the Catia software. The kinematics of the studied structure is presented and the system modeling prerequisites are created through the Adams program. Importing the system into Adams simulates the kinematics of the proposed structure. The state of stresses and deformations of the prehensor module is determined and presented. The prehensor module of this type is designed for the reliability of the prehensor.

1. Introduction

The role of the prehensor in a robot is to achieve the prehension of various objects with specific design. Depending on the shape of the prehensat object, modeling of the constructive form of the prehensor occurs so that the phases of prehension are fulfilled. The griping of the object must comply with the minimum prehension conditions. The quality of prehension is given by the degree of sliding of the prehensat object. In order to achieve good prehension it is necessary to perform the self-adaptive control function of the robot, respectively of the prehensor. Ever since the design phase, there must be a close link between the elements of the prehensor structure and the the sensory system to highlight the quality of prehension by self-adaptive control. The geometric shape of the prehensat object imposes the prehensor structure. The paper presents the geometric and constructive form of an anthropomorph with 5 flanks. From the anthropomorphic prehension analysis results different types of prehensors depending on the number of fingers, the number of phalanges, the type of command and control system adopted.

Mechanical prehensors with 5 fingers are complex systems that imitate a human hand, and the hand is an ideal prehensor. The prehensors have the role of transporting objects in a controlled and well-determined manner in a defined space. The required conditions of this type of prehensor are the planarity to the prehensat object and to the position on the work table, and the movement must be very precise with 1 Kg of prehensat, '[1]'.

2. The schematic concept establishment of the prehensor module

To establish a concept, the necessary details are needed in order to achieve it. In the case of a prehensor module it is required to establish the principle and the functioning mode. Under the conditions imposed, the prehensor system created fulfills the parallelism function of the prehensat object and the function of parallelism to the work table on which the object is located.

This type of prehensor is very similar to functionality with that of a griper, only the prehensor system exerts forces on the controlled object.

2.1. The kinematic scheme of the prehensor module

According to the conditions, the dynamic scheme of the prehensor module was created, "Figure 1", where the structure and its functioning are represented.

The prehensor module is made of 8 kinematic joints where the joint (A) is the primary rotation joinit that engages the secondary joint (B) and the joint (C). This rotation chain allows the freedom to perform a translational movement in (E) and (H). With the help of the segments (1), (2), and (3), which are connected by the passive joints (B), (C) and (D), the translational movements of (E) and (H) are assured, without this system blocking angularly due to the length of the levers. By the angular modification (a0) which is described by the horizontal of point (A) and segment (1), the variable angles and distances (d1), (d2) si (d3) change between the joints (C) and (E), (C) and (F), respectively (E) and (H), to move the point (M) to the axis (O).

This prehensor mode works on the principle of the parallelogram where the active side (4) is the side trained by the joint (C), the passive sides (3) and (5) are equal and parallel, as are the sides (4) and (6). It also fulfills the angular condition where the angle (a4) and the angle (a5) are equal. Under these conditions, it it assured the planarity of the prehensor module against the prehensat object, '[2]'.

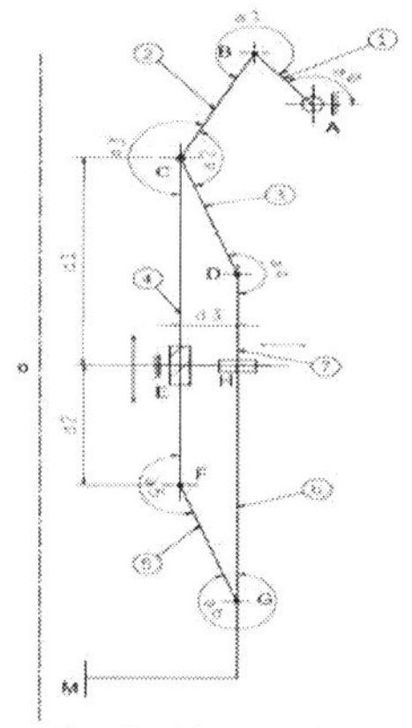

Figure 1. The kinematic scheme.

3. 3D modeling and execution of prehensor modul kinematic in the Catia software

3.1. The stages of execution the 3D model of the prehensor module:

- Sketcher Module

It is the module in which the geometric description of the desired piece is created in plan, and the geometric shapes of the pieces can be determined or undetermined. For making the shape, the commands in "Figure 2" are needed, where we can use the "Profile" commands and we can dimension or constrain using commands from "Constraint".

Figure 2. Sketcher tools.

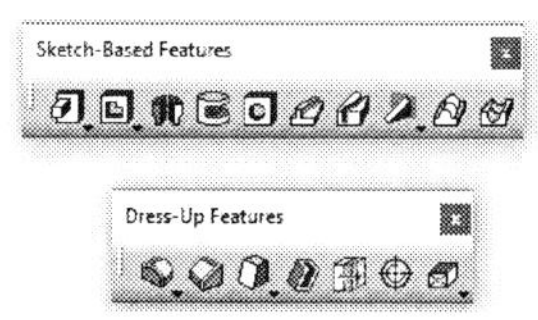

Figure 3. Part Design Module tools.

- Part Design Module

It can only be used after it has been dimensioned and described the piese in plan in Sketcher Mode, then 3D modeling is realised in Part Desing module. In the making of the 3D model of a piece there are considered many details regarding the processing and assembling. To make a 3D shape of a piece the commands in "Figure 3" are used.

- Assambly Design Module

In Assambly mode, all the pieces are inserted after they have been created to make the prehensor module. The pieces will be mounted according to the degree of freedom required to achieve the proper movement of the prehensor module without simulating it. Initially, the fixed pieces are positioned and declared, and then the other pieces are moved. To achieve the constraints and general structure of the prehensor module, the commands in "Figure 4" were used.

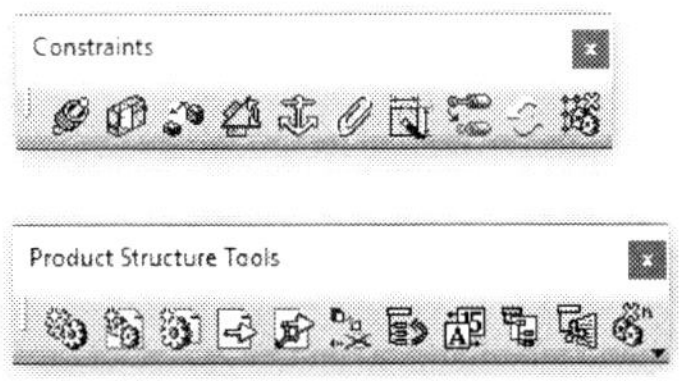

Figure 4. Assambly Design tools.

After completing 3D modeling and respecting the kinematic conditions, the prehensor module was created. In "Figure 5" is represented the kinematic scheme with the mechanical respecting and realization so that kinematic joints have a necessary degree of freedom. The rotation joints (A), (B), (C), (D), (F) and (G) are made of a bearing type between the parts that form the lever arms and the radial-axial bearings, to reduce the friction from the movement of the prehensor module.[8,9]
The translational movements are performed by means of linear ball guides (E) and (H), which allow a very good continuity in the movement of the prehensor module. With roller bearings and guides we get a friction very small between joints, which helps us in 3D simulation. Whereas simulation is ideal, it means friction is neglected.

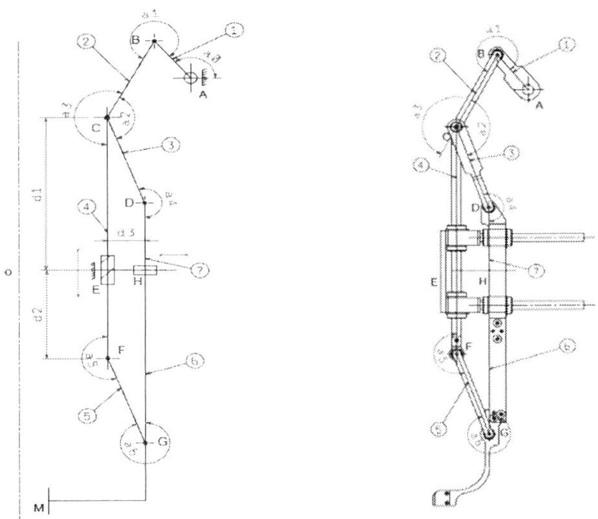

Figure 5. Schematic representation of the kinematic scheme.

3.2. Kinematics execution in the Catia software
To achieve cinematics in the Catia software, it is necessary to enter DMU Kinematics mode, where constraints and joints are required to implement. As in the kinematic scheme, in Catia there is a primary joint and the rest of the secondary joints. The primary joint exerts the movement, and the secondaries ones follows it until it stops or is in a moment of mechanical limitation. In "Figure 6" are represented the kinematic stages of the prehensor module, '[4]'.

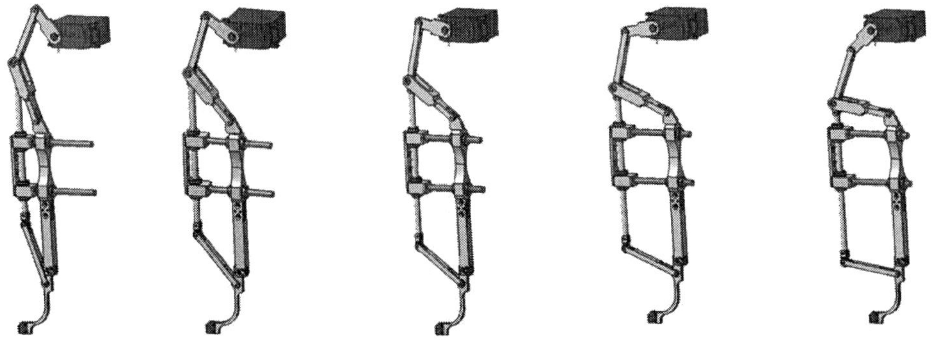

Figure 6. The kinematic stages of the prehensor module.

3.3. *Structural analysis generating of the prehensor module*
This type of analysis is done when we want to structurally analyze deformations of bodies or assemblies when applying a force. This analysis is done in the Generative Structural Analysis module and is used to verify the structure, the time reliability of the mechanical components and to determine the solicited areas. By determining the solicited areas, we can establish the sensors implementation areas.

Figure 7. Structural analysis of the prehensor module.

"Figure 7" shows the mechanic deformation of the prehensor module, where this analysis is performed in order to implement the sensors. Solicited areas are represented in red and least solicited in blue. The prehensor module is subjected to a 5 N force both at closing and at opening.

4. Prehensor module importation and simulation in the Adams software

4.1. *Importing the prehensor module into Adams*
After executing the design in a software like Catia, the parts are imported into Adams. Depending on the version of Adams, the prehensor module can be directly imported. Once imported the prehensor

IOP Conf. Series: Materials Science and Engineering **514** (2019) 012024 doi:10.1088/1757-899X/514/1/012024

module, the type of material for each piece is declared in order to apply the constraints. The first constraint is the fixation of the fixed objects, then the constraint of the following pieces with the tools in "Figure 8".

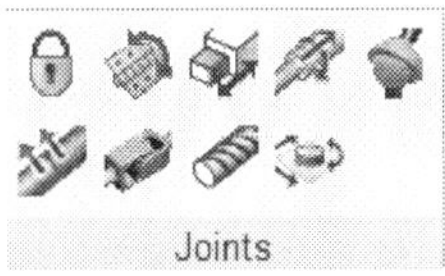

Figure 8. Tools for constraint of bodies.

With these tools, all the components of the prehensor module have been constrained, "Figure 9", so that the the movement can be realised. Following constraints, the primary joint and the secondary joints must be declared, '[3]'.

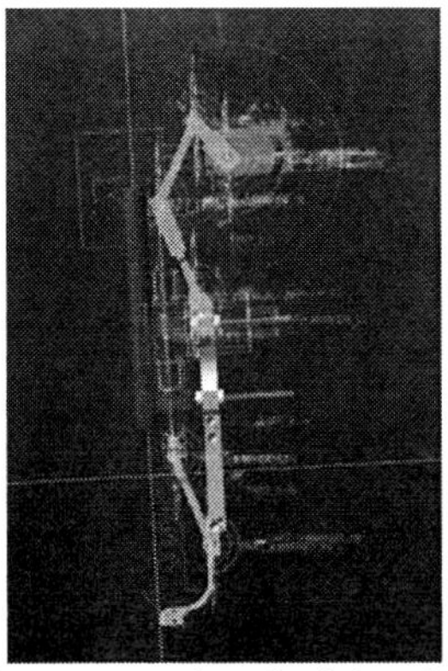

Figure 9. Representing the joints in Adams.

4.2. Performing simulation in Adams

Once all the joints have been applied to the components of the prehensor module, the forces required to move have been implemented. One of these forces is constant, namely the gravitational force. By fulfilling all the functional conditions, the 3D simulation of the prehensor module can be applied. In "Figure 10" is presented the simulation window, where we can control the number of steps in a given time, '[9]'.

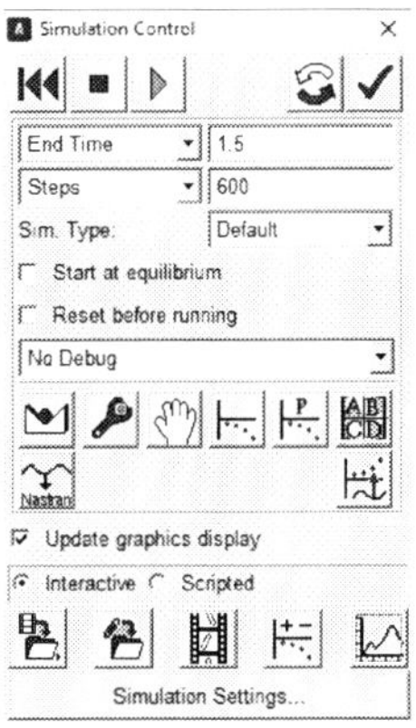

Figure 10. Control panel.

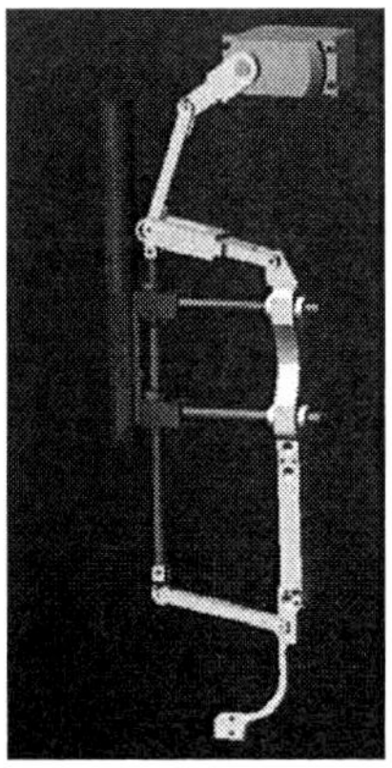

Figure 11. Kinematic execution in Adams.

After determining the time and steps required to generate the complete movement of the prehensor module, it set to start and the prehensor module will move to the forward position relative to the origin, as in "Figure 11".

Once with the movement of the prehensor module, acceleration sensors and force sensors can be introduced. As it can be seen in the graph in "Figure 12", the angular acceleration fluctuations are influenced by the applied force.

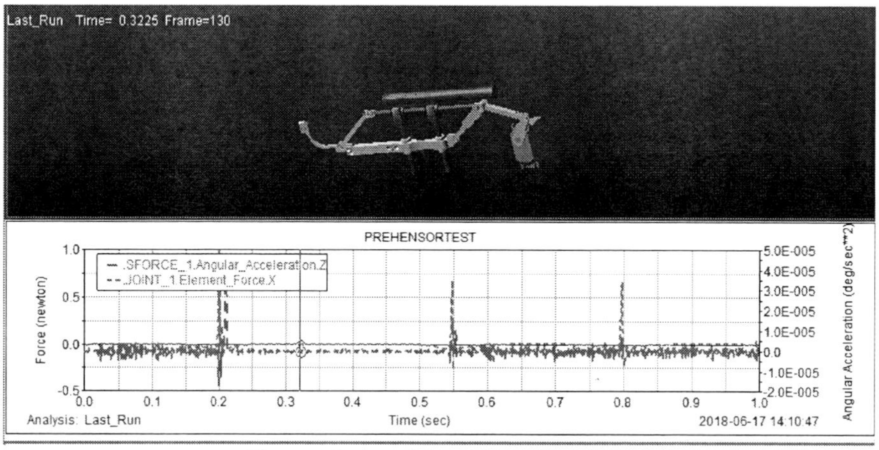

Figure 12. Angular acceleration graph.

5. Conclusions

This prehensor module is designed to create a complex prehensor with 3 or 5 modules. The prehensor allows the capture of small objects and objects with a larger gauge. This particularity is due to the parallelogram system and the translational movements the prehensor has.

6. References

[1] Staretu I 1996 *Sisteme de Prehensiune* (Brasov: Lux Libris)

[2] Dudita F, Staretu I 1988 *Sinteza si sistematizarea structural a mecanismelor elementare plane cu bare articulate utilizabile ca robotomecanismelor de prehensiune* (Cluj-napoca: Buletinul celui de al VIII-lea Simpozion National de I R)

[3] Alexandru C 2000 *Simularea pe calculator a sistemelor mecanice articulate* (Brasov: Lux Libris)

[4] Ghionea I 2007 *Proiectare asistata CATIA V5. Elemente teoretice si aplicatii* (Bucuresti: BREN)

[5] Ghionea I 2009 *CATIA V5. Aplicatii in ingineria mecanica* (Bucuresti: BREN)

[6] Alexandru C 2009 *Softuare Platform for Analyzing and Optimizing the Mechanical System* (Brasov: Springer Sience&Business Media)

[7] Wang H, Zhang J 2004 *The parameter modelling and analysis of Virtual prototype basing on the ADAMS* (Michigan: Mechanical Manufacture)

[8] Barbu, I., V*irtual prototyping for an mechatronics device used for medical rehabilitation* COMEC 2009, 29-30 October 2009, Brasov, Romania; ISBN 978-973-598-572-1.

[9] Barbu, I., *Modeling and optimization of a knee orthosis using virtual prototyping* Bulletin of the *Transilvania* University of Braşov • Vol. 7 (56) No. 2 - 2014

[10] *Getting Starting Using ADAMS* (Mechanical Design Inc. 2013)

IOP Conf. Series: Materials Science and Engineering **514** (2019) 012025 doi:10.1088/1757-899X/514/1/012025

Nonstandard teeth profile description and justification of precessional gear parameters selection

I Bostan[1] and V Dulgheru[1]

[1]Basics of Machine Design Department, Technical University of Moldova, Republic of Moldova

E-mail: valeriu.dulgheru@bpm.utm.md

Abstract. Teeth profiles have an important role in the efficient transformation of motion in the precessional transmissions. Multiple precessional gear theory, previously developed, did not take into consideration the influence of the diagram error of the linking mechanism in the processing device for gear wheel on the teeth profile. Functioning under the multiplication regime, these errors have major influence, which can lead to instant blocking of gear and to power losses. With this purpose, a thorough analysis was conducted on the motion development mechanism under multiplication, and on the teeth profile error generating source. On the basis of fundamental theory of multiple precessional gear, previously developed, a new gear with modified teeth profile and the technology for its industrial manufacturing was proposed and patented. The elaboration of the mathematic model of the modified teeth profile is based integrally on the mathematic model of teeth profile, previously developed by the authors. With this purpose it is necessary to present the detailed description of teeth profile without modification and, then, to present of the description of modified profile peculiarities.

1. Introduction

The diversity of beneficiaries' requirements concern mechanical transmissions is reduced, specially, to the increase of reliability, efficiency and to the decrease of mass and dimensions. It becomes more and more difficult to satisfy the mentions requirements by updating partially the traditional transmissions. This problem can be solved by using new types of mechanical transmissions - planetary precessional transmissions [1,2]. The absolute multiplicity of the precessional gear (up to 100% simultaneously engaged teeth pairs) provides high bearing capacity and increased kinematic accuracy, reduced dimensions and mass. In addition, large kinematic possibilities (-8...-3600), reduced acoustic emission and solution of all technological issues open advantages that are very important for the utilization of planetary precessional transmissions in various areas of machine building. The authors elaborated a large number of diagrams concern planetary precessional transmissions (for reducers, multipliers and differentials), multiple gears for power and kinematic transmissions, gear processing methods and their control, the majority of works being patented with about ***190 patents.*** Know-how in the elaboration of the multicouple precessional gear, manufacturing technology and control methods, and a range of precessional transmission diagrams belong to research team from the Technical University of Moldova.

The specific character of sphere-spatial (precessional motions of the precessional transmissions pinion makes impossible the utilisation of teeth classical involute profiles. This fact requires the elaboration of new profiles adequate to the sphere-spatial motion of pinion which would ensure high performances to the precessional transmission.

2. Elaboration of diagram of precessional transmissions

The elaboration of working machines driving mechanisms is based on the diagram of precessional transmissions, presented in figure 1 [1]. The rotating motion of the crank shaft 1 is transformed into sphere-spatial motion of the block pinion 2 with two teethed crowns 3 and 4, which are rolling without sliding on the immovable and driven toothed wheel teeth 5 and 6. Due to the minimum difference between the number of teeth $Z_5 = Z_3 - 1, Z_6 = Z_4 - 1$, $Z_3 = Z_4 + 1,2,3...$ the transmission ratio is:

$$i = \pm \frac{Z_3 Z_6}{Z_5 Z_4 - Z_3 Z_6}. \quad (1)$$

The teeth of crowns 3 and 4 are manufactured in the shape of conical rollers installed on axis having the possibility to rotate round them, and the teeth of central wheel 5 and 6 have non-standard convex-concave profile.

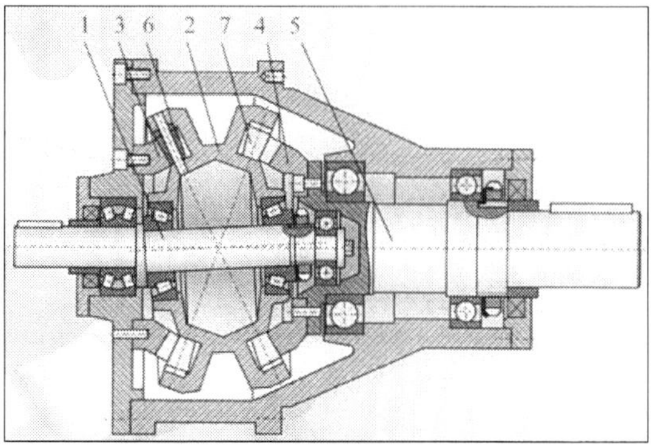

Figure 1. Diagram of the planetary precessional transmission.

3. Parametrical equation of the tooth profile

In precessional gear the planet gear performs spherical-spatial motion around one fixed point [3,4]. It is known [Euler], that the body that performs spherical motion has three degrees of freedom. In theoretical mechanics, usually, the position of the body that performs precessional motion is defined by Euler angles.

The coordinates of the modified point $\boldsymbol{D^m}$ are:

$$\begin{aligned} X_D^m &= -\sin\delta \sin\left[Y_C^m \sin\theta + Z_C^m (1-\cos\theta)\cos\psi\right]; \\ Y_D^m &= -Y_C^m \cos\delta + Z_C^m \sin\delta\left[\cos^2\psi + \cos\theta \sin^2\psi\right]; \\ Z_D^m &= -Y_C^m \sin\delta\left(\cos^2\psi + \cos\theta \sin^2\psi\right) - Z_C^m \cos\delta. \end{aligned} \quad (2)$$

The motion of point $\boldsymbol{D^m}$ related to the movable system linked rigidly to the semiproduct is described by the following formulas

$$\begin{aligned} X_{1D}^m &= X_D^m \cos\frac{\psi}{Z_1} - Y_D^m \sin\frac{\psi}{Z_1}; \\ Y_{1D}^m &= X_D^m \sin\frac{\psi}{Z_1} + Y_D^m \cos\frac{\psi}{Z_1}; \\ Z_{1D}^m &= Z_D^m. \end{aligned} \quad (3)$$

The coordinates of point $\boldsymbol{E^m}$ on the sphere is calculated according to the formulas:

$$\begin{aligned} X_{1E}^m &= k_2^m Z_{1E}^m + d_2^m; \\ Y_{1E}^m &= k_1^m Z_{1E}^m - d_1^m; \\ Z_{1E}^m &= \frac{(k_1^m d_1^m - k_2^m d_2^m)}{k_1^{m2} + k_2^{m2} + 1} - \\ &- \frac{\sqrt{(k_1^m d_1^m - k_2^m d_2^m)^2 + (k_1^{m2} + k_2^{m2} + 1)\cdot(R_D^2 - d_1^{m2} - d_2^{m2})}}{k_1^{m2} + k_2^{m2} + 1}, \end{aligned} \quad (4)$$

where

$$k_1^m = \frac{X_{1D}^m \left(X_{1D}^m \dot{X}_{1D}^m + Y_{1D}^m \dot{Y}_{1D}^m \right) + Z_{1D}^{m\,2} \dot{X}_{1D}^m}{Z_{1D}^m \left(X_{1D}^m \dot{Y}_{1D}^m - Y_{1D}^m \dot{X}_{1D}^m \right)};$$

$$k_2^m = -\frac{\left(k_1^m Y_{1D}^m + Z_{1D}^m\right)}{X_{1D}^m};$$

$$d_1^m = \frac{R_D^2 \cos\beta \, \dot{X}_{1D}^m}{\left(X_{1D}^m \dot{Y}_{1D}^m - \dot{X}_{1D}^m Y_{1D}^m \right)};$$

$$d_2^m = \frac{\left(R_D^2 \cos\beta + d_1^m Y_{1D}^m\right)}{X_{1D}^m}.$$

Point E^m projection on tooth sectional plane has the following coordinates:

$$X_E^{''m} = \varepsilon^m \cdot X_{1E}^m,$$
$$Y_E^{''m} = \varepsilon^m \cdot Y_{1E}^{''m},$$
$$Z_E^{''m} = \varepsilon^m \cdot Z_{1E}^{''m}, \qquad (5)$$

where

$$\varepsilon^m = \frac{D}{AX_{1E}^m + BY_{1E}^m + CZ_{1E}^m}.$$

The modified profile of the tooth is described by the equations:

$$\xi^m = X_E^{''m} cos\frac{\pi}{Z_1} + \left[R_D cos(\delta+\theta+\beta) + Y_E^{''m}\right] sin\frac{\pi}{Z_1};$$
$$\zeta^m = X_E^{''m} sin\gamma sin\frac{\pi}{Z_1} - \left[R_D cos(\delta+\theta+\beta) + Y_E^{''m}\right] sin\gamma cos\frac{\pi}{Z_1} + \qquad (6)$$
$$+\left[R_D sin(\delta+\theta+\beta) + Z_E^{''m}\right] cos\gamma.$$

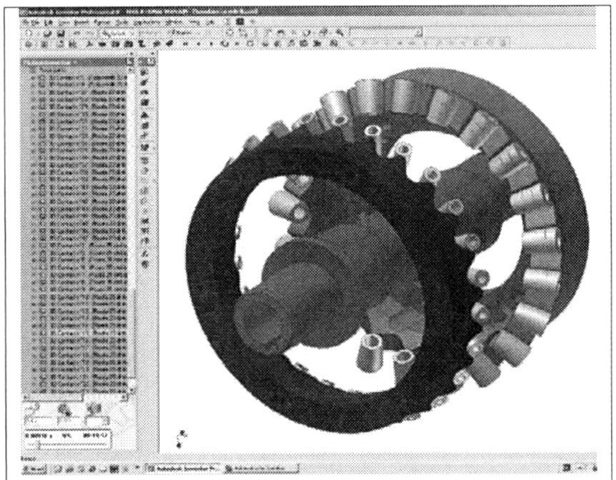

Figure 2. Computerised model of the planetary precessional gear.

The computerised model of the planetary precessional gear on the figure 2 is presented.
Based on the analytical description of teeth profiles by a system of parametric equations on spherical surface the profilograms of the selected gearing teeth have been designed in MatchCAD. The obtained profilograms are shown in figure 3, 4, 5. The analysis of the obtained teeth profiles, based on the fundamental conditions of gearing selection (high bearing capacity due to gearing multiplicity, small dimensions and mass, technology, etc.) has allowed the selection of the teeth profiles parameters.

This continuing dramatically change in available computational resources offers new options in gear design for gear manufacturing processes. They include the complete 3D model of the whole gear, including the gear body and al gear flanks, obtained by using of modelling system *CATIA V5R7*.

Constructions peculiarities and high multiplicity of gear create favourable premises for the improvement of precessional transmissions kinematics accuracy. Within these activities we elaborated:

- theoretical basis for the identification of kinematics error generated by various primary error (frontal and radial knocking), on the basis of error independent action principle by fulfilling computer assisted mathematics experiment;
- compensation method for manufacturing and assembling errors;

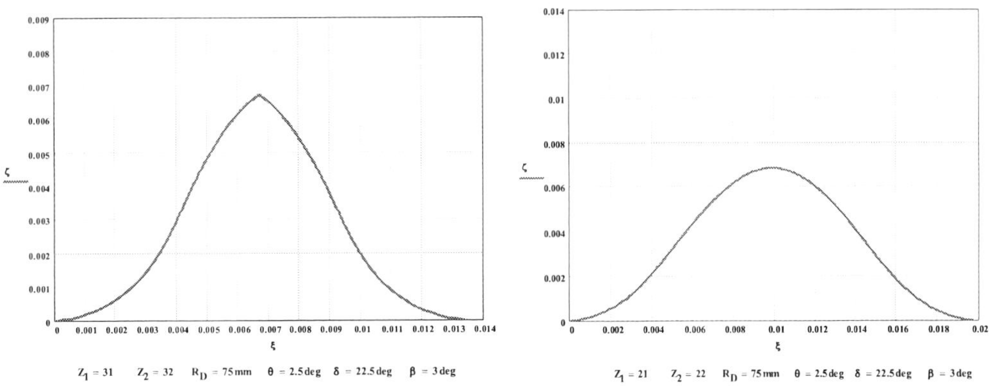

Figure 3. The influence of the number of teeth on the form of teeth profile.

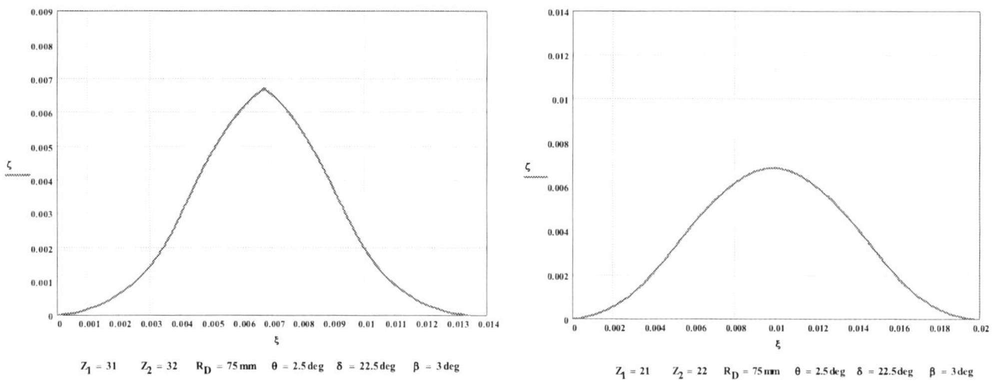

Figure 4. The influence of the number of teeth on the form of teeth profile.

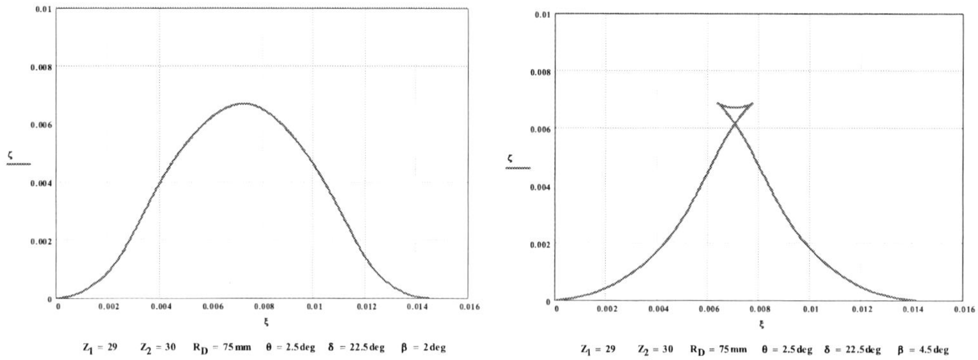

Figure 5. The influence of the taper angle of the rollers, **β** on the form of teeth profile.

- method of determination of probable limit error for precessional reducers with account of the stochastic character of manufacturing and assembling errors.

Special attention was paid to precessional reducers experimental research. For this purpose two laboratories were set up: 1) for mechanical tests and; 2) working technology for gear wheels. The laboratories are equipped with stands for testing and with control and modern measuring devices.

Know-how in the elaboration of the multicouple precessional gear, manufacturing technology and control methods, and a range of precessional transmission diagrams belong to research team from the Technical University of Moldova. During the last 35 years the team patented about 190 inventions.

References:

[1] Bostan, I., Dulgheru, V., Sochirean A., Babaian I. Anthology of inventions: planetary precessional transmissions. Vol.1. – Ch.: S.n., 2011. 593p. ISBN 978-9975-4100-9-0.

[2] Bostan, I. Precessional transmissions with multicouple gear, Stiinta, Chişinău. 1992, ISBN5-376-01005-8.

[3] Bostan, I., Dulgheru V., Vaculenco M. Precessional gear-engagement. Patent nr. 1886MD, 2001.

[4] Bostan, I., Babaian, I. Precessional gear-engagement. Patent nr.1455094 RU (patent nr. 560 MD), 1996.

Acknowledgments

Ion Bostan, PhD.Dr.Sc.Prof., academician, Technical University of Moldova, Department "*Basics of Machine Design*", E-mail: ion.bostan@cnts.utm.md Office Phone: 00373 22 509986. Home Address: Lidia Istrati 20, Str. Home Phone: 00373 22 31 92 63.

Valeriu Dulgheru, PhD.Dr.Sc.Prof., Head of Department, Technical University of Moldova, Department "*Basics of Machine Design*", E-mail: valeriu.dulgheru@bpm.utm.md Office Phone: 00373 22 509939. Home Address: Lidia Istrati 23, Str. Home Phone: 00373 22 31 92 05.

IOP Conf. Series: Materials Science and Engineering **514** (2019) 012026 doi:10.1088/1757-899X/514/1/012026

Cinetostatic analysis of planetary precessional multiplier

I Bostan[1], V Dulgheru[1] and R Ciobanu[1]

[1]Basic of Machine Design Department, Technical University of Moldova, Republic of Moldova

E-mail: valeriu.dulgheru@bpm.utm.md

Abstract. The majority planetary precessional transmissions diagrams developed previously operate efficiently in reducer's regime. Teeth profiles have an important role in the efficient transformation of motion in the precessional transmissions that operate as multipliers. Multiple precessional gear theory, previously developed, did not take into consideration the influence of the diagram error of the linking mechanism in the processing device for gear wheel on the teeth profile. Functioning under the multiplication regime, the errors of the linking mechanism have major influence, which can lead to instant blocking of gear and to power losses. With this purpose, a thorough analysis was conducted on the motion development mechanism under multiplication, and on the teeth profile error generating source. This paper describes also the precessional transmission diagrams operate efficiently in multiplier's regime, and cinetostatic analysis of precessional multipliers.

1. Introduction

Depending on the structural diagram, precessional transmissions fall into two main types – *K-H-V* and *2K-H*, from which a wide range of constructive solutions with wide kinematical and functional options that operate in multiplier regime. The kinematical diagram of the precessional transmission *K-H-V* (fig. 1 [1]) comprises five basic elements: planet career *H*, satellite gear *g*, two central wheels *b* with the same number of teeth, controlling mechanism *W* and the body (frame). The roller rim of the satellite gear *g* gears internally with the sun wheels *b*, and their teeth generators cross in a point, so-called the centre of precession. The satellite gear *g* is mounted on the planet (wheel) career *H*, designed in the form of a sloped crank, which axis forms some angle with the central wheel axis *θ*.

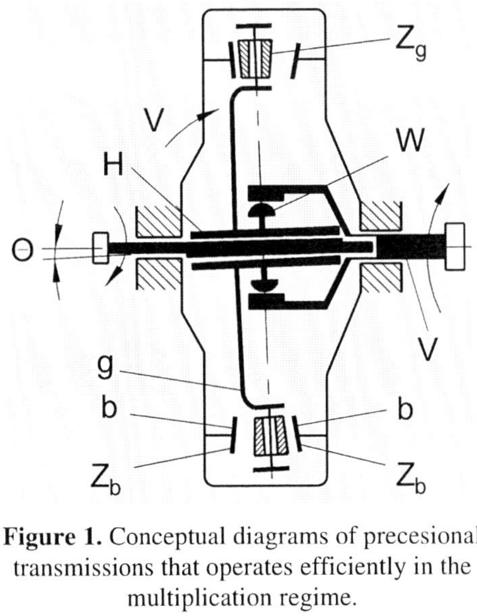

Figure 1. Conceptual diagrams of precesional transmissions that operates efficiently in the multiplication regime.

Revolving, the sloped crank *H* transmits sphero-spatial motion to the satellite wheel regarding the ball hinge installed in the centre of precession. For the transmission with the controlling mechanism designed as clutch coupling (fig.1.141, a), the gear ratio (gear reduction rate) varies in the limits:

$$i_{HV}^{g} = -\frac{z_g \cos\Theta - z_b}{z_b};\ i_{HV}^{g} = -\frac{z_g \cos\Theta - z_b}{z_b \cos\Theta}, \quad (1)$$

reaching the extreme values of 4 times for each revolution of the crank ***H***. If necessary, this shortcoming can be eliminated using as a controlling mechanism the constant Cardan joint (Hooke's joint), the ball synchronous couplings, etc.

$$i_{HVmed}^{g} = -\frac{z_g - z_b}{z_b}. \quad (2)$$

For $z_g = z_b+1$, $i_{HV}^{g} = -\frac{1}{z_b}$, the driving and driven shafts have opposite directions.

For $z_g = z_b-1$, $i_{HV}^{g} = \frac{1}{z_b}$, the shafts revolve in the same direction.

This kinematical diagram of the precessional transmission ensures a range of gear ratios $i = 8...60$, but in the multiplication regime it operates efficiently only for the range of gear ratios $i = 8...25$. As well, in the controlling mechanism W, that operates with pitch angles of the semi couplings up to 3°, power losses occur reducing the efficiency of the multiplier on the whole.

To avoid power losses in the multiplier and to widen the kinematical options (fig. 2) [2] the conceptual diagram of the precessional multiplier with wide kinematical options was designed.

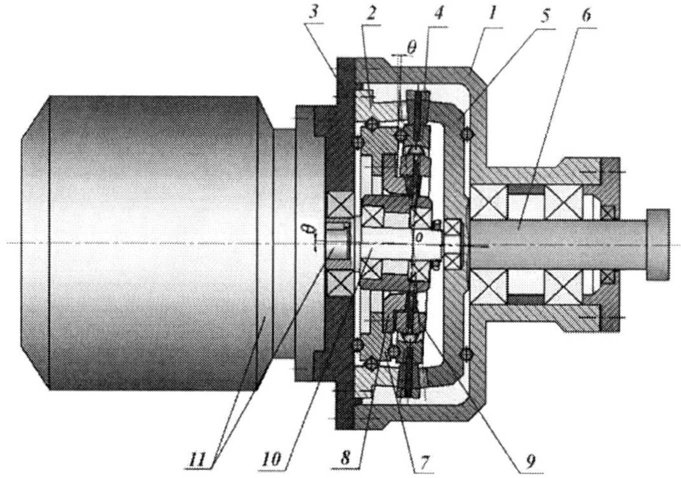

Figure 2. Planetary precessional multiplier with satellite gear mounted radially.

The planetary precessional multiplier comprises the following units: the housing 1, inside which the fixed sun wheel 2 is placed and connected rigidly to the housing cover 3, exterior satellite wheel 4 with the teeth in the shape of rollers, movable sun wheel 5, linked rigidly to the input shaft 6. The satellite wheel 3 is connected kinematically with the sloped flange of the disk 7, connected rigidly with the sun wheel 8, that gears with the interior satellite wheel 9 mounted unbound on the output crank shaft 10, and linked rigidly to the rotor generator 11. The exterior satellite wheel 4 is mounted void on gear bodies on exterior spherical surfaces of the interior wheel 9. The pitch angle of the output crank shaft 10 axis and of the sloped flange is equal to . The rotational motion of the input shaft 6 is transmitted to the movable sun wheel 5. Due to the difference in the number of teeth of wheel 5 and of the exterior satellite wheel (gear) 4 ($Z_6 = Z_5 \pm 1$), the last will have to carry out a precession motion around the fixed point O (centre of precession). Precessional motion around its axis is excluded as the number of teeth of the sun wheel 2 is equal to the number of rollers of the satellite gear 4 ($Z_2 = Z_4$). The precessional motion of the exterior satellite gear 4 is transformed, by means of an inclined flange of the disc 7, into rotational motion around the disc axis 7 that will revolve by the degree of multiplication

$$i_7 = -\frac{z_5}{z_4 - z_5}, \tag{3}$$

where Z_4 is the number of rollers of the exterior satellite gear 4;

Z_5 – is the number of teeth of the movable sun wheel 5.

The multiplied rotational motion of the disc 7 is transformed into multiplied precessional motion of the interior satellite gear 9, due to the difference in the number of teeth $Z_8 = Z_9 \pm 1$. At the rotation of disc 7 of the sun wheel 8 at an angle equal to the teeth angular pitch, the interior satellite gear will perform a complete precession cycle around point „O". The precessional motion of the interior satellite gear 9 is transformed by means of output crank shaft 10 into rotational motion, multiplied to the degree of multiplication

$$i_{10} = -\frac{z_8}{z_9 - z_8}, \tag{4}$$

where Z_8 is the number of teeth of the sun wheel 8,

Z_9 – is the number of rollers of the satellite gear 9.

The multiplied rotational motion of the output crank shaft 10 is transmitted to the generator rotor 11.

IOP Conf. Series: Materials Science and Engineering **514** (2019) 012026 doi:10.1088/1757-899X/514/1/012026

2. Analytic description of teeth profile and justification of precessional gear parameters selection

Teeth profiles have an important role in the efficient transformation of motion in the precessional transmissions that operate as multiplier. Multiple precessional gear theory, previously developed, did not take into consideration the influence of the diagram error of the linking mechanism in the processing device for gear wheel on the teeth profile.

Functioning under the multiplication regime, these errors have major influence, which can lead to instant blocking of gear and to power losses. With this purpose, a thorough analysis was conducted on the motion development mechanism under multiplication, and on the teeth profile error generating source. On the basis of fundamental theory of multiple precessional gear, previously developed, a new gear with modified teeth profile and the technology for its industrial manufacturing was proposed and patented [3].

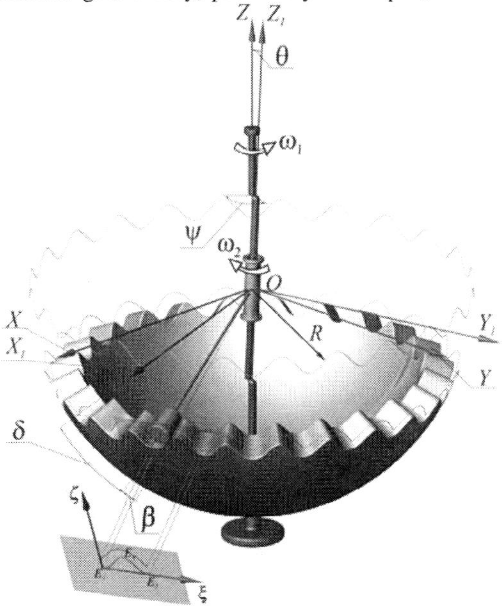

Figure 3. Tooth profile in normal section.

To ensure continuity of the transfer function and to improve the performances of precessional transmission under multiplication it is necessary to modify teeth profile with the diagram error value ψ_3 by communicating supplementary motion to the tool. In this case the momentary transmission ratio of the manufactured gear will be constant. Usually, in theoretical mechanics the position of the body making spherical-spatial motion is described by Euler angles.

The mobile coordinate system $\boldsymbol{OX_1Y_1Z_1}$ is connected rigidly with the satellite wheel, which origin coincides with the centre of precession $\boldsymbol{0}$ (fig. 3) and performs spherical-spatial motion together with the satellite wheel relative to the motionless coordinate system ***OXYZ.***

The elaboration of the mathematic model of the modified teeth profile is based integrally on the mathematic model of teeth profile, previously developed by the authors. With this purpose it is necessary to present the detailed description of teeth profile without modification and, then, to present of the description of modified profile peculiarities.

3. Description of teeth profile designed on sphere

An arbitrary point D of the tool axis describes a trajectory relative to the fixed system according to the equations:

$$\begin{aligned} X_D^m &= -\sin\delta \sin\left[Y_C^m \sin\theta + Z_C^m (1-\cos\theta)\cos\psi\right]; \\ Y_D^m &= -Y_C^m \cos\delta + Z_C^m \sin\delta\left[\cos^2\psi + \cos\theta \sin^2\psi\right]; \\ Z_D^m &= -Y_C^m \sin\delta\left(\cos^2\psi + \cos\theta \sin^2\psi\right) - Z_C^m \cos\delta. \end{aligned} \tag{5}$$

Index ***m*** means „*modified*".

The motion of point $\boldsymbol{D^m}$ compared to the movable system connected rigidly to the semi product is described by formulas:

$$\begin{aligned} X_{1D}^m &= X_D^m \cos\frac{\psi}{Z_1} - Y_D^m \sin\frac{\psi}{Z_1}; \\ Y_{1D}^m &= X_D^m \sin\frac{\psi}{Z_1} + Y_D^m \cos\frac{\psi}{Z_1}; \\ Z_{1D}^m &= Z_D^m. \end{aligned} \tag{6}$$

The projections of point $\boldsymbol{D^m}$ velocities is expressed by formulas:

$$\dot{X}_D^m = -\sin\delta\cos\psi\left[Y_C^m\sin\theta + Z_C^m(1-\cos\theta)\cos\psi\right]\dot{\psi} -$$
$$-\sin\delta\sin\psi\left[\dot{Y}_C^m\sin\theta + \dot{Z}_C^m(1-\cos\theta)\cos\psi - Z_C^m(1-\cos\theta)\sin\psi\cdot\dot{\psi}\right]; \qquad (7)$$
$$\dot{Y}_D^m = -\dot{Y}_C^m\cos\delta + \dot{Z}_C^m\sin\delta\left[\cos^2\psi + \cos\theta\sin^2\psi\right] +$$
$$+Z_C^m\sin\delta\left[-2\cos\psi\sin\psi + 2\cos\theta\sin\psi\cos\psi\right]\dot{\psi};$$
$$\dot{X}_{1D}^m = \dot{X}_D^m\cos\frac{\psi}{Z_1} - \frac{\dot{\psi}}{Z_1}X_D^m\sin\frac{\psi}{Z_1} - \dot{Y}_D^m\sin\frac{\psi}{Z_1} - \frac{\dot{\psi}}{Z_1}Y_D^m\cos\frac{\psi}{Z_1};$$
$$\dot{Y}_{1D}^m = \dot{X}_D^m\sin\frac{\psi}{Z_1} + \frac{\dot{\psi}}{Z_1}X_D^m\cos\frac{\psi}{Z_1} + \dot{Y}_D^m\cos\frac{\psi}{Z_1} - \frac{\dot{\psi}}{Z_1}Y_D^m\sin\frac{\psi}{Z_1}.$$

The coordinates of point $\boldsymbol{E^m}$ on the sphere is calculated by formulas:

$$X_{1E}^m = k_2^m Z_{1E}^m + d_2^m;$$
$$Y_{1E}^m = k_1^m Z_{1E}^m - d_1^m; \qquad (8)$$
$$Z_{1E}^m = \frac{(k_1^m d_1^m - k_2^m d_2^m) - \sqrt{(k_1^m d_1^m - k_2^m d_2^m)^2 + (k_1^{m2} + k_2^{m2} + 1)\cdot(R_D^2 - d_1^{m2} - d_2^{m2})}}{k_1^{m2} + k_2^{m2} + 1},$$

where:

$$k_1^m = \frac{X_{1D}^m\left(X_{1D}^m\dot{X}_{1D}^m + Y_{1D}^m\dot{Y}_{1D}^m\right) + Z_{1D}^{m\,2}\dot{X}_{1D}^m}{Z_{1D}^m\left(X_{1D}^m\dot{Y}_{1D}^m - Y_{1D}^m\dot{X}_{1D}^m\right)}; \quad k_2^m = -\frac{\left(k_1^m Y_{1D}^m + Z_{1D}^m\right)}{X_{1D}^m};$$

$$d_1^m = \frac{R_D^2\cos\beta\,\dot{X}_{1D}^m}{\left(X_{1D}^m\dot{Y}_{1D}^m - \dot{X}_{1D}^m Y_{1D}^m\right)}; \quad d_2^m = \frac{\left(R_D^2\cos\beta + d_1^m Y_{1D}^m\right)}{X_{1D}^m}.$$

According to the obtained analytical relations a soft for the calculation and generation of teeth was developed in CATIA V5R7 modelling system that allowed obtaining the modified trajectories of points E^m_e and E^m_i on the spherical front surfaces, both exterior and interior ones, by which the teeth surface was generated (fig. 4).

Projection of point E^m on the tooth transversal plane has the following coordinates:

$$X_E^{\prime\prime m} = \varepsilon^m\cdot X_{1E}^m, \quad Y_E^{\prime\prime m} = \varepsilon^m\cdot Y_{1E}^m, \quad Z_E^{\prime\prime m} = \varepsilon^m\cdot Z_{1E}^m, \qquad (9)$$

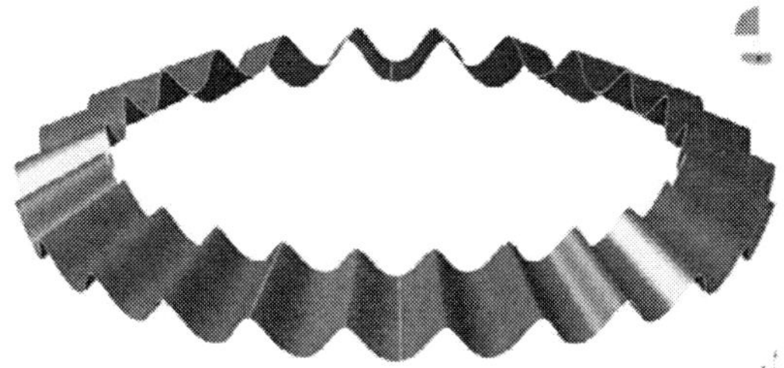

Figure 4. Teeth generating surface.

where $\varepsilon^m = -\dfrac{D}{AX_{1E}^m + BY_{1E}^m + CZ_{1E}^m}$.

The modified teeth profile in plane is described by the equations:

$$\xi^m = X_E^{\prime\prime m}\cos\frac{\pi}{Z_1} + \left[R_D\cos(\delta+\theta+\beta) + Y_E^{\prime\prime m}\right]\sin\frac{\pi}{Z_1};$$
$$\zeta^m = X_E^{\prime\prime m}\sin\gamma\sin\frac{\pi}{Z_1} - \left[R_D\cos(\delta+\theta+\beta) + Y_E^{\prime\prime m}\right]\sin\gamma\cos\frac{\pi}{Z_1} + \qquad (10)$$
$$+\left[R_D\sin(\delta+\theta+\beta) + Z_E^{\prime\prime m}\right]\cos\gamma.$$

A wide range of modified teeth profiles with different geometrical parameters were generated in MathCAD 2010 Professional software (fig. 5). The solid model of a gear wheel is shown in fig. 6.

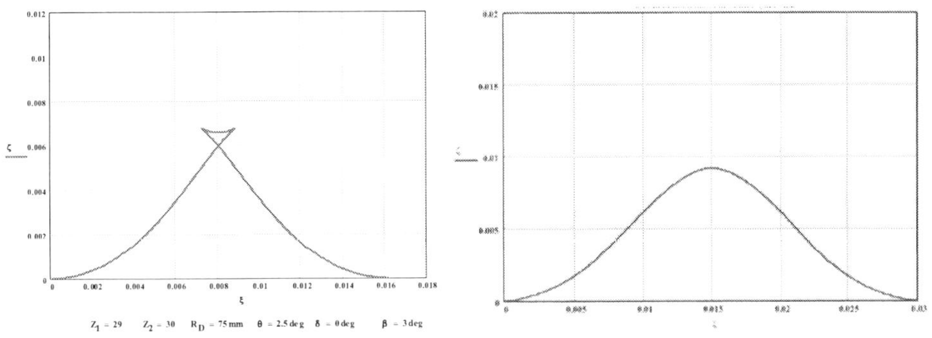

Figure 5. Teeth profiles for multipliers.

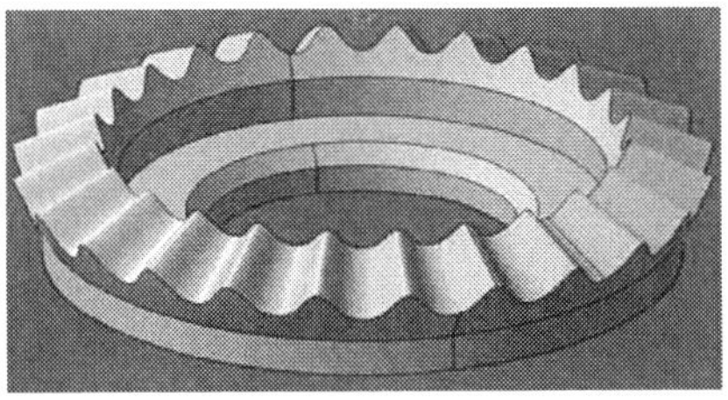

Figure 6. Computerised model of the sun gear.

Based on the carried out research it was established that from the point of view of decreasing energy losses in gearing, in the multiplication mode of operation, the gearing angle should be $\alpha > 45^0$, and the nutation angle (the pitch angle of the crank shaft) should be – $\theta \leq 2,5^0$. This is dictated by the reverse principle of movement in the multipliers compared to the reducers: the axial component of the normal force in gear must be maximal to drive the crank shaft in the rotation movement through the satellite wheel.

References

[1] Bostan I, Dulgheru V, Sochirean A., Babaian I. Antologia invențiilor, Vol. 1. Transmisii planetare precesionale. Chișinău (Bons Offices), 2011. 594p. ISBN 978-9975-4100-9-0.

[2] Bostan I., Dulgheru V., Ciobanu R., Ciobanu O. Patent No. 3153 B1 (MD), CIB F16H1/32. Planetary precessional multiplier. Publ. BOPI - 2006. - No. 9.

[3] Bostan I., Țopa M., Dulgheru V., Vaculenco M. Patent No. 1886 B1 (MD), CIB F16H1/32; B23F9/06. Precessional gearing and procedure for its fulfilment / T.U.M. – No.2000 – 071. Publ. BOPI – 2002. – No.3.

Acknowledgments

Ion Bostan, PhD.Dr.Sc.Prof., academician, Technical University of Moldova, Department "*Basics of Machine Design*", E-mail: ion.bostan@cnts.utm.md Office Phone: 00373 22 509986. Home Address: Lidia Istrati 20, Str. Home Phone: 00373 22 31 92 63.

Valeriu Dulgheru, PhD.Dr.Sc.Prof., Head of Department, Technical University of Moldova, Department "*Basics of Machine Design*", E-mail: valeriu.dulgheru@bpm.utm.md Office Phone: 00373 22 509939. Home Address: Lidia Istrati 23, Str. Home Phone: 00373 22 31 92 05.

Radu Ciobanu, PhD. Assoc.prof. Technical University of Moldova, Department "*Basics of Machine Design*", E-mail: radu.ciobanu@bpm.utm.md Office Phone: 00373 22 509988. Home Address: Studenților, 3/1,Str.

IOP Conf. Series: Materials Science and Engineering **514** (2019) 012027 doi:10.1088/1757-899X/514/1/012027

A survey on gear meshing quality based on tooth contact analysis

M C Radu[1], L Andrei[1] and G Andrei[1]

[1]Mechanical Engineering Department, Faculty of Engineering, *Dunărea de Jos* University of Galați, România

E-mail: radumihaicat@gmail.com

Abstract. Gears are common mechanical elements in power transmission systems. In order to avoid failure of transmission systems, it is necessary to inspect the gear meshing condition with respect to kinematic and dynamic parameters. Tooth contact analysis (TCA) method represents a useful tool for designing and analysing gear performance. It represents an important method to examine gear tooth contact points, contact ratios and kinematic errors. The primary objective of TCA is to establish the contact trail on gear tooth surfaces, and also to determine the transmission error functions generated by gear misalignments. TCA has several branches such as the loaded tooth contact analysis, unloaded tooth contact analysis and tooth contact analysis with errors. The determination of the initial contact point is necessary in order to achieve an accurate solution. The present survey allows a new insight into the improvement of the gear meshing quality, based on tooth contact analysis and transmission error.

1. Introduction

Gears are common mechanical elements in power transmission systems. Numerous industries benefit from their use. Their main advantages are robustness, small dimension, constant transmission ratio, high performance etc. In order to avoid failure of transmission systems, it is necessary to inspect the gear meshing condition in terms of kinematic and dynamic parameters. Gears also demand high accuracy of machining and assembly process which may influence gear meshing system by spreading the vibration and noise. The transmission error is considered the main cause for noise occurrence. Gearing noise is one of the main concerns for mechanical engineers.

Walker [1] was the first to study noise and vibration by analyzing the modifications of spur gear tooth profile. Gear noise has three important elements: gear whine, gear clatter and gear rattle. Gear whine is a result of time-varying mesh stiffness and is regarded as a clean tonal noise [2]. Whining noise represent the noise produced by the meshing gears under load [3]. Its existence is an effect of the static transmission error resulted from the meshing process [4,5]. Gear rattling occurs as a consequence of gear backlash and may be present when the meshing gears are under a slight load [6]. Gear clatter is generated by torsional oscillations when the gear mesh is under load.

Tooth contact analysis (TCA) algorithm is a useful tool for designing and analysing gears performance. It represents an important method to examine gear tooth contact points, contact ratios and kinematic errors. Involute tooth surfaces are usually used to achieve linear contact and lower transmission error functions. Usually, in order to reduce noise and vibration in transmission systems, high contact ratio gears are used. The primary objective of TCA is to establish the contact path on gear tooth surfaces and also to determine the transmission error functions generated by gear misalignments. Gear meshing quality can be assessed through a number of elements, like the transmission error and contact pattern.

This paper is a state-of-the-art on the theory of TCA algorithm that determines the gear meshing.

2. Condition of continuous tangency

Three stiff independent coordinate systems (S_1,S_2,S_f) interacting with both gears and the framework are considered. In order to reproduce the misalignment, an extra settled coordinate system is connected (S_q). In S_1 and S_2 coordinate systems, two tooth surfaces (Σ_1,Σ_2) are examined. The first gear with the corresponding tooth surface (Σ_1) spin about a settled axis situated in S_f. Therefore, the coordinate system S_f contains a group of gear tooth surfaces. The second gear with the corresponding tooth surface (Σ_2) rotates about another settled axis situated in S_q. The position and direction of the coordinate system S_q, considering Sf, reproduces the misalignment of the pinion [7]. When the location vectors and the normal fit at any moment, it is considered that contacting surfaces are in continuous tangency (Fig. 1).

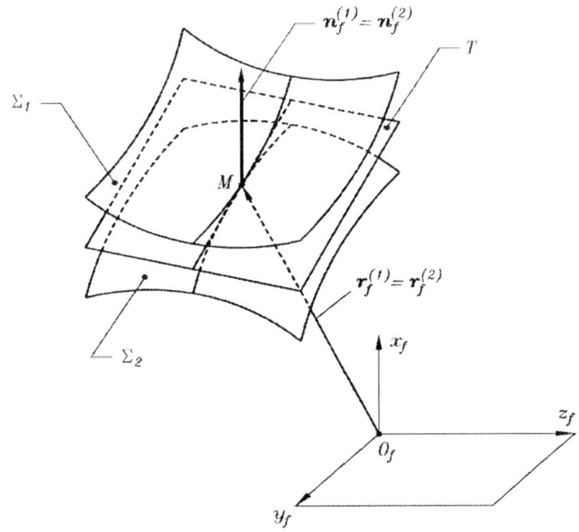

Figure 1. Tangency of surfaces in ideal gear train [7]

The equation of continuous tangency of pinion and gear surfaces is reached when establishing the vector conditions:

$$r_f^{(1)}(u_1,\theta_1,\phi_1) - r_f^{(2)}(u_2,\theta_2,\phi_2) = 0; \quad n_f^{(1)}(u_1,\theta_1,\phi_1) - n_f^{(2)}(u_2,\theta_2,\phi_2) = 0 \qquad (1)$$

where r_f represent the position normal vector, n_f is the unit normal vector, u_i and θ_i are the parameters of the gear surface and ϕ_i represent the angle of gear revolution.

3. Mathematical pattern of surface contact analysis

The involute curve represents the most common gear profile and also the ideal shape of the unmodified tooth flank. When the modified tooth flank case is considered, a section of the pressure line is kept unchanged during the gear meshing, in order to maintain the involute curve [8].

$$L_{Ca} = 0.5 \times (g_\gamma - p_e) \qquad (2)$$

where L_{Ca} is the length of tip and root relief, g_γ represent the length of the pressure line and p_e is the base pitch length.

Tooth flank is set by covering the curvature family of the cutting route. Tooth flank geometry can be described using a universal generation model which is developed as a widespread machine setting. Usually, eight fundamental generic machine tools settings are well-known. Using conversion matrices for symbolizing the kinematic chain, tooth flank generic mathematical model was achieved [9]:

$$\begin{cases} r_b(\mu,\theta,\phi) = M_{bc}(R_a,S_r,E_M,X_D,X_B,\gamma_m,\sigma,\zeta;\phi) \times r_c(\mu,\theta) \\ n_b(\mu,\theta,\phi) = M_{bc}(R_a,S_r,E_M,X_D,X_B,\gamma_m,\sigma,\zeta;\phi) \times n_c(\mu,\theta) \\ n_b(\mu,\theta,\phi) \times v(\mu,\theta,\phi) = 0 \end{cases} \qquad (3)$$

where $n_b \times v = 0$ form the gearing theory, μ and θ are the cutter blade parameters, also known as Gaussian parameters and ϕ is a motion parameter. M_{bc} represents the transformation matrix. $R_a, S_r, E_M, X_D, X_B, \gamma_m, \sigma, \zeta$ are the fundamental machine tool settings used to change the polynomial terms through the manufacturing process.

The cutter blade usually has three types of blade shape: straight, Top-Rem and parabolic profile.

3.1. Determining the points of contact

Cao et al. [10] presented in their article the mathematical pattern for estimating the edge contact, tooth contact area and depicted the tooth contact analysis ease-off numerical procedure.

The common algorithm presents the following contact conditions: the unit normal vectors are aligned at the point of contact and the surface position vectors are coincidences. The equations can be presented as [10]:

$$\begin{cases} r_h^1(u_1, v_1, \varphi_1) = r_h^2(u_2, v_2, \varphi_2) \\ n_h^1(u_1, v_1, \varphi_1) = n_h^2(u_2, v_2, \varphi_2) \end{cases} \tag{4}$$

where $\mathbf{r}_i$ are the position vectors of pinion and gear, $\mathbf{u}_i$ and $\mathbf{v}_i$ are the face generating characteristics and $\boldsymbol{\varphi}_i$ represent the angular position.

When the $\mathbf{r}_1$ and $\mathbf{r}_2$ functions are considered, a string of instantaneous contact points are created considering each angular position of the gear drive. The equivalent transmission error function can be expressed as [10]:

$$\Delta\varphi_2(\varphi_1) = \varphi_2 - (z_1/z_2) \times \varphi_1 \tag{5}$$

where the Transmission Error is considered to be the gap between the current position of the rotation angle of the output gear and the theoretical one if the meshing gears are ideally conjugate [11,12].

The common method for the estimation of the instantaneous contact point ensures a good precision as a result of the necessity for an answer to solve the nonlinear equations.

3.2. Transmission Error and the separation of the point of contact

A function H(j) is considered to be the difference between the contact curve (CC) and the changed curve at the considered mating point taking into account the gear drive angular position [10]. The instantaneous unloaded TE represents the minimum difference Δ(i) between the instantaneous conjugate CC and the changed curve. The unloaded TE can be written as [10]:

$$TE(i) = [\Delta(i)/(-z_1^* n_{y1}^* + y_1^* n_{z1}^*)](z_1/z_2) \tag{6}$$

where $\mathbf{n}_1^*$ and $\mathbf{r}_1^*$ represent the unit normal and position vectors of the point p_1^*, and the length deviation Δ(i) had to be transformed into angular deviation.

4. Contact characteristics

Considering shaft misalignment and tip corner contact, it is necessary to inspect the impact of the flank changes on the contact features [8]. Lately, numerous researches have been conducted to investigate the contact characteristics. Pedrero determines the surface resistance and tooth bending strength by analysing a load distribution sample developed on the minimum resilient potential principle. Munro [13] proposed an analytical method for estimating the transmission error of spur gears heaving high contact ratio and corner contact.

Zhuo et al. [14] used spiral and hypoid bevel gears to obtain a method for the optimization of the transmission error and tooth contact pattern. Their method includes optimization objectives, a genetic algorithm and control parameters in order to found the pinion tooth face. They determined the correlation between the curvature elements and the pitch cone and achieved a new list of basic equations for the hypoid gear pitch cone elements. A perfect geometry for the tooth contact pattern cannot be established using standard machine adjustments and cutter specifications for the fabrication of hypoid and spiral bevel gear controls. Tooth contact pattern may present errors related to shape, location or size. Many types of control parameters exist, such as pressure and spiral angle, pitch angle and pitch cone distance, root angle or the coefficient of contact zone length. Still, all these parameters

IOP Conf. Series: Materials Science and Engineering **514** (2019) 012027 doi:10.1088/1757-899X/514/1/012027

are not appropriate to use for the design of meshing features. Changes in machine-tool settings lead to different types of transmission errors. The shape of the tooth surface is influenced by the machine-tool parameters and also by the type of the cutter blade. Curvature characteristics are determined by the cutter blade section while the adjustment parameters have a strong correlation with the pitch cone elements.

The need to reduce the gear noise and sensitivity to misalignment represents the main issue for the aerospace and automotive industries. The original contact point had to be identified at the beginning of the process in order to acquire an exact response. The initial contact point affects the dimension of the contact pattern, the contact pressure and the amplitude of TE [15]. Huang et al. article [16] present the modification of the contact pressure produced between the meshing teeth. The contact stress changes are investigated by the FEM method.

Jabbour and Asmar [17] determined a method for the estimation of contact stress throughout each contact line of spur and helical gears. In other paper [18] they created a pattern for the simulation of the conduct of a plastic helical gear.

Liang et al. [19] present the estimation strategy for the deviation of tooth areas of the conjugate-curve pinion. Zhou [20] described the micro-contact characteristics of gear tooth shape using an elastic-plastic roughness contact pattern considering or not the friction case. The micro-contact characteristics of different rough surfaces are very important to friction, fatigue, wear or to the transmission performances.

Fernandez del Rincon et al. [21] proposed a connection between a regional and a global term to establish the deformation of each contact point. The local term was established using the Hertzian contact method, and the FEM analysis was used to determine the global term. Winter et al. [22] study the load distribution when profile changes are considered. Sanchez et al. [23] also had approached the problem of a load distribution model for inward gear transmission who used the minimum elastic potential energy. Their model established the boundary load regime and the level of the contact and bending stress. The elastic potential energy can be described as the sum of the compressive part with the shear component and the bending component. The contact stress can be determined using the Hertz's equation. Contact stress was also investigated by Chen et al. [24] who studied the case of concave conical involute meshing gears having a non-parallel axis.

In the signal-vibration study for the improvement of gearbox diagnostic models, it is necessary to determine the impact of gear faults in order to estimate the errors and boundary operation conditions [25]. Tharmakulasingam [2] used numerical methods and FEA to highlight the result of tooth profile changes on the TE in spur gears. Sheveleva et al. [26] suggested a way for tracking the contact path where the relative location among tooth surface pairs was important and the revolution of a gear wheel was established when the contact was acquired.

Kawasaki et al. [27] used Klingelnberg cyclo-palloid model to present the design process, the determination of the contact path and TE and the effect of assembly errors on gear meshing properties.

Digital image correlation represents an efficient instrument for the measurement of in-plane deformation considering a planar object area. It specifically gives full-field displacements with a very good precision and strains by looking at the computerized pictures of a test subject area obtained before and after the deformation [28].

4.1. Basic contact conditions for meshing teeth

The engaging teeth having the specific displacements (δ_1,δ_2) come into contact at the point Q_y. The contact situation of two different points on the mating teeth, having a partition distance h, is described by the following equations [29]:

$$h + w_{k1} + w_{k2} = \delta \text{, when the two points are in contact} \quad (7)$$

$$h + w_{k1} + w_{k2} > \delta \text{, when the two points are out of contact} \quad (8)$$

where w_{k1} and w_{k2} represent the specific deformation.

4.2. Discretization of the contact zone of the meshing tooth

The contact surface of the mating teeth is usually discretized into a large number of units on the mutual tangential plane. The influence coefficient method is used to determine the deformations having the assigned stresses on the separated elements. As shown in fig.2 the deformation w_{kj} of the medium point P_k of a single element, which is generated by a stress p_j that influence another spot P_j, can be interpreted using the influence coefficient $f_{k,j}$ [29].

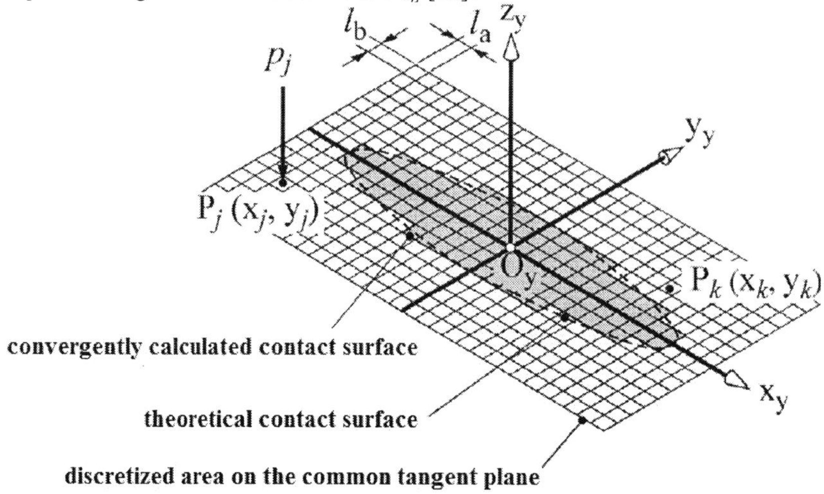

Figure 2. Gear meshing on the mutual tangent plane [29]

4.3. Corner contact

In their work Yu and Mechefske [30] developed an analytical procedure to indicate the impact of corner contact on the static transmission error and TVMS of loaded spur gears. They used FEM to calculate the dynamic response of meshing gears and the static transmission error (STE).

Corner contact represents the contact which is external to the standard contact length and it takes place as a result of the resilient deviation of the loaded teeth. This situation is considered to be the cause for early meshing. The contact point is influenced by the implemented torque. The elastic deformation grows with the increase of the transmitted load. Gear teeth partition distance represents the length between mating teeth without connection when there is no resilient deformation. The contact situation is established by comparing the distance along the pressure line with the LSTE. Lin et al. [31] presented the link between transmission error and corner contact without taking into consideration gear tooth profile faults.

For spur gears having a standard contact ratio, corner contact appears when the approached tooth couple arrive at the presumed starting point of the meshing and the separating tooth couple is leaving the finish point of the meshing considered. Yu et al. [30] made the following assumptions: corner contact concerns only the gear mesh at the beginning and the end of a singular tooth couple part in meshing and have no effect on the double engagement.

4.4. Tip relief optimization

In Fuentes et al. paper [32] a parabolic tip relief is used by changing the cross-section geometry of the generating tool.

Spur and helical gear use involute tooth faces which ensure linear contact and a reduced function of TE considering normal meshing situation. Spur gears can be considered a special case of the helical gears when the helix angle is nil. Tip relief is generally necessary in order to achieve better contact conditions throughout gear meshing. Tip relief represents the permanent elimination of the material from the top edge of the tooth face. It can be implemented using linear and parabolic functions of deflection of the tooth profile. In the absence of tip relief, different surfaces of high contact stress arise on the top edge of the tooth face. The use of an ideal level of tip relief allows for improvement of contact solicitation throughout the entire meshing period and bypass the occurrence of the different surfaces with intense contact solicitation.

As a result of the tip relief changes, bending stress rises slightly influenced by the decrease of the actual contact ratio.

5. Tooth contact analysis

TCA software programs predict meshing and contact of gear tooth areas taking into account the localized bearing contact, which provides the surface contact point. Using the TCA algorithm the following objectives are set [7]:

a) the transmission errors generated by the misalignment between gear axes;

b) the contact route on pinion tooth surfaces;

c) bearing contact as the arrangement of momentary contact ellipses.

Considering misalignment, the place and direction of gear axes are known and gear tooth surfaces are perfectly determined.

TCA technique represents an important instrument for achieving spiral and hypoid gears and also determines gear tooth contact performances [9]. TCA outcome is strongly influenced by the original contact point. TCA equations are established when the considered original values that affect the original contact point are acceptable, otherwise the computational procedure is stopped or the TCA outcome is incorrect.

There are currently several ramifications of the tooth contact analysis technology, such as the loaded tooth contact analysis (LTCA), the unloaded tooth contact analysis (UTCA) or the tooth contact analysis with errors (ETCA).

Vogel et al. [33] used a so-called "ease-off topography" to estimate the tooth contact analysis of hypoid bevel gears. In their research, the transmission error, the curvature properties and the path of contact parameters were obtained. A formulation of the ease-off was given by Stadtfeld [34] who also determined the instantaneous contact between two areas.

In Su Jinzhan et al. paper [35] it was achieved a seventh-order polynomial function of transmission error for spiral bevel gears. They used both TCA and LTCA algorithms to compare the seventh-order polynomial function and the parabolic function of transmission error. The Finite Element Method, mathematical programming and flexibility matrixes are part of the numerical simulation LTCA technique. A reverse TCA procedure is set for the derivation factor of the changed roll.

Jiang Jinke and Fang Zongde [36] presented a solution to diminish noise and vibration of cylindrical spur gears by changing the tooth face using a verifiable higher sequence polynomial function of TE (H-TE) and also described the process of achieving the mathematical pattern. They optimized the loaded transmission error minimal amplitude (ALTE) using TCA and LTCA of pinions and used particle swarm optimization technique in order to achieve the polynomial coefficients of H-TE. The TCA algorithm represents a solution to analyse the profile of transmission errors and the changes of contact pattern caused by manufacturing and assembly errors. Under lightly load tooth meshing features and TE can be established using TCA technics. A derivation of gear drive tooth surface and additional rotation factors were created using a reverse TCA pattern.

TCA algorithm and FEA software are the methods by which computer simulation of contact pattern and gear meshing is performed. Bearing contact among different surface pairs do not influence the TCA procedure. The estimation of TCA does not require solving nonlinear equations, but it has to take into account the impact of the neighbouring tooth couple on the contact pattern [32].

Vilmos Simon [37] described a procedure for computer-aided TCA in inconsistent spiral bevel gears. In his work he presented the contact path and also the possible contact lines. His procedure considers the minimization of a particular function that causes the detachment of the tooth surfaces throughout the face width. The possible contact lines are represented by the points having this minimum detachment. The method was created using a software program. The use of this software enables the determination of the tooth faces that separate throughout the contact lines, the contact path, the angular displacement and the changing of the angular speed ratio over a meshing period.

Chao et al. [38] developed a TCA algorithm for spherical gear sets. TCA outcome delivers important information about the path of contact, kinematic errors and the contact ellipses. From the geometrical point of view, two types of spherical gear tooth shape are known: concave and convex. Unlike helical or spur gears, spherical gears enable the changeable shaft angle and the existence of misalignment without having gear interference throughout the meshing cycle.

The gear tooth presents a number of errors among which we can list: pressure angle errors, lead angle errors, tooth profile errors etc. The most common assembly errors are the horizontal and vertical axial misalignment and the centre distance errors. Location and unit normal vectors are considered inside a coordinate system when the TCA algorithm is applied. Thus, the unit normal vectors are collinear. Tooth surface contact point is located inside an elliptical area, due to the fact that gear tooth surfaces exhibits elasticity. TCA outcomes allow the estimation of the immediate contact point of the gear mesh. The consideration of continuous tangency of the tooth faces gear drive represent the hypothesis on which the TCA procedure is based. Tooth surface contact pattern is obtained from the succession of the contact ellipses which are generated by the motion of the gear drives. The methods used for the investigation of the contact pattern are divided into two main categories: rigid body and the elastic body method. The best-known elastic body method is the FEM which is used to examine the contact surface with respect to the resilient deformation of tooth areas. The curvature research process belongs to the contact pattern investigation which represents a rigid body method [38].

Hsu and Su [39] analyze TCA and TE of a gear meshing pair for the case of a crowned pinion and a common involute gear.

Daniele Vecchiato [40] creates a TCA algorithm for misaligned planetary gear train. He establishes that changes for the contacting teeth area are necessary in order to bypass edge contact and also gives the parabolic function of TE.

Guan et al. [41] study the manufacturing procedures for crown gear coupling having misalignment. They located the contact points and examine the crown gear coupling having misalignment using a precise solution based on the TCA algorithm.

5.1. Loaded tooth contact analysis

The influence coefficient model is used to determine the LTCA algorithm with respect to the interconnection between the deformation and the loading on a contact region. According to Yu et al. [8] research contact stress had a powerful correlation with the separation distances. In their LTCA algorithm, the influence coefficients used for deformations are subject to tooth bending, shear loading and Hertzian contact of the engaged teeth. Real models are also affected by the twisted shafts due to the existing torque on the contact pressure.

The LTCA algorithm used by Jinke and Zonde was developed by Zhang and Fang [42,43]. During the meshing cycle, a couple of mating tooth (indicated as I and II) are in contact at some point, as presented in fig. 3 [36].

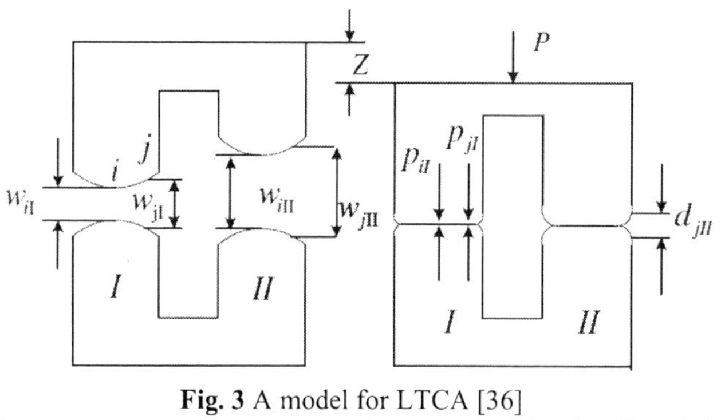

Fig. 3 A model for LTCA [36]

The contact load among two reaching tooth surfaces is considered to be initiated throughout the relative main direction. The intersection between the normal plane throughout the relative main direction and the tooth surface represents the tooth curvature. The clearance among a pair of mating tooth surfaces, when tooth deformation is not taken into account, can be described by the following set of equations [36]:

$$\begin{cases} F_k p_k + w_k = Z + d_k, \ldots k = I, II \\ \sum_{j=1}^{n} p_{jI} + \sum_{j=1}^{n} p_{jII} = P \\ d_{jk} = 0, \ldots (p_{jk} > 0) \\ d_{jk} > 0, \ldots (p_{jk} = 0) \end{cases} \tag{9}$$

where *ik* represent the contact point, *jk* indicates a point throughout the relative main direction, p_k represent the column matrix, F_k signify the assembled adaptability matrix of contact place conditioned by the FEM analysis, w_k represent a column matrix that includes the underlying tooth difference from TCA, and Z indicates the tooth approximation throughout the pressure line.

According to Li [44], LTCA of spur gears had been determined by implementing FEM and mathematical programming method.

Considering element deformation and flank topology, LTCA pattern can reproduce the loaded contact particularities [29]. Among the methods used for establishing LTCA, the FEM is the most useful for analysing the contact between gear teeth. High precision and the deformations of the elements being analysed are the main FEM advantages. Influence coefficient method was used by Wu and Tsai [45] to create a computerized LTCA.

Ye and Tsai [8] created an effective computerized instrument using the effect of misalignment and tip corner contact to establish the LTCA.

In Sanchez-Marin et al. work [46] a new geometric approximation was suggested for the TCA.

5.2. Unloaded tooth contact analysis

In Astoul et al. [47] article a numerical algorithm for UTCA is presented. His suggested algorithm is considered steady and offered a decent estimation of the contact design area. In order to obtain a higher robustness, a lower complexity level was chosen. Mesh density influences the precision of the contact point place and also the computing time.

In Fong's research [48] was developed a model to make less demanding the improvement of a general hypoid gear generator. An adjustment of Fong's model [49] was realized by Shih. He was able to reproduce different face-hobbing processes.

5.3. Tooth contact analysis with errors

ETCA technology had been used to reproduce the gear meshing operation [50]. Using ETCA and considering the benefit of numerical techniques, the contact points of the gear meshing process can be established. The most important element of the calculation process represents the precise and efficient determination of the primary contact point. This first contact point immediately influences the dimension and course throughout the contact length of the tooth flank. The original contact point is assessed by various numerical models, such as the improved Powell hybrid algorithm [51], the Newton-Raphson algorithm (52) or FEM. In order to determine the initial contact point, it is necessary to introduce a string of possible primary values [50]. In other TCA researches, the original contact point value is established as an estimated value that leads to human errors. The examination of the original point value having minimum distance requires many iterations and some coefficients must be randomly considered. Litvin et al. (51) presented an iterative procedure to obtain the estimated values, taking into consideration the minimum deviation of the normal vectors. However, this solution is not very efficient as a large number of operations are required to compute a set of complex nonlinear equations. Considering ETCA, errors mainly arise from the assembly errors of the gear meshing and the geometric and kinematic errors of the machine tool.

Han et al. [50] presented a number of noteworthy improvements, like the subdivision procedure, the counting of matching tooth face points or the coordinating methodology. In their work, the actual tooth flank with errors is represented using a generic term of the ease-off. Manufacturing errors that are specific to the active flank are generated by the effects of different faults. The ease-off topography can be described by the following equations [50]:

$$\varepsilon_i^j = [r_i^{*j}(q_i,\theta_i) - r_i^j(q_i,\theta_i)] \times n_i^j(q_i,\theta_i) \quad (i = 1, 2; j \in [1, n \times m]) \tag{10}$$

where r_i represents the tooth flank vector, n_i is the tooth flank normal vector, i is considered the drive or driven gear and q and θ are the basic design parameters of the tooth flank.

Using the Coordinate Measuring Machine (CMM), it was possible to obtain the proper coordinate parameters of the calculated locations on the tooth flank. The nonlinear ETCA equations are determined by the tooth flank, cutter blade and the tooth contact conditions. The contact pattern is made from all existing contact ellipses. The contact ellipses can be determined for each contact point using Euler and Bertrand formula. The occurrence of the transmission error is caused by the fact that the transmission ratio is different in each contact point [50].

6. Time-varying mesh stiffness and meshing simulation

The vibration study of the gear meshing can be determined by analysing the time-varying mesh stiffness (TVMS), which represents an internal excitation of the gears [53]. The TVMS calculation can be realized using several methods, among which we list: analytical, FEM and experimental methods. The most effective way to determine TVMS represents the use of the FEM method, which takes into account geometric and assembly errors. However, the FEM method presents an important drawback as the processing time may be long.

In Astoul et al. article [47] the algorithm used for the meshing simulation considers the relative displacement of the two sections. Relative displacement presents four separate elements: the master part axial displacement, the slave part axial displacement, the shaft angle and the offset displacement. Tooth flank shape affects the relative displacement and the kinematic and geometric conduct of the contact pattern. The Gleason principle is used to estimate the value and sign, as well as the displacement direction. The tooth flank creation procedure is the only parameter that influences the stability of the algorithm. On the master flank, an ordinary mesh is created and its nodes are projected on the slave flank. The driven gear may have the master part functionality. The slave flank is considered to be stationary, while the master flank spins. In order to detect if the contact point is inside the contact pattern, it was necessary to compare the lowest calculated angle value at any place in the gear mesh with the value calculated at the previous step. If the first value was smaller, then the contact point was present in the meshing area, otherwise it does not exist in the specified area.

In Chen et al. research [54] a refined analytical TVMS pattern is presented for gear tooth root break having deep spread irregular along tooth width.

Analytical computation of an individual tooth pair can be separated into a number of elements, such as [30]:

a) the narrow deformation of the tooth determined by the Hertzian contact;

b) the tooth beam deviation having the route of the load, considering gear tooth as an irregular cantilever when the tooth basis is entirely stiff;

c) the tooth deviation having the route of the load provoked by the elasticity of the basis when tooth is considered to be entirely stiff.

Mesh stiffness influences the load distribution of a tooth pair, which varies at each contact point [23]. Load measured per every unit of length also differs at each point of the contact line and contact length.

Wu et al. [55] described a FEA method for the dynamic contact, taking into account the loaded contact deformation, sliding friction and changes of the tooth meshing couple. Dynamic Transmission Error (DTE) is determined when the investigation is conducted at low speed. Dynamic meshing features are influenced by the sliding friction which generates a pitch point effect. As a result of this pitch point effect, the dynamic contact force decreases immediately. When the investigation takes place at high speed the dynamic meshing features with an existing TVMS is analysed by the described method.

7. Contact fatigue

Shen et al. [56] determined a solution to estimate the contact fatigue life by studying the properties of a gear pump. Their method merges the Finite Element Method with the nominal stress method pursuing the Miner cumulative damage hypothesis. A gear pump was used as it represents the fundamental energy source of a hydraulic device. Tooth face wear and fatigue resulted from the high contact strain that gear tooth face had to support.

According to Miner cumulative damage assumption, the damage produced by cyclic stress at every stage is self-sufficiency and permanence!when is under a periodic load and fatigue failure of a component take place when an accumulated damage threshold level is reached. When a nominal stress method is observed, it is necessary to take into account many material elements such as surface processing coefficient, size factor or the actual stress concentration coefficient. It had been noticed that gears present a lower noise level when the pressure angle is greater than 20 degrees and the contact ratio is greater than two. Gear noise and vibration and tooth fatigue are influenced by dynamic gear loads. To increase gear efficiency, lower the noise levels, avoid tooth breaking and eliminate the pitting effect, it is necessary to reduce the dynamic gear loads. The change of the pressure angle affects the meshing properties [57].

Gear mesh stiffness changes during the meshing process. It represents the main source of excitation for noise and vibration. When a crack or other fault is present in the gear tooth then gear mesh stiffness is also altered [58].

Many kinds of research were made for the investigation of the crack propagation path. Researches outcomes are applied to prevent unexpected failure, to monitor the health status by calculating the dynamic vibration response or to diagnose gear faults [59]. Considering the case of bevel gears, the maximum bending stress takes place on the highest point of singular tooth contact and maximum contact fatigue solicitation is in the proximity of the pitch line [60].

8. Conclusion

The article reviews the basic types of tooth contact analysis and also the mathematical pattern for the surface contact analysis as well as gear mesh contact features.

Tooth contact analysis algorithm represents an important instrument for designing and examining the gear meshing quality. One of the most important purposes of TCA is to determine the contact path on gear tooth faces, and also to establish the transmission error functions produced by gear misalignments.

The profile of transmission error function is affected by the TCA algorithm.

A precise geometry of the tooth contact pattern cannot be determined using common machine adjustments and cutter specifications for the fabrication of hypoid and spiral bevel gear controls. Tooth contact model features errors associated with size, location or shape.

The identification of the initial contact point is very important for achieving an accurate response. Initial contact point affects contact stress, transmission error amplitude and the size of the contact shape. The investigation of the contact stress changes is accomplished using the Finite Element Method. Some authors suggested a union between a global term and a local term in order to determine the deformation of each contact point. The contact point is also affected by the applying torque.

The contact conditions are established by comparing the distance considered along the pressure line with the loaded static transmission error. Tip relief is usually required to achieve better contact conditions throughout the meshing.

TCA software programs predict meshing and contact of gear tooth areas taking into account the localized bearing contact, which provides the surface contact point. TCA equations can be determined if the considered original values that affect the initial contact point are admissible, if not, the computational process will be terminated or the TCA result will be wrong. Tooth contact analysis method represents an answer for the investigation of the transmission error shape and for the modification of the contact model induced by manufacturing and assembling errors. TCA result provides significant information about the contact path, contact ellipses, kinematic errors and the evaluation of the instantaneous contact point of the meshing gear. The LTCA method is determined using the influence coefficient considering the interconnection between the deformation and the loading on a contact region. In the ETCA algorithm, the initial contact point can be assessed using the improved Powell hybrid algorithm, the Newton-Raphson algorithm and the FEM.

This paper represents the introductive part for an in-deep research. A number of analytical, numerical and optimization procedures will be developed in order to achieve the transmission error functions as well as the proper TCA algorithm. All these methods will be implemented in non-circular gears.

9. References

[1] Walker H 1938 Gear Tooth Deflection and Profile Modification *Engineer* **166** pp 410-435
[2] Tharmakulasingam R 2009 *Transmission Error in Spur Gears: Static and Dynamic Finite-Element Modeling and Design Optimization* (London: School of Engineering and Design Brunel University – Doctoral Thesis) pp 6-50

[3] Fietkau P, Bertsche B 2013 Influence of tribological and geometrical parameters on lubrication conditions and noise of gear transmissions *Mechanism and Machine Theory* **69** pp 303–320
[4] Gregory R W, Harris S L, Munro R G 1963 Dynamic behaviour of spur gears *Proceedings of the Institution of Mechanical Engineers* **178** pp 207-218
[5] Garambois P, Perret-Liaudet J, Rigaud E 2017 NVH robust optimization of gear macro and microgeometries using an efficient tooth contact model *Mechanism and Machine Theory* **117** pp 78–95
[6] Ottewill J R, Neild S A, Wilson R E 2010 An investigation into the effect of tooth profile errors on gear rattle *Journal of Sound and Vibration* **329** pp 3495–3506
[7] Litvin F L, Fuentes A 2009 *Gear Geometry an Applied Theory – Second Edition* (New York: Cambridge University Press) Chapter 9.4
[8] Ye S-Y, Tsai S-J 2016 A computerized method for loaded tooth contact analysis of high-contact-ratio spur gears with or without flank modification considering tip corner contact and shaft misalignment *Mechanism and Machine Theory* **97** pp 190-214
[9] He D, Ding H 2018 A new analytical identification approach to the tooth contact points considering misalignments for spiral bevel or hypoid gears *Mechanism and Machine Theory* **121** pp 785-803
[10] Cao X, Deng X, Wei B 2018 A novel method for gear tooth contact analysis and experimental validation *Mechanism and Machine Theory* **126** pp 1-13
[11] Welbourn D B 1979 Fundamental Knowledge of Gear Noise *Proceedings of the Institution of Mechanical Engineers* pp 9-14
[12] Gregory R W, Harris S L, Munro R G 1963 A Method of Measuring Transmission Error in Spur Gears of 1:1 Ratio, *Journal of Scientific Instruments* **40** pp 5-9
[13] Munro R G, Morrish L, Palmer D 1999 Gear transmission error outside the normal path of contact due to corner and top contact *Proceedings of the Institution of Mechanical Engineers, Part C: Journal of Mechanical Engineering Science* **213** pp 389-400
[14] Zhuo Y-B, Xiang X-Y, Zhou X-J, LV H-L, Teng G-Y 2017 A method for the global optimization of the tooth contact pattern and transmission error of spiral bevel and hypoid gears *Journal of Zhejiang University-Science A (Applied Physics & Engineering)* **18**(5) pp 377-392
[15] Mermoz E, Astoul J, Sartor M, Linares J M, Bernard A 2013 A new methodology to optimize spiral bevel gear topography *CIRP Annals – Manufacturing Technology* **62** pp 119-122
[16] Hwang S-C, Lee J-H, Lee D-H, Han S-H, Lee K-H 2013 Contact stress analysis for a pair of mating gears *Mathematical and Computer Modelling* **57** pp 40-49
[17] Jabbour T, Asmar G 2015 Tooth stress calculation of metal spur and helical gears *Mechanism and Machine Theory* **92** pp 375-390
[18] Jabbour T, Asmar G 2009 Stress calculation for plastic helical gears under a real transverse contact ratio *Mechanism and Machine Theory* **44** pp 2236-2247
[19] Liang D, Chen B, Peng S, Hua C, Liao R 2017 Deviation Calculation and Analysis of Tooth Surface of Conjugate-Curve Gear Drive *International Journal of Precision Engineering and Manufacturing* **18** (5) pp 689-696
[20] Zhou C, Huang F, Han X, Gu Y 2017 An elastic–plastic asperity contact model and its application for micro-contact analysis of gear tooth profiles *Int J Mech Mater Des* **13** pp 335-345
[21] Fernandez del Rincon A, Viadero F, Iglesias M, Garcia P, de-Juan A, Sancibrian R 2013 A model for the study of meshing stiffness in spur gear transmissions *Mechanism and Machine Theory* **61** pp 30-58
[22] Winter H, Placzek T 1991 Load distribution and topological flank modification of helical and double helical gears *European Journal of Mechanical Engineering* **36**
[23] Sanchez M B, Pleguezuelos M, Pedrero J I 2016 Calculation of tooth bending strength and surface durability of internal spur gear drives *Mechanism and Machine Theory* **95** pp 102-113
[24] Chen Y-C, Liu C-C 2011 Contact stress analysis of concave conical involute gear pairs with non-parallel axes *Finite Elements in Analysis and Design* **47** pp 443-452
[25] Lee D-H, Moon K-H, Lee W-Y 2016 Characteristics of transmission error and vibration of

broken tooth contact *Journal of Mechanical Science and Technology* **30**(12) pp 5547-5553
[26] Sheveleva G I, Volkov A E, Medvedev V I 2007 Algorithms for analysis of meshing and contact of spiral bevel gears *Mechanism and Machine Theory* **42** pp 198-215
[27] Kawasaki K, Tamura H, Iwamoto Y 1999 Klingelnberg spiral bevel gears with small spiral angles *Proceedings 4th World Congress on Gearing and Power Transmissions* pp 697-703
[28] Raghuwanshi N K, Parey A 2017 Experimental measurement of spur gear mesh stiffness using digital image correlation technique *Measurement* **111** pp 93-104
[29] Tsai S-J, Ye S-Y 2018 A computerized approach for loaded tooth contact analysis of planetary gear drives considering relevant deformations *Mechanism and Machine Theory* **122** pp 252-278
[30] Yu W, Mechefske C K 2016 Analytical modeling of spur gear corner contact effects *Mechanism and Machine Theory* **96** pp 146-164
[31] Lin H H, Wang J, Oswald F B, Coy J J 1994 Effect of extended tooth contact on the modeling of spur gear transmissions *Gear Technology* **July/August** pp 18-25
[32] Fuentes-Aznar A, Ruiz-Orzaez R, Gonzalez-Perez I 2017 Comparison of spur, helical and curvilinear gear drives by means of stress and tooth contact analyses *Meccanica* **52** pp 1721-1738
[33] Vogel O, Griewank A, Bar G 2002 Direct gear tooth contact analysis for hypoid bevel gears *Computer Methods in Applied Mechanics and Engineering* **191** pp 3965-3982
[34] Stadtfeld H. J. 1993 Handbook of Bevel and Hypoid Gears (Rochester Institute of Technology)
[35] Jinzhan S, Zongde F, Xiangwei C 2013 Design and analysis of spiral bevel gears with seventh-order function of transmission error *Chinese Journal of Aeronautics* **26** 1310-1316
[36] Jinke J, Zongde F 2015 Design and analysis of modified cylindrical gears with a higher-order transmission error *Mechanism and Machine Theory* **88** pp 141-152
[37] Vilmos S 2007 Computer simulation of tooth contact analysis of mismatched spiral bevel gears *Mechanism and Machine Theory* **42** pp 365-381
[38] Chao L C, Tsay C B 2008 Contact characteristics of spherical gears *Mechanism and Machine Theory* **43** pp 1317-1331
[39] Hsu R-H, SU H-H 2014 Tooth contact analysis for helical gear pairs generated by a modified hob with variable tooth thickness *Mechanism and Machine Theory* **71** pp 40-51
[40] Vecchiato D 2006 Tooth contact analysis of a misaligned isostatic planetary gear train *Mechanism and Machine Theory* **41** pp 617-631
[41] Guan Y, Fang Z, Yang X, Chen G 2018 Tooth contact analysis of crowned gear coupling with misalignment *Mechanism and Machine Theory* **126** pp 295-311
[42] Zhang Y, Fang Z 1997 Analysis of transmission errors under load of helical gears with modified tooth surfaces *J. Mech. Des.* **119** pp 120-126
[43] Zhang Y, Fang Z 1999 Analysis of tooth contact and load distribution of helical gears with crossed axes *Mechanism and Machine Theory* **34** pp 41-57
[44] Li S 2008 Effect of addendum on contact strength, bending strength and basic performance parameters of a pair of spur gears *Mechanism and Machine Theory* **43** pp 1557–1584
[45] Wu S-H, Tsai S-J 2009 Contact stress analysis of skew conical involute gear drives in approximate line contact *Mechanism and Machine Theory* **44** pp 1658-1676
[46] Sanchez-Marin F, Iserte J L, Roda-Casanova V 2016 Numerical tooth contact analysis of gear transmissions through the discretization and adaptive refinement of the contact surfaces **101** pp 75-94
[47] Astoul J, Geneix J, Mermoz E, Sartor M 2013 A simple and robust method for spiral bevel gear generation and tooth contact analysis *Int J Interact Des Manuf* **7** pp 37-49
[48] Fong Z-H 2000 Mathematical model of universal hypoid generator with supplemental kinematic flank correction motions *J. Mech. Des.* **122** pp 136-142
[49] Shih Y-P, Fong Z-H, Lin G C 2006 Mathematical model for a universal face hobbing hypoid gear generator *J. Mech. Des.* **129** pp 38-47
[50] Ding H, Zhou Y, Tang J, Zhong J, Zhou Z, Wan G 2017 A novel operation approach to determine initial contact point for tooth contact analysis with errors of spiral bevel and hypoid gear *Mechanism and Machine Theory* **109** pp 155-170

[51] Litvin F L, Vecchiato D, Fuentes A, Gonzalez-Perez I 2004 Automatic determination of guess values for simulation of meshing of gear drives *Computer Methods in Applied Mechanics and Engineering* **193** pp 3745-3758
[52] Litvin F L, Sheveleva G I, Vecchiato D, Gonzalez-Perez I, Fuentes A 2005 Modified approach for tooth contact analysis of gear drives and automatic determination of guess values *Computer Methods in Applied Mechanics and Engineering* **194** pp 2927-2946
[53] Ma H, Zeng J, Feng R, Pang X, Wen B 2016 An improved analytical method for mesh stiffness calculation of spur gears with tip relief *Mechanism and Machine Theory* **98** pp 64-80
[54] Chen Z, Zhai W, Shao Y, Wang K, Sun G 2016 Analytical model for mesh stiffness calculation of spur gear pair with non-uniformly distributed tooth root crack *Engineering Failure Analysis* **66** pp 502-514
[55] Wu Y-J, Wang J-J, Han Q-K 2012 Contact finite element method for dynamic meshing characteristics analysis of continuous engaged gear drives *Journal of Mechanical Science and Technology* **26** pp 1671-1685
[56] Shen H, Li Z, Qi L, Qiao L 2018 A method for gear fatigue life prediction considering the internal flow field of the gear pump *Mechanical Systems ad Signal Processing* **99** pp 921-929
[57] Gupta K, Chatterjee S 2018 Analysis of Design and Material Selection of a Spur gear pair for Solar Tracking Application *Material Today: Proceedings* **5** pp 789-795
[58] Raghuwanshi N K, Parey A 2016 Experimental measurement of gear mesh stiffness of cracked spur gear by strain gauge technique *Measurement* **86** pp 266-275
[59] Pandya Y, Parey A 2013 Crack behavior in a high contact ratio spur gear tooth and its effects on mesh stiffness *Engineering Failure Analysis* **34** pp 69-78
[60] Song D, Lin H, Xing-hui H, Song H 2013 Finite element analysis of contact fatigue and bending fatigue of a theoretical assembling straight bevel gear pair *J. Cent. South Univ* **20** pp 279-292

Acknowledgments
The research was funded by the Ministry of National Education from Romania.

IOP Conf. Series: Materials Science and Engineering **514** (2019) 012028 doi:10.1088/1757-899X/514/1/012028

The effect of urban buildings on the implementation of small wind turbines

D Ciobanu[1], M Gheorghe[1], R Saulescu[1], B Midoi[2] and C Jaliu[1]

[1]Product Design, Mechatronics and Environment Department, Transilvania University of Brasov, Romania
[2]CANAM Steel, Brasov, Romania

E-mail: daniela.ciobanu@unitbv.ro

Abstract. The paper presents the main factors that can influence the location of a wind turbine (building shape, construction height, direction and wind speed) aiming to increase its performance. Thus, three types of buildings with various shapes and heights, which can be found in the urban environment (houses, offices and hotels) are presented in a case study, numerical simulations being carried out using the ANSYS software. The aerodynamic tunnel is customized for each building while maintaining a proportion of the building and the space needed to analyse it. All buildings are rotated to the wind direction and not to the theoretical North. The objective of this paper is to identify the area of the building in which an increase of the airflow is obtained, which allows to increase the performances of the wind turbine. The results allow formulating recommendations for the implementation location and types of units that should be installed.

1. Introduction

The literature suggests that, theoretically, the wind potential exceeds, in a significant way, the global energy demand; however approx. 2-3% of global electricity demand is obtained from wind power, despite the fact that, over the last two decades, the generation capacity of wind energy has increased by 27% [1].

The energy performance of buildings is an increasingly pressing need of the present times, as reflected in the 2013/31 / EU Directive, which stipulates that, starting from 2021, the buildings will be Nearly Zero Energy Buildings (nZEB) through the efficient use of renewable energy resources (including wind energy) [1].

Although many countries have an increasing interest in producing energy on renewable basis, in some countries the electricity generation from wind power still has a negative perception due to the variability of wind potential and its low value that limits the implementation of wind systems in the urban environment. Along with these impediments, the built environment has also a major influence on the good functioning of a wind system. For the architectural acceptance of the wind systems implemented in the urban environment, it is necessary to collaborate with the architects to build buildings that have integrated wind turbines (WT) into their structure. Concepts of this type are already implemented at Council House 2 - Melbourne [2], Mercy Housing - Chicago [3] or Wrigley Hall - Arizona [4], which produce electric power at lower wind speeds by using turbulent airflows [5, 6, 7].An incorrect setup of the wind system on the roof can lead to a considerable decrease in WT power even under favourable wind conditions, a statement already proved in the literature [8, 9].

The methodology, which is based on common shapes of symmetrical building roofs [10], involves the modelling of the architectural forms specific to the implementation area and the analysis of the airflows dynamics; therefore, the impact of building architecture on the wind potential behaviour has an important role in this analysis, CFD software being usually used to identify the disturbing areas and to optimally place a wind system.

As presented in [11], a wind system integrated in the urban environment can be a cheap option for energy generation, the renewable energy source being directly influenced by the building shape and the intensity of the wind turbulences [12].

This paper addresses four factors that can influence the implementation location of a low capacity WT (building shape and height, wind direction and speed) by using CFD software to analyse the airflow behaviour in the building vicinity. Thus, the optimal location of the wind system is identified in order to get a full capacity operation. The study is based on the weather data presented in [13]. The conclusions include recommendations for the identification of the WT optimal location under specific wind potential and buildings with different shapes.

2. Problem formulation

The paper analyses the influence of the main factors on the positioning of a low capacity wind turbine in order to obtain higher WT performances. These factors refer to the type and shape of the building, the height of construction and the wind speed; another factor would be the wind direction that will be neglected in this study as the building will be placed in the wind tunnel towards wind.

The paper presents as a case study three types of urban buildings placed under a known wind speed. The analysis will be carried out on buildings with various shapes and heights that can be found in almost any type of urban environment. These buildings will be imported [14] into the wind tunnel and placed under an airflow of minimum 2 m/s speed (the speed at which a small wind system usually starts to produce energy) and maximum 6 m/s [13], values that cover the wind potential from Brasov County at 12 m from the ground. The optimal WT location of the wind systems is proposed based on the numerical simulations with the ANSYS software.

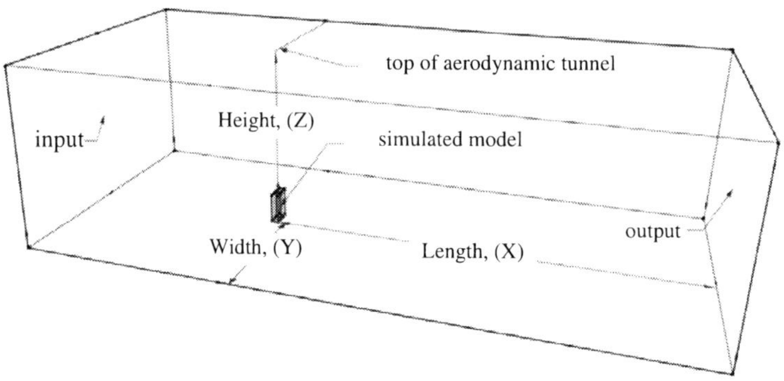

Figure 1. Graphical representation of the wind tunnel

The aerodynamic tunnel was customized for each building, keeping a proportion between the building and the space needed to analyse it (Figure 1), ensuring a proper airflow around the building. All buildings were rotated in the direction of the wind rather than towards the theoretical North. As the WT will be mounted on the roof, the behaviour of the air currents in the XZ plane (in a section along the X axis) will be further analysed.

The objective of this paper is to identify the area of the building in which an increase of the airflow is obtained, with less roughness, allowing the wind turbine performance to be increased. The results will allow to formulate recommendations for the implementation location, the types and number of WT that should be installed.

3. Airflow formulation and recommendations regarding the implementation location

As presented in Figure 1, the air flows from left to right, the buildings being oriented to be analysed according to specific weather data, with a wind direction from N-NNW to SSW-S (see [13]). The buildings placed in the wind tunnel have different destinations: office buildings, family homes, hotels.

To achieve the aerodynamic tunnel, proportional values are assigned to the buildings dimensions. The results are based on the assumption of no roughness, the buildings being isolated.

3.1. Office building

The aerodynamic tunnel for the analysis of an office building (Figure 2) with the dimensions at the base of X=40 m, Y=20 m and Z=80 m has the length X=720 m, width Y=240 m, and height Z=240 m.

Figure 2. Example of an office building [14]

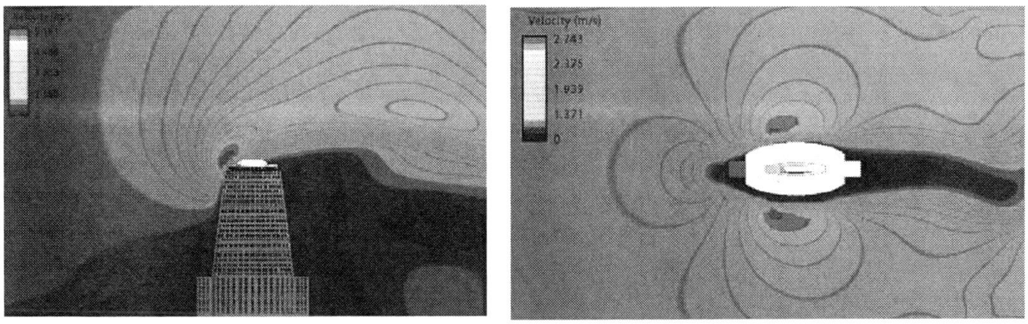

a) b)

Figure 3. The behaviour of the airflows around the office building in the aerodynamic tunnel at a wind speed of 2 m/s: view in the ZX plane and b) view in the XY plane, at 5 m from the ground

Considering a wind speed of 2 m/s, it can be observed that, in the roof area (Figure 3a), there are turbulences created by the shape of the roof, which amplifies the wind speed of approx. 2.5 times. In this case, it can be also noticed that turbulent airflows appear in the side parts of the building (Figure 3b), although the facade has a circular surface.

Due to the reduced roof dimensions, it is recommended to implement wind turbines with the rotor diameter of up to 5 m, on a tower of approx. 7 - 9 m.

Figure 4. Example of an apartment building [14]

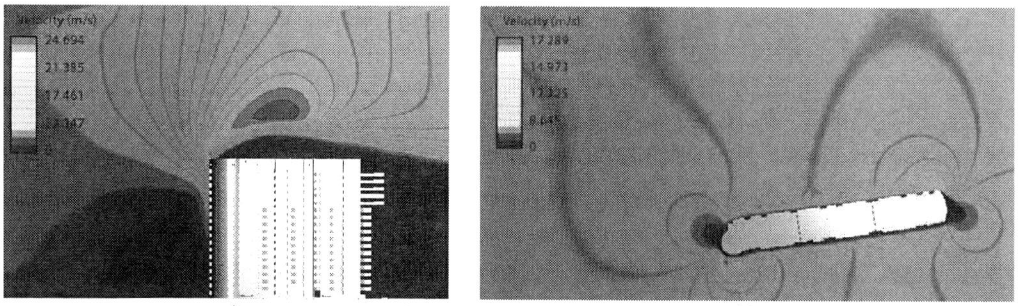

a) b)

Figure 5. The behaviour of the airflows around the apartment building in the aerodynamic tunnel at a wind speed of 6 m/s: a) view in the ZX plane and b) view in the XY plane at 17 m above the ground

3.2. Apartment building

A building with apartments of a parallelepiped shape (Figure 4) is further considered [14], having the following dimensions X=70 m, Y=20 m and Z=50 m. The aerodynamic tunnel used in the CFD analysis is X=150 m long, Y=100 m wide and Z=150 m high.

According to Figure 5, the wind speed is amplified by approx. 4 times in the terrace area, the wind speed being of 6 m/s. Therefore, it is recommended to use wind turbines with vertical axis, with a diameter of up to 4 m, and a 5 m tower (reaching a height of 9 - 10 m from the roof terrace). Besides, the vertical axis wind turbines are preferred in the urban environment to the horizontal ones as their noise during operation is smaller.

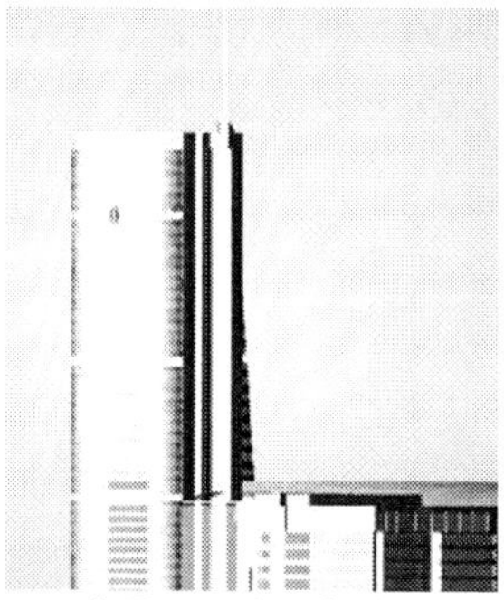

Figure 6. Example of a hotel [14].

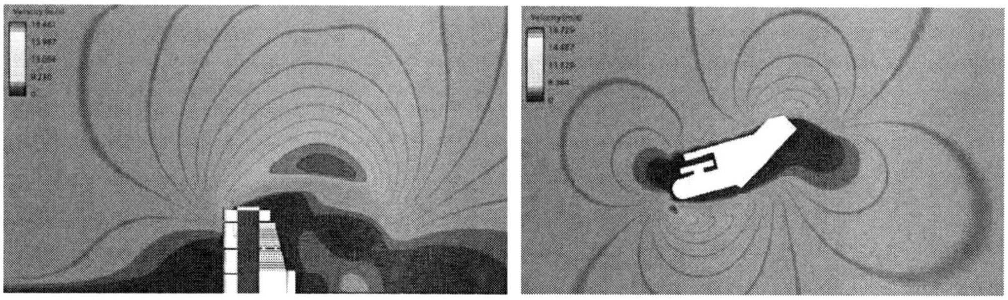

a) b)

Figure 7. The behaviour of the airflows around the hotel in the aerodynamic tunnel at a wind speed of 6 m/s: a) view in the ZX plane and b) view in the XY plane at 50 m above the ground

3.3. Hotel

The hotel from Figure 6 [14] has a geometric footprint on the ground made up of triangles and rectangles, with straight and oblique external walls, having the dimensions X=40 m, Y=60 m and Z=100 m. The building is placed into an aerodynamic tunnel with the length X=300 m, width Y=180 m, and height Z=250 m.

This type of building amplifies the wind about 3 times (Figure 7), the considered wind speed being of 6 m/s.

For this type of building it is recommended to use vertical axis wind turbines of 5 kW nominal power, whose cut-in speed is of 1 m/s, WT that can be mounted on the edge of the hotel upper terrace.

4. Conclusions and discussions

The optimal location for mounting wind systems on buildings depends on the variables affecting the wind flow. The effect of the shape and height of the building as well as the wind velocity on the behaviour of the air currents is analysed in the paper at the contact with the considered building in order to highlight the optimal location for the wind turbines.

In all cases, the area with the peak intensity of the air turbulences extends above the roof to about 1/4 of the building height, which means that the wind systems should be positioned so as to benefit of wind amplification; however, the support tower should not be overly high to avoid the occurrence of vibrations in the anchorage.

The shapes of all investigated buildings have an influence on wind speed, amplifying the wind potential. It can be concluded that the height of the building, on which the wind turbines are installed, is important, thus eliminating or diminishing the effect of the surrounding buildings on the wind potential.

For the buildings whose side walls are vertical, the wind has turbulences at the edges, where wind systems with vertical axes are recommended to be installed; if the building side wall is oblique, a higher wind is gained to the inside of the building, where wind systems with vertical axes should be installed.

Regardless of the building shape, the wind has a higher speed on the building terrace / roof than on its sides.

The circular shapes, compared to the straight ones, allow a smoother wind flow and, therefore, an easier wind amplification.

Due to the complexity of the urban environment, these studies must be conducted in a simplified manner by considering fixed values for all variables, the only variable that can alter the results being the locally wind potential.

The analysis highlights the influence of the building on the amount of captured wind energy, depending on the location of wind systems on the roof or its sides.

The main conclusions of the paper can be formulated as follows:

- buildings can influence through their various shapes the amount of energy captured by the turbines installed on them; therefore, they should be designed from the project stage to allow wind potential amplification;
- vertical and / or horizontal axes wind turbines can be installed on almost any type of building, taking into account the tower height on which they are mounted that allows the WT operation at nominal power;
- vertical axes wind turbines are recommended to be installed on the sides of the buildings, wherever allowed; these turbines are silent, start at lower wind speeds and are easier to be architecturally accepted;
- wind systems of different types and sizes, can be mounted on buildings of different shapes while maintaining the environment aesthetics;
- both the architectural shapes in the installation area and the wind potential of that area are important for the maximization of the energy potential of a wind system installed in the urban environment.

5. References

[1] Mostafaeipour A, Jadidi M, Mohammadi K, Sedaghat A 2014 An analysis of wind energy potential and economic evaluation in Zahedan, Iran. *Renewable and Sustainable Energy Reviews* **30** pp 641-650.

[2] http://commons.wikimedia.org/wiki/File:Council_House_2_Melbourne_east_and_turbine_detai l. JPG (accessed on 7.09.2017)

[3] http://www.aerotecture.com/projects_mlh.html (accessed on 1.09.2017)

[4] http://schoolofsustainability.asu.edu/media/image-gallery.php (accessed on 4.09.2017)

[5] Lubitz W D 2014 Impact of ambient turbulence on performance of a small wind turbine. *Renewable Energy* **61** pp 69-73.

[6] Sunderland K, Woolmington T, Blackledge J, Conlon M 2013 Small wind turbines in turbulent (urban) environments: A consideration of normal and Weibull distributions for power prediction *Journal of Wind Engineering and Industrial Aerodynamics* **121** pp 70–81

[7] Toja-Silva F, Colmenar-Santos A, Castro-Gil M 2013 Urban wind energy exploitation systems: Behaviour under multidirectional flow conditions - Opportunities and challenges *Renewable and Sustainable Energy Reviews* **24** pp 364-378

[8] Blackmore P 2008 *Siting micro-wind turbines on house roofs* (Watford: BRE)

[9] Blackmore P 2010 *Building-mounted micro-wind turbines on high-rise and commercial buildings* (Watford: BRE)

[10] Abohela I, Hamza N, Dudek S 2013 Effect of roof shape, wind direction, building height and urban configuration on the energy yield and positioning of roof mounted wind turbines *Renewable Energy* **50** pp 1106-1118

[11] Stankovic S, Campbell N, Harries A 2009 *Urban windenergy* (London: Earthscan)

[12] Rafailidis S 1997 *Influence of building areal density and roof shape on the wind characteristics above a town* Boundary-Layer Meteorology; 85: 255-71

[13] Saulescu R, Climescu O, Jaliu C 2014 Assessment of Wind Energy Resources in Communities. Case Study: Brasov, Romania. *Proceedings of the Conference for Sustainable Energy CSE 2014* pp 151-166

[14] http://www.qookar.com/type/ArchitecturalAppearance/CommercialOffice.html (accessed on 02.09.2017)

IOP Conf. Series: Materials Science and Engineering **514** (2019) 012029 doi:10.1088/1757-899X/514/1/012029

Kinematic Modelling of Contact Point between Chain Bush and Sprocket

R Velicu[1], R Săulescu[1] and C C Gavrilă[1]

[1]Product Design, Mechatronics and Environment Department, Transilvania University of Braşov, Romania

E-mail: rsaulescu@unitbv.ro

Abstract. As is well known, in the automotive industry there is a tendency to obtain the most efficient, reliable and compact systems. Because the belt drives used are subject to excessive elongation and wear in most applications, since high performance engines are desired, distribution chains are preferred. The dynamic behavior of a chain drive during operation is directly influenced by its geometry and, implicitly, by the chain contact with the sprocket. Since the application of normal and transverse forces occurring in a chain transmission depends directly on their application point and thus on the point of contact between the chain and the sprocket, the kinematic modeling of the point of contact between the chain bushings and the sprocket is proposed in present paper.

1. Introduction

Chain transmissions used in the automotive industry must be efficient, reliable and compact. Along with these requirements, they also have the following advantages: they are capable of transmitting a wide range of speeds and powers with a negligible range; have a long lifetime due to oil lubrication and chain-to-chain contacting through the bushings (making rolling friction); can operate in hostile environments (oily areas or high temperatures). In spite of these advantages, because of the differences in the profiles of the sprockets, the dynamic behavior is treated with enough reluctance in the literature. The optimization of the chain tooth profile in the design [1] influences directly its geometry, respecting the contact surfaces that can influence the wear of the contacts between the chain bush and the sprocket [2]. A graphical analysis of the contact between the bush and the sprocket is compared for two types of profiles in [3], and in [4] a method of modeling the contact between them, based on the use of geometric parameters characteristic of a contact sprocket - chain bush [5]. In the paper [6] there is presented a kinematic model of chain link displacement according to the angle of rotation of the wheel, a model continued with variation of the specific chain elongation according to the number of teeth in contact [7]. Following these steps, it becomes necessary the modeling of the contact point between the chain bush and the sprocket.

An optimized transmission structure, engine speed and dynamic forces (such as normal and transverse forces or friction forces) can have a significant effect on its optimal functioning.

Depending on the shape of the sprocket teeth and the number of teeth in the chain-wheel contact, the chain transmission forces are differentially distributed. These contact forces can be determined on the basis of the power balance between the chain and the sprocket and are directly dependent on the contact angle variation.

Knowing the variations in the contact angle and the point of contact (the point of application of the forces in the transmission) allow optimization of the design of the transmission; therefore in this paper it is proposed a calculation model for determining the geometric position of the contact point between the sleeve and the sprocket.

Then, the model presented is applied to a case study and the results are validated with graphical representations.

2. Problem formulation

The proposed mathematical model for determining the contact point is based on the determination of the contact angle, presented in [8], for a chain transmission with a unitary transmission ratio. The considered transmission is a standardized one [9], being a short-bush chain, known for the number of sprocket teeth and geometric elements of the transmission. Based on these parameters, will result the contact angle (α_i) and then the contact point coordinates (x and y). For verification, another method is used to determine the coordinates of the contact point, on the basis of which the angle of contact is determined. These values are then compared with corresponding graphical or tabular representations.

In the case study, knowing the chain pitch (p), the diameter of the chain bush (d_b) and the number of teeth (z), the mathematical model allows the parameters of the bushed chain transmission and the numerical values of the contact point coordinates (x_{ci}, y_{ci}) and also the contact angle (α_i).

3. Determining the contact point and contact angle

3.1. Method I

This first method consists in determining the contact angle (α_i) for a number of contacts (*i*) between the chain bush and the sprocket. Based on this, the contact point coordinates (x_i and y_i) are calculated for the same number of contacts.

The mathematical model for calculating the contact angle is based on the mounting position (see Figure 1), where tangential forces are assumed to be zero, and also that, for i = 0 result the first contact between the chain bush, and, the wheel position is vertical, where the notations considered to be with index 0 and $\alpha_i = 0°$.

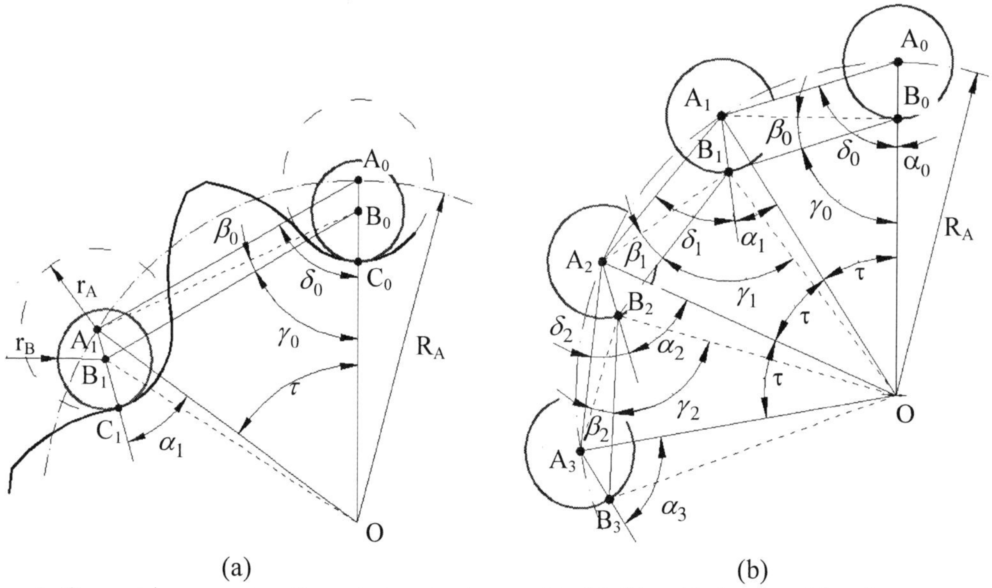

Figure 1. Geometric parameters for a chain-wheel ensemble (a) [8], with details of first four contacts (b).

For a general case of chain-wheel ensemble, is considering i = 0, 1, 2..., respectively j = i+1, the relation for determining the curves center position for sprocket (A_i, A_j) and also for chain-bush center

(B_i, B_j) can be determined:

$$A_iB_i = r_A - r_B,\ A_iA_j = 2R_A\, sin\left(\frac{\tau}{2}\right),\ OA_i = R_A,\ B_iB_j = l = p + x \text{, where } x = 0.2\% \text{ from } p, \quad (1)$$

for $i = 0$ (mounting position) result $\alpha_0 = 0$, $\delta_0 = 90 - \frac{\tau}{2}$, respectively $OB_0 = R_A - r_A + r_B$.

The contact angle (α_i) can be determined using the Figure 2, where the generalised Pithagora theorema is applied.

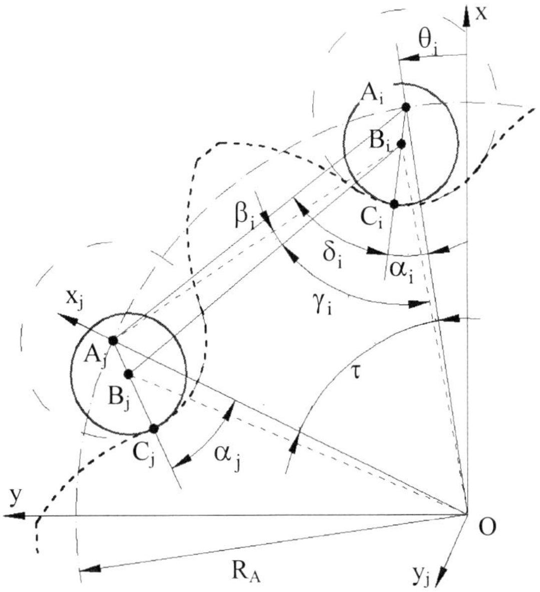

Figure 2. The parameters for successive contact points, i and j, between the chain bush and sprocket

The general algorithm [8] for contact angle calculus, result as follows:
from $\Delta A_iA_jB_i$

$$A_jB_j^2 = A_iB_i^2 + A_iA_j^2 - 2A_iB_i \cdot A_iA_j\, cos\,\delta_i$$

$$A_jB_i^2 = (r_A - r_B)^2 + \left(2R_A\, sin\frac{\tau}{2}\right)^2 - 4R_A(r_A - r_B)\, sin\left(\frac{\tau}{2}\right) cos\,\delta_i$$

$$\delta_i = \delta_0 - \alpha_i; \quad (2)$$

from $\Delta A_jB_iB_j$

$$cos\,\beta_i = \frac{B_iB_j^2 + A_jB_i^2 - A_iB_i^2}{2B_iB_j \cdot A_jB_i}$$

$$\beta_i = arccos\left(\frac{B_iB_j^2 + A_jB_i^2 - A_iB_i^2}{2B_iB_j \cdot A_jB_i}\right); \quad (3)$$

from ΔOB_iA_j

$$cos(\gamma_i + \beta_i) = \frac{OB_i^2 + A_jB_i^2 - OA_i^2}{2OB_i \cdot A_jB_i}$$

$$\gamma_i = arccos\left(\frac{OB_i^2 + A_jB_i^2 - OA_i^2}{2OB_i \cdot A_jB_i}\right) - \beta_i; \tag{4}$$

from ΔOB_iB_j

$$OB_j^2 = B_iB_j^2 + OB_i^2 - 2B_iB_j \cdot OB_i \cos\gamma_i;$$

from ΔOA_jB_j

$$\cos\alpha_j = \frac{OA_j^2 + A_iB_i^2 - OB_j^2}{2OA_j \cdot A_iB_i}$$

$$\alpha_j = arccos\left(\frac{OA_j^2 + A_iB_i^2 - OB_j^2}{2OA_j \cdot A_iB_i}\right). \tag{5}$$

The contact angle led to the contact point coordinates. For a known teeth number of sprocket, (z), and the wheel angular pitch (τ=360/z), result the contact points coordinates (x_{cj}, y_{cj}).

Taking account by Figure 2 and relation $\theta_j = \theta_i + \tau$, result

$$\begin{bmatrix} x_jC_j \\ y_jC_j \end{bmatrix} = \begin{bmatrix} R_A - r_A\cos\alpha_j \\ r_A\sin\alpha_j \end{bmatrix} \text{ and } T_{XjYj} = \begin{bmatrix} \cos\theta_j & -\sin\theta_j \\ \sin\theta_j & \cos\theta_j \end{bmatrix}. \tag{6}$$

The contact point coordinates x_{cj} şi y_{cj} result as relation

$$\begin{bmatrix} x_{Cj} \\ y_{Cj} \end{bmatrix} = T\begin{bmatrix} x_jC_j \\ y_jC_j \end{bmatrix} = \begin{bmatrix} (R_A - r_A\cos\alpha_j)\cos\theta_j - r_A\sin\alpha_j\sin\theta_j \\ (R_A - r_A\cos\alpha_j)\sin\theta_j + r_A\sin\alpha_j\cos\theta_j \end{bmatrix}. \tag{7}$$

By transforming x_jy_j in xy($T_{XjYj\to xy}$) result

$$T_{XjYj\to xy} = \begin{bmatrix} \cos(\theta_i + \tau) & -\sin(\theta_i + \tau) \\ \sin(\theta_i + \tau) & \cos(\theta_i + \tau) \end{bmatrix}. \tag{8}$$

Also, the relation (6) and (7) led to

$$\begin{bmatrix} x_{Cj} \\ y_{Cj} \end{bmatrix} = \begin{bmatrix} (R_A - r_A\cos\alpha_j)\cos(\theta_i + \tau) - r_A\sin\alpha_j\sin(\theta_i + \tau) \\ (R_A - r_A\cos\alpha_j)\sin(\theta_i + \tau) + r_A\sin\alpha_j\cos(\theta_i + \tau) \end{bmatrix}. \tag{9}$$

The previous mathematical model is applied to a real case and led to results presented in Table 1.

3.2. Method II

The previous mathematical model for determining the contact angle (α) and also the contact points coordinates (x_{cj}, y_{cj}), can be verified using the second method.

In Figure 3, the chain link and sprocket ensemble is considered rotated counter clockwise with angle θ^o relative to the mounting position; so, the first position is for $A_0B_0C_0$ and α_0 when the angle θ^o is imposed.

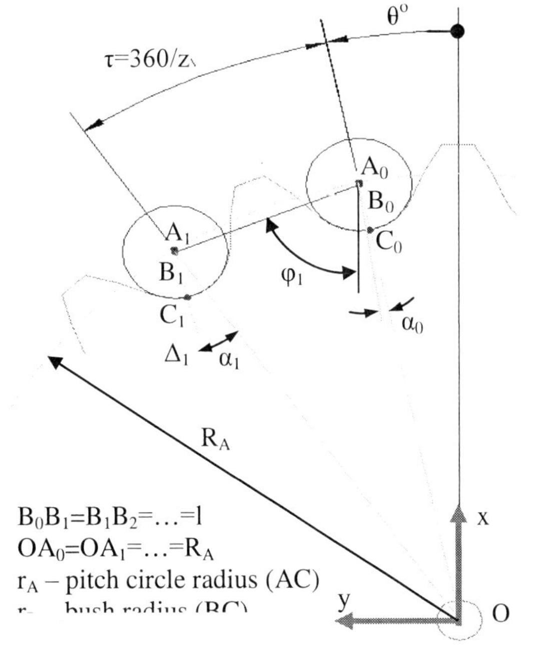

Figure 3. Contact points and angles (C_0& α_0=0 şi C_1& α_1) for first two contacts, 0 and 1.

Following the Figure 3, the center coordinates for curve radius (A_0,A_1) and bush center (B_0, B_1) can be written as

$$A_0: \begin{cases} x_{A0} = R_A \cos\theta \\ y_{A0} = R_A \sin\theta \end{cases}; \quad B_0: \begin{cases} x_{B0} = (R_A - r_A + r_B)\cos\theta \\ y_{B0} = (R_A - r_A + r_B)\sin\theta \end{cases}, \tag{10}$$

$$A_1: \begin{cases} x_{A1} = R_A \cos(\theta + 1\tau) \\ y_{A1} = R_A \sin(\theta + 1\tau) \end{cases}; \quad B_1: \begin{cases} x_{B1} = x_{B0} - l\cos\varphi_1 \\ y_{B1} = y_{B0} + l\sin\varphi_1 \end{cases}. \tag{11}$$

Using the right line ecuation Δ_1 for point C_1 result

$$-\frac{x_{C1} - x_{A1}}{y_{C1} - y_{A1}} = -\frac{x_{B1} - x_{A1}}{y_{B1} - y_{A1}}. \tag{12}$$

Using the circle ecuations for radius r_A and r_B, for point C_1, result

$$(x_{C1} - x_{A1})^2 + (y_{C1} - y_{A1})^2 = r_A^2, \tag{13}$$

$$(x_{C1} - x_{B1})^2 + (y_{C1} - y_{B1})^2 = r_B^2. \tag{14}$$

Also, from relation (13) and (14) result

$$(x_{A1} - x_{B1})(2x_{C1} - x_{A1} - x_{B1}) + (y_{A1} - y_{B1})(2y_{C1} - y_{A1} - y_{B1}) = r_B^2 - r_A^2. \tag{15}$$

Taking account by relation (12), result the coordinates x_{c1} and y_{c1}

$$x_{C1} = x_{A1} + (y_{C1} - y_{A1})\frac{x_{B1} - x_{A1}}{y_{B1} - y_{A1}}, \tag{16}$$

$$y_{C1}=\frac{(y_{A1}-y_{B1})[(x_{A1}^2-x_{B1}^2)+(y_{A1}^2-y_{B1}^2)]}{2[(x_{A1}-x_{B1})^2+(y_{A1}-y_{B1})]}+\frac{2y_{A1}(x_{A1}-x_{B1})^2-(r_A^2-r_B^2)(y_{A1}-y_{B1})-2x_{A1}(x_{A1}-x_{B1})(y_{A1}-y_{B1})}{2[(x_{A1}-x_{B1})^2+(y_{A1}-y_{B1})]}. \quad (17)$$

Replacing the relations (16) and (17) in (13), and considering it equal by zero, result the numeric value for angle φ_1

$$F(\varphi)=(x_{C1}-x_{A1})^2+(y_{C1}-y_{A1})^2-r_A^2=0\ . \quad (18)$$

For contact 2 coordinates, for the centers A_2 and B_2, result x_{c2} and y_{c2}

$$A_2:\begin{cases}x_{A2}=R_A\cos(\theta+2\tau)\\ y_{A2}=R_A\sin(\theta+2\tau)\end{cases},\quad B_2:\begin{cases}x_{B2}=x_{B1}-l\cos\varphi_2\\ y_{B2}=y_{B1}+l\sin\varphi_2\end{cases}, \quad (19)$$

$$y_{C2}=\frac{(y_{A2}-y_{B2})[(x_{A2}^2-x_{B2}^2)+(y_{A2}^2-y_{B2}^2)]}{2[(x_{A2}-x_{B2})^2+(y_{A2}-y_{B2})]}+\frac{2y_{A2}(x_{A2}-x_{B2})^2-(r_A^2-r_B^2)(y_{A2}-y_{B2})-2x_{A2}(x_{A2}-x_{B2})(y_{A2}-y_{B2})}{2[(x_{A2}-x_{B2})^2+(y_{A2}-y_{B2})]}, \quad (20)$$

$$x_{C2}=x_{A2}+(y_{C2}-y_{A2})\frac{x_{B2}-x_{A2}}{y_{B2}-y_{A2}}\ . \quad (21)$$

Similar as for contact 1, from relation (22) result the angle φ_2:

$$F(\varphi)=(x_{C2}-x_{A2})^2+(y_{C2}-y_{A2})^2-r_A^2=0\ . \quad (22)$$

Generalizing for a plane circle, from relation (13) result the following relation for the angle φ_i which led to the point coordinates B_i and C_i

$$F(\varphi)=(x_{Ci}-x_{Ai})^2+(y_{Ci}-y_{Ai})^2-r_A^2=0, \quad (23)$$

where

$$A_i:\begin{cases}x_{Ai}=R_A\cos(\theta+i\tau)\\ y_{Ai}=R_A\sin(\theta+i\tau)\end{cases},\ B_i:\begin{cases}x_{Bi}=x_{Bi-1}-l\cos\varphi_i\\ y_{Bi}=y_{Bi-1}+l\sin\varphi_i\end{cases}, \quad (24)$$

and

$$y_{Ci}=\frac{(y_{Ai}-y_{Bi})[(x_{Ai}^2-x_{Bi}^2)+(y_{Ai}^2-y_{Bi}^2)]}{2[(x_{Ai}-x_{Bi})^2+(y_{Ai}-y_{B2}i)]}+\frac{2y_{Ai}(x_{Ai}-x_{Bi})^2-(r_A^2-r_B^2)(y_{Ai}-y_{Bi})-2x_{Ai}(x_{Ai}-x_{Bi})(y_{Ai}-y_{Bi})}{2[(x_{Ai}-x_{Bi})^2+(y_{Ai}-y_{B2}i)]}, \quad (25)$$

$$x_{Ci}=x_{Ai}+(y_{Ci}-y_{Ai})\frac{x_{Bi}-x_{Ai}}{y_{Bi}-y_{Ai}}, \quad (26)$$

Using the sprocket teeth number (z) the angle θ_{max} can be calculated. This angle can be equal with the angular pitch (τ). Also, depending by this angle, the angle φ_i for $F(\varphi)=0$ can be represented.

For determining the angle (α_i) between the right line Δ_i (A_iC_i) and A_iO the scalar multipliying is used

$$\cos\alpha_i=\bar{e}_{A_iC_i}\cdot\bar{e}_{A_iO}=\frac{\overline{A_iC_i}}{|A_iC_i|}\cdot\frac{\overline{A_iO}}{|A_iO|}, \quad (27)$$

$$cos\,\alpha_i = \frac{(x_{Ci} - x_{Ai})(x_O - x_{Ai}) + (y_{Ci} - y_{Ai})(y_O - y_{Ai})}{\sqrt{(x_{Ci} - x_{Ai})^2 + (y_{Ci} - y_{Ai})^2} \cdot \sqrt{(x_O - x_{Ai})^2 + (y_O - y_{Ai})^2}}. \tag{28}$$

4. Case Study

For considered chain teeth number (z), bush diameter (d_b) and chain pitch (p), the contact points coordinates between the chain bush and sprocket, which from a short-bush chain transmission, will be determined. So, for a standardized chain transmission with unitary ratio, are considered the teeth number z=16, the bush diameter d_b=5.08 [mm] and also the chain pitch p= 9.525 [mm].

The transmission parameters (Figure 1a) result as follow [7]:

$$D_d = \frac{p}{sin\frac{180}{z}} = 48.82 \text{ mm}$$

$$D_f = D_d - d_b = 43.74 \text{ mm}$$

$$R_{1max} = 0.505 \cdot d_b + 0.069\frac{d_b}{3} = 2.68 \text{ mm}$$

$$R_A = \frac{D_f}{2} + R_{1max} = 24.55 \text{ mm}$$

$$r_A = R_{1max} = 2.68 \text{ mm}$$

$$r_B = \frac{d_b}{2} = 2.54 \text{ mm}.$$

With the previous parameters and the first method presented, there are determined the contact angle (α) depending by the contact number (i) between the chain bush and sprocket. For contact angle known values (Figure 2), the contact points coordinates (x_{ci}, y_{ci}) can be determined, when the sprocket is rotating fron 0 to the angular pitch ($\theta \in [0,\tau]$). In table 1 the different values of coordinates x_{ci}, y_{ci} (i=0...2) and α_i (i=0...2), depending by rotation angle θ values, are presented.

Table 1. Contact point (x_i și y_i) and contact angle (α_i) values depending by rotation angle (θ), according to the first method.

θ_i [deg]	x_{c0} [mm]	y_{c0} [mm]	x_{c1} [mm]	y_{c1} [mm]	x_{c2} [mm]	y_{c2} [mm]	α_0 [deg]	α_1 [deg]	α_2 [deg]
0	21.87	0	20.21	8.37	15.47	15.47	0	7.65	15.03
4.5	21.8	1.72	19.35	9.93	14.37	16.45	1.67	9.36	16.39
9	21.6	3.42	18.58	11.35	12.95	17.48	3.21	10.72	17.57
11.25	21.46	4.27	18.19	12.15	12.15	18.19	4.07	11.49	18.33
13.5	21.27	5.11	17.65	13.06	11.48	18.82	4.77	12.33	18.97
18	20.8	6.79	16.52	14.39	10.06	19.89	6.28	13.77	20.27
22.5	20.21	8.37	15.47	15.47	8.87	20.91	7.65	15.03	21.9

Applying the second method, for known geometrical parameters and wheel rotation angle by mounting position (θ_i = 0;4.5;11.25;22.5°) and also the angle given by the precedent chain bush (φ_i), for $F(\varphi) = 0$, result the contact point coordinates and also the contact angles, for the first three contacts (i=0, 1, 2). For easy comparison, in tables 1 to 5 these values (x_{ci}, y_{ci}) and (α_i) are highlighted.

Table 2. The values for the contact angle and contact point coordinates, for θ_i = 0°, according by second method

i=0	x_{A0} [mm]	y_{A0} [mm]	x_{B0} [mm]	y_{B0} [mm]

	24.55		0.00		24.41			0.00
i=1	x_{A1} [mm]	y_{A1} [mm]	x_{B1} [mm]	y_{B1} [mm]	y_{C1} [mm]	x_{C1} [mm]	φ_1 [deg]	α_1 [deg]
	22.69	9.40	22.55	9.34	8.37	20.21	78.75	7.65
i=2	x_{A2} [mm]	y_{A2} [mm]	x_{B2} [mm]	y_{B2} [mm]	y_{C2} [mm]	x_{C2} [mm]	φ_2 [deg]	α_2 [deg]
	17.36	17.36	17.26	17.26	15.47	15.47	56.30	15.03

Table 3. The values for the contact angle and contact point coordinates, for $\theta_i = 4.5^\circ$, according by second method

i=0	x_{A0} [mm]		y_{A0} [mm]		x_{B0} [mm]			y_{B0} [mm]
	24.48		1.93		24.34			1.92
i=1	x_{A1} [mm]	y_{A1} [mm]	x_{B1} [mm]	y_{B1} [mm]	y_{C1} [mm]	x_{C1} [mm]	φ_1 [deg]	α_1 [deg]
	21.87	11.15	21.75	11.08	9.93	19.35	74.25	9.36
i=2	x_{A2} [mm]	y_{A2} [mm]	x_{B2} [mm]	y_{B2} [mm]	y_{C2} [mm]	x_{C2} [mm]	φ_2 [deg]	α_2 [deg]
	15.95	18.67	15.86	18.57	16.15	14.37	51.80	16.39

Table 4. The values for the contact angle and contact point coordinates, for $\theta_i = 11.25^\circ$, according by second method

i=0	x_{A0} [mm]		y_{A0} [mm]		x_{B0} [mm]			y_{B0} [mm]
	24.08		4.79		23.94			4.76
i=1	x_{A1} [mm]	y_{A1} [mm]	x_{B1} [mm]	y_{B1} [mm]	y_{C1} [mm]	x_{C1} [mm]	φ_1 [deg]	α_1 [deg]
	20.42	13.64	20.30	13.56	12.15	18.19	67.50	11.49
i=2	x_{A2} [mm]	y_{A2} [mm]	x_{B2} [mm]	y_{B2} [mm]	y_{C2} [mm]	x_{C2} [mm]	φ_2 [deg]	α_2 [deg]
	13.64	20.42	13.56	20.30	18.19	12.15	45.00	18.33

Table 5. The values for the contact angle and contact point coordinates, for $\theta_i = 22.5^\circ$, according by second method

i=0	x_{A0} [mm]		y_{A0} [mm]		x_{B0} [mm]			y_{B0} [mm]
	22.69		9.40		22.55			9.34
i=1	x_{A1} [mm]	y_{A1} [mm]	x_{B1} [mm]	y_{B1} [mm]	y_{C1} [mm]	x_{C1} [mm]	φ_1 [deg]	α_1 [deg]
	17.36	17.36	17.26	17.26	15.47	15.47	56.30	15.03
i=2	x_{A2} [mm]	y_{A2} [mm]	x_{B2} [mm]	y_{B2} [mm]	y_{C2} [mm]	x_{C2} [mm]	φ_2 [deg]	α_2 [deg]
	9.40	22.69	9.34	22.56	20.91	8.87	33.56	21.9

5. Results and Discussion

Based on the proposed mathematical methods and studying a real case of a chain transmission, the contact angle between chain bush and sprocket are determined. In the considered case the wheel teeth number is minimum and the transmission ratio is unitary.

According by first presented method, in Figure 4 the variation of the contact angle (α) is presented. In Figures 5 and 6 the variation of the contact point coordinates (x and y), for different values of the sprocket rotation angle (between 0 and its angular pitch (τ) are presented.

Analyzing the Figures 4, 5 and 6 can be observed that:

- the contact angle increases linear with the increasing of the sprocket rotation angle;
- the contact point coordinates on x axis decreases with the increasing of the sprocket rotation angle; the accentuate decrease is at contacts considered in increasing order;
- the contact point coordinates on y axis increases with the increasing of the sprocket rotation angle; the accentuate increase is at contacts considered in increasing order.

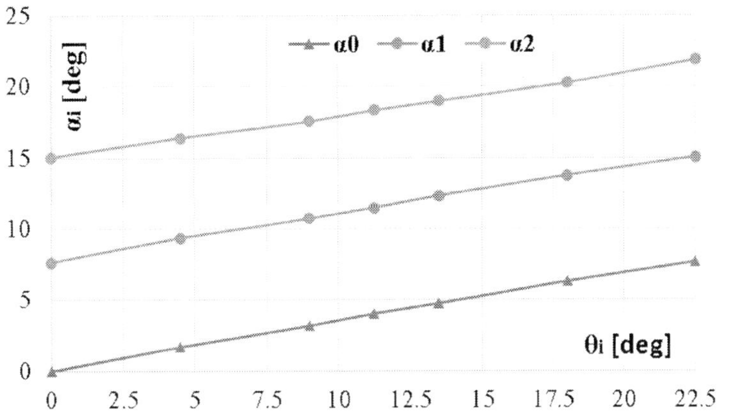

Figure 4. Contact angle variation (α_i), depending by chain wheel rotation angle (θ_i), for sprocket angular pitch (τ), according by first method.

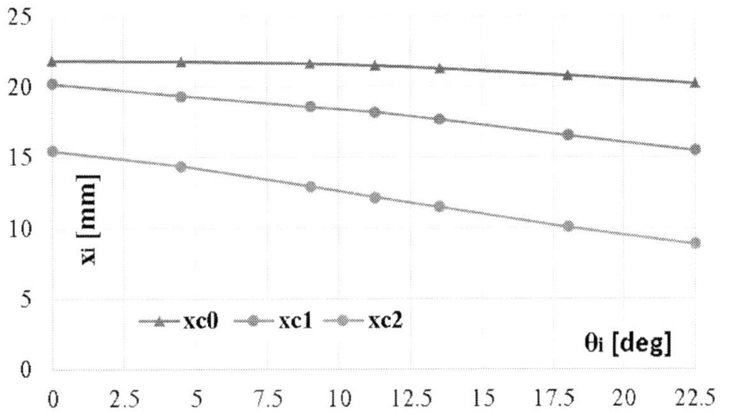

Figure 5. Contact point variation for coordinate (x_i), depending by the sprocket rotation angle (θ_i), for sprocket angular pitch (τ), according by first method.

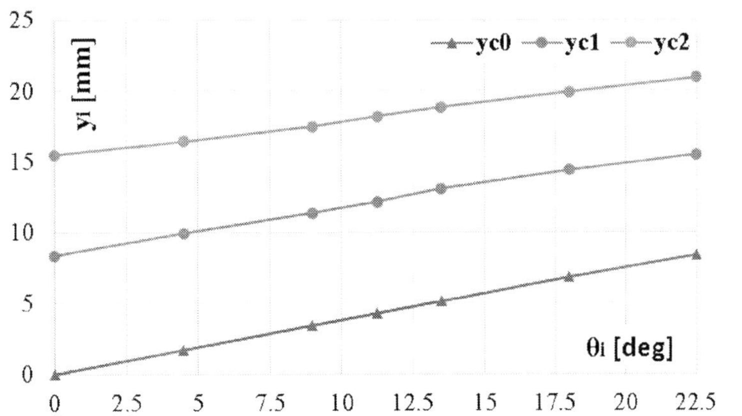

Figure 6. Contact point variation for coordinate (y_i), depending by the sprocket rotation angle (θ_i), for sprocket angular pitch (τ), according by first method.

For results comparison, in Table 6 are presented the point coordinates and also contact angles values. It is considered that, the first contact starts from initial position on the vertical axis, $\alpha_0 = 0$, (Figure 1) and vary along the sprocket angular pitch, τ (Figure 3).

Table 6. The contact point coordinates (x_i și y_i) and contact angle (α_i) values, depending by the chain wheel rotation angle (θ_i), according by first method and checked by the second method (highlighted in Tables 1...5)

θ_i [deg]	x_{c0} [mm]	y_{c0} [mm]	x_{c1} [mm]	y_{c1} [mm]	x_{c2} [mm]	y_{c2} [mm]	α_0 [deg]	α_1 [deg]	α_2 [deg]
0	21.87	0	20.21	8.37	15.47	15.47	0	7.65	15.03
4.5	21.8	1.72	19.35	9.93	14.37	16.45	1.67	9.36	16.39
11.25	21.46	4.27	18.19	12.15	12.15	18.19	4.07	11.49	18.33
22.5	20.21	8.37	15.47	15.47	8.87	20.91	7.65	15.03	21.9

6. Conclusions

The correct functioning of a bush chain transmission is given, among other functional parameters, by the contact point / angle between the bushing and the sprocket, knowing that with the increase of the contact angle the friction surface between them increases.

The design of a chain drive with bushings must also take into account the contact angle or contact point between the bushing and the sprocket as they influence the transmission dynamics (vibrations and wear) and can compensate for deviations induced by assembly or manufacturing.

Knowing accurately the contact angle and the contact point allows correct determination of the application of the transversal forces during operation.

The number of contacts between the bush and the sprocket influences directly, on the one hand, the contact angle (which is limited according to the sprocket geometry) and on the other hand the load on each tooth. A larger number of contacts implies lower transmission efficiency with a lower tooth load, so a higher lifetime.

The use of such a mathematical model allows the determination of the areas of application of the transverse forces of the transmission and thus an essential step in the dynamic calculation of the transmission which contributes significantly to the optimization of such a transmission.

7. References

[1] Sergeev S A, Moskalev D V 2009 Parametric Optimization of Chain-Transmission Sprockets *Russian Engineering Research* **29** pp 452–455

[2] Metil'kov S A, Berezhnoi S B, Yunin V V 2008 Wear of Hinges in Roller Drive Chain *Russian Engineering Research* **28** pp 839–844

[3] Todi-Eftimie A, Velicu R, Saulescu R, Jaliu C 2013 Geometric modelling of power joints from bush chain drives *Mechanisms and Machine Science* **18** pp 471-479 DOI 10.1007/978-3-319-01845-4_47

[4] Pedersen S L 2005 Model of contact between rollers and sprockets in chain-drive systems *Archive of Applied Mechanics* **74** pp 489-508 DOI 10.1007/s00419-004-0363-4

[5] Lates M T 2012 Bush chains design process *Annals of the Oradea University, Fascicle of Management and Technological Engineering* **XI (XXI)** Nr.2 pp 2.51-2.55

[6] Velicu R, Saulescu R, Jurj L 2016 Contact point of bush – sprocket tooth depending on pitch differences of bush chain transmissions *IOP Conference Series: Materials Science and Engineering* **147** (1) 012039 DOI 10.1088/1757-899X/147/1/012039

[7] Saulescu R, Velicu R, Lates M 2016 Geometric modelling of the contact point between the bushing and sprocket in chain drives *IOP Conference Series: Materials Science and Engineering* **174** 012049 DOI 10.1088/1757-899X/174/1/012049

[8] Velicu R, Saulescu R, Jurj L 2018 The Influence of the Bush-Bushes Pocket Geometry on the Bush Contact Angle *Applied Mechanics and Materials* **880** pp 15-20 DOI 10.4028/www.scientific.net/AMM.880.15

[9] https://www.iso.org/standard/38940.html ISO 606:2004 Short pitch transmission precision bush chains and sprockets *International Organization for Standardization* (accessed on March 2018)

IOP Conf. Series: Materials Science and Engineering **514** (2019) 012030 doi:10.1088/1757-899X/514/1/012030

Thermally coupled structural analysis of a connecting rod obtained by powder metallurgy

[1]A Didu, N Dumitru[1] and A M Calangiu[2]

[1] Faculty of Mechanics, University of Craiova, România
[2] FORD Romania S.A., Craiova, Romania

E-mail: nicolae_dtru@yahoo.com , acalangi@ford.com

Abstract:The paper is structured in two parts, the first part is dedicated to modeling and numerical simulation in dynamic regime of the motor mechanism from a 4-stroke combustion engine and the second part, to finite element method analysis in thermally coupled structural behavior of sliding bearings. The bearings in the connecting rod structure are obtained by Powder Metallurgy. It is described the procedure for the experimental determination of the material properties of the parts made of metallic powders, which are required as input data for modeling in an integrated system of the sintering process coupled with the structural analysis of the engine assembly.

1.Introduction

This paper aimes to analyze in dynamic regime a mechanism from a 4-stroke internal combustion engine. There are identified the laws of variation in time for the kinematic and dynamic parameters by virtual prototyping. The main purpose of this paper is to identify the answer of dynamic loads of the connecting rods' mechanism. The two sliding bearings composing the connecting rod subassembly are made of metallic powders obtained by pressing and sintering. To identify bearings behavior, the connecting rod subassembly is analyzed with the finite element method in thermally coupled regime. At the same time, the modeling and simulation with finite elements in thermal regime of the bearings behavior during the sintering process is pursued. The kinematic and dynamic analysis of the motor mechanism is studied in several works, respectively [1], [2].

There are several attempts for modeling with the finite element method of the sintering process of parts obtained through powder metallurgy.

Connecting rods made of sintered powders with a complex shape show problems related to non-uniform distribution of density. The optimal quality of the connecting rod is conditioned by the physical properties of the preform. The obtained results show that the stress flow increases with decreasing temperature and increasing strain rate. [3]

The invention presented by Lynn concerns an improved manufacturing method of a powder metal connecting rod with a stress riser crease formed in a side face by forging a sintered preform with a generally V-shaped notch in the side face. [5]

Many powder metallurgy processing techniques are available for manufacturing different components, ranging from pressing and sintering to hot isostatic pressing, powder forging, metal injection molding and rapid prototyping; these are discussed in various publications. [4]

2. Dynamic analysis of the mechanism of an internal combustion engine

IOP Conf. Series: Materials Science and Engineering **514** (2019) 012030 doi:10.1088/1757-899X/514/1/012030

2.1 Modeling with Finite Element Method

In order to analyze with finite elements in dynamic regime, the following steps are taken:

- The geometric model of the assembly crankshaft, connecting rod, piston.is imported from Solid Works;
- There are defined the material properties for each component in the mechanical system;
- The kinematic joints between the mobile elements of the assembly are defined as 3D models with reference systems and contact surfaces;
- There are identified and defined the dynamic parameters for the two types of contacts , respectively:
- ▪ The contact between elements without relative movement;
- ▪ The link between elements with relative motion;
- We choose the type of finite element and mesh each component in the model;
- There are defined the pressure distribution laws on the four pistons(figure 2);
- There are defined the parameters for dynamic analysis (direct method, integration step, control algorithm, etc.);
- There are defined the local reference systems for monitoring the movement of the mechanical system in dynamic regime;
- There are identified the time-varying laws for dynamic system parameters or component elements, including strains, stresses or displacements resulting from the combination of rigid solid movement and the one of deformable solids. [8]

2.2 Input data for dynamic modeling of the mechanism:

The law of pressure variation on the piston, experimentally determined:

$$p = -4.930 + 3203.386t - 1.631 \cdot 10^{7} t^{2} + 4.110 \cdot 10^{9} t^{3} - 4.504 \cdot 10^{11} t^{4} + 2.229 \cdot 10^{13} t^{5} - -4.102 \cdot 10^{14} t^{6} \quad (1)$$

- the law of variation of the crankshaft rotation angle.

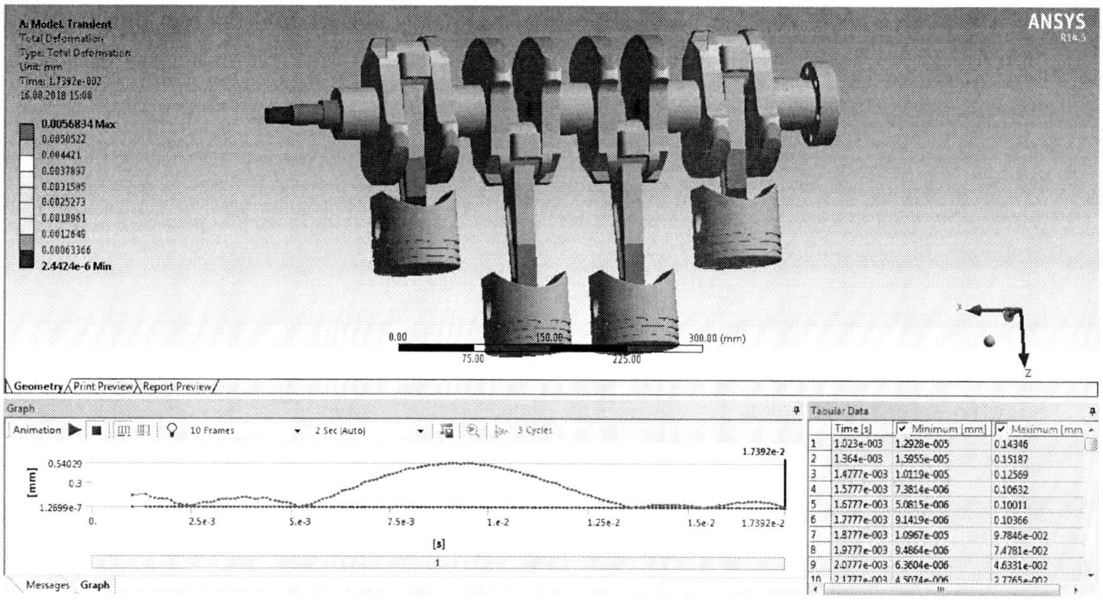

Figure 1. Distribution of the resultant deformations on the assembly

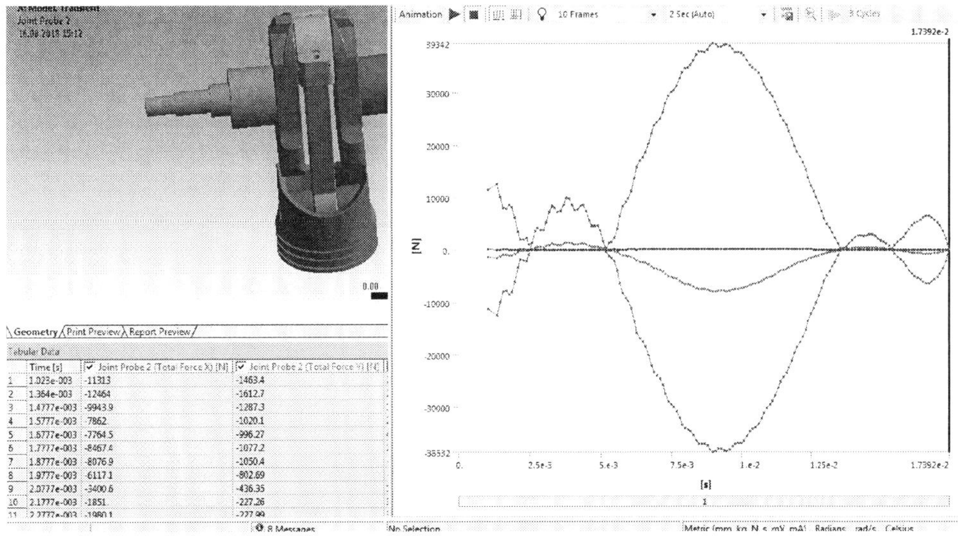

Figure 2. The distribution of connecting forces between the connecting rod and the bolt

3. Thermally coupled structural analysis of the connecting rod - bearing assembly.

3.1. Technology for obtaining bearings using metallic powders

Bearings obtained from sintered metallic powders

The connecting rod subassembly in the motor mechanism structure consists of two plain bearings made of pressed and sintered metallic powders.

Sintering was carried out at 1150° C, the samples being maintained at this temperature for 60 minutes. Heating was done slowly (60 min. at 350 degrees, 20 min. at 650 degrees) to maintain the temperature for the homogenization of heat in the mass of the pieces, as it is shown in the cyclogram from figure 6.

3.2. Characterization of the obtained specimens

After pressing and sintering, the densities of the specimens, their porosities and their contractions were determined.

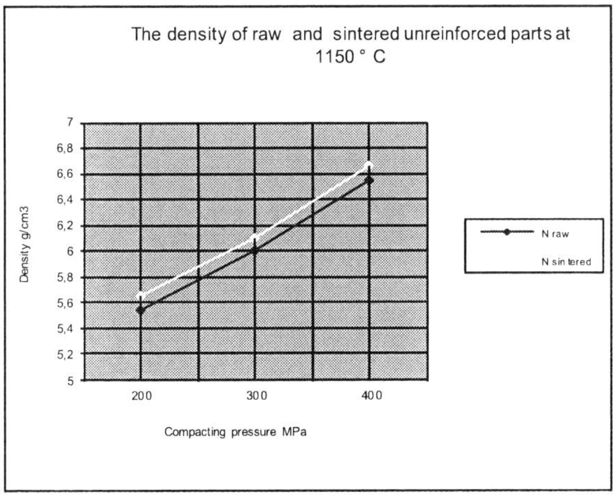

Figure 3. Density variation of unreinforced parts before and after sintering

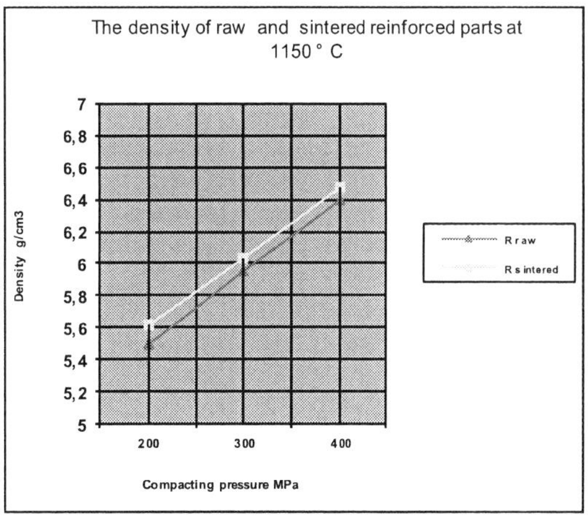

Figure 4. Density variation of reinforced parts before and after sintering

The curve of the density variation dependent on compaction pressure, before and after sintering for the unreinforced specimens (N) can be seen in Figure 3 and in Figure 4 for the reinforced ones (R).

By overlaying the two diagrams, we can compare the two types of specimens (Figure 5).

From the diagram study (Figure 5), it is noticed that the unreinforced specimens are denser than the reinforced ones. This is due to the fact that the reinforcement prevents the free flow of iron powder particles; it is possible that in the area of the wire to have more porous areas.

After sintering, the density of the pieces increases and the dimensions decrease.

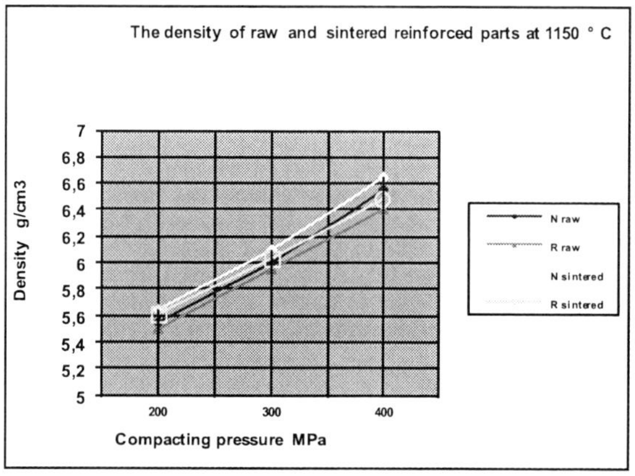

Figure 5. Density variation of reinforced and unreinforced parts before and after sintering

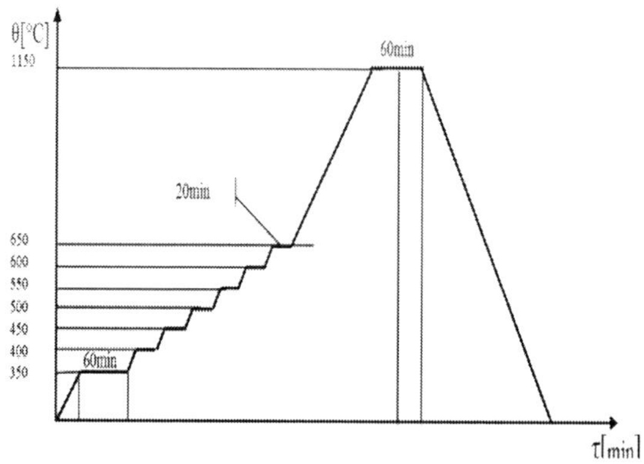

Figure 6. Sintering cyclogram of the samples

Analyzing contraction variation charts, it is noted that the highest contractions are for B (0.8% for both unreinforced and reinforced parts pressed with 200 MPa pressures).The contraction value increases with the decrease of compaction pressure for width B, both for reinforced and unreinforced parts. [7] The contraction in height has the same tendency to decrease as the compaction pressure increases for both types of specimens. The L-length contraction value increases with increasing compaction pressure for unreinforced parts and has a random distribution for the reinforced ones (increases for a pressure of 300 MPa versus 200 MPa, then decreases to 400 MPa). [6]

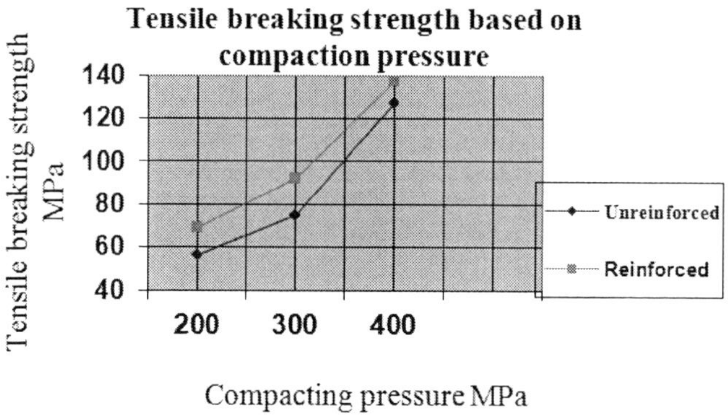

Figure 7. Variation of tensile breaking strength depending on the compaction pressure of the standard parts

It is determined by experimental testing of tensile strength for different compaction pressures. In order to analyze the dynamic behavior of the bearings in the structure of the connecting rod subassembly, the data obtained on the models are at a compaction pressure of 400 MPa.

IOP Conf. Series: Materials Science and Engineering **514** (2019) 012030 doi:10.1088/1757-899X/514/1/012030

From the obtained force-strain diagrams analysis, it was found that the tensile breaking strength is directly proportional to the density of the parts (implicitly with the compaction pressure). It is also clear that the tensile breaking strength is higher for the reinforced parts than for the unreinforced ones.

Using the previous data, can plot the variation of the tensile strength resistance of the unreinforced calibrator parts and the reinforced ones according to the compacting pressure which can be seen in Figure 7.

3.3. Finite element analysis in structural thermally coupled regime

For the finite element analysis of a bearing, the following steps were taken:

- Developing the geometric model;
- Defining the type of finite elements and material properties;
- Model meshing;
- Specification of loads and contouring conditions;
- Thermally coupled structural analysis;
- Processing the results.

For identifying the critical values of thermal or structural parameters were selected the nodes from the interest areas, obtaining the diagrams of variation for thermal flux, the thermal gradient and respectively the equivalent stresses von Mises (Figure 11).

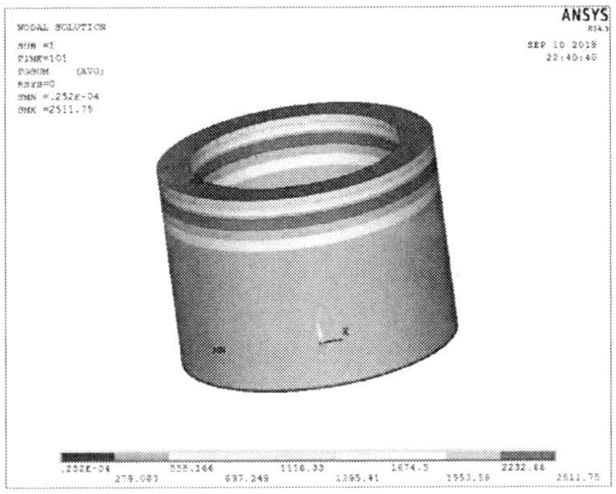

Figure 8. Distribution of the resultant gradient on the bearing

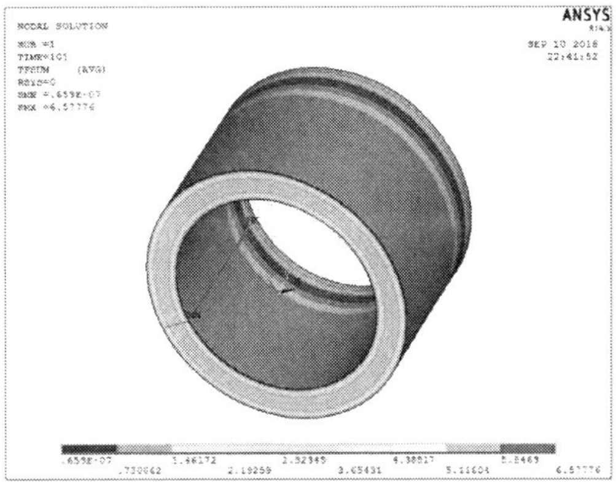

Figure 9. Distribution of the resultant thermal heat flux on the bearing

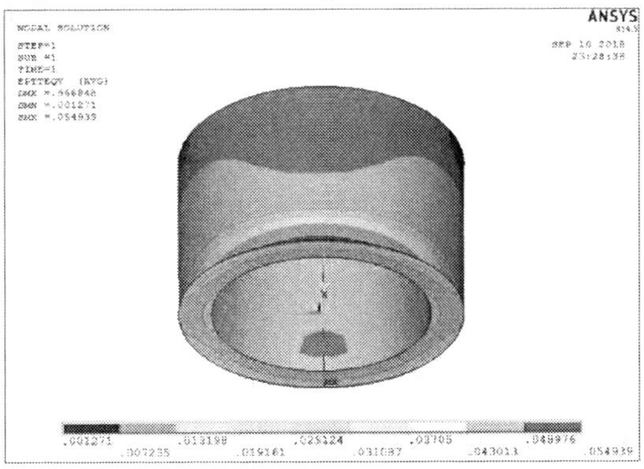

Figure 10. The equivalent strain von Mises thermally -structural coupled

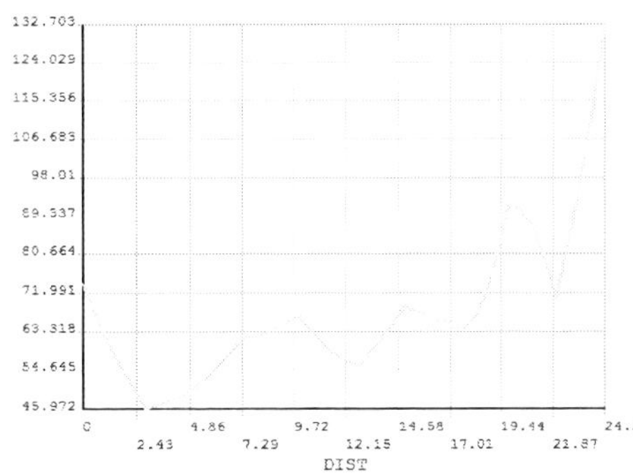

Figure 11. The resulting von Mises stresses in distributed on the bearing generatrix.

4. Results and disscutions

From analyzing this paper, there can be established the following aspects:

- The mathematical model for determining the angular velocity and the motor moment of the actuation mechanism has been realized;
- The dynamic model with finite elements of the entire engine assembly with the ANSYS program was built;
- There were identified the laws of variation in time for the connecting forces in the two sliding bearings in the connecting rod structure;
- It is described the procedure for the experimental determination of the mechanical characteristics of the two bushings obtained by powder metallurgy (PM);
- The material properties were established on unreinforced and reinforced specimens, before and after sintering, by comparison;
- There were realized samples for more compaction pressures, the dynamic study being carried out on the 400 MPa sample without reinforcement;

By laboratory tests were established the modulus of elasticity, the Poisson ratio and other characteristics necessary for the analysis with finite elements in thermally coupled thermal regime;

5. Conclusions

It was found that the stresses, deformations and displacements monitored by the finite element method are in normal values for the two bearings in the connecting rod structure;

The obtained results recommend the use of parts made by powder metallurgy in the construction of the internal combustion engine drive mechanisms, which have an optimal thermal and structural behavior.

Mainly the thermal transient analysis of the studied bearings highlights the following results:

- distribution of heat flux in the bearing body (figure 9);
- distribution of the thermal gradient (figure 8);
- distribution of friction forces on the bearing in cylindrical coordinates;
- the distribution of equivalent strain (figure 10) and equivalent stresses (figure 11) in a thermally - structural coupled span.

The above-mentioned laws are materialized in the form of diagrams. The elaborated models fit into an integrated system that allows the extension of the procedure for the study of different types of sliding bearings in the structure of which are used groups of materials with different properties, allowing a complete verification in terms of static and dynamic resistance calculation, as well as thermal transfer.

6. References

[1] Charles E., Wilson J., Sadler P. 1993 *Kinematics and Dynamic of Machinery* Harper Collins College Publishers
[2] Norton R. L. 2008 Design of Machinery McGraw-Hill
[3] Ramakrishnan P. 2013 *Automotive applications of powder metallurgy* Woodhead Publishing Limited pp 493-519
[4] Fengxian L., Jianhong Y., Jurgen E. 2017 *Metallurgical and Materials Transactions A The Minerals* Metals & Materials Society and ASM International pp 2971-2978
[5] Lynn - et al v. FCA U.S., LLC et al, No. 1:2016mc02034 patent
[6] De Mello J. D. B., Klein A., Binder R. and Hutchings I.M. 2001 *Effect of compaction pressure and powder grade on microstructure and hardness of steam oxidised sintered iron* Powder Metallurgy **44** pp 53-61
[7] Didu A.,Didu M., Radu St. 2010 *Physical-Mechanical Characteristics Dependence on Compacting Pressure for Reinforced with Steel Yarns Pieces* Annals of DAAAM for 2010 & Procedings of the 21 st International DAAAM Symposium **21** pp 1501-1502
[8] Moaveni S. 2003 *Finite Element Analysis* Pearson Education Prentice Hall.

IOP Conf. Series: Materials Science and Engineering **514** (2019) 012031 doi:10.1088/1757-899X/514/1/012031

Double Flank Gear Roll Inspection of Spur Gears Manufactured using Fused Deposition Modeling

C I Rizescu[1], R R Ciobanu[1] and D Rizescu[1]

[1]Mechatronics and Precision Engineering Department, POLITEHNICA of Bucharest University, Romania

E-mail: dana.rizescu@upb.ro

Abstract. The paper presents an experimental study that has as main objective the evaluation of the integration capacity of the spur gears obtained by FDM in high precision mechanisms. The development of precision mechanical systems required the use of the additive layer manufacturing technology in order to manufacture the gears that have specific particularities for the applications in which gears are integrated. Take these aspects into consideration, the paper presents the results achieved by testing on a ZWP-06 machine the gears obtained by fused deposition modeling technology. Gears were obtained on an Anet A2, semiprofessional 3D printer. To obtain a superior quality of toothed wheels, both process parameters: the material properties, and the features of the 3D printer were taken into account. The results, obtained from the test, revealed the level of influence of the execution parameters of the toothed wheels by FDM on their quality. The correlation of the process parameters remains the most important aspect for obtaining components with a high accuracy class, which complies with the resistance standards under the conditions of a shorter execution time.

1. Double flank gear roll inspection

The double flank gear roll inspection is a simple method to test the functional accuracy easily and quickly for spur gears [1, 2]. The result is the sum of all given deviations. No matter which deviation the specimen has, the double flank gear roll inspection will detect it. The result of the double flank gear roll inspection is a summery error. That means, that it is not possible to determine the single errors causing the deviation. If the detected summary errors are within tolerance, it is assumed that also the single errors are within tolerance. If the deviation detected by the roll inspection is too big, the cause study is the operator's job or the specimen has to be measured by a machine which is able to determine single deviations. The authors developed a research for different gears materials on a measuring setup for the double flank gear rolling inspection ZWP 06 from FRENCO. During the double flank gear rolling inspection two gears are rolled together free from backlash. ZWP 06, the double flank gear roll inspection machine, is specially designed for small high-precision gears. It is also suitable for plastic gears. The measuring force can be lowered to 0 N. The sophisticated design is extraordinarily precise and sensitive. The measuring carriage is supported free from backlash on four leaf springs [3, 4].

This so-called parallelogram suspension is very sensitive and registers even the smallest change in centre distance. The measurement process is motor driven by default. We used the FRENCO Software 'FGI pro'. The sketch of the experimental setup is represented in figure 1. There are two gears: the master gear and the workpiece (specimen). The driver gear is the master gear. Master gear is manufactured from steel.

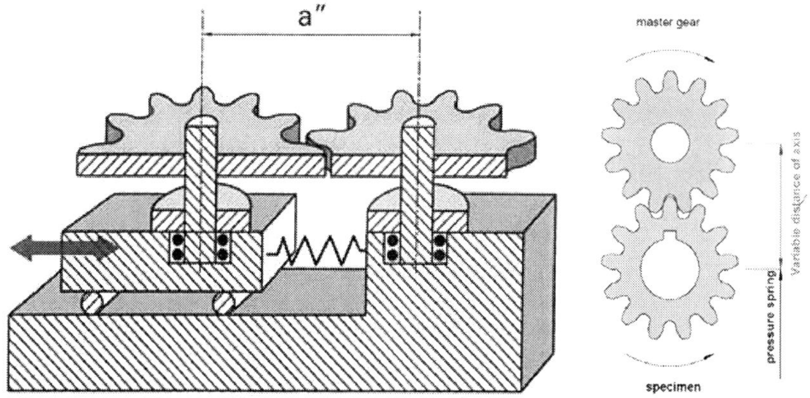

Figure 1. Experimental setup ZWP 06

The software FGI pro includes both, the control of the drive and the evaluation of data. The software is in-house developed and programmed by developers for applications software. With the actual values being marked in colour, the specimen (workpiece) can quickly be evaluated as 'Pass' (green) or 'Fail' (red). Basically the software determines the following values:

- total radial composite deviation Fi"
- tooth-to-tooth radial composite deviation, fi";
- runout deviation by composite test, Fr";
- short-wave component, fk".

2. Operating mode

Usually, gears transfer rotations and torques. Like all machine components, they are afflicted with production caused deviations. The challenge to produce a constant quality requires quick and easy inspection methods, which can be smoothly integrated in the production process. For a precise assessment of the gears' operational behaviour, manufacturers make again use of the well tried double flank gear rolling inspection [2]. This method enables a quick assessment of gears in terms of total deviations.

It determines gear roll variations, tooth to tooth composite deviations, composite runout deviations and the short wave part and displays them clearly. Immediately after the measurement it can be seen whether the gear is acceptable or not. By the colouring the visual comparison of the actual values and the reference value is made easier. For each deviation there are used different colours, operator can easily observe the wrong deviations. The software creates a clearly laid out record showing (presented in figure 2):

Double flank rolling deviation Fi"

Fi" is the difference between the maximum and minimum values of the working centre distance, **a"**, which occurs during a radial (double flank) composite test, when the product gear with its right and left flank simultaneously in tight mesh contact with those of a master gear, is rotated through one complete revolution. (DIN3960/3963)

Double flank rolling tooth to tooth deviation f_i"

fi" is the value of the tooth radial composite deviation corresponding to one pitch, 360°/z, during one complete cycle of engagement of all the product gear teeth (z = number of gear teeth).

Radial Runout Fr''

F_r" is the value of radial runout of the gear between the maximum and the minimum radial distance from the gear axis as observed by removing the short-term or undulation pitch deviations and analysing the long -term sinusoidal waveform.

IOP Conf. Series: Materials Science and Engineering **514** (2019) 012031 doi:10.1088/1757-899X/514/1/012031

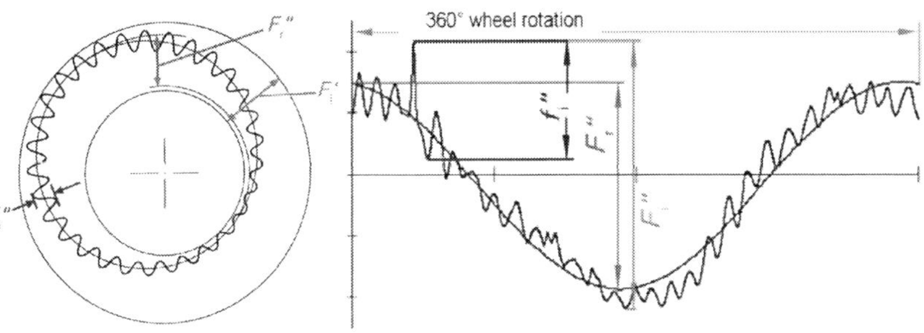

Figure 2. Representing deviations F"$_i$, f"$_i$, F"$_r$

3. Materials and methods for FDM technology

The process of rapid prototyping by FDM has developed spectacular, having as main advantage of obtaining robust parts with a relative low cost [5, 6]. However, a major disadvantage of this technology is represented by the accuracy and the roughness of the parts obtained.

Considering the advantages of rapid prototyping FDM, for the development of the specific control systems, at the global level there is a strong support through the open source applications type [5]. From this point of view, to assess the quality of the parts obtained by FDM there are a number of studies and research on the common software solutions and the properties of materials used [7, 8].

In the literature there are a number of researches which had as main objective optimize part surfaces obtained by FDM, addressing solutions that: application of a metallic paint on the surfaces of the printed, the application of any chemical treatments or post processing technologies of adjustment controls. These solutions have the main disadvantage to enter a new stage in the process of obtaining parts, stage which increases the time and cost of production.

Additional manufacturing processes are required for obtaining complex components in a short time, at a cost of production decreased. In order to fulfil these requirements, FDM process parameters must be established for each application [9]. The final quality is determined by the properties of the material, the parameters of the process and the characteristics of the machine [9, 10].

Researching the influence of the characteristics of the material on the quality of the components obtained by FDM there was used a 3D printer ANET A2, where the table movements run on X-Y axis, and the print head runs along the Z axis. System extruder material, using a system consisting of two gear-driven by a stepper motor. The printer has been calibrated resulting an accuracy of 0.1 mm for components which are simple forms.

The gears have been designed in SOLIDWORKS, and for obtaining the G-Code we used: v2.05 Repetier Cura Engine Slicer software. Materials characteristics are presented in table 1, for both materials: ABS and PLA.

Table 1 Materials characteristics: ABS and PLA

ITEM	ABS	PLA
Density[kg/m^3]	1.01	1.20-1.25
Melt Point [C]	220-260	190-220
Melt Flow Index [g /10 min]	1.43	7.8
Tensile Yield Strength [MPa]	40.96	62.63
Elongation at Break [%]	20.86	4.43
Flexural Strength [MPa]	45.44	65.02
Flexural Modulus [MPa]	1948.45	2504.4
Impact Strength [MPa]	22.11	4.28

Also, for both gears have been keep the same process parameters, being changed only temperature of the extruder and the temperature of the printer platform. In figure 3 there are presented the 3D printer (left) and the designed gear laying on the printing table (right).

Figure 3. 3D printer and Repetier Cura Engine Slicer

4. Experimental results

There were considered three gears manufactured from steel, ABS1 and ABS2. All three gears have the same modulus, m = 1mm, number of teeth, z = 58, and the same thickness b = 12 mm.

Steel gear was manufactured on a special gear cutting drill machine. Acrylonitrile Butadiene Styrene (ABS) is a 3D printing plastic and it was used for the yellow gear. It is a versatile petroleum-based material that belongs to a family of thermoplastic polymers. There are different brands and grades of ABS filament, which makes them unequal in terms of usage and design.

PLA is the most commonly used filament. It's biodegradable, easy to print, and very strong material and was used for the black gear. The perfect choice for printing large objects thanks to its low thermal expansion (little to no warping) and for printing tiny parts because of its low melting temperature. All three gears (steel, ABS and PLA) are shown in figure 4. In figure 5 is presented the inspection of each gear on ZWP 06 machine. The results are presented in table 2.

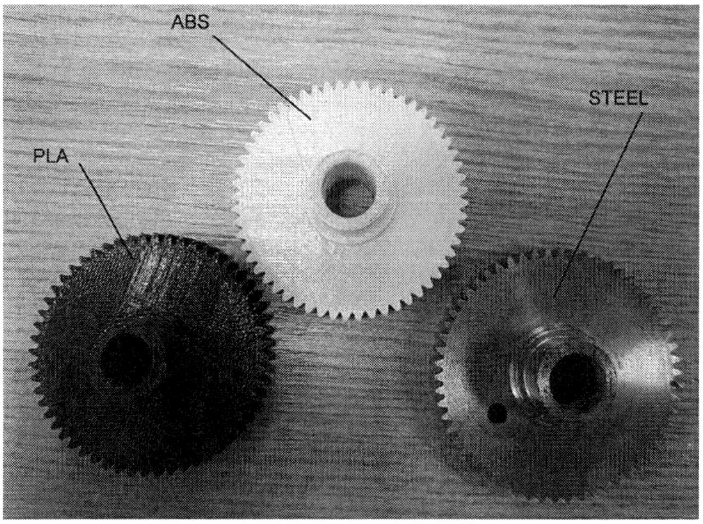

Figure 4 ABS, PLA and STEEL gears

Figure 5 Gears inspection on ZWP machine

Table 2

	Linear chart	Polar chart
Steel gear	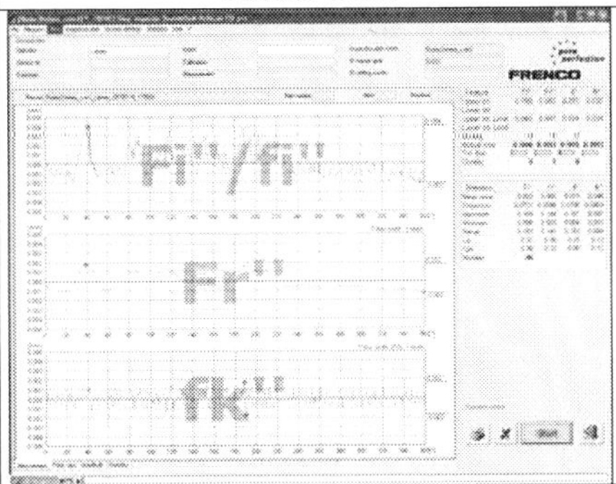	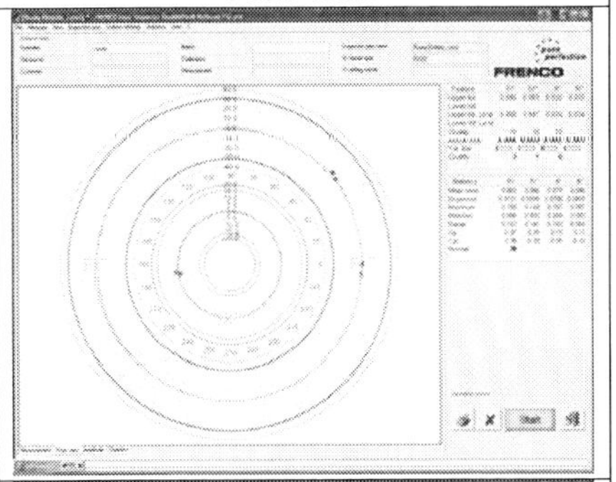
ABS gear	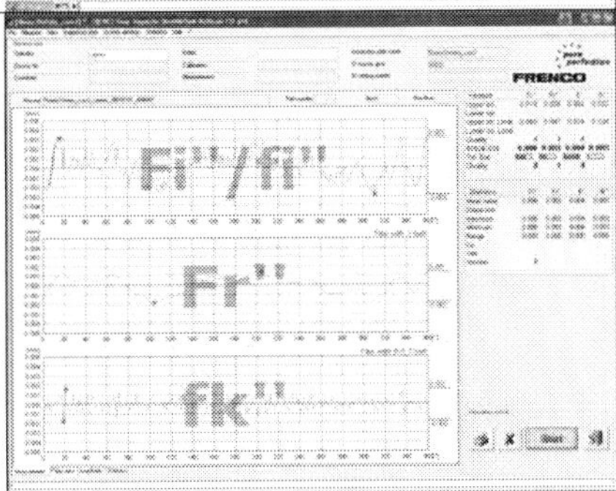	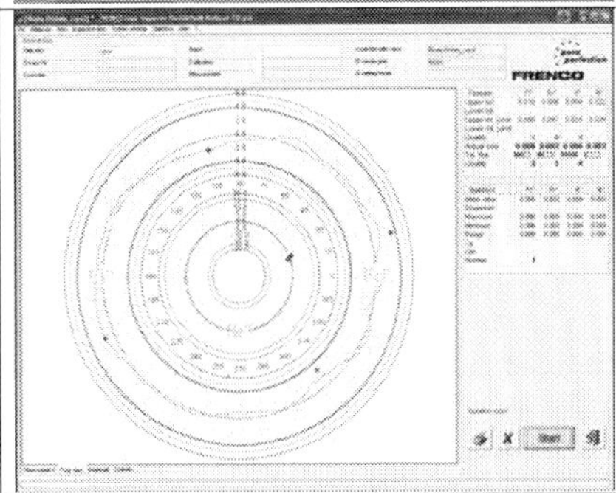
PLA gear	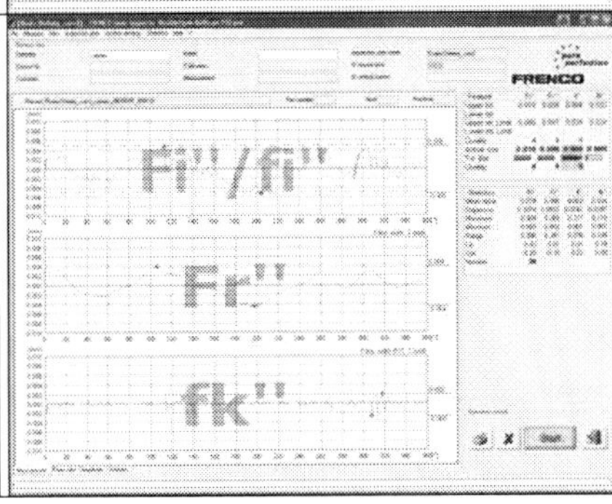	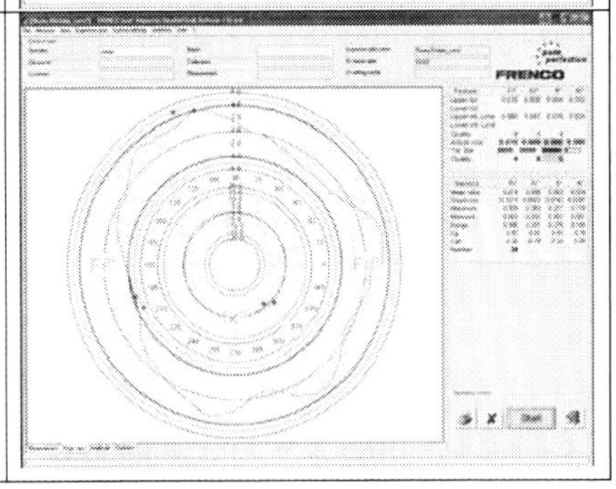

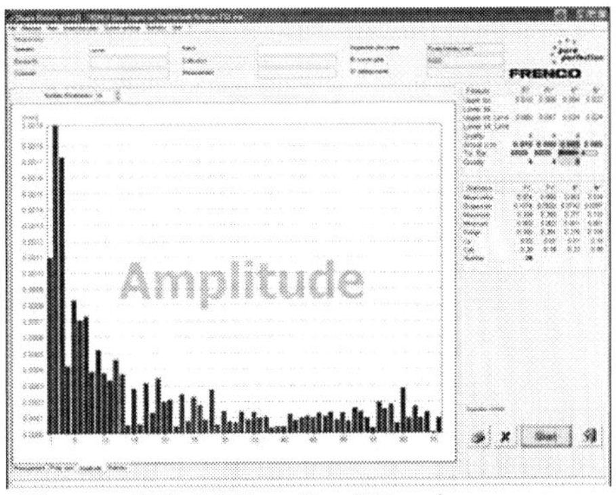

Figure 6 PLA Visualized Fourier spectrum

The Fournier spectrum of the measurement is displayed very clearly (figure 6). The frequency of the teeth is coloured. Beside that the software provides further features, like a comfortable filing, statistical evaluations, automatic detection of damages, measurement of tooth segments, qs-STAT port, etc.

5. Conclusions

The ABS and PLA gears were left in the environmental temperature in our laboratory and they were measured after three days from their manufacturing. For both printed gears we used the same fill factor of 20%.

The experimental results show that:

- The steel gear could be considered for Quality range = 4; (DIN3960/3963)
- The ABS gear could be considered for Quality range = 4;(DIN3960/3963)
- The PLA gear could be considered for Quality range = 5;(DIN3960/3963)

Considering steel gear and ABS gear we found that **double flank rolling tooth to tooth deviation, f_i", is better for ABS than steel gear.** PLA gear is almost of the same quality like the steel gear (except double flank rolling tooth to tooth deviation $f"_i$).

The paper presents an experimental study that has as main objective the evaluation of the integration capacity of the spur gears obtained by FDM in high precision mechanisms.

The tests results reveal a superior precision of ABS gear in comparison with PLA gear.

A future research direction is represented by fatigue and strength tests of ABS gear wheel.

The authors developed several studies in order to evaluate the mechanical behavior of printed machine elements like: springs, gears, clutches, bearings, etc.

References

[1] G. Budzik, The use of the rapid prototyping method for the manufacture and examination of gear wheels, in M.E. Hoque (Ed.), Advanced Applications of Rapid Prototyping Technology in Modern Engineering, (InTech publisher, 2011) 339-364.

[2] Rizescu C, Elemente si mecanisme de mecanică fină-Partea I, *Editura Printech 2013*, pp 41 -55.

[3] https://www.frenco.de/en/double-flank-gear-inspection-machines/

[4] Voicu A, Gheorghe G and Bădița L L, Studies and Research on the measurement and 3D integrated control of complex components from automotive industry by laser scanning, *International conference ICNAAM, 22-28 september 2014*, Rodhos, Greece.

[5] Besnea D, Constantin V, Dontu O, Spanu A, and Cioboată D, Additive Manufacturing Concepts and Design for Advanced Composites Materials, *International Journal of Materials, Mechanics and Manufacturing-IJMMM* **6(3)**, pp 187-190.

[6] Ludmila Novakova-Marcincinova, Ivan Kuric "Basic and Advanced Materials for Fused Deposition Modeling Rapid Prototyping Technology" Manuf. and Ind. Eng., 11(1), 2012, ISSN 1338-6549.

[7] Vishal Sharma and Sharanjit Sing "Rapid Prototyping: Process advantage comparison and application" International Journal of Computational Intelligence Research ISSN 0973-1873 Volume 12, Number 1 (2016), pp. 55-61;

[8] O.A. Mohamed, S.H. Masood, and J.L. Bhowmik, Optimization of fused deposition modeling process parameters: A review of current research and future prospects, Adv. Manuf., 3, 2015, 42–53.

[9] P. Ravi, P.S. Shiakolas, and A.D. Thorat, Analyzing the effects of temperature, nozzlebed distance, and their interactions on the width of fused deposition modeled struts using statistical techniques toward precision scaffold fabrication, Journal of Manufacturing Science and Engineering, 139, 2017.

[10] Pearce, J. M., Morris Blair, C., Laciak, K. J., Andrews, R., Nosrat, A. Zelenika-Zovko, I., 2010. 3-D Printing of Open Source Appropriate Technologies for Self-Directed Sustainable Development", Journal of sustainable development, Vol. 3(4): 17- 29.

IOP Conf. Series: Materials Science and Engineering **514** (2019) 012032 doi:10.1088/1757-899X/514/1/012032

Study behavior of Geneva mechanism using 3D printing technology

D Rizescu[1], D Besnea[1], C I Rizescu[1] and E Moraru[1]

[1]Mechatronics and Precision Engineering Department, POLITEHNICA of Bucharest University, Romania

E-mail: dana.rizescu@upb.ro

Abstract. The authors developed a study for behaviour of Geneva mechanism considering different materials and technologies. Generally, there are three basic types of Geneva motion mechanisms named external, internal and spherical. In this paper the authors considered external Geneva mechanism. The design and machining of a conventional Geneva mechanism are generally well known, but there are some particularities, if we consider 3D printing technologies. Conventional, Geneva mechanisms are used to low-speed applications, or to those in which noise and vibration are of no importance. Using PLA+Copper or ABS materials noise and vibrations are reduced considerably. Also, the entire mechanism has a reduced weight. There were considered different fill factors for 3D printed parts. The experimental setup allows fast changing of mechanism parts. This research aims the study of behaviour of 3D Printed machine elements like: springs, bearings, clutches gears, bellows, diaphragms, bushes, brakes, sliders, etc developed by authors.

1. Introduction

There is well known that indexing mechanisms are of great practical importance in many applications, including weaving looms, precision measurement instruments, automated packaging and printing machinery, film movie projectors and movie cameras. One of the most common forms of indexing mechanism is the Geneva mechanism or Maltese cross mechanism. The design and machining of a conventional Geneva mechanism is generally straightforward since its structure consists primarily of no more than a driving crank and a wheel with straight slots which is Maltese Cross. However, in such mechanisms, a significant impact load is produced at the initial and final stages of the indexing operation as the roller enters and exits the slot, respectively. Consequently, conventional Geneva mechanisms are confined to low-speed applications, or to those in which noise and vibration are of no importance.

The authors developed a study for behaviour of Geneva mechanism considering different materials and technology. The design and machining of a conventional Geneva mechanism are generally well known but there are some particularities if we consider 3D printing technologies.

2. Geneva mechanism

The authors developed five Geneva mechanisms manufactured with different technologies. These mechanisms consist of five Maltese crosses, with the same dimensions, and five driving cranks with pins. There were considered five different materials as it follows:

1. Plexiglas. Plexiglas is a type of acrylic and someone will often find the two words (plexiglass and

acrylic) to be interchangeable;
2. PLA. Poly (lactic acid) or polylactic acid or polylactide;
3. ABS. Acrylonitrile Butadiene Styrene;
4. PLA + Copper;
5. PLA + Bronze.

The Geneva mechanism translates continuous motion to intermittent motion through the driving wheel whose crank pin is shown in figure 1. It drives the Geneva wheel as it slides into and out of the slot of the Geneva wheel. It thus advances it - one step at a time when engaged. The driver wheel usually consists of both the crank and a raised circular blocking disk that locks the Geneva wheel in position between steps to avoid excessive vibration while rotating. The driver angle of the pin is labelled with φ_1 and the driven angle of the Maltese Cross is labelled with φ_2. Using notations from figure 1 the relation between angles φ_2 and φ_1 is:

$$\mathrm{tg}\,\varphi_2=\frac{R_1\cdot\sin\varphi_1}{L-R_1\cdot\cos\varphi_1}=\frac{\lambda\cdot\sin\varphi_1}{1-\lambda\cdot\cos\varphi_1} \tag{1}$$

where $$\lambda=\frac{R_1}{L}=\sin\varphi_2=\sin\frac{\pi}{z} \tag{2}$$

z = number of slots from Maltese cross.

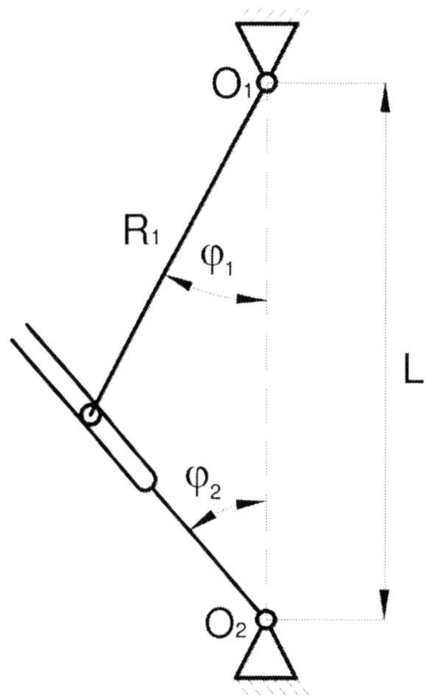

Figure 1. Geneva mechanism

All five crosses, from experimental setups, have six slots, but a theoretical study was developed using Matlab environment. There were considered different Geneva mechanisms with z = 3,4 and 6 slots. The driven angle φ_2 which represents the rotation of Maltese cross is represented in figure 2, depending on driver angle φ_1 which represents the rotation of driver crank. For example, the driver angle, $\varphi_1 = 2160°$ for Geneva mechanism with 6 slots and $\varphi_1 = 1080°$ for Geneva mechanism with 3 slots. Using Matlab function, **diff**, one can calculate the relative angular speed, ω_2/ω_1, if there is assumed that $\omega_1 = d(\varphi_1)/d(t)$ and $\omega_2 = d(\varphi_2)/d(t)$. This is a numerical derivation and the results are presented in figure 2, considering the same number of slots, z = 3, 4 and 6, for different Geneva mechanisms.

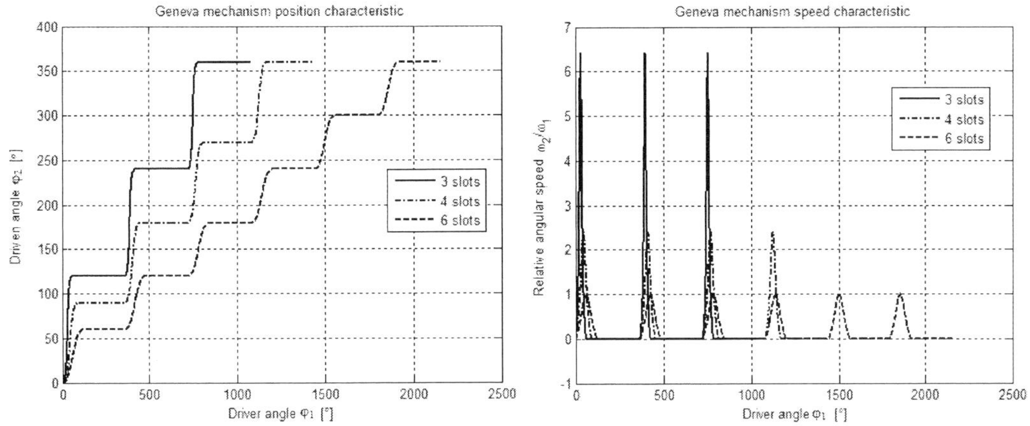

Figure 2. Geneva mechanisms position and speed characteristics with different number of slots

3. Experimental setups

The authors developed experimental setups in order de determine cinematic parameters of Geneva mechanisms. There were designed five Geneva mechanism. Considering only weight of cross and crank slider (neglecting shafts, bearings) the weights are presented in table 1.

Table 1. Weight of Geneva mechanisms

Material	Weight [g]
Plexiglas	55
PLA	33
ABS	27
PLA + Copper	68
PLA + Bronze	65

In order to determine the time between two successive slots the authors uses reflective sensors. In figure 3 is presented the operating mode of a reflective sensor. When the slot of the Maltese cross is in front of the reflective sensor the reflected beam is interrupted. Then, the cross is rotating with an angle of 60°. The cross passes in front of the reflective sensor and the time is recorded until the next (successive) slot is reached. When the cross is in front of the reflective sensor the beam is reflected.

There are six time data to be recorded for one cross rotation. These data are stored in an excel file and then read with a Matlab script file [2].

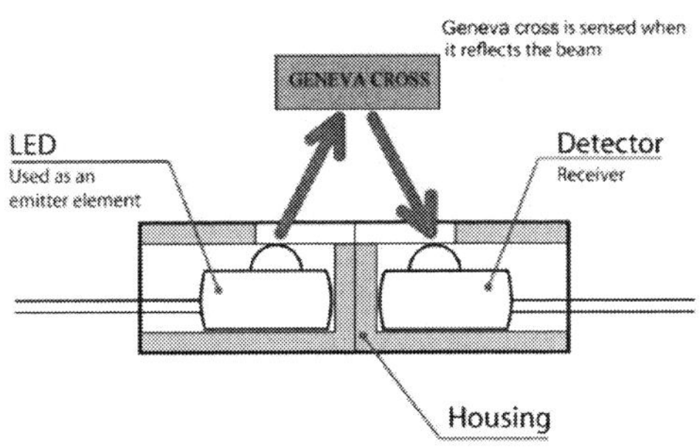

Figure 3. Operating principles of optical sensors

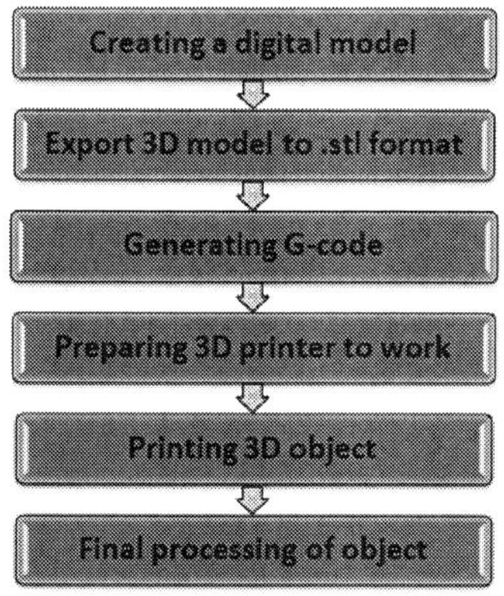

Figure 4. Steps of the FDM process

4. Experimental results

The experimental test includes different angular rotation speeds for driver crank. There were used: ω_1 = 24 rpm and also ω_1' = 66 rpm. These speeds allow studying the cinematic and dynamic behaviour of Geneva mechanisms.

Thermoplastic extrusion technology begins like other rapid prototyping processes by designing the virtual image of the future object in a 3D editor or CAD program. In the case of this study, the three-dimensional model that was based was a Geneva mechanism. When 3D modelling is completed, the resulting file is transferred to .stl format, which is recognized by most modern 3D printers. The STL file with future restoration is processed by a special slicing program, which translates it into a control code G for the FDM printer. The slicing program with which the file was processed is ReplicatorG.

3D printing technology is a well-planned and ready process for turning virtual models into physical objects. The steps of FDM technology are shown in the Fig.4. The most important elements of the 3D printer are the work platform and the printhead. The finished workpiece is formed on the work platform. During operation, the platform moves up and down the Z axis. The printhead extrudes a melted polymer thread onto the work platform layer after layer forming the finished structure. The printhead of the 3D printer moves horizontally and vertically (X, Y axis). The process of 3D printing itself is quite simple. The print head extrudes the first layer of molten material into the work area, after which the platform is lowered to the thickness of the layer, and the next overlapping layer begins to form.

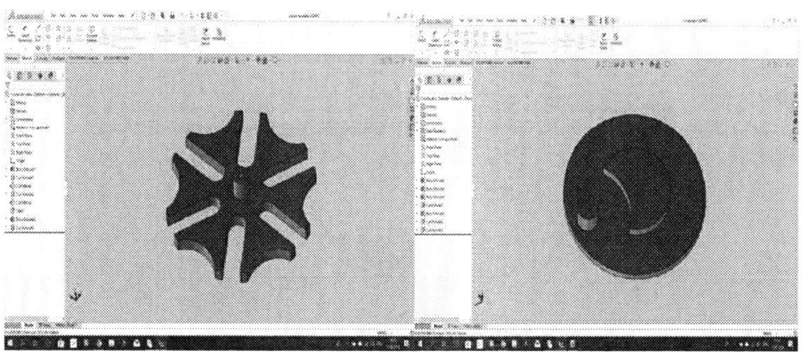

Figure 5. CAD models of Maltese cross

After completion of the process, the sacrificial layers are removed and result the final pieces [3].
In figure 6 are two of experimental setups with Geneva mechanisms: with PLA material (left) and Plexiglas material (right). These experimental setups are connected to PC through a USB interface.

In figure 7 are represented all five setups at it follows: a) Plexiglas (PMMA) , b) PLA, c) ABS, d)PLA + copper and e) PLA + bronze.

Figure 6. Experimental setups PLA (left) and PLEXI (right)

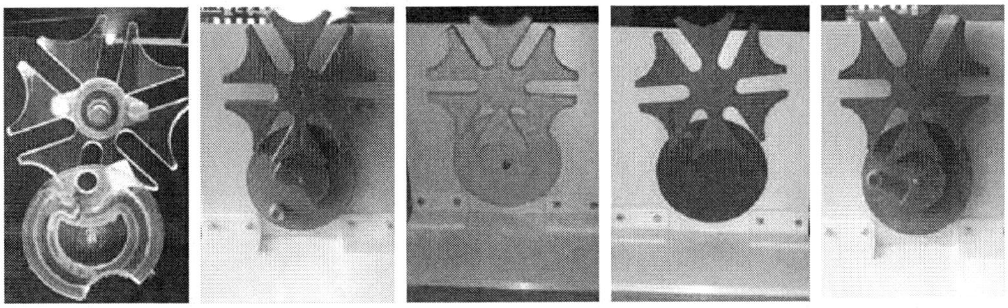

a b c d e

Figure 7. Geneva mechanisms from different material:

a) PMMA b) PLA c) ABS d) PLA+copper e) PLA+bronze

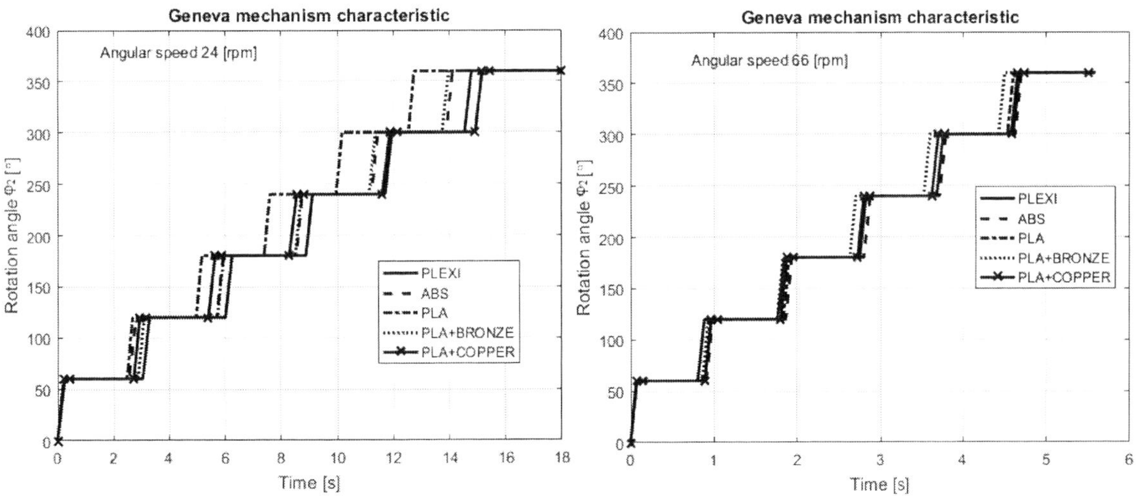

Figure 8. Geneva mechanism characteristic in function of rotation angle (24 rpm and 66 rpm)

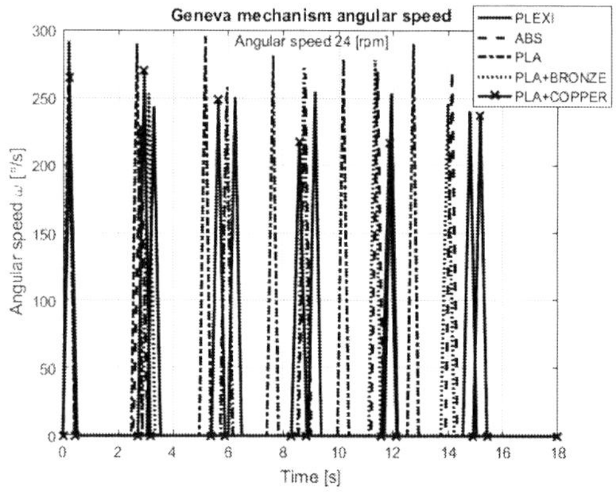

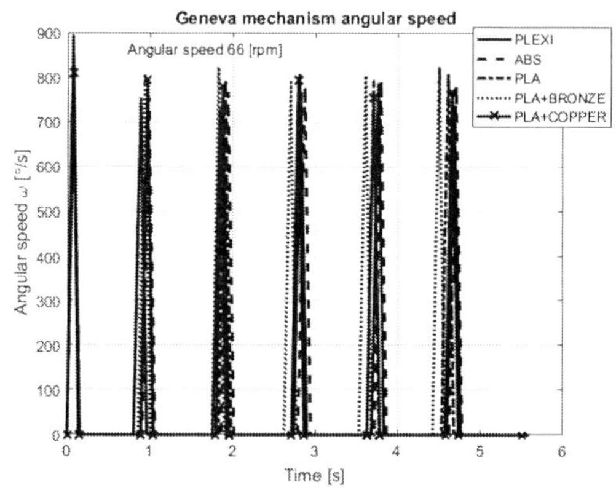

Figure 9. Geneva mechanism characteristic in function of angular speed (24 rpm and 66 rpm)

5. Conclusions

In this paper a Geneva mechanism was studied and successfully realized with the help of additive technologies from five different materials. The models obtained are used as demonstration stands used in didactic applications. Of the five investigated materials, ABS and PLA + Copper have been proved to have properties and characteristics close to mechanical engineering applications and these results are demonstrated by graphs. ABS also has the advantage of lighter weight, which makes it suitable for applications where low mass is needed. It can also be concluded that the angular speed of the drive element is inversely proportional to the difference between the angular speeds of the Malta crosses.

The latest years have brought spectacular developments to these new 3D printing technologies, which makes it possible today to find a wide variety of products manufactured by these new technologies and many in development, making them more and more competitive in wider areas.

Applications of AM technologies in this area are gaining more and more importance. Challenges are related to finding and approving new materials that can be put into desirable physical forms with the most appropriate mechanical, increasing dimensional precision and surface quality, making parts from different materials with a functional gradient.

References

[1] Bickford J.H, Geneva Mechanisms, Mechanism for Intermittent Motion, *New York: Industrial Press Inc., 1972*, p. 264.

[2] Rizescu D, Miu S, Rizescu C.I, Complemente de mecatronică, *Editura Printech, Bucuresti, 2000*, p.7.

[3] Moraru E, Dontu O, Besnea D, Constantin V. Study and realization of prosthetic dental models by additive technologies *ACME 2018*, Iasi, Romania.

[4] Spanu A, Constantin V. The using of additive manufacturing for prototype production of moulds, *International Journal of Mechatronics and Applied Mechanics*, 2017(1), p.7-16

Acknowledgments

This work has been funded by University POLITEHNICA of Bucharest, through the "Excellence Research Grants", Program UPB-GEX 2017. Identify: UPB-GEX2017, Grant no. 48/25.09.2017, ME 14-17-05, ID98.

IOP Conf. Series: Materials Science and Engineering **514** (2019) 012033 doi:10.1088/1757-899X/514/1/012033

Identification and viewing of singularities for a new type of 6RSS parallel mechanism

L Milica[1], A Năstase [1] and G Andrei [1]

[1] Mechanical Engineering Department, "Dunarea de Jos University" of Galati, Galati, Romania

E-mail: milica.lucian@ugal.ro

Abstract. This paper presents a method for determining the singular configurations of a 6RSS parallel mechanism based on its kinematic model. The parallel mechanism consists of a fixed plate and a mobile platform connected through six independent kinematic chains. Each of the six kinematic chains has an actuating rotational joint and two spherical joints. The kinematic analysis is based on the kinematic screw method. The resultant motion can thus be represented by composing the kinematic screws of the component motions. Starting from the two Jacobian matrices of the kinematic model, the two types of singularities encountered in parallel mechanisms are identified. Singular configurations are determined by an algorithm which is further used to produce three-dimensional representations of the two matrices determinants. Also, with the same program, it is possible to accurately establish the existence or inexistence of a mechanism for a given position and orientation.

1. Introduction

Parallel mechanisms have attracted the attention of many researchers due to the high degree of structural resistance that makes them less susceptible to input errors than their serial counterparts [1-5]. These qualities make parallel mechanisms find their applicability in various fields: flight simulators and fine positioning devices and fast packing in high-speed milling machines [6, 7].

Because of the kinematic complexity of these mechanisms, numerous researches have been undertaken on this issue [8-13].

An important role in the kinematic control of the parallel mechanisms is the determination of the singular configurations associated with the direct Jacobian matrix and the inverse one, respectively. Determining the singlular configurations is a central issue for the kinematic model of the robot .

Identifying them means finding those critical configurations that cause important changes in the kinematic performance of a manipulator. The study of singularities is the main tool in order to avoid these critical positions.

The development of the robotic industry and the emergence of parallel mechanisms with increasingly complex movements imply the development of general methods that contribute to the analysis of singularities, thus improving the process of designing and programming the robot. Planning robot movements, taking into account the identification of all singular configurations, is the major objective to be pursued.

In these particular positions, the degree of spatial mobility of the effector element decreases therefore some of the input parameters no longer have an effect on its movement. The issue of critical positions is subject to special studies on parallel mechanisms [14-16].

In order to avoid singularities the researches undertaken so far have revealed different means in their analysis. Some researchers have developed a specific technique based on treating the geometric place determined by singularities as a set of obstacles [17]. Using three-dimensional representations, other researchers have determined the singular configurations for a Stewart parallel mechanism, for a certain orientation and position of the mobile platform [18].

2. The kinematics of the 6RSS parallel mechanism

The kinematic analysis of a mechanism establishes the implicit relationships between the infinitesimal motion of the actuated kinematic joints and the effector element. The present paper performs the kinematic analysis of a six-degree parallel manipulator 6RSS (rotational joint, spherical joint, spherical joint) from figure 1, using the kinematic screw method.

The parallel mechanism is represented in figure 1 in 3D view. The 6RSS parallel manipulator consists of a fixed plate to which six actuators are fitted, and a mobile platform on which is fitted the seventh actuator to drive a milling tool. One of the six kinematic k chain of the 6RSS parallel manipulator has the kinematic rotational joints at points O_k and the kinematic spherical joints at points A_k and B_k respectively. The drive arms O_kA_k are rotated at an angle θ around the axis of the vertex $\mathbf{u_k}$ passing through the point O_k.

The mechanism has the following features:

- axis of rotational active joints that generate circular trajectories has an $\mathbf{u_k}$ unit vector and are coplanar, two-by-two coincident and central-symmetrically disposed ;
- the actuating arms O_kA_k are of equal length;
- A_kB_k rods are of equal length;
- spherical joints of the mobile platform are coplanar and their centres are located on the vertices of a regular hexagon.

Determining the kinematics of parallel mechanisms involves establishing the equations of the resultant motion of a rigid body having a relative motion in relation to a mobile coordinates system. The resultant movement can be represented through the kinematic screws by composing the screws of all the components in motion.

In figure 2 we have attached to a rigid body a coordinate system with center in point P (identical to the origin of the system O) which belongs to this.

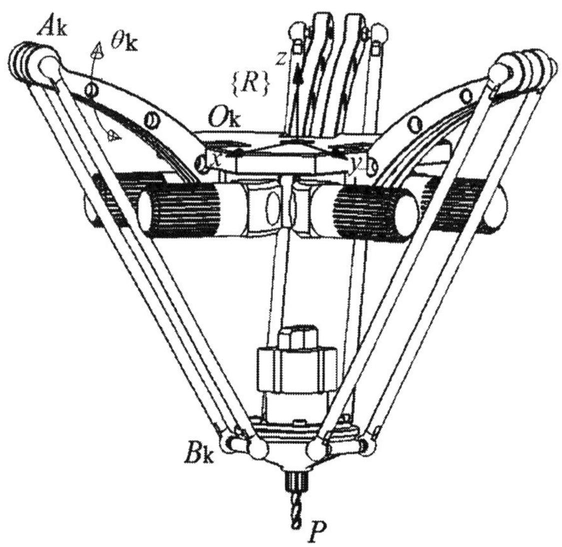

Figure 1. The new 6RSS parallel manipulator.

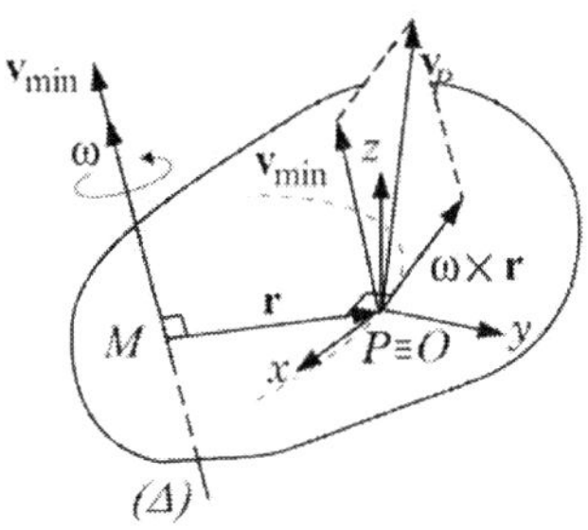

Figure 2. Elements of the kinematic screw.

Consider the rigid body in an instantaneous motion relative to a roto-translational axis (Δ). ω is the angular velocity of the body, r is the distance from point *P* relative to the roto-translational axis (Δ) and λ is the screw parameter given by the relation:

$$\lambda = \frac{\mathbf{v}_{\min}}{\omega} \tag{1}$$

The screw parameter λ has the length dimension and characterizes the movement geometrically. We have thus expressed two particular cases:

- when λ = 0 results a pure rotation;
- when λ = ∞ results a pure translation.

We have the following expressions:

$$\boldsymbol{\omega} = \omega \cdot \mathbf{u} \tag{2}$$

$$\mathbf{k} = \mathbf{r} \times \mathbf{u} + \lambda \cdot \mathbf{u} \tag{3}$$

where **u** is the unit vector of the instantaneous axis of rotation. When this axis is coincident with the axes of the reference system, the expressions of the unit vector $\mathbf{u}_x$, $\mathbf{u}_y$, $\mathbf{u}_z$ of the homologous axes can be determinate easily.

Relationships (2) and (3) define the geometric screw **$** whose general formula is:

$$\$ = \left[\mathbf{u} \cdot \mathbf{k}\right]^{T} \tag{4}$$

If for the expression (3) we make the notation:

$$\mathbf{w}_j = \mathbf{r}_j \times \mathbf{u}_j \text{ și } \mathbf{r}_j = \mathbf{M}_j\mathbf{O}_j \tag{5}$$

we have the following two cases:

- for the rotational joint:

$$\$ = \left[\mathbf{u}_j \cdot \mathbf{w}_j\right]^{T} \tag{6}$$

- for the translational joint:

$$\$ = \left[\mathbf{0} \cdot \mathbf{u}_j\right]^{T} \tag{7}$$

By notating the kinematic screw with the **Ω**, the expression becomes:

$$\boldsymbol{\Omega} = \omega \cdot \$ \tag{8}$$

in which $\omega = |\boldsymbol{\omega}|$.

In the case of the parallel mechanism analyzed, by composing a number of kinematic screws of a kinematic chain, the resulting motion will be described by the kinematic screw obtained by summing up the *n* kinematic screws:

$$\boldsymbol{\Omega} = \boldsymbol{\Omega}_1 + \boldsymbol{\Omega}_2 + \ldots + \boldsymbol{\Omega}_n \tag{9}$$

Let us examine one of the kinematic chains k of the parallel mechanism 6RSS in figure 3. Each actuated arm has a local mobility given by the rotation around the axis A_kB_k, independent movement relative to the other possible movements of the mechanism. To simplify the calculations, we eliminated these six isolated nobilities and replaced the spherical joint from point *B* with a fourth-class joint that does not affect the degree of mobility of the mechanism.

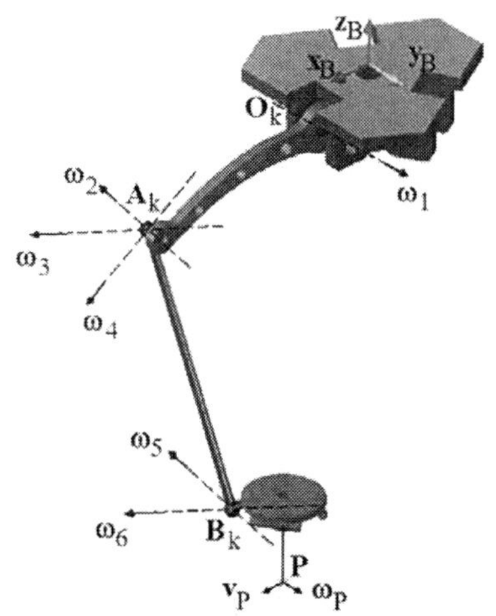

Figure 3. Angular velocities of kinematic joints belonging to a *k*-chain.

We have attached to a rigid body a coordinate system with center in point *P* (identical to the origin *O* of the system) which belongs to this.

At each angular velocity $\omega_1,\ldots,\omega_P$ corresponds a kinematic screw $\mathbf{\Omega_1,\ldots,\Omega_P}$ so that for the six kinematic chains of the mechanism will result the system of equations:

$$S=\begin{cases}\mathbf{\Omega}_1^{(1)}+\mathbf{\Omega}_2^{(1)}+\ldots+\mathbf{\Omega}_6^{(1)}=\mathbf{\Omega}_P\\ \mathbf{\Omega}_1^{(2)}+\mathbf{\Omega}_2^{(2)}+\ldots+\mathbf{\Omega}_6^{(2)}=\mathbf{\Omega}_P\\ \vdots\\ \mathbf{\Omega}_1^{(6)}+\mathbf{\Omega}_2^{(6)}+\ldots+\mathbf{\Omega}_6^{(6)}=\mathbf{\Omega}_P\end{cases}\tag{10}$$

in which it is known that:

$$\mathbf{\Omega}_P=\begin{bmatrix}\mathbf{\Omega}_1^P & \mathbf{\Omega}_2^P & \mathbf{\Omega}_3^P & \mathbf{\Omega}_4^P & \mathbf{\Omega}_5^P & \mathbf{\Omega}_6^P\end{bmatrix}^T\tag{11}$$

Taking into account the relation (8) above system of equations is:

$$S=\begin{cases}\omega_1^{(1)}\cdot\mathbf{\$}_1^{(1)}+\omega_2^{(1)}\cdot\mathbf{\$}_2^{(1)}+\omega_3^{(1)}\cdot\mathbf{\$}_3^{(1)}+\omega_4^{(1)}\cdot\mathbf{\$}_4^{(1)}+\omega_5^{(1)}\cdot\mathbf{\$}_5^{(1)}+\omega_6^{(1)}\cdot\mathbf{\$}_6^{(1)}=\mathbf{\Omega}_P\\ \omega_1^{(2)}\cdot\mathbf{\$}_1^{(2)}+\omega_2^{(2)}\cdot\mathbf{\$}_2^{(2)}+\omega_3^{(2)}\cdot\mathbf{\$}_3^{(2)}+\omega_4^{(2)}\cdot\mathbf{\$}_4^{(2)}+\omega_5^{(2)}\cdot\mathbf{\$}_5^{(2)}+\omega_6^{(2)}\cdot\mathbf{\$}_6^{(2)}=\mathbf{\Omega}_P\\ \vdots\\ \omega_1^{(6)}\cdot\mathbf{\$}_1^{(6)}+\omega_2^{(6)}\cdot\mathbf{\$}_2^{(6)}+\omega_3^{(6)}\cdot\mathbf{\$}_3^{(6)}+\omega_4^{(6)}\cdot\mathbf{\$}_4^{(6)}+\omega_5^{(6)}\cdot\mathbf{\$}_5^{(6)}+\omega_6^{(6)}\cdot\mathbf{\$}_6^{(6)}=\mathbf{\Omega}_P\end{cases}\tag{12}$$

Taking into account expression (6), the expression of screws for this joint is:

$$\mathbf{\$}_i^{(k)}=\begin{bmatrix}\mathbf{u}_i^{(k)}\cdot\mathbf{r}_i^{(k)}\times\mathbf{u}_i^{(k)}\end{bmatrix}^T\tag{13}$$

It is known that for any five geometric screws $\$_j$, (j = 1…5) there is a reciprocal screw $\mathbf{\$}_r$, so that the product reciprocally between it and any of the five screws is 0:

$$\begin{aligned}&\mathbf{\$}_1\circ\mathbf{\$}_r=\mathbf{0}\\ &\vdots\\ &\mathbf{\$}_5\circ\mathbf{\$}_r=\mathbf{0}\end{aligned}\tag{14}$$

Considering two geometric screws $\mathbf{\$}_1=[\mathbf{u}_1\cdot\mathbf{w}_1]^T$ and $\mathbf{\$}_2=[\mathbf{u}_2\cdot\mathbf{w}_2]^T$ we have the relationship (15) that defines the expression of their reciprocal product:

$$\mathbf{\$}_1\circ\mathbf{\$}_2=\mathbf{u}_1\cdot\mathbf{w}_2+\mathbf{u}_2\cdot\mathbf{w}_1\tag{15}$$

The components of the reciprocal screws $\mathbf{\$}_{2,6}^{\mathbf{r(k)}} = \left[\mathbf{u}_{\mathbf{r}}^{(\mathbf{k})} \cdot \mathbf{w}_{\mathbf{r}}^{(\mathbf{k})}\right]^{\mathrm{T}}$ for the k kinematic chains are determined with condition (14) and through expression (15) , the six systems will have the form:

$$S = \begin{cases} \mathbf{\$}_2^{(\mathbf{k})} \circ \mathbf{\$}_{2,6}^{\mathbf{r(k)}} = 0 \\ \mathbf{\$}_3^{(\mathbf{k})} \circ \mathbf{\$}_{2,6}^{\mathbf{r(k)}} = 0 \\ \mathbf{\$}_4^{(\mathbf{k})} \circ \mathbf{\$}_{2,6}^{\mathbf{r(k)}} = 0 \\ \mathbf{\$}_5^{(\mathbf{k})} \circ \mathbf{\$}_{2,6}^{\mathbf{r(k)}} = 0 \\ \mathbf{\$}_6^{(\mathbf{k})} \circ \mathbf{\$}_{2,6}^{\mathbf{r(k)}} = 0 \\ \mathrm{u}_{\mathrm{r_x}}^2 + \mathrm{u}_{\mathrm{r_y}}^2 + \mathrm{u}_{\mathrm{r_z}}^2 = 1 \end{cases} \rightarrow S = \begin{cases} \mathbf{u}_2^{(\mathbf{k})} \cdot \mathbf{w}_{\mathbf{r}}^{(\mathbf{k})} + \mathbf{u}_{\mathbf{r}}^{(\mathbf{k})} \cdot \mathbf{w}_2^{(\mathbf{k})} = 0 \\ \mathbf{u}_3^{(\mathbf{k})} \cdot \mathbf{w}_{\mathbf{r}}^{(\mathbf{k})} + \mathbf{u}_{\mathbf{r}}^{(\mathbf{k})} \cdot \mathbf{w}_3^{(\mathbf{k})} = 0 \\ \mathbf{u}_4^{(\mathbf{k})} \cdot \mathbf{w}_{\mathbf{r}}^{(\mathbf{k})} + \mathbf{u}_{\mathbf{r}}^{(\mathbf{k})} \cdot \mathbf{w}_4^{(\mathbf{k})} = 0 \\ \mathbf{u}_5^{(\mathbf{k})} \cdot \mathbf{w}_{\mathbf{r}}^{(\mathbf{k})} + \mathbf{u}_{\mathbf{r}}^{(\mathbf{k})} \cdot \mathbf{w}_5^{(\mathbf{k})} = 0 \\ \mathbf{u}_6^{(\mathbf{k})} \cdot \mathbf{w}_{\mathbf{r}}^{(\mathbf{k})} + \mathbf{u}_{\mathbf{r}}^{(\mathbf{k})} \cdot \mathbf{w}_6^{(\mathbf{k})} = 0 \\ \mathrm{u}_{\mathrm{r_x}}^2 + \mathrm{u}_{\mathrm{r_y}}^2 + \mathrm{u}_{\mathrm{r_z}}^2 = 1 \end{cases} \tag{16}$$

where $\mathrm{u}_{\mathrm{r_x}}^2, \mathrm{u}_{\mathrm{r_y}}^2, \mathrm{u}_{\mathrm{r_z}}^2$ are the components on the three directions of the vector $\mathbf{u}_{\mathbf{r}}^{(\mathbf{k})}$. Solving each of the six systems (16) results in the components of the six reciprocal screw $\mathbf{\$}_{2,6}^{\mathbf{r(k)}}$.

Multiplying each system equation (16) by the reciprocal screw, it results:

$$S = \begin{cases} \omega_1^{(1)} \cdot \mathbf{\$}_1^{(1)} \circ \mathbf{\$}_{\mathbf{2,6}}^{\mathbf{r(1)}} = \mathbf{\Omega}_{\mathbf{P}} \circ \mathbf{\$}_{\mathbf{2,6}}^{\mathbf{r(1)}} \\ \omega_1^{(2)} \cdot \mathbf{\$}_1^{(2)} \circ \mathbf{\$}_{\mathbf{2,6}}^{\mathbf{r(2)}} = \mathbf{\Omega}_{\mathbf{P}} \circ \mathbf{\$}_{\mathbf{2,6}}^{\mathbf{r(2)}} \\ \vdots \\ \omega_1^{(6)} \cdot \mathbf{\$}_1^{(6)} \circ \mathbf{\$}_{\mathbf{2,6}}^{\mathbf{r(6)}} = \mathbf{\Omega}_{\mathbf{P}} \circ \mathbf{\$}_{\mathbf{2,6}}^{\mathbf{r(6)}} \end{cases} \tag{17}$$

The reciprocal products $\mathbf{\Omega}_{\mathbf{P}} \circ \mathbf{\$}_{\mathbf{2,6}}^{\mathbf{r(k)}}$ will have the following expressions:

$$\mathbf{\Omega}_{\mathbf{P}} \circ \mathbf{\$}_{\mathbf{2,6}}^{\mathbf{r(k)}} = \begin{bmatrix} \Omega_1^{\mathrm{P}} & \Omega_2^{\mathrm{P}} & \Omega_3^{\mathrm{P}} \end{bmatrix}^{\mathrm{T}} \cdot \mathbf{w}_{2,6}^{(\mathbf{k})} + \mathbf{u}_{2,6}^{(\mathbf{k})} \cdot \begin{bmatrix} \Omega_4^{\mathrm{P}} & \Omega_5^{\mathrm{P}} & \Omega_6^{\mathrm{P}} \end{bmatrix}^{\mathrm{T}} \tag{18}$$

The system (17) becomes:

$$S = \begin{cases} \omega_1^{(1)} \cdot \mathbf{\$}_1^{(1)} \circ \mathbf{\$}_{\mathbf{2,6}}^{\mathbf{r(1)}} = \begin{bmatrix} \Omega_1^{\mathrm{P}} & \Omega_2^{\mathrm{P}} & \Omega_3^{\mathrm{P}} \end{bmatrix}^{\mathrm{T}} \cdot \mathbf{w}_{2,6}^{(1)} + \mathbf{u}_{2,6}^{(1)} \cdot \begin{bmatrix} \Omega_4^{\mathrm{P}} & \Omega_5^{\mathrm{P}} & \Omega_6^{\mathrm{P}} \end{bmatrix}^{\mathrm{T}} \\ \omega_1^{(2)} \cdot \mathbf{\$}_1^{(2)} \circ \mathbf{\$}_{\mathbf{2,6}}^{\mathbf{r(2)}} = \begin{bmatrix} \Omega_1^{\mathrm{P}} & \Omega_2^{\mathrm{P}} & \Omega_3^{\mathrm{P}} \end{bmatrix}^{\mathrm{T}} \cdot \mathbf{w}_{2,6}^{(2)} + \mathbf{u}_{2,6}^{(2)} \cdot \begin{bmatrix} \Omega_4^{\mathrm{P}} & \Omega_5^{\mathrm{P}} & \Omega_6^{\mathrm{P}} \end{bmatrix}^{\mathrm{T}} \\ \vdots \\ \omega_1^{(6)} \cdot \mathbf{\$}_1^{(6)} \circ \mathbf{\$}_{\mathbf{2,6}}^{\mathbf{r(6)}} = \begin{bmatrix} \Omega_1^{\mathrm{P}} & \Omega_2^{\mathrm{P}} & \Omega_3^{\mathrm{P}} \end{bmatrix}^{\mathrm{T}} \cdot \mathbf{w}_{2,6}^{(6)} + \mathbf{u}_{2,6}^{(6)} \cdot \begin{bmatrix} \Omega_4^{\mathrm{P}} & \Omega_5^{\mathrm{P}} & \Omega_6^{\mathrm{P}} \end{bmatrix}^{\mathrm{T}} \end{cases} \tag{19}$$

By denoting with $r_1^{(k)}$ the scalar of the reciprocal product $\mathbf{\$}_1^{(\mathbf{k})} \circ \mathbf{\$}_{\mathbf{2,6}}^{\mathbf{r(k)}}$, ($k$ = 1,…,6) from (19), its matrix expression becomes:

$$\begin{bmatrix} r_1^{(1)} & \cdots & 0 \\ \vdots & \ddots & \vdots \\ 0 & \cdots & r_1^{(6)} \end{bmatrix}_{(6\times6)} \cdot \begin{bmatrix} \omega_1^{(1)} \\ \vdots \\ \omega_1^{(6)} \end{bmatrix}_{(6\times1)} = \begin{bmatrix} \mathbf{w}_{2,6}^{(1)} & \cdots & \mathbf{w}_{2,6}^{(1)} & \mathbf{u}_{2,6}^{(1)} & \cdots & \mathbf{u}_{2,6}^{(1)} \\ \vdots & & \vdots & \vdots & & \vdots \\ \mathbf{w}_{2,6}^{(6)} & \cdots & \mathbf{w}_{2,6}^{(6)} & \mathbf{u}_{2,6}^{(6)} & \cdots & \mathbf{u}_{2,6}^{(6)} \end{bmatrix}_{(6\times6)} \cdot \begin{bmatrix} \Omega_1^{\mathrm{P}} \\ \vdots \\ \Omega_6^{\mathrm{P}} \end{bmatrix}_{(6\times1)} \tag{20}$$

The left-components $r_1^{(k)}$ of matrix determine the direct Jacobian [B] and the right-components $\mathbf{w}_{2,6}^{(\mathbf{k})}, \mathbf{u}_{2,6}^{(\mathbf{k})}$ of the matrix determine the inverse Jacobian [A].

The equation (20) written in matrix form becomes:

$$[\mathrm{B}] \cdot \begin{bmatrix} \omega_1^{(1)} \\ \vdots \\ \omega_1^{(6)} \end{bmatrix} = [\mathrm{A}] \cdot \begin{bmatrix} \Omega_1^{\mathrm{P}} \\ \vdots \\ \Omega_6^{\mathrm{P}} \end{bmatrix} \tag{21}$$

The expression of the kinematic screw of the platform is:

$$\mathbf{\Omega}_{\mathbf{P}} = \omega_P \cdot \mathbf{\$}_{\mathbf{P}} = \omega_P \cdot \begin{bmatrix} \mathbf{u}_{\mathbf{P}} & \mathbf{r}_{\mathbf{P}} \times \mathbf{u}_{\mathbf{P}} + \lambda_P \cdot \mathbf{u}_{\mathbf{P}} \end{bmatrix}^{\mathrm{T}} = \begin{bmatrix} \omega_P \cdot \mathbf{u}_{\mathbf{P}} & \omega_P \cdot (\mathbf{r}_{\mathbf{P}} \times \mathbf{u}_{\mathbf{P}}) + \lambda_P \cdot \mathbf{u}_{\mathbf{P}} \cdot \omega_P \end{bmatrix}^{\mathrm{T}} \tag{22}$$

Where, based on the relationships below:

$$\lambda_P = \mathbf{u} \cdot \mathbf{w} \tag{23}$$

$$\mathbf{r_P} = \mathbf{u} \times \mathbf{w} \tag{24}$$

we get the expressions of ω_P, λ_P and $\mathbf{r_P}$ that fully describe the movement of the platform:

$$\omega_P = \sqrt{\left(\Omega_1^P\right)^2 + \left(\Omega_2^P\right)^2 + \left(\Omega_3^P\right)^2} \tag{25}$$

$$\lambda_P = \Omega_1^P \cdot \Omega_4^P + \Omega_2^P \cdot \Omega_5^P + \Omega_3^P \cdot \Omega_6^P \tag{26}$$

$$\mathbf{r_P} = \left(\Omega_1^P \cdot \mathbf{i} + \Omega_2^P \cdot \mathbf{j} + \Omega_3^P \cdot \mathbf{k}\right) \times \left(\Omega_4^P \cdot \mathbf{i} + \Omega_5^P \cdot \mathbf{j} + \Omega_6^P \cdot \mathbf{k}\right) \tag{27}$$

3. Determining the singular configurations of the 6RSS parallel manipulator

Based on the two Jacobian matrices, we can define the two types of singularities encountered in parallel mechanisms. These singularities interfere when the mechanism loses one or more degrees of freedom in certain configurations. Thus, when the determinant of the matrix [A] tends to infinitely we have type I singularity:

$$\det\left[\mathrm{A}\right] = \pm\infty \text{ or } \det\left[\mathrm{B}\right] = 0 \tag{28}$$

In the case of this type of singularity, the mobile platform remains motionless while the actuated arms move. When matrix determinant [A] is zero, we have singularity of type II:

$$\det\left[\mathrm{A}\right] = 0 \tag{29}$$

This type of singularity is characterized by the fact that the mobile platform continues to move while the actuated arms are locked.

In the case of the 6RSS parallel manipulator, these singular positions can be geometrically expressed by the following two conditions:

- when point *B* belongs to the plane [P_1]determined by the unit vector **u** of the rotational joint and point *A* (figure 4);
- when point *A* belongs to the plan [P_2] of the mobile platform (figure 5).

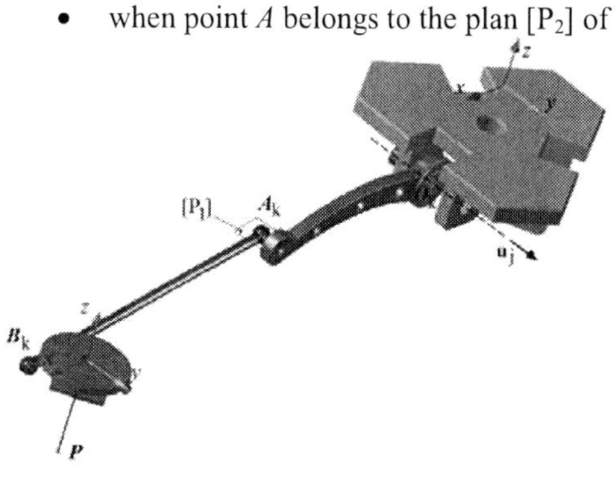

Figure 4. Singularity of type I, point *B* belongs to the plane [P_1]

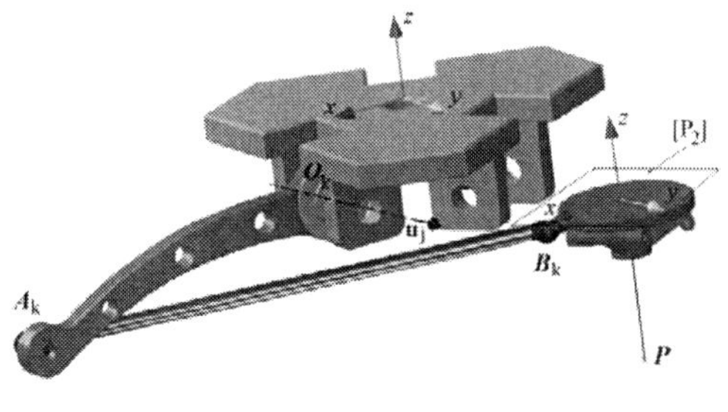

Figure 5. Singularity of type II, point *A* belongs to the plane [P_2]

To determine these singular configurations, a program was created to produce three-dimensional representations of the functions of determinants $\det\left[\mathrm{B}\right]$ and $\det\left[\mathrm{A}\right]$ of the direct and inverse Jacobian matrix from equation (21). The two Jacobian matrices determine the kinematic model of the parallel mechanism. The analysis of singular configurations has been achieved for cases where the mobile platform operates at a certain z-position relative to the fixed reference system attached to the base (figures 6-13).

IOP Conf. Series: Materials Science and Engineering **514** (2019) 012033 doi:10.1088/1757-899X/514/1/012033

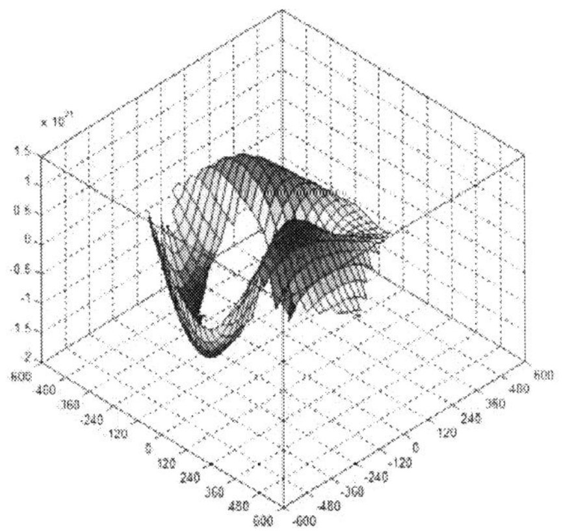

Figure 6. Function $\det[A]$, z = -200mm

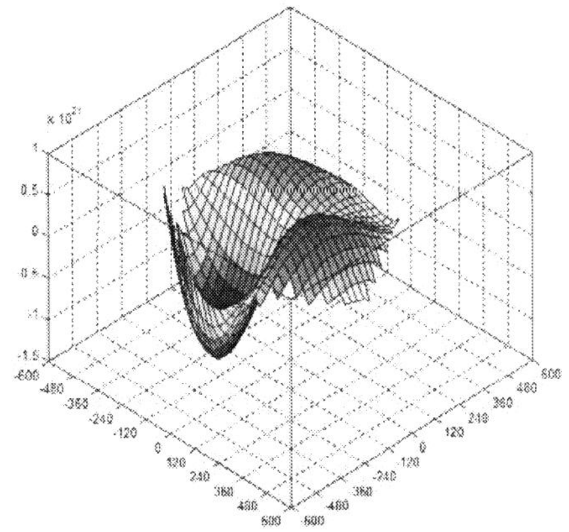

Figure 7. Function $\det[A]$,z = -315mm

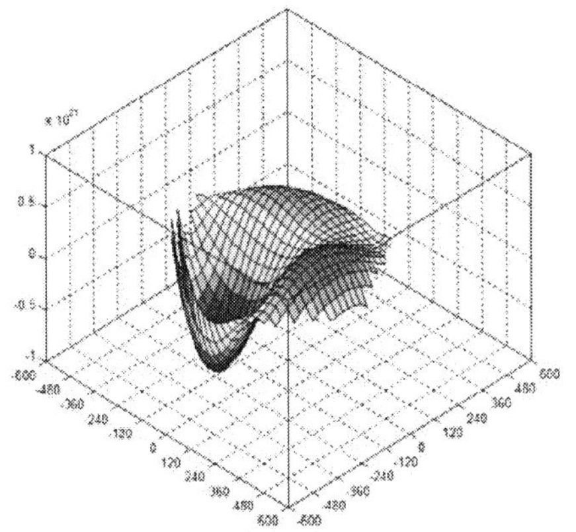

Figure 8. Function $\det[A]$, z = -400mm

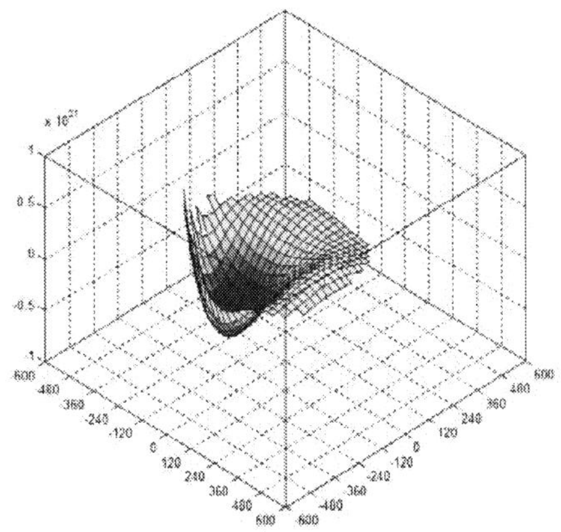

Figure 9. Function $\det[A]$,z = -495mm

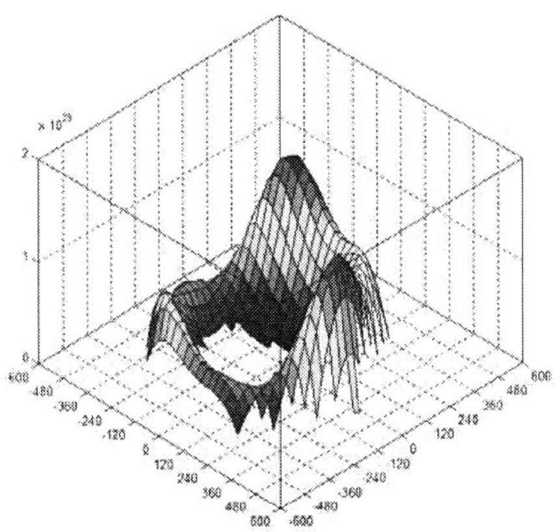

Figure 10. Function $\det[B]$, z = -200mm

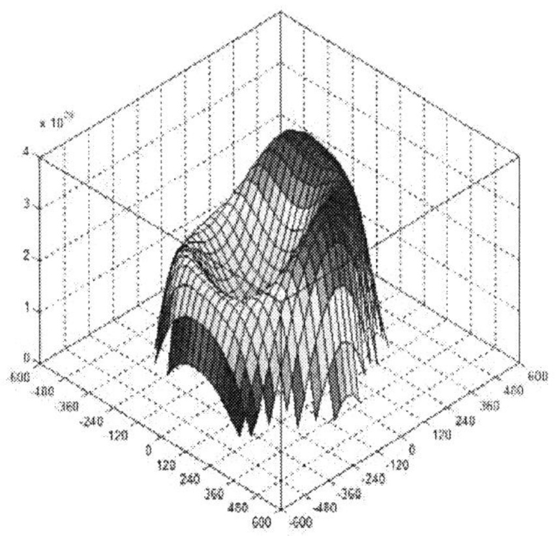

Figure 11. Function $\det[B]$, z = -315mm

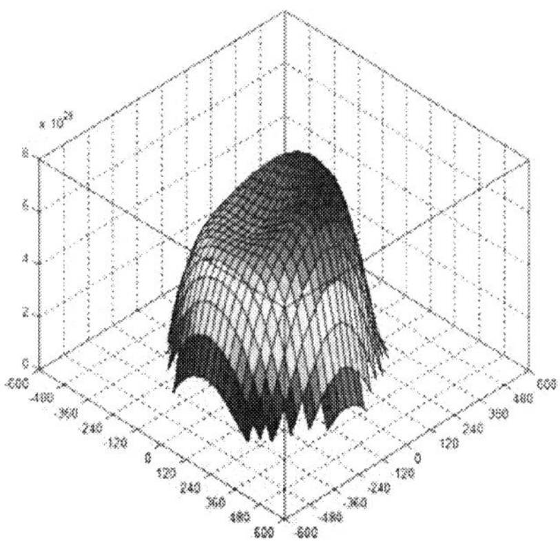

Figure 12. Function $\det[B]$, z = -400mm

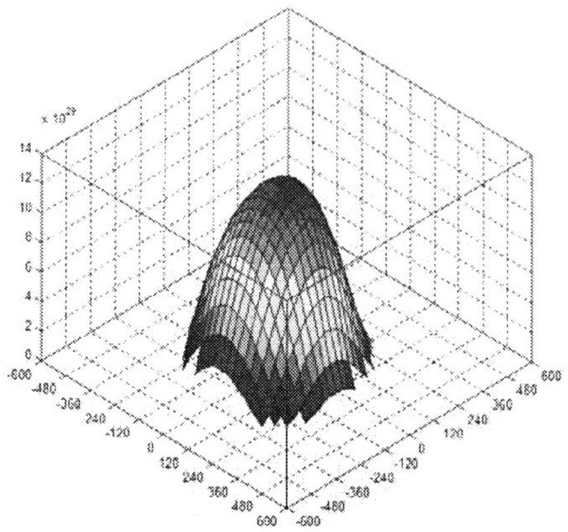

Figure 13. Function $\det[B]$, z = -495mm

Based on these three-dimensional representations, sections with a plane corresponding to the zero value of the two functions $\det[B]$ and $\det[A]$ were done. These sections highlight those critical curves determined by the set of singular positions fulfilling one of the two conditions mentioned above (figures 14, 15). Also, with the help of the same program, it was possible to establish precisely the existence or the inexistence of the mechanism for a given position and orientation.

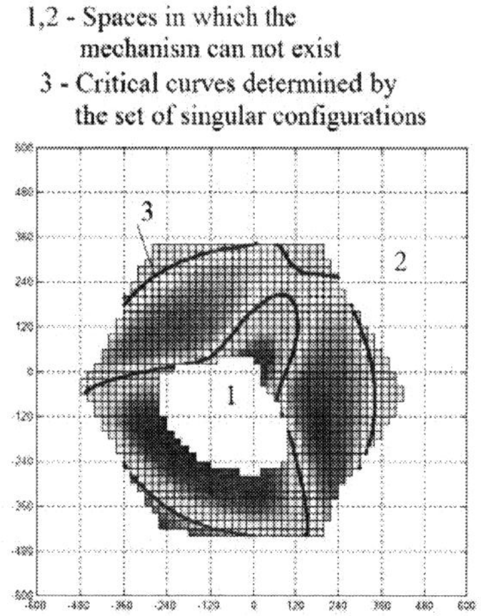

Figure 14. Section with a plane corresponding to the value $\det[A]=0$, for z = -200mm

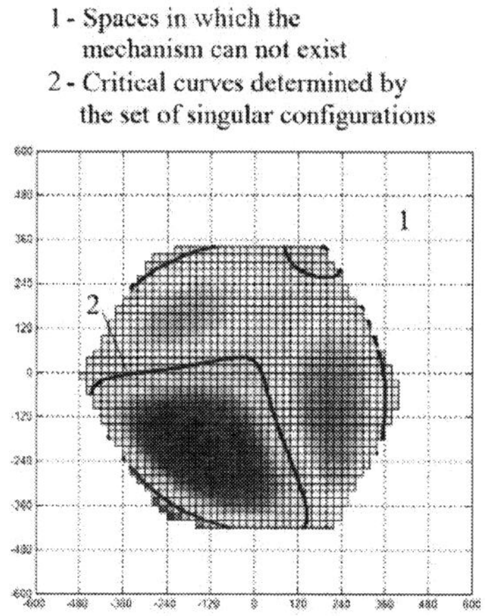

Figure 15. Section with a plane corresponding to the value $\det[A]=0$, for z = -315mm

4. Conclusions

This work contributes to an original approach to avoiding the single configurations of the new 6RSS parallel mechanism. The issue of singularities was highlighted based on the general kinematic model of the parallel manipulator determined by the kinematic screws. The mechanism presented in this paper is part of the manipulators with complex architectures that appear in the current robotics, characterized by new mechanisms of increasingly complex movements.

Based on the two Jacobian matrices, the two types of singularities encountered in parallel mechanisms were defined. A program was created to make three-dimensional representations of the determinate functions $\det[B]$ and $\det[A]$ of the direct and inverse Jacobian matrix. The developed method thus allows the determination of the singular configurations of the parallel mechanism for a particular position and orientation of the mobile platform. Viewing these sections with precise details on those trajectories that generate singular positions allows the motion capabilities of the manipulator to be determined. These diagrams can provide a map on which to identify those trajectories that do not cross singularity, in other words they are safe in terms of the robot's motion capabilities.

5. References

[1] Briot S and Bonev I A 2007 Are parallel robots more accurate than serial robots *Transactions of the* Canadian *Society for Mechanical Engineering* **31** pp 445–455
[2] Brogardh T 2007 Present and future robot control development–An industrial perspective *Annual Reviews in Control* **31** pp 69–79
[3] Weck M and Staimer D 2002 Parallel kinematic machine tools-current state and future potentials *CIRP Annals–Manufacturing Technology* **51** pp 671–683
[4] Merlet J P Parallel Robots 2006 Kluwer Academic Publishers (Norwell USA)
[5] Grigore G 2008 Structural Synthesis of Parallel Robots. Part 1: Methodology Springer (Netherlands)
[6] Huang Z, Cao Y 2005 Property identification of the singularity loci of a class of Gough-Stewart manipulators *International Journal of Robotics Research* **24** pp 675–685
[7] Pierrot F, Reynaud C, Fournier A 1990 Delta: A simple and efficient parallel robot *Robotica* **8** pp 105–109
[8] Merlet J P 1990 Les robot parallèles Ed. Hermes (Paris)
[9] Aleshin A K, Glazunov V A , Rashoyan G V 2016 Analysis of Kinematic Screws That Determine the Topology of Singular Zones of Parallel-Structure Robots *Journal of Machinery Manufacture and Reliability* **45** pp 291–296
[10] Kong X, Gosselin C M 2004 Type synthesis of 3T1R 4-DOF parallel manipulators based on screw theory *IEEE Transactions on Robotics and Automation* **20** pp 181–190
[11] Guo S, Fang Y, Qu H 2012 Type synthesis of 4-DOF non overconstrained parallel mechanisms based on screw theory *Robotica* **30** pp 31–37
[12] Xu Q, Li Y 2008 An investigation on mobility and stiff-ness of a 3-DOF translational parallel manipulator via screwtheory, *Robotics and Computer-Integrated Manufacturing* **24** pp 402–414
[13] Kong X, Gosselin C M 2004 Type synthesis of 3-DOF translational parallel manipulators based on screw theory *Journal of Mechanical Design* **126** pp 83–92
[14] Bonev I A 2002 Geometric Analysis of Parallel Mechanisms *PhD thesis, Faculté des Sciences et de Génie Université de Laval*
[15] St-Onge B M and Gosselin C M 2000 Singularity analysis and representation of the general Gough-Stewart platform *International Journal of Robotics Research* **19** pp 271–288
[16] Sen S, Dasgupta B and Mallik A K 2003 Variational approach for singularity-free pathplanning of parallel manipulators *Mechanism* and *Machine Theory* **38** pp 1165–1183
[17] Dash A K, Chen I, Yeo S H and Yang G 2005 Workspace generation and planning singularity-free path for parallel manipulators *Mechanism* and *Machine Theory* **40** pp 776–805
[18] Li H, Gosselin C M, Richard M J and St-Onge B M 2006 Analytic form of the six dimensional singularity locus of the general Gough-Stewart platform *ASME Journal of Mechanical Design* **128** pp 279–287

Acknowledgments
The research was funded by the Ministry of National Education from Romania.

IOP Conf. Series: Materials Science and Engineering **514** (2019) 012034 doi:10.1088/1757-899X/514/1/012034

Mono-axial tracking system for parabolic trough collector

D Ciobanu[1], M Enescu[2] and C Jaliu[1]

[1]Renewable Energy System and Recycling Research Centre, Transilvania University of Brasov, Romania
[2]Design of Mechanical Elements and Systems Research Centre, Transilvania University of Braşov, Romania

E-mail: daniela.ciobanu@unitbv.ro; monicaenescu@unitbv.ro; cjaliu@unitbv.ro

Abstract. Increased living standards in the developed countries lead to increased energy consumption. By using renewable energy, the fossil fuel consumption and greenhouse gas effect are reduced. The main source of renewable energy is the sun that can be used to produce the thermal energy or electric energy. The concentrating collectors are used to obtain both heat and electricity. The most common and used concentrating collectors are the parabolic trough collectors due to their advantages, such as: lower price, use of mono-axial tracking systems, simple maintenance. Parabolic trough collectors can have orientation around the N-S axis or E-W axis. In the case of N-S orientation, the angular stroke is higher than 120^0, and in case of E-W orientation, the angular stroke that the tracking system has to ensure is less than 100^0. The mechanisms used to orientate these collectors are using gears, cam or chain transmission. The paper presents a new type of mono-axial tracking system with linear actuators that has the advantages of assuring the required angular stroke, a high orientation accuracy and lower price.

1. Introduction

The 2017 reports inform that 70% of the energy was produced from renewable energy sources, but also that the amount of CO_2 emissions has risen by 1.4%, after three years of stagnation [1]. This increase is due to the cuts in fossil fuel prices, to the global economic growth (+ 3.7%), but also due to the reduced efforts to increase energy efficiency.

Worldwide, the new annual investments reached USD 279.8 billion at the end of 2017, 179 countries having as target the use of renewable energies, while 57 countries are targeting by 2020 to generate the whole amount of electric energy from renewable sources.

In 2017, 48% of the globally consumed energy was used for heating, 27% of it being produced from renewable energy (RE), 32% for transportation (3% from RE), and 20% for energy (25% from RE) (Figure 1).

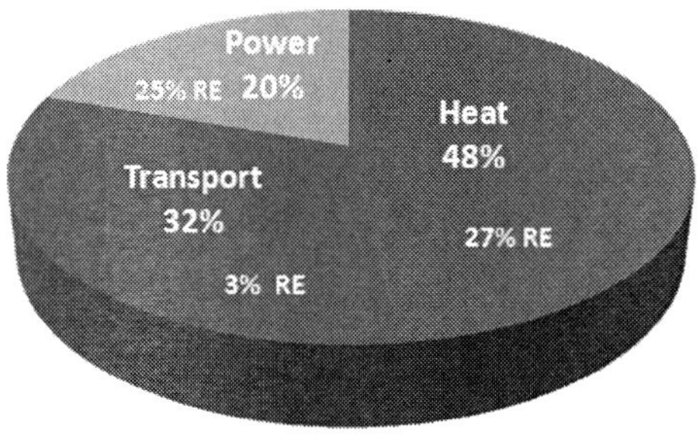

Figure 1. Renewable energy in the total energy consumption in 2017

The main source of renewable energy is the sun. Solar energy can be used both to produce heat using solar collectors and to generate electricity by using solar panels or through thermal conversion by using concentrating collectors.

In the case of solar thermal systems, the temperature of the agent is approx. 60^0C, while for concentrating collectors is over 100^0C. A large amount of solar radiation should be captured to obtain temperatures of the thermal agent higher than 100^0C. A way to capture more radiation is by using reflective surfaces that concentrate the solar radiation [2].

The main types of collectors that concentrate solar radiation are: trough-type parabolic collectors, dish-type collectors and central tower collectors [3]. The most used concentrating collectors are the trough type, due to the use of a tracking system, an easier maintenance and lower cost than the other two variants [1, 2]. Thus, to capture the solar radiation, the parabolic trough collectors are equipped with mono-axial tracking systems. These systems have to meet the following functional conditions: to provide a good orientation (the orientation error should be below 0.5^0), to ensure a high rotation angle of the reflector, and a good stability of the whole structure and to allow the control of the reflector position under critical conditions (high wind speeds, etc.)

Two main variants of mono-axial tracking systems for trough-type parabolic collectors are on the market today: an E-W orientation system and a N-S orientation system. In the case of the E-W orientation, the angular stroke of the collector is up to 90^0, being obtained by using either linkages or gears [4]. In the case of N-S tracking systems, the angular stroke that has to be achieved is higher than 100^0. The mechanisms that can provide these angular strokes and which are found in the construction of the trough collectors are: cam mechanisms, gears and chain / belt transmissions [5].

The paper is proposing a new variant of tracking mechanism for the parabolic trough collectors. The linkage based system has the advantage of achieving large angular strokes and a lower price.

2. Conceptual design of the tracking system

The parabolic trough collectors consist of: reflector, receiver and a mono-axial tracking system. The operating temperature of these collectors is high due to the reflection of the solar radiation from the receiver to the reflector. In order to reflect the solar radiation, the reflector must follow the sun motion on the sky (see Figure 2).

The angles that determine the position of the sun on the sky and which the tracking system of the trough collector has to ensure are given by the following relations [2]:

$$\alpha = \sin^{-1}(\sin\delta\sin\varphi + \cos\delta\cos\varphi\cos\omega) \tag{1}$$

$$\psi = (\text{sgn}\omega)\cos^{-1}\left(\frac{\sin\alpha\sin\varphi - \sin\delta}{\cos\alpha\cos\varphi}\right) \tag{2}$$

where α [^{0}] is the altitude angle (the angle between the sunray and the observer plane [6]), ψ [^{0}] is the azimuth angle (it is the compass direction from which the sunlight is coming [6]), δ [^{0}] is the declination angle (the angle between the sunray and the equatorial plane [6]), ω [^{0}] – the hour angle (the angle that describes the Earth's rotation around its polar axis [6]) and φ [^{0}] is the latitude angle (the angle that describes the observer's position on the Earth [6]).

For Brasov (the latitude angle $\varphi = 45.583^0$) and, for the four important days of the year (day no N = 82 - spring equinox, N = 172 - summer solstice, N = 266 - autumn equinox, and N = 355 - winter solstice) the altitude angle has the values presented in Figure 3, while the azimuth angle variation is illustrated in Figure 4, both depending on the solar hour.

The azimuth angle variation for different latitude angle values is presented in Figure. 6. The charts highlight that the azimuth angle varies from $\pm$ 120^0 for a latitude angle of 15^0 North to $\pm$ 123.8^0 for the latitude angle of 65^0.

In order to concentrate the solar radiation, the reflector of the parabolic collector has to perform an angle whose values are between 240^0-248^0 (see Figure 5). As the solar radiation intensity is reduced in the mornings and evenings, it will be further considered that the collector orientation will begin at 8:00 and will end at 17:00, which will reduce the energy consumption required for orientation. During this period of time, the angular stroke the collector must achieve is $\pm$ 90^0 (see Figure 4).

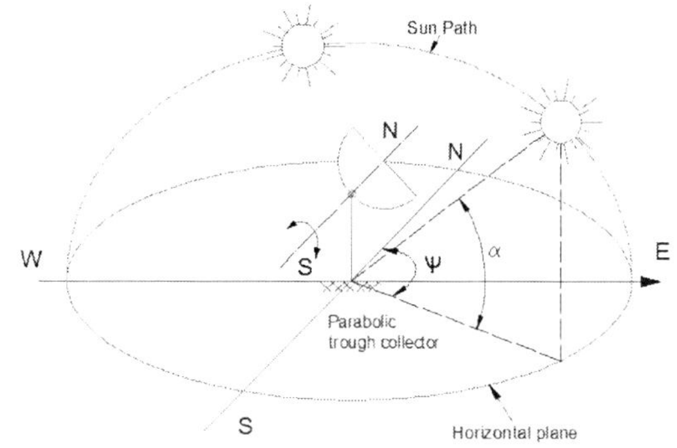

Figure 2. The Sun path on the sky

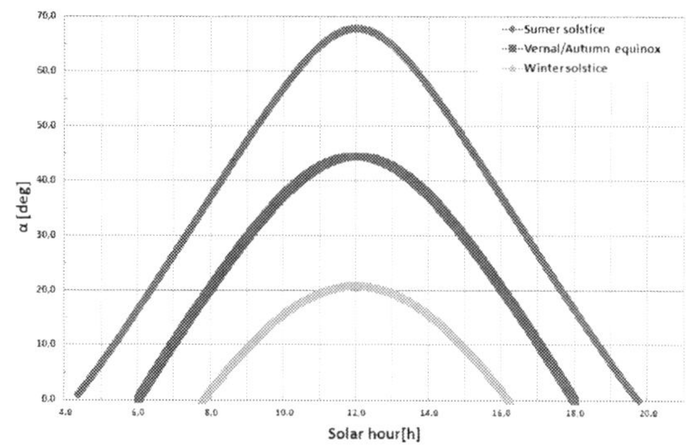

Figure 3. The altitude angle variation

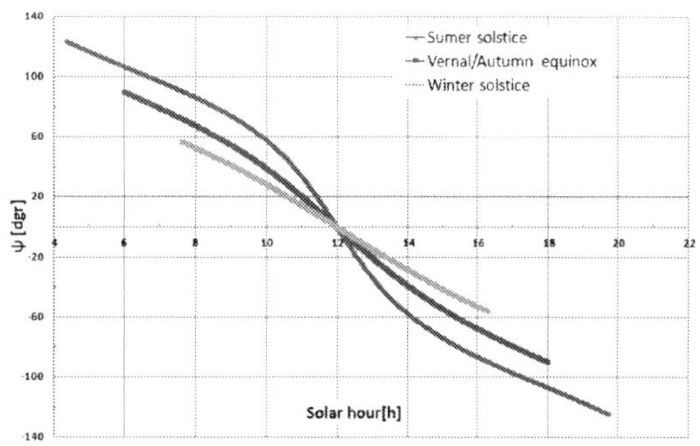

Figure 4. The azimuth angle variation

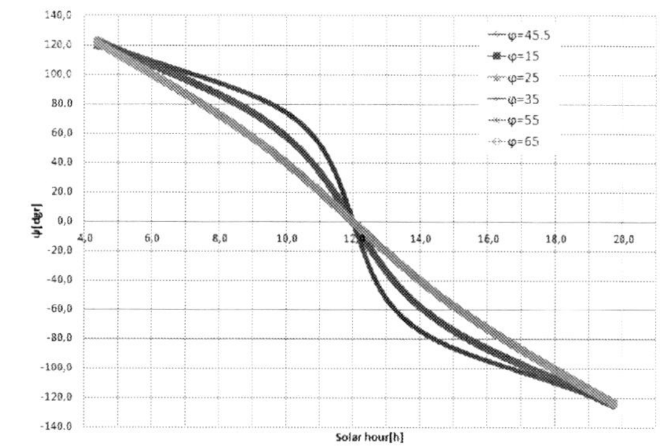

Figure 5. The azimuth angle variation for different latitude angles

To ensure this angular stroke, gears or linkages driven by two linear actuators can be used. These mechanisms allow a reduced orientation accuracy. The accuracy can be increased and a large angular stroke can be achieved by using the tracking system presented in Figure 6.

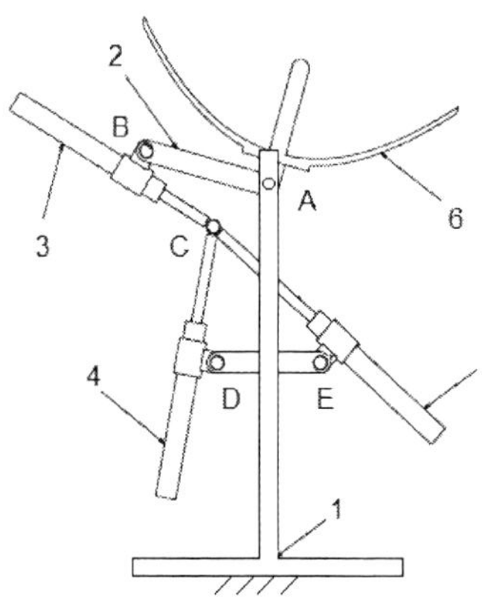

Figure 6. Mono-axial tracking system

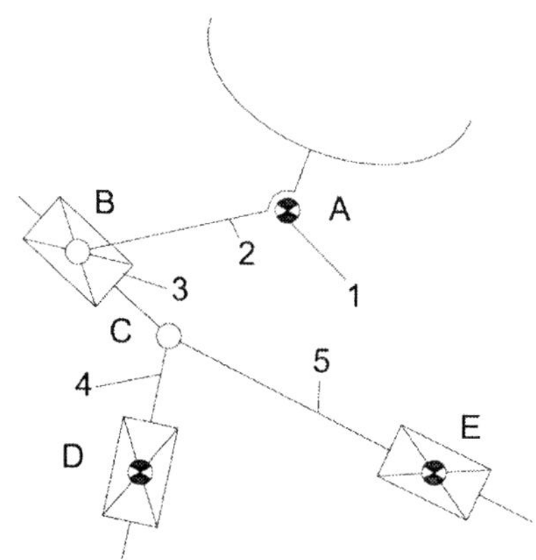

Figure 7. Structural scheme of the tracking mechanism

The mono-axial tracking system from Figure 6 is composed of: 1 – the collector base ; 2 - lever that is solidar to the parabolic collector; 3, 4, 5 - linear actuators that provide a large angular stroke (higher than 180^0) of the trough collector; 6 – parabolic trough collector.

The linear actuators 4 and 5, which are disposed triangularly and connected to the base through the revolute joints D and E, move the parabolic trough collector with a large angular stroke, but with a low accuracy [7]. The orientation accuracy is ensured by actuating the actuator 3.

Kinematic analysis

The tracking mechanism is presented in Figure 7 in structural scheme, being defined as a multibody system with a minimum number of bodies $n_b = 5$ bodies: 1 - fixed body; 2 - output body; 3,4,5 - other bodies. The number of driving constraints can be determined according to Eq. (3):

$$M = S \cdot (n_b - 1) - \sum c_i = 3 \cdot (5 - 1) - 9 = 3, \quad (3)$$

where: S is the mechanism spaciality; n_b - number of bodies; c_i - number of geometrical constraints. The type and number of geometrical constraints between each two bodies are presented in Table 1, while the geometrical models of the bodies are systematized in Table 2.

Tabel 1. Geometrical constraints

Bodies	Geometrical constraints	Place	Σc_i
1-2	R	A	2
1-4	RT	D	1
1-5	RT	E	1
2-3	RT	B	1
3-4	R	C	2
3-5	R	C	2

R- rotation, T – translation,

Table 2. Geometry of bodies

Body 1

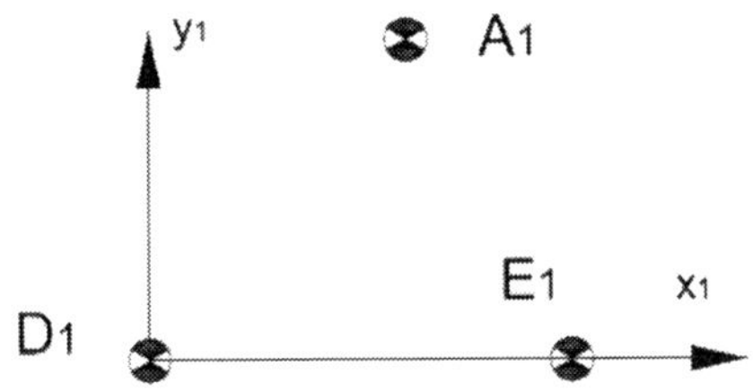

x_{A_1} =-505.9 mm; y_{A_1} = 876.3 mm;
x_{D_1} = 0 mm; y_{D_1} = 0 mm;
x_{E_1} = 1011mm; y_{E_1} =0 mm;

Body 2

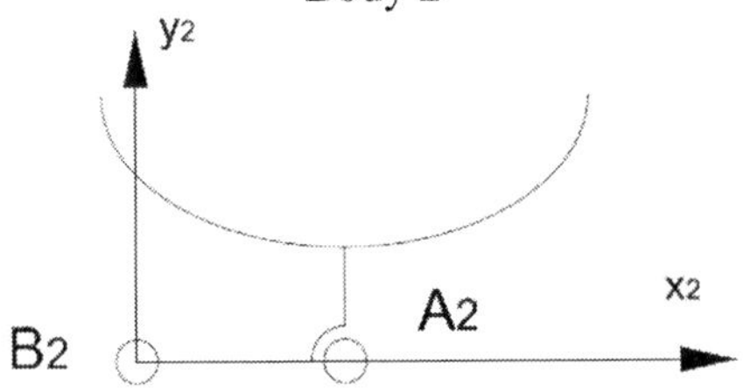

$x_{B_2}^{(2)}$ = 0 mm; $y_{B_2}^{(2)}$ = 0 mm;

$x_{A_2}^{(2)}$ = 409.2 mm; $y_{A_2}^{(2)}$ = 0 mm;

Body 3

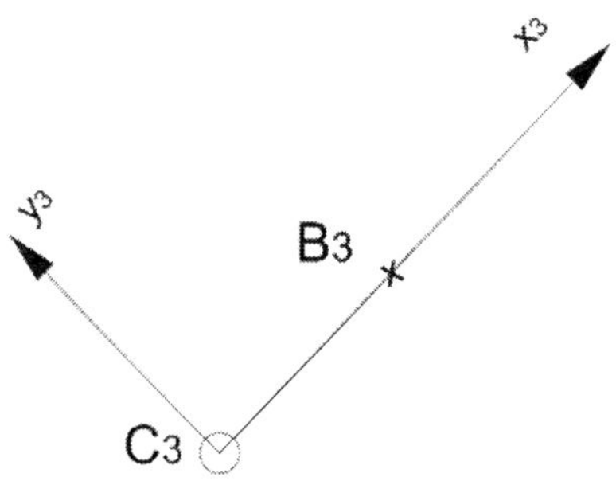

$x_{B_3}^{(3)}$ = 422 mm; $y_{B_3}^{(3)}$ = 0 mm;
$x_{C_3}^{(3)}$ = 0 mm; $y_{C_3}^{(3)}$ = 0 mm;

Table 2. Geometry of bodies (continuation)

Body 4

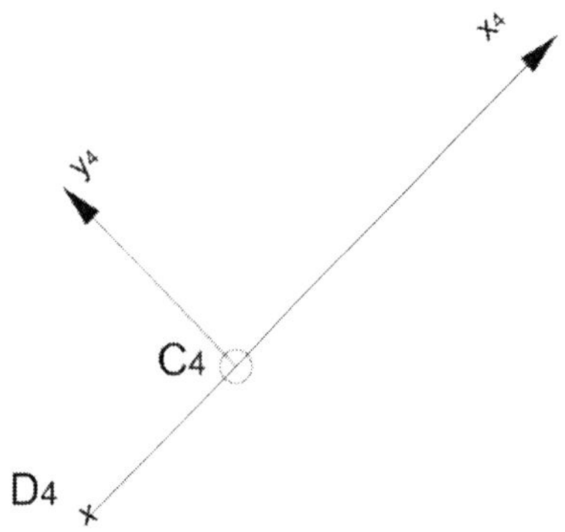

$x_{C_4}^{(4)} = 0\text{mm};\ y_{C_4}^{(4)} = 0\text{ mm};$
$x_{D_4}^{(4)} = 1288\text{ mm};\ y_{D_4}^{(4)} = 0\text{mm};$

Body 5

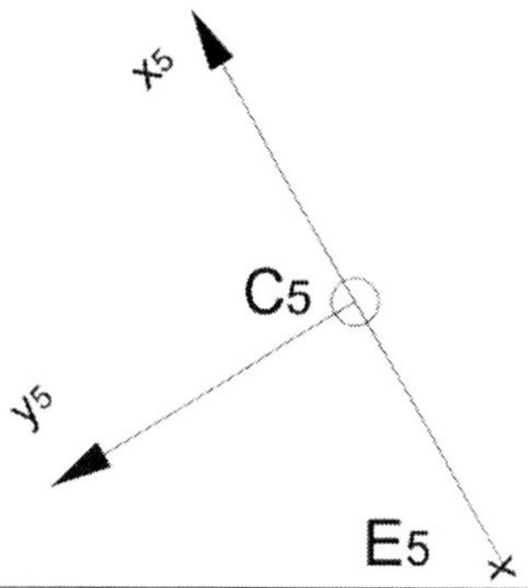

$x_{C_5}^{(5)} = 0\text{ mm};\ y_{C_5}^{(5)} = 0\text{ mm};$
$x_{E_5}^{(5)} = 1585\text{ mm};\ y_{E_5}^{(5)} = 0\text{ mm};$

The 12 generalized coordinates (x_{O2}, y_{O2}, φ_2; x_{O3}, y_{O3}, φ_3; x_{O4}, y_{O4}, φ_4; x_{O5}, y_{O5}, φ_5) from Table 2 can be obtained from the system of the position functions, consisting of 9 equations for the geometrical restrictions (Eq. 4…12) and 3 equations for the kinematic restrictions (Eq.13….15):

$1-2 \quad R \quad A_1 = A_2$

$$F_1: x_{A_1} - x_{O_2} - x_{A_2}^{(2)} cos\varphi_2 + y_{A_2}^{(2)} sin\varphi_2 = 0 \quad (4)$$

$$F_2: y_{A_1} - y_{O_2} - x_{A_2}^{(2)} sin\varphi_2 - y_{A_2}^{(2)} cos\varphi_2 = 0 \quad (5)$$

$1-4 \quad RT \quad D_1$

$$F_3: \left(x_{D_1} - x_{O_4} - x_{D_4}^{(4)} cos\varphi_4 + y_{D_4}^{(4)} sin\varphi_4\right)\left(y_{O_4} + x_{D_4}^{(4)} sin\varphi_4 + y_{D_4}^{(4)} cos\varphi_4 - y_{O_4} - x_{C_4}^{(4)} sin\varphi_4 - y_{C_4}^{(4)} cos\varphi_4\right) - \left(y_{D_1} - y_{O_4} - x_{D_4}^{(4)} sin\varphi_4 - y_{D_4}^{(4)} cos\varphi_4\right)\left(x_{O_4} + x_{D_4}^{(4)} cos\varphi_4 - y_{D_4}^{(4)} sin\varphi_4 - x_{O_4} - x_{C_4}^{(4)} cos\varphi_4 + y_{C_4}^{(4)} sin\varphi_4\right) = 0 \quad (6)$$

$1-5 \quad RT \quad E_1 \in E_5C_5$

$$F_4: \left(x_{E_1} - x_{O_5} - x_{E_5}^{(5)} cos\varphi_5 + y_{E_5}^{(5)} sin\varphi_5\right)\left(y_{O_5} + x_{E_5}^{(5)} sin\varphi_5 + y_{E_5}^{(5)} cos\varphi_5 - y_{O_5} - x_{C_5}^{(5)} sin\varphi_5 - y_{C_5}^{(5)} cos\varphi_5\right) - \left(y_{E_1} - y_{O_5} - x_{E_5}^{(5)} sin\varphi_5 - y_{E_5}^{(5)} cos\varphi_5\right)\left(x_{O_5} + x_{E_5}^{(5)} cos\varphi_5 - y_{E_5}^{(5)} sin\varphi_5 - x_{O_5} - x_{C_5}^{(5)} cos\varphi_5 + y_{C_5}^{(5)} sin\varphi_5\right) = 0 \quad (7)$$

$2-3 \quad RT \quad B_2 \in B_3C_3$

$$F_5: \left(x_{O_2} + x_{B_2}^{(2)} cos\varphi_2 - y_{B_2}^{(2)} sin\varphi_2 - x_{O_3} - x_{B_3}^{(3)} cos\varphi_3 + y_{B_3}^{(3)} sin\varphi_3\right)\left(y_{O_3} + x_{B_3}^{(3)} sin\varphi_3 + y_{B_3}^{(3)} cos\varphi_3 - y_{O_3} - x_{C_3}^{(3)} sin\varphi_3 - y_{C_3}^{(3)} cos\varphi_3\right) - \left(y_{O_2} + x_{B_2}^{(2)} sin\varphi_2 + y_{B_2}^{(2)} cos\varphi_2 - y_{O_3} - x_{B_3}^{(3)} sin\varphi_3 - y_{B_3}^{(3)} cos\varphi_3\right)\left(x_{O_3} + x_{B_3}^{(3)} cos\varphi_3 - y_{B_3}^{(3)} sin\varphi_3 - x_{O_3} - x_{C_3}^{(3)} sin\varphi_3 + y_{C_3}^{(3)} cos\varphi_3\right) = 0 \quad (8)$$

$3-4 \quad R \quad C_3 = C_4$

$$F_6: x_{O_3} + x_{C_3}^{(3)} cos\varphi_3 - y_{C_3}^{(3)} sin\varphi_3 - x_{O_4} - x_{C_4}^{(4)} cos\varphi_4 + y_{C_4}^{(4)} sin\varphi_4 = 0 \quad (9)$$

$$F_7: y_{O_3} + x_{C_3}^{(3)} sin\varphi_3 + y_{C_3}^{(3)} cos\varphi_3 - y_{O_4} - x_{C_4}^{(4)} sin\varphi_4 + y_{C_4}^{(4)} cos\varphi_4 = 0 \quad (10)$$

$$3-5 \quad R \quad C_3 = C_5$$

$$F_8: x_{O_3} + x_{C_3}^{(3)} cos\varphi_3 - y_{C_3}^{(3)} sin\varphi_3 - x_{O_5} - x_{C_5}^{(5)} cos\varphi_5 + y_{C_5}^{(5)} sin\varphi_5 = 0 \quad (11)$$

$$F_9: y_{O_3} + x_{C_3}^{(3)} sin\varphi_3 + y_{C_3}^{(3)} cos\varphi_3 - y_{O_5} - x_{C_5}^{(5)} sin\varphi_5 - y_{C_5}^{(5)} cos\varphi_5 = 0 \quad (12)$$

$$F_{10}: d_{CD} = f(t) \quad (13)$$

$$F_{11}: d_{CE} = f(t) \quad (14)$$

$$F_{12}: d_{CB} = f(t) \quad (15)$$

Equations (4) and (5) represent the geometrical constraint of R type from joint A, equation (6) - the geometrical restriction of RT type from joint D, equation (7) – the geometrical restriction of RT type from joint E, equation (8) – the geometrical restriction of RT type from joint B, equations (9) - (12) – the restrictions of R type from C, and equations (13)...(15) – the kinematic restrictions.
The displacement function of the tracking mechanism (Figure 9) can be obtained by solving numerically the system (4)...(15) in Maple 5 software.

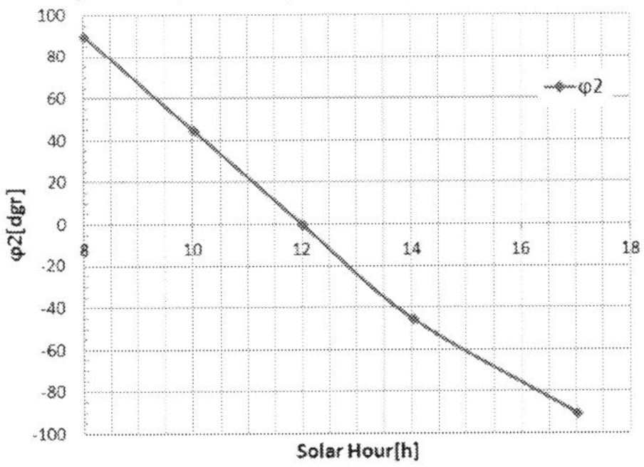

Figure 8. Angular stroke for parabolic trough collector

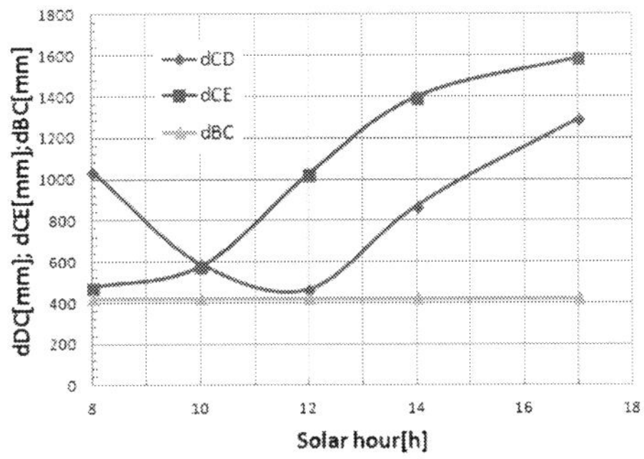

Figure 9. Linear actuator strokes

Conclusions

The proposed mono-axial tracking mechanism has the advantage of achieving a large angular stroke through the two actuators in a triangular arrangement, articulated to the base at their fixed ends. In order to ensure a smooth correction of the orientation and to increase the collector stroke, a third linear actuator, having its cylinder solidary to the trough collector, is used.

The variation of the angular stroke of the parabolic trough collector (Figure 9) and the strokes of the three linear actuators (Figure 10) can be obtained from the system of equations (4 ... 15) for the numerical data presented in Table 2, by using MAPLE 5 software.

The analysis of the mechanism dynamic behaviour can be peformed based on the kinematic analysis of the tracking system.

References

[1] http://www.ren21.net/gsr-2018/ (last access on 20 of June)
[2] Visa I, Jaliu C, Duta A, Neagoe M, Comsit M, Moldovan M, Ciobanu D, Burduhos B, Saulescu R 2015 *The Role of Mechanisms in Sustainable Energy System* (Transilvania University Publishing House)
[3] Duffie A and Beckman W A 2006 *Solar engineering of thermal processes* (John Willey&Sons Inc.)
[4] Desai N B and Bandyopadhyay S Optimization of concentrating solar thermal power plant based on parabolic trough collector *Journal of Clean Production* **89** pp 262-271
[5] Mousazadeh H et. al. A review of principle and Sun tracking methods for maximizing solar systems output *Renewable and Sustainable Energy Reviews* **13** pp 1800-18018
[6] http://www.powerfromthesun.net/book.html (last access on May 15)
[7] Patent No. RO131808(A0): Highly Accurate Uniaxial Orientation Mechanism

Mechanisms that generate Gerono's lemniscate

I Popescu[1], L Luca[2] and S S Ghimisi[3]

[1]Mechanical Faculty, University of Craiova, Craiova, Romania
[2]Engineering Faculty, University Constantin Brancusi of Targu-Jiu, Targu-Jiu, Romania
[3]Engineering Faculty, University Constantin Brancusi of Targu-Jiu, Targu-Jiu, Romania

E-mail: lylyanaluca@yahoo.com

Abstract. It is presented Gerono's lemniscate with bibliographic references on the geometrical aspects of this curve vastly studied in literature. A mechanism of Artobolevski that draws this curve is studied. The desired lemniscate was obtained. We synthesize a new, original mechanism that generates this curve, starting from two references showing a geometric construction and animation. The mechanism obtained is of the type R-RPP-RPR-PRP, so that the relations of the positions are written by the contouring method and the desired lemniscate is obtained. There are changes in the mechanism resulting in similar but deformed curves.

1. Introduction

Gerono's lemniscate (the eight-shaped curve) is a curve constructed by the French mathematician Camille-Christophe Gerono, who lived between 1799 and 1891, a curve that is part of the group of polyosomal curves, i.e. the median curves of two conical curves, having a common axis, a parallel and a perpendicular to this axis, the name being given by Cayley in 1868 [1]. By customization the curves are obtained: double- heart curve, slider- crank curves by Bérard, regular bifolium, quartic by Bernoulli, Gerono's lemniscate. In [2], Gerono's lemniscate is detailed, is given the construction with animation and equations are given in different forms, the curve being a quart, and in [3] there are many such curves obtained by the transformation of Newton. This curve is further mathematically detailed in [4] and [5]. It is also shown in [6] how to construct this curve (figure 1): From a point P on a circle it goes horizontally until it cuts the parallel with the ordinate tangent to the circle at the point M. The intersection of the OM straight with the order passing through P gives the Q point that generates Gerono's lemniscate.

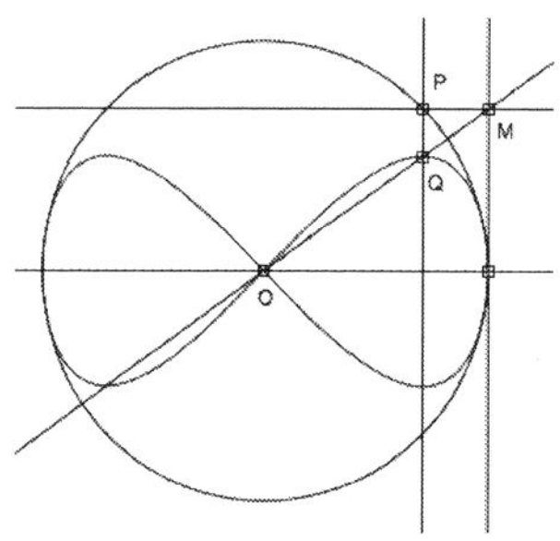

Figure 1. The geometric construction [6]

In [7] there is a demonstration (on Youtube) how to get this curve mathematically. Other calculations on the geometry of this curve are given in [8]. In [9] there are also mathematical calculations for this curve, also called Huygens lemniscate. By rotating this curve, [10] gives an interesting spatial surface. In the [11] is given a mechanism that generates this curve. Below, the mechanism in [11] is analyzed and an original mechanism generating this curve is constructed, starting from figure 1 and the animation in [2].

In [12, 13, 14, 15, 16] there are presented many original mechanisms that generate mathematical curves with aesthetic shapes. The structure and kinematics of some aesthetic curves were studied in papers [17, 18, 19].

This paper presents a new mechanism, original, that generates this curve (lemniscate). The mechanism can be applied to machines in the textile industry, to mechanical toys and to the industrial drawing of this curve.

2. Artobolevskii's mechanism

The Artobolevskii's mechanism is shown in Figure 2.

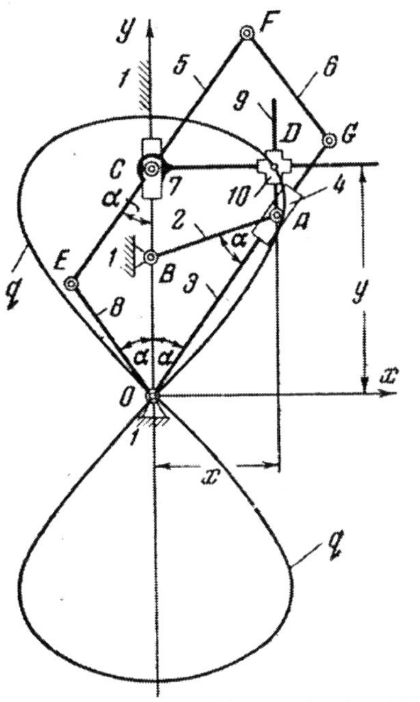

Figure 2. Artobolevskii's mechanism [11]

For geometrical reasons, the author has set the following conditions:

$$OE = EC = FG = AB = OB = \frac{a}{2};\ EF = OG \tag{1}$$

and relationships:

$$x = OA\sin\alpha = 2OB\cos\alpha\sin\alpha \tag{2}$$

$$y = 2OE\cos\alpha \tag{3}$$

In relations (1) the values of the sides are expressed according to a length "a". For a = 10 mm, using the relations (2) and (3), the Gerono's lemniscate of Figure 3 was obtained, and for a = 30 mm the one in Figure 4, the dimensions increasing to increase "a".

For the α = 0 ... 180 degrees, the half of the curve is obtained and this is shown in Figure 5.

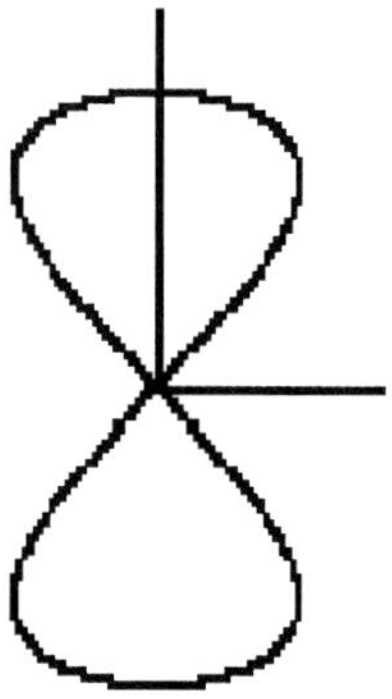

Figure 3. Curve, a=10

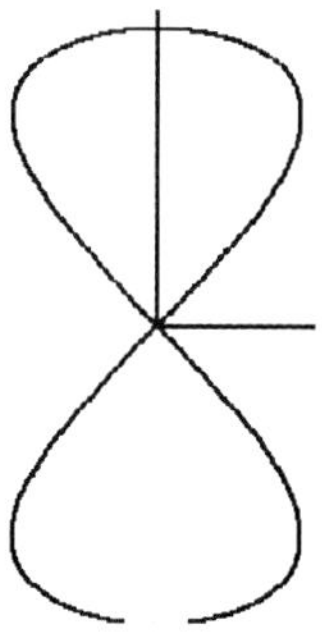

Figure 4. Curve, a=30

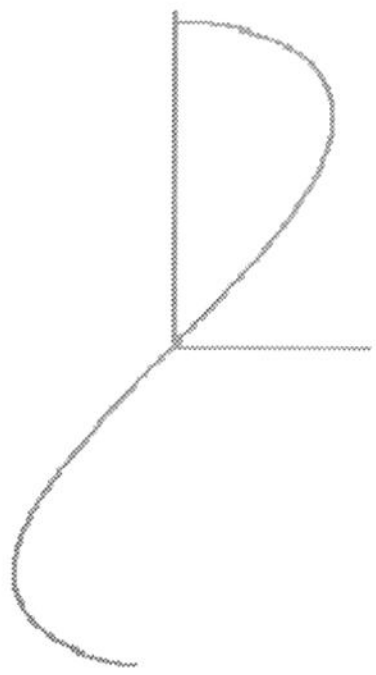

Figure 5. The curve, α =0...180

3. Synthesizing a new mechanism

It has gone from Figure 1 and from the animation in [2], constructing the kinematic scheme of the mechanism in Figure 6. The circle with the radius AB was drawn and the tangent to this parallel with the ordered, considered straight line. The right angle DBC was constructed with the variable length sides, taking point B somewhere in the original circle. The point of intersection of the BC side of the right angle with the tangent to the circle is C. This point is variable on the BC, therefore C was introduced in slider 4, but it must also move on the tangent, therefore it is a double slider of type PP (P comes from prismatic). It joins A with C, its length is variable, which requires another slide 5 in C, connected by a rotation joint to the element 4. The position of D is at the intersection of the BD side of the right angle with the variable radius AC so that the point D slides to the right BD, so slide 7 into C, but C also slips to the AC radius, that is, the sliding 6, connected to 7 by a rotation joint. Point D will shoot Gerono's lemniscate.

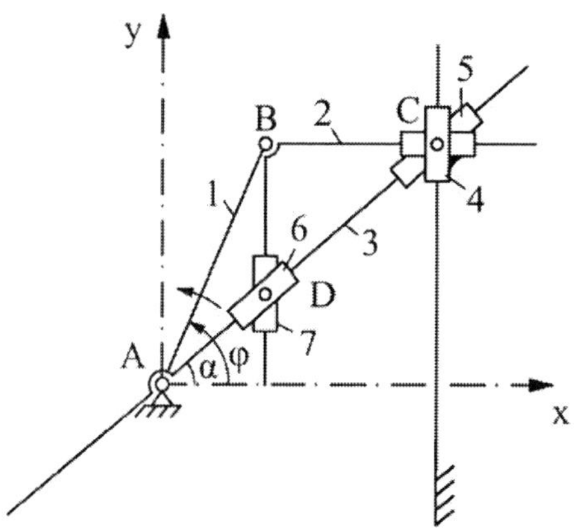

Figure 6. The synthesized mechanism

The mechanism is quite complicated, as can be seen from the structural analysis of Figure 7. It is found that the mechanism consists of the leading element AB, dyad BC type RPP (R comes from rotation), dyad AC type RPR and dyad 6-7 of D, type PRP.

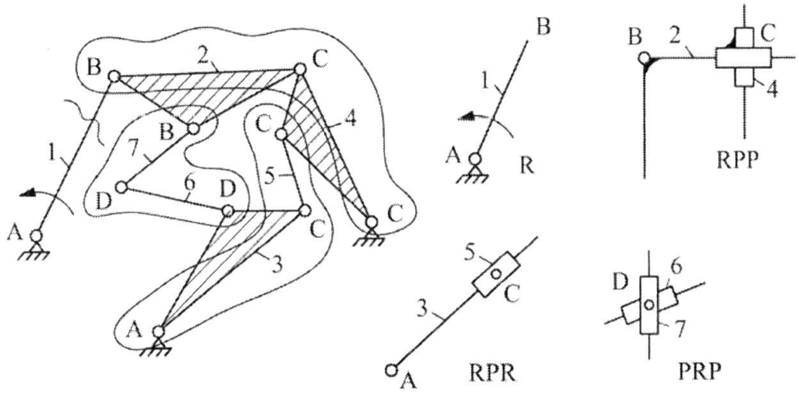

Figure 7. The structural analysis

4. Relations and results

Based on Figure 6 the following equations are written:

$$x_B = AB\cos\varphi;\quad y_B = AB\sin\varphi \tag{4}$$

$$x_C = const \tag{5}$$

$$BC = x_C - x_B \tag{6}$$

$$y_C = y_B \tag{7}$$

$$AC^2 = x_C^2 + y_C^2 \tag{8}$$

$$\cos\alpha = \frac{x_C}{AC} \tag{9}$$

$$x_D = x_B + BD\cos 270 = AD\cos\alpha = x_B \tag{10}$$

$$y_D = y_B + BD\sin 270 = AD\sin\alpha \tag{11}$$

From (4) equation results the position of B, and from (5), (6), (7) equations the position of C is obtained. With (8) equation it calculated the length of the AC, and with (9) equation it calculated the angle α. From (10) equation it obtained AD and x_D, and from (11) equation, it obtained y_D.

Figure 8 shows the mechanism obtained with the written program based on the above relations for a position, and Figure 9 shows the mechanism in successive positions.

The curve is shown in Figure 10. Only half of the curve was obtained, the one for the (+) sign in front of the radical by which it is calculated α from equation (9). The other half of the curve results for the case where the adopted sign is (-), Figure 11.

The entire curve is obtained for the signs $\pm$ in front of the radical (Figure 12). In the case of the real mechanism, it generates half a curve when it is initially positioned above the axis and the other half when it is positioned under the x-axis.

Figure 13 shows the variations of the coordinates of D for the sign (+), and Figure 14 for the sign (-). The curves are continuous, i.e. there are no breaks in the operation of the mechanism, the curve of x_D is the same in both charts, and the curve of y_D is inverted to the (-) sign.

IOP Conf. Series: Materials Science and Engineering **514** (2019) 012035 doi:10.1088/1757-899X/514/1/012035

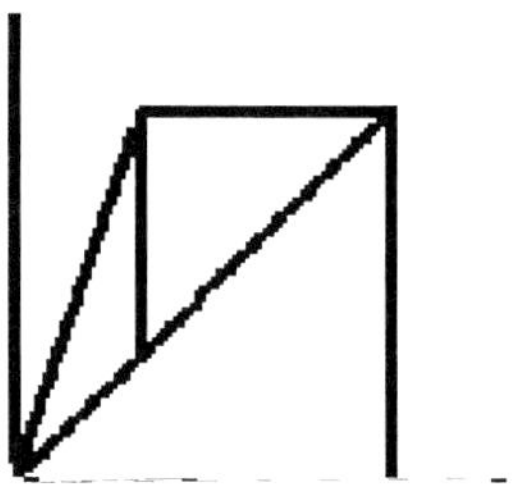

Figure 8. The mechanism in a position

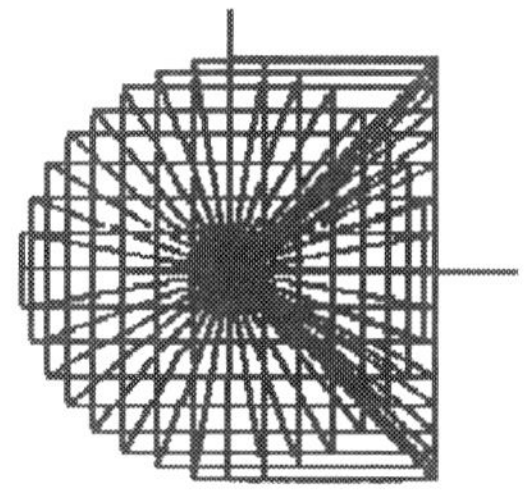

Figure 9. The mechanism in successive positions

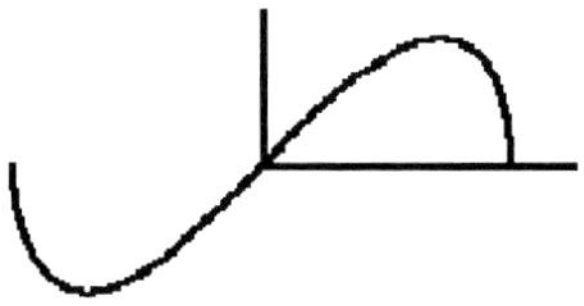

Figure 10. Curve for the sign (+)

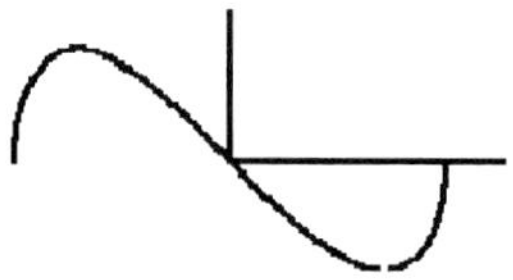

Figure 11. Curve for the sign (–)

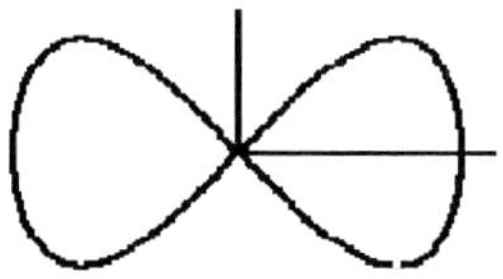

Figure 12. The full curve

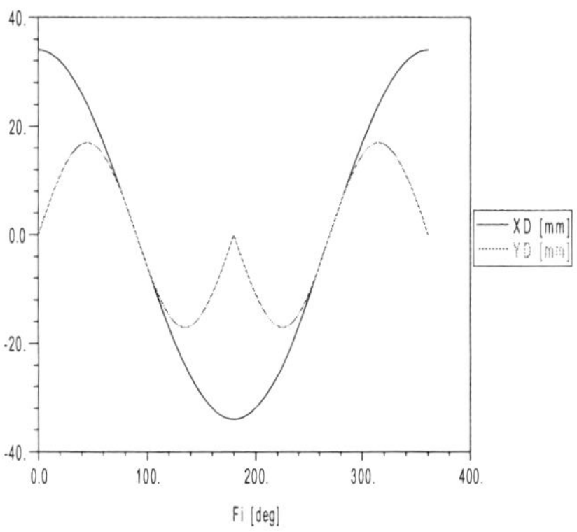

Figure 13. Curves for D at the sign (+)

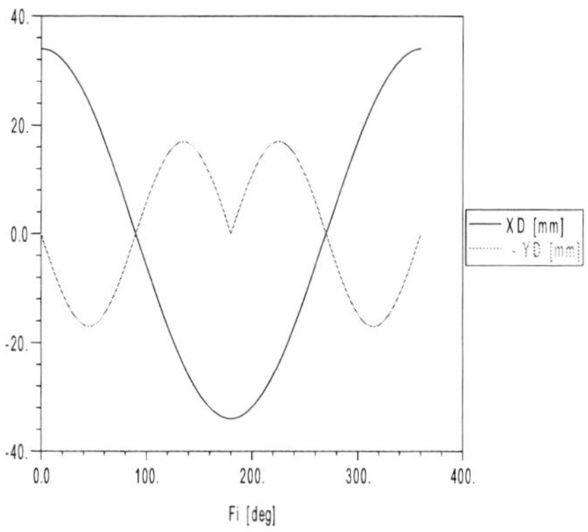

Figure 14. Curves for D at the sign (–)

It was also attempted to change the position of the tangent by C, considering a line parallel to the order that is no longer tangent to the circle with the radius AB. Thus, for x_C = 10 mm, was obtained the curve of Figure 15, for x_C = 25 mm was obtained the curve of Figure 16, and for x_C = 50 mm was obtained the curve of Figure 17, i.e. the same kind of curves, but deformed in height. The curves from Figures 15, 16 and 17 result for the same values of xc but negative, i.e. C is positioned to the left of the y-axis".

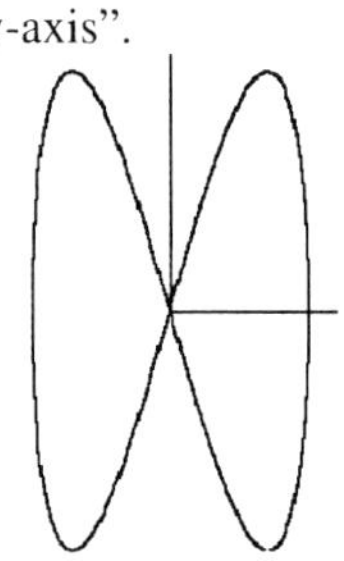

Figure 15. Curve for x_C=10

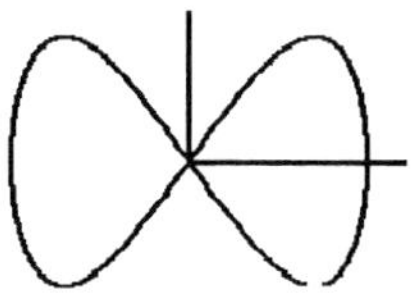

Figure 16. Curve for x_C=25

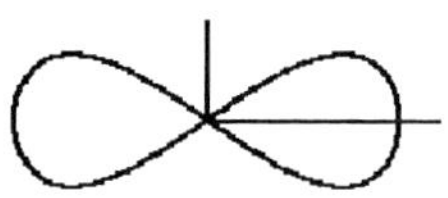

Figure 17. Curve for x_C=50.

5. Conclusions

It has departed from geometric considerations on Gerono's curve, with many works showing equations and other geometric features in the literature. A mechanism of Artobolevskii was also studied, generating the curve of Gerono. Next, a geometric solution and an animation, given in the literature, made the synthesis of an original generating mechanism of this lemniscate. The structure of this new R-RPP-RPR-PRP mechanism was studied, the position relationships were plotted by the contouring method and the desired curve was obtained. Equations provide two solutions, each generating a half of the curve. From the diagrams obtained it is found that the movement of the mechanism is continuous. The mechanism has been modified by positioning a line parallel to the y-axis so that it is no longer tangent to the initial circle, obtaining similar but deformed curves. This original mechanism, obtained through structural and geometric synthesis, is added to similar mechanisms in the literature and can be used to build machinery from the textile industry, mechanical toys, when cutting parts.

6. References

[1] Ferreol R 2008 *Mathcurve. Polyzomal curve.* [http://www.mathcurve.com/courbes2d/polyzomale/polyzomale.shtml]

[2] Ferreol R and Mandonnet J 2017 *Lemniscate of Gerono - Mathcurve.* [http://www.mathcurve.com/courbes2d.gb/gerono/gerono.shtml]

[3] Ferreol R 2017 Hyperbolism and antihyperbolism of a curve newton transformation. *Mathcurve.* [http://www.mathcurve.com/courbes2d.gb/hyperbolisme/hyperbolisme.shtml]

[4] https://en.wikipedia.org/wiki/Lemniscate_of_Gerono

[5] http://mathworld.wolfram.com/EightCurve.html

[6] http://xahlee.info/SpecialPlaneCurves_dir/LemniscateOfGerono_dir/lemniscateOfGerono.html

[7] [https://www.youtube.com/watch?v=E7oR_JBgUzA]

[8] Weisstein E W 1999 *Lemniscate of Gerono* [http://archive.lib.msu.edu/crcmath/math/math/e/e039.htm]

[9] https://www.revolvy.com/main/index.php?s=Lemniscate%20of%20Gerono

[10] https://www.geogebra.org/m/X2U9Jv3K

[11] Artobolevskii I I 1959 *Teoria mehanizmov dlia vosproizvedenia ploskih crivâh.* Izd. Academii Nauk, SSSR, Moskva.

[12] Luca L and Popescu I 2015 Paths and Laws of Motion of a Mechanism with Two Successive Conductive Elements and a Triad *Applied Mechanics and Materials* **772** Trans Tech Publications, Switzerland pp 344-349

[13] Luca L, Popescu I and Ghimisi S 2012 Studies regarding generation of aesthetics surfaces with mechanisms *Proceedings of the 3-rd International Conference on Design and Product Development* Montreux Elvetia pp 249-254

[14] Luca L, Ghimisi S and Popescu I 2012 Studies regarding the movement on the cochleoid *BookSeries: Advanced Materials Research* **463-464** pp 147-150

[15] Luca L and Popescu I 2012 Generation of aesthetic surfaces through trammel mechanism *Fiability& Durability/FiabilitatesiDurabilitate* **1**supplement pp 55-61

[16] Luca L and Popescu I 2013 Curves and aesthetic surfaces generated by the R-R-RTR mechanism *Fiability& Durability/FiabilitatesiDurabilitate* **1** pp28-34

[17] Popescu I, Luca L and Cherciu M 2011 *Trajectories and motion laws of some mechanisms (Traiectorii si legi de miscare ale unor mecanisme)* Sitech Publishing House Craiova

[18] Popescu I, Luca L and Mitsi S 2011 *Geometry, structure and kinematics of some mechanisms (Geometria , structura si cinematica unor mecanisme)* Sitech Publishing House Craiova

[19] Popescu I, Luca L and Cherciu M 2013 *Structure and kinematics of mechanisms. Applications (Structura si cinematica mecanismelor. Aplicatii)* Sitech Publishing House Craiova

IOP Conf. Series: Materials Science and Engineering **514** (2019) 012036 doi:10.1088/1757-899X/514/1/012036

Modeling and simulation of a dual-axis solar tracker for PV modules

A Marcu[1], C Alexandru[1] and I Barbu[1]

[1]Product Design, Mechatronics and Environment Department, Transilvania University of Braşov, Romania

E-mail: calex@unitbv.ro

Abstract. The paper deals with the modeling and simulation in virtual prototyping environment of a mechatronic solar tracker used for photovoltaic systems, with the aim to increase the energetic efficiency, by maximizing the rate of incoming incident solar radiation. The solar tracker in study is an equatorial dual-axis mechanism, which allows the adjustment of the diurnal and seasonal angles of the PV module in accordance with a predefined tracking program, the actuating sources being linear actuators. The modeling and simulation of the tracking system is carried out by using the MBS (Multi-Body System) commercial software solution ADAMS (Automatic Dynamic Analysis of the Mechanical Systems).

1. Introduction

This paper addresses a topic from a very important field: renewable sources for the generation of electricity - increasing the efficiency of conversion in solar panels. Research in this area represents a global priority as it provides viable alternatives to a series of major problems the mankind is facing: the limiting and polluting nature of conventional sources, global warming, or greenhouse effect.

The sun represents the most important source of renewable energy, the current systems allowing the conversion of the solar energy into electricity or heat. The energy efficiency of the solar panels depends largely on the solar energy capture, which can be maximized by using systems for the orientation of the panels towards the sun. The orientation principle of the solar panels is based on the data regarding the position of the sun on the celestial sphere. To ensure the greatest possible efficiency of the solar energy conversion into electricity, the sun rays must fall normally on the surface of the receiver, so it is necessary that the system periodically modifies its position in such a way as to maintain the relationship between the sun rays and the panel [1-3]. On the one hand, over the course of one year, the Earth describes an elliptical rotation movement around the Sun, and on the other hand, over the course of a day, it performs a complete rotation around its own axis (a movement that generates sunrises and sunsets). Therefore, for the design of the solar tracking systems, both rotating movements have to be taken into consideration: the diurnal movement and the seasonal (elevation) movement [4, 5].

As the complexity and competitiveness of the development of the tracking systems grows, the design and production times are reduced, conditions in which the creation and testing of physical prototypes become major impediments to launching new products. Therefore, it is necessary to implement techniques based on virtual prototyping that can deliver higher performance and product quality using only a fraction of the time and cost required in the traditional creation and testing approaches [6-11].

IOP Conf. Series: Materials Science and Engineering **514** (2019) 012036 doi:10.1088/1757-899X/514/1/012036

2. Modeling the solar tracking mechanism

For this paper, an equatorial dual-axis tracking mechanism was considered, whereby the necessary movements are generated by means of two linear actuators. Its operation is done in steps, the movement/positioning laws of the panel being modeled to achieve a minimum number of actuations (for energy efficiency purposes). The kinematic model of the tracking system concerned, having two degrees of freedom, contains the bodies in the system, connected by kinematic joints (geometric constraints), and the geometric parameters specific to the mechanism (locations/orientation of the joints). The input (operation) is made by kinematic constraints (motion generators) applied in the translation joints between the pistons and the linear actuator cylinders.

The tracking mechanism considered includes two kinematic chains to generate the two necessary movements, namely the kinematic chain for diurnal movement (ABCD) and the kinematic chains for the seasonal movement (EFGH) respectively. The geometric constraints in A, C, D and E, G, H respectively, are modeled by revolute joints, while the B and F are modeled by translation joints. In order to achieve the diurnal movement, the panel rotates relative to the support element around the axis D-D', while the seasonal movement of the panel is made by rotation of the support element relative to the supporting pillar around the H joint axis.

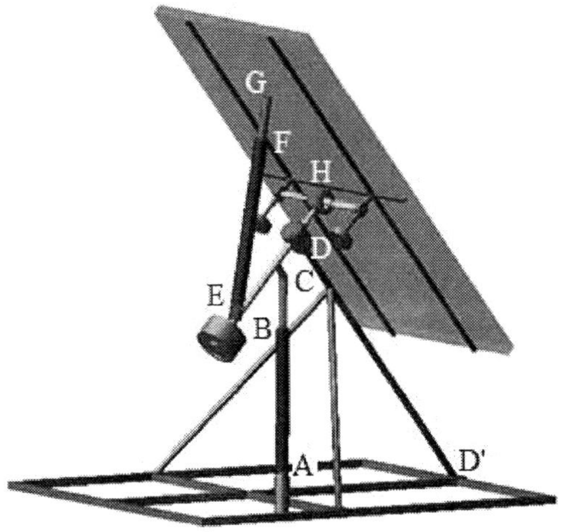

Figure 1. The MBS model of the equatorial dual-axis tracking system.

The angular movements of the panel can be done continuously or in steps. For energy efficiency reasons of the system, the movement is chosen to be made in steps, the movement laws of the panel, for both rotations, being set so as to achieve a minimum number of actuations; the simulation is made considering the data specific to the summer solstice and the latitude of the Brașov area, which actually determine the orientation of the panel relative to the position of the sun on the celestial sphere.

In terms of diurnal movement, the total angle achieved by the panel is 120° (-60°, ..., + 60°), with the zero position (at noon), while for the seasonal position the elevation angle is adjusted from 11° to 22.05°. In the virtual model created by using the MBS ADAMS software, the movement laws of the panel were modeled by summing a series of STEP time functions as follows:

- for the diurnal movement:
  STEP(time, 0.0, 0.0d, 4.26, 0.0d)+STEP(time, 4.26, 0.0d, 7.91, 0.0d)+STEP(time, 7.91, 0.0d, 8.01, 7.5d)+STEP(time, 8.91, 0.0d, 9.01, 15.0d)+STEP(time, 9.91, 0.0d, 10.01, 15.0d)+STEP(time, 10.91, 0.0d, 11.01, 15.0d)+STEP(time, 11.91, 0.0d, 12.01, 15.0d)+STEP(time, 12.91, 0.0d, 13.01, 15.0d)+STEP(time, 13.91, 0.0d, 14.01, 15.0d)+STEP(time, 14.91, 0.0d, 15.01, 15.0d)+STEP(time, 15.91, 0.0d, 16.01, 7.5d)+STEP(time, 16.01, 0.0d, 18.46, 0.0d)+STEP(time, 18.46, 0.0d, 22.0, 0.0d)+STEP(time, 22.0, 0.0d, 22.2, -120.0d)+STEP(time, 22.2, 0.0d, 24.0, 0.0d);

- for the seasonal (elevation) movement:
  STEP(time, 0.0, 0.0d, 4.26, 0.0d)+STEP(time, 4.26, 0.0d, 8.91, 0.0d)+STEP(time, 8.91, 0.0d, 9.01, 11.05d)+STEP(time, 9.01, 0.0d, 14.91, 0.0d)+STEP(time, 14.91, 0.0d,15.01,-11.05d)+STEP(time, 15.01, 0.0d, 18.46, 0.0d)+STEP (time, 18.46, 0.0d, 24.0, 0.0d).

In these terms, the time variation laws of the diurnal and seasonal angles are presented in Figure 2 (relative to the initial position, at 4.26 hours, where $\varphi_d = 60°$, $\varphi_s = 11°$). On the basis of these laws applied to the revolute joints in D (diurnal movement) and H (seasonal movement), according to the notations in Figure 1, the transposed/equivalents for the linear displacements of the actuators were obtained (the relative movement of the pistons relative to the cylinders), actually the kinematic constraints applied to the translation joints in B (diurnal movement) and H (seasonal movement), for which time variation diagrams are shown in Figure 3 (a - diurnal motion, b - elevation motion).

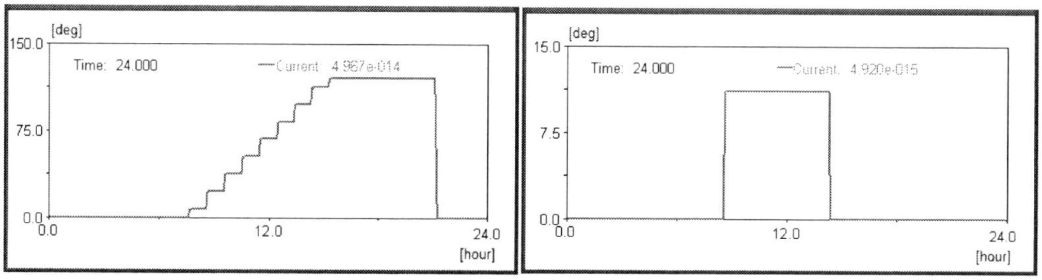

a. b.

Figure 2. The time-history variations of the daily (a) and elevation (b) angles.

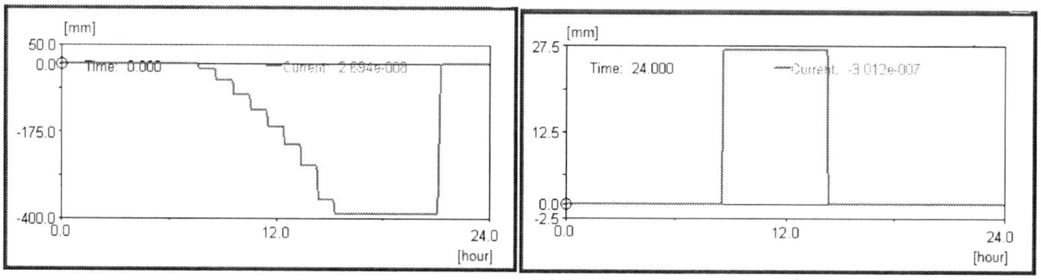

a. b.

Figure 3. The time-history variations of the actuators strokes.

3. Results and conclusions

The kinematic model processing is done by launching the ADAMS/Solver module, which auto-formulates and solves the kinematic motion equations. Following the model analysis, the behavior of the tracking system is determined, for whose evaluation the post-processing of the results is further carried out. Post-processing consists in plotting variation diagrams for the values of interest (positions, velocities, accelerations), graphic animations, and others. For example, Figure 4 shows the angular velocities (a - diurnal motion, b - elevation) and accelerations (c - diurnal motion, d - elevation) for the two panel movements, and Figure 5 shows the linear velocities (a - diurnal motion, b - elevation) and accelerations (c - diurnal motion, d - elevation) in the actuators.

The results obtained demonstrate that the designed tracking system fulfills the movement functions, ensuring the necessary positioning of the panel for the whole functioning cycle (no self-lockings of the mechanism show up). Further on, by means of the reverse dynamic analysis, the driving forces are determined, which, when applied to the leading elements of tracking system charged with forces, they generate the prescribed kinematic behavior. The driving forces represent the input data into the direct dynamic model, an then the tracking system behavior under the action of the forces is to be assessed.

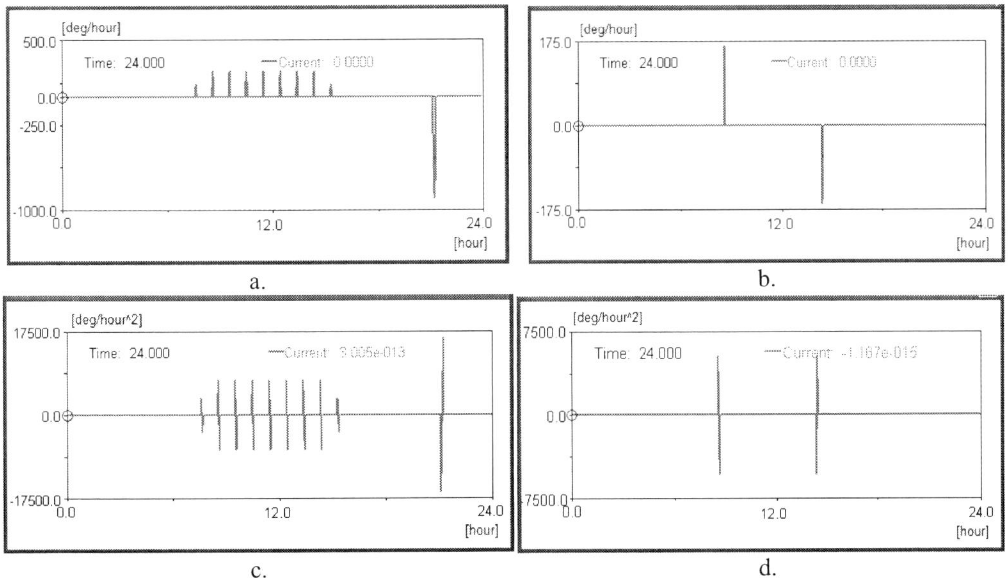

Figure 4. The time-history variations of the angular velocities (a, b) and accelerations (c, d).

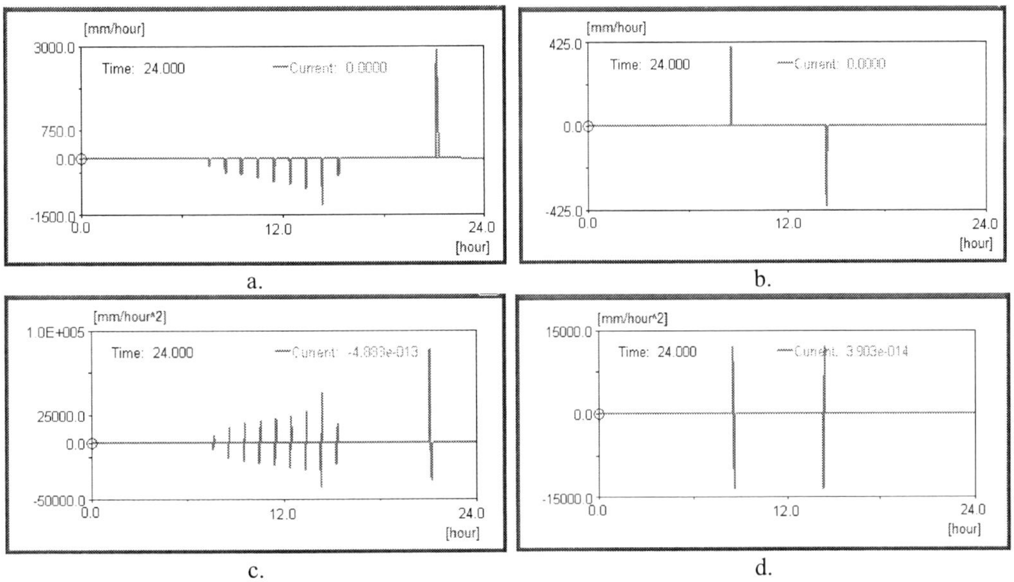

Figure 5. The time-history variations of the linear velocities (a, b) and accelerations (c, d).

The tracking system concerned has two degrees of freedom, namely the linear displacements in the actuators (which generate the two movements of the panel), which in the kinematic model are controlled (imposed) by kinematic constraints of the form S(t) whose variation laws have been shown in Figure 3. In the case of the dynamic model, the kinematic constraints are replaced by driving forces, so that the system has two generalized independent coordinates.

The inverse dynamic model of the tracking system includes, on the one hand, the components specific to the kinematic model (the bodies, the links between them, the specific geometric parameters) and, on the other hand, the external and internal system of forces which engage the mechanism. The charging of the tracking system concerned is done by mass forces only (other forces may also be considered, such as the action of the wind, etc.). Thus, the values of the driving forces in the linear actuators are obtained, the time variations being shown in the diagrams in Figure 6 (a - diurnal motion, b - elevation). The driving forces thus obtained represent inputs into the direct dynamic model of the tracking system, replacing the kinematic constraints in the kinematic model.

The dynamic model evaluates the behavior of the system under the action of the forces. As an example, Figure 7 shows the results of the dynamic analysis on the magnitude of the reaction forces in the revolute joints A (Fig. 7, a) and E (Fig. 7, b) of the actuator cylinders to the adjacent elements (see Figure 1). Such reactions are necessary for the drawing of the engagement schemes, for the dimensioning of the joints and of the elements of the mechanism.

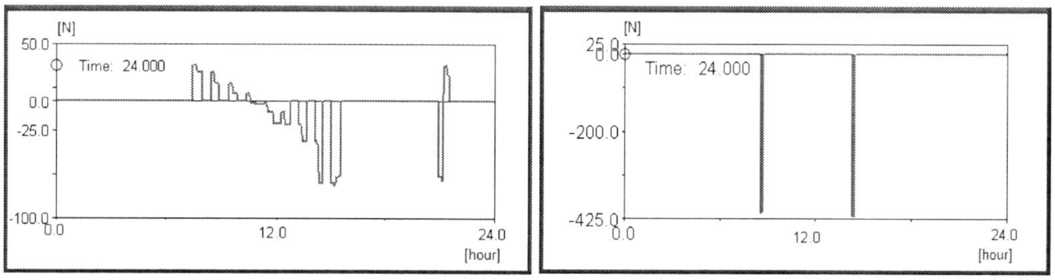

a. b.

Figure 6. The motor forces in actuators.

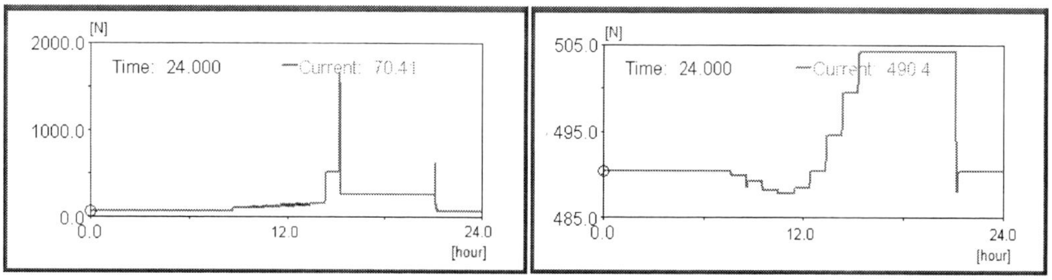

a. b.

Figure 7. The reaction forces in the revolute joints A and E.

In the case of the joints in the kinematic chain for the diurnal movement (ABCD contour), there was a large increase in the reactions caused by the swing/ sudden rotation of the actuator within a very short interval, for kinematic reasons. Except for this interval, the variation is relatively low, as it can be observed from the time variation diagram of the A joint reaction (similar to the other joints). The reported problem will be subject to dynamic optimization of the tracking system, which is presented in another paper submitted to the Xth Product Design, Robotics, Advanced Mechanical & Mechatronic Systems and Innovation Conference - PRASIC [12].

On the other hand, Figure 8 shows the time variation diagrams for power consumption in order to achieve the desired movements (diurnal and seasonal/elevation), a very important parameter for assessing the energy efficiency of the tracking system. Maximum consumption occurs on the return area through continuous movement (without steps) to the initial position at the end of the day. Except for this area, the maximum occurs over the overlapping of the two movements/rotations in the 14.91-15.01 time slot, as well as over the sudden tilting/rotation of the diurnal actuator (problem which was reported for the reactions in the joints).

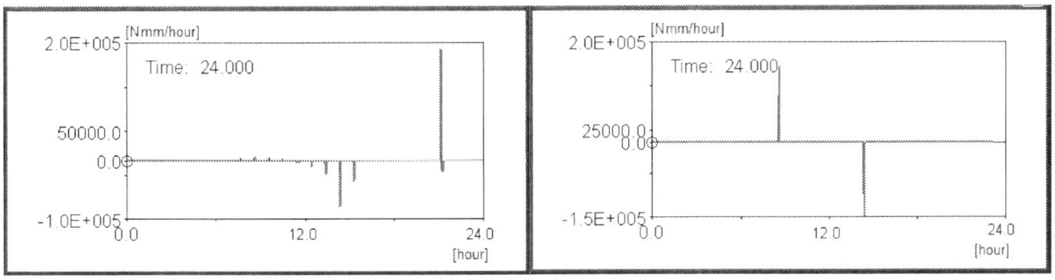

a. b.

Figure 8. The power consumption for the diurnal (a) and elevation (b) motions.

Along with the development of the MBS method and software, the design of the tracking systems evolved in a new direction: the design and analysis of complex theoretical models, close to actual models both in terms of system structure as well as operating conditions. True virtual prototypes for the tracking systems can be designed with the help of the MBS software platforms, considering reducing to the minimum the expenditures that would result from designing physical prototypes as well as the time required for testing. Regarding the equatorial tracking system concerned, although from a kinematic point of view, its behavior followed the required movement functions, some dynamic problems (strong reactions in joints, high power consumption) have been identified, which are to be eliminated based on a study of dynamic optimization [12].

References

[1] Alexandru C 2009 The design and optimization of a photovoltaic tracking mechanism *Proceedings of the 2nd IEEE International Conference on Power Engineering, Energy and Electrical Drives POWERENG 2009* pp 436-441

[2] Alexandru C and Tatu I N 2013 Optimal design of the solar tracker used for a photovoltaic string *Journal of Renewable and Sustainable Energy* **5** pp 1-16

[3] Altuzarra O, Macho E, Aginaga J and Petuya V 2015 Design of a solar tracking parallel mechanism with low energy consumption *Proceedings of the Institution of Mechanical Engineers, Part C: Journal of Mechanical Engineering Science* **229** pp 566-579

[4] Edgar R, Stachurski Z and Cochard S 2016 Optimisation direct normal insolation of rectangular PV platforms *Solar Energy* **136** pp 166-173

[5] Ioniță M and Alexandru C 2012 Dynamic optimization of the tracking system for a pseudo-azimuthal photovoltaic platform *Journal of Renewable and Sustainable Energy* **4** pp 1-15

[6] Alexandru C and Pozna, C 2009 Dynamic modeling and control of the windshield wiper mechanisms *WSEAS Transactions on Systems* **8** pp 825-834

[7] Alexandru C and Alexandru, P 2010 The virtual prototype of a mechatronic suspension system with active force control *WSEAS Transactions on Systems* **9** pp 927-936

[8] Alexandru C and Alexandru, P 2011 A comparative analysis between the vehicles' passive and active suspensions *International Journal of Mechanics* **5** pp 371-378

[9] Alexandru P, Macaveiu D and Alexandru C 2012 Design and simulation of a steering gearbox with variable transmission ratio *Proceedings of the Institution of Mechanical Engineers, Part C: Journal of Mechanical Engineering Science* **226** pp 2538-2548

[10] Bernard A 2005 Virtual engineering: methods and tools *Journal of Engineering Manufacture* **219** pp 413-421

[11] Tarniță D, Marghitu D 2013 Analysis of a hand arm system *Robotics and Computer-Integrated Manufacturing* **29**, pp 493-501

[12] Marcu A, Alexandru C and Barbu I 2018 Dynamic optimization of a dual-axis solar tracker for PV modules *Proceedings of the Xth Product Design, Robotics, Advanced Mechanical & Mechatronic Systems and Innovation Conference PRASIC 2018*

IOP Conf. Series: Materials Science and Engineering **514** (2019) 012037 doi:10.1088/1757-899X/514/1/012037

Dynamic optimization of a dual-axis solar tracker for PV modules

A Marcu[1], C Alexandru[1] and I Barbu[1]

[1]Product Design, Mechatronics and Environment Department, Transilvania University of Braşov, Romania

E-mail: calex@unitbv.ro

Abstract. This paper deals with the dynamic optimization of the dual-axis tracking mechanism, which is used for increasing the energetic efficiency of a photovoltaic module, by maximizing the rate of incident solar radiation. The study is based on the optimal design algorithm included in the commercial MBS (Multi-Body Systems) software solution ADAMS, through the parameterization of the virtual model, which offers the possibility to perform a series of studies aiming at identifying the design parameters that decisively influence the dynamic behavior of the tracking systems. The global coordinates of the design points (i.e. the mounting points of the actuating sources) are defined as design variables for the optimization process, while the design objectives refer to the power consumptions of the actuating sources.

1. Introduction

The continuous diminution of fossil fuel resources has led the energy market to turn to the renewable resource systems industry, which can obtain electricity and heat through environmentally-friendly and non-polluting methods. So, the sun, water, wind, geothermal energy and biomass are increasingly used to generate energy. Solar energy is the most important source of renewable energy and offers enormous potential to generate green energy without emissions of pollutant gases into the atmosphere. A viable solution for increasing the efficiency of solar energy conversion systems (photovoltaic or thermal panels) is the use of tracking mechanisms that change the position of the panel to maximize the amount of incident solar radiation.

Depending on the two axes of motion of the earth, the tracking systems may be bi-axial and mono-axial. The dual-axis tracking systems provide precise positioning by performing the diurnal and seasonal movements [1-5]. The mono-axis tracking systems are frequently used for the diurnal motion of the panel, the seasonal position being fixed at an optimal angle predefined to the location (i.e. the elevation angle) [6-9]. The mono-axis systems have the advantage of low cost due to lower number of elements and drive with a single motor, but have the disadvantage of lower energy efficiency relative to the dual-axis systems.

Given the working principle, two basic types of tracking systems can be identified: passive and active. The operation of passive orientation systems is usually based on the thermal expansion of a Freon-based fluid from one corner of the system to the other because of the heat sensitivity of the working fluid. The active tracking systems are mechatronic systems based on electrically driven positioning devices, including motors (rotary or linear actuators), speed reducers, linkages, couplings, etc. The orientation of solar panels by active tracking systems can increase the efficiency of the conversion system with values between 20% and 50% relative to the fixed panel situation [10-16].

By using MBS commercial software (such as ADAMS - Automatic Dynamic Analysis of the Mechanical Systems), given the computing power and the facilities they provide, as pointed out in [17-23], true virtual prototypes for the tracking systems can be designed, with a view to obtaining products that function properly to meet the high market demands. This means that faithful modeling of both the components of the tracking system and its operating conditions can be achieved, which allows for rapid testing of numerous geometric - constructive variants in order to optimize the system.

The use of a virtual prototype platform allows for the possibility that, once the digital prototype is designed, the optimization of the mechanical tracking system can be easily done by modifying some specific features, mainly geometrical. Thus, a multitude of constructive variants can be tested regarding the location and the characteristics of the tracking system elements, mass distribution and so on, at reduced costs compared to the optimization based on the physical prototype construction and testing.

This paper deals with an optimization model based on the optimal design algorithm included in the ADAMS software, through the parameterization of the virtual model, which offers the possibility to perform a series of studies aiming at identifying the design parameters that decisively influence the dynamic behavior of the tracking systems. The case study is developed by considering an equatorial dual-axis tracking system, which was modeled - analyzed in virtual prototyping environment in another paper submitted to the Xth Product Design, Robotics, Advanced Mechanical & Mechatronic Systems and Innovation Conference - PRASIC [24]. The tracking mechanism includes two kinematic loops (Figure 1): ABCD - for the diurnal movement, and EFGH - for the seasonal (elevation) movement. The geometric constraints in A, C, D, E, G and H are revolute joints, while the B and F connections are modeled by translational joints.

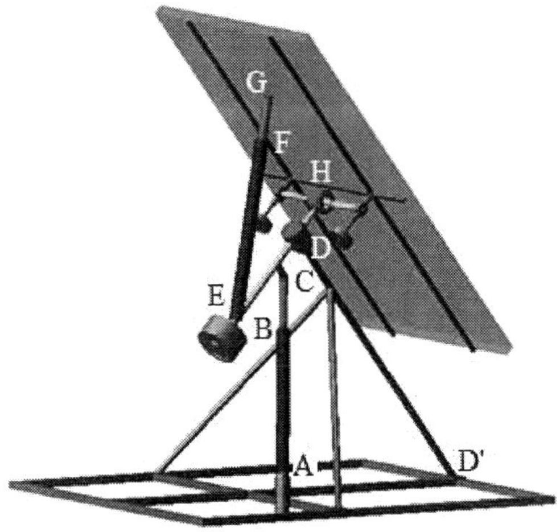

Figure 1. The virtual model of the dual-axis tracking mechanism.

2. The optimization algorithm

As stated in [24], the objective of optimizing the dynamic behavior of the tracking system is to reduce the power consumption of the motors of the actuators (for energy efficiency purposes) while reducing the magnitude of the reactions in the joints of the mechanism (in this case, by avoiding that reaction leap/peak caused by the sudden swing of the actuator for the diurnal movement).

Under these conditions, the optimization of the dynamic behavior of the tracking systems is based on taking the following steps: parameterization of the model; defining design variables; defining the objective functions for optimization and the design constraints; carrying out the analysis of design sensitivities, in order to identify the main design variables (having a significant influence on the objective functions); the actual optimization of the system on the basis of the main variables.

The parameterization of the dual-axis tracking system shown in Figure 1 for the purpose of dynamic optimization is based on the use of the points defining the structural scheme of the mechanism, in this case the points defining the locations of the joints between the elements. For this study, only the connection points of the linear actuators' cylinders relative to the adjacent elements have been considered, namely the joint A to the fixed support - for the diurnal movement, and the joint E to the intermediary support - for the seasonal movement. The coordinates of the other points are considered to be established solely on constructive considerations relative to the way the system is mounted.

As additional conditions, both the orientation of the actuators (thus the positioning of the axes of the translational joints between the cylinders and the pistons) as well as the orientation of the axes of the revolute joints between the actuators' cylinders and the elements to which they are connected (the fixed support - for the diurnal actuator, respectively the intermediary support - for the seasonal actuator), are preserved in the modeling position (the initial one, from the beginning of the simulation), changing only the locations of the joints. In addition, in the case of the seasonal actuator, as the seasonal movement takes place in the XZ plane, the global Y_E coordinate remains unchanged (at the initial value) during the optimization study.

In this way, five independent design variables, which control the model during the optimization process, have been obtained, as follows: DV_1 $\rightarrow X_A$, DV_2 $\rightarrow Y_A$, DV_3 $\rightarrow Z_A$, DV_4 $\rightarrow X_E$, DV_5 $\rightarrow Z_E$. Each design variable is defined by an initial value (corresponding to the initial design) and a variation range relative to the initial value, which has been set with the view of maintaining the tracking mechanism within acceptable construction limits.

The objective of the optimization is to identify the values of the design parameters for which the system behavior is optimal. In ADAMS, the objective function is defined by a measure, which tracks the time variation of the dynamic parameter of interest. The measure is a specific ADAMS object that allows investigating the behavior of a (predefined or user-defined) feature of the model during or after carrying out the analysis. The objective function (which is also called design objective, functional cost or performance index) is a numerical representation of the quality, efficiency, cost, or stability of the system, representing a numerical quantification that distinguishes/evaluates multiple design variants; optimal design is ensured when the objective function is, as the case may be, minimized or maximized.

At the same time, a number of design constraints can be defined, which impose restrictions on the structure and/or operation of the tracking system, for example, the system has to fall within a certain working space or certain parameters do not exceed predetermined limit values. The optimization study improves, as far as possible, the objective function without violating the design constraints. Constraints are frontiers that directly or indirectly eliminate the unacceptable system design, often taking the form of additional goals in the design of the tracking system. Each design constraint generates a relationship of inequality, while during the optimization process the constraint value is kept under or equal to zero.

The objective of optimizing the dynamic behavior of the tracking system in study is to minimize the power consumption of the two linear actuators (for energy efficiency purposes) while reducing the magnitude of the joint reactions (more specifically, to avoid the reaction peak caused by the swing of the actuator for the diurnal movement [24]).

The objective of the parametric design study is to examine design sensitivities, i.e. to identify the design variables with a major influence on the objective functions to be optimized; only these variables are taken into account in the optimization process of the tracking system (the secondary variables, having a low/insignificant influence, will be disregarded).

3. Results and conclusions

As mentioned previously, for the equatorial tracking system concerned, the design variables consist of the overall coordinates of the points of articulation where the actuator cylinders are placed on the panel support (for the diurnal movement), or on the supporting pillar, respectively (for the seasonal movement). The objectives to be analyzed during the parametric design studies have been defined on the basis of the measures by which the reaction in joint A and the power consumptions for the diurnal and seasonal movements are modeled, with the aim to observe the maximum value of these measures during the dynamic simulation/analysis of the tracking system.

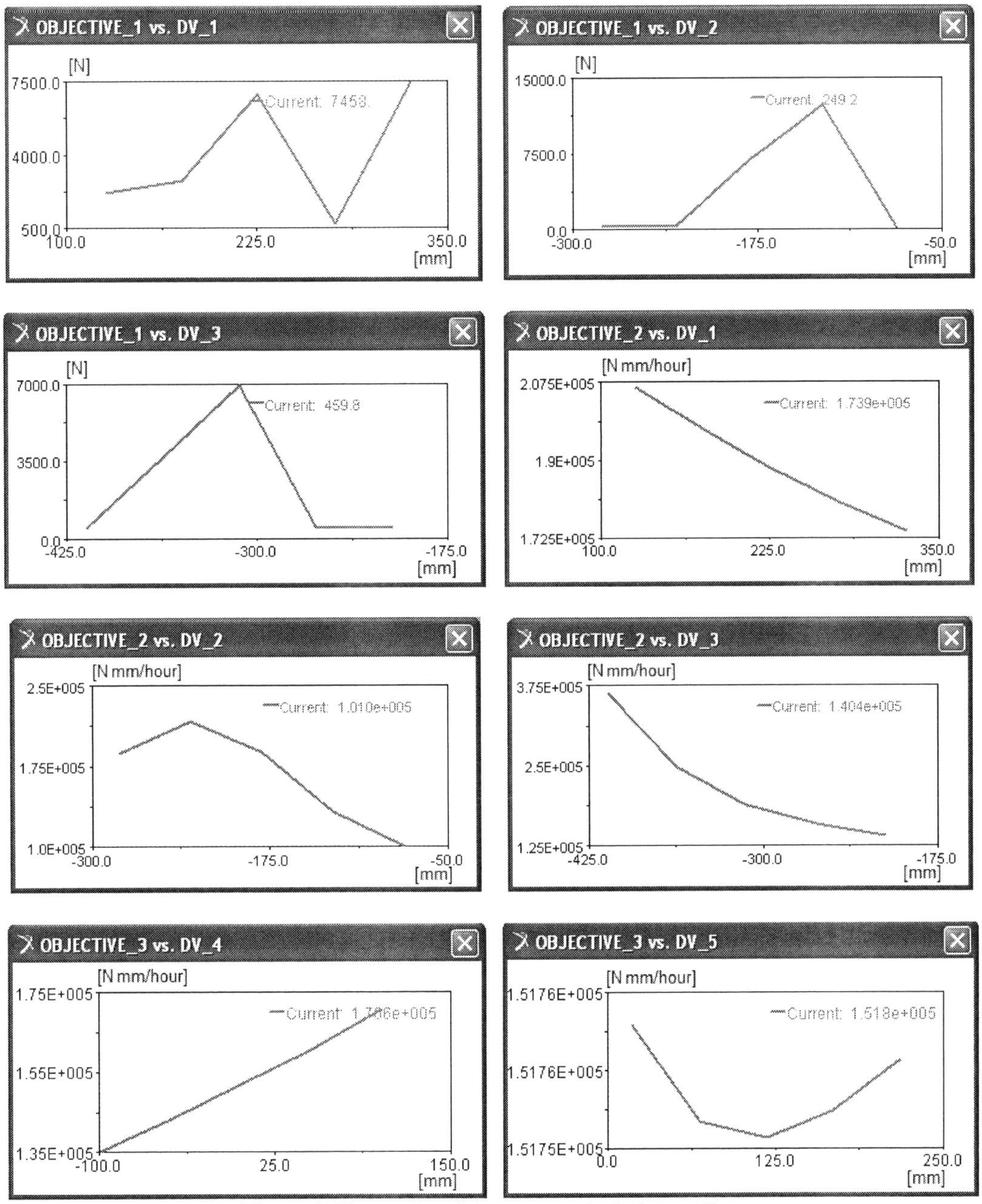

Figure 2. The results of the parametric design studies.

The design study is performed for each design variable individually, in its value range. Thus, using the selected design variables and the indicated objectives, the diagrams in Figure 2 were obtained, based on which the sensitivity of the objectives to the variation of the design variables is evaluated (in diagrams, on the ordinate there are the values of the design variable, and on the abscissa there are the values of the objective function).

Following the study of parametric design, the following main design variables resulted: for objective 1 (reaction in joint A): DV_1, DV_2, DV_3; for objective 2 (power consumption of the linear actuator for the diurnal movement): DV_2, DV_3; for objective 3 (power consumption of the linear actuator for the seasonal movement): DV_4. The dynamic optimization of the tracking system behavior will be performed based on these variables, giving up the secondary variable DV_5.

The actual optimization is done progressively for each objective, with the specific main design variables, starting with objective 1; this means that the optimal option for an tracking system resulting from the optimization study for objective 1 becomes the initial option for the optimization of objective 2 and so on.

The results obtained based on optimization are shown in Figure 3, as follows: the reaction force in the revolute joint A (a), and the power consumptions for the diurnal (b) & elevation (c) motions. Compared to the results obtained for the initial model (before optimization) presented in [24], a significant reduction in the power consumptions and coupling reactions can be observed, while maintaining the system within rational construction limits, which proves the viability of the optimization algorithm implemented in this research.

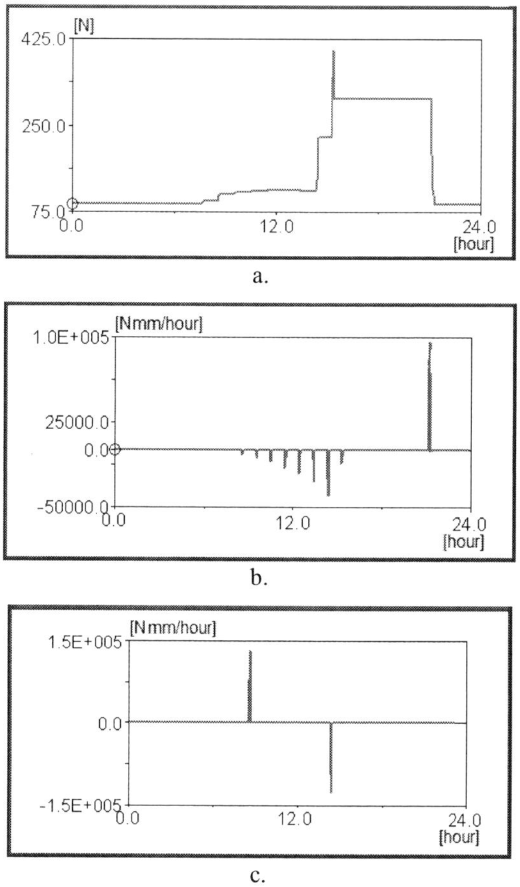

Figure 3. The results of the optimization process.

References

[1] Alexandru C and Comşiţ M 2008 The energy balance of the photovoltaic tracking systems using virtual prototyping platform *Proceedings of the 5th IEEE International Conference on the European Electricity Market EEM 2008* pp 253-258

[2] Alexandru C and Pozna C 2008 Virtual prototype of a dual-axis tracking system used for photovoltaic panels *Proceedings of the IEEE International Symposium on Industrial Electronics ISIE 2008* pp 1598-1603

[3] Alexandru C 2009 The design and optimization of a photovoltaic tracking mechanism *Proceedings of the 2nd IEEE International Conference on Power Engineering, Energy and Electrical Drives POWERENG 2009* pp 436-441

[4] Flores-Hernández D A, Palomino-Resendiz S, Lozada-Castilloc N and Luviano-Juárez A 2017 Mechatronic design and implementation of a two axes sun tracking photovoltaic system driven by a robotic sensor *Mechatronics* **47** pp 148-159

[5] Hoffmann F M, Molz R F, Kothe J V, Nara E O B, Pablo L and Tedesco C 2018 Monthly profile analysis based on a two-axis solar tracker proposal for photovoltaic panels *Renewable Energy* **115** pp 750-759
[6] Alexandru C and Tatu I N 2013 Optimal design of the solar tracker used for a photovoltaic string *Journal of Renewable and Sustainable Energy* **5** pp 1-16
[7] Çinar S M, Hocaoğlu F O and Orhun M 2014 A remotely accessible solar tracker system design *Journal of Renewable and Sustainable Energy* **6** pp 1-12
[8] Ioniță M and Alexandru C 2012 Dynamic optimization of the tracking system for a pseudo-azimuthal photovoltaic platform *Journal of Renewable and Sustainable Energy* **4** pp 1-15
[9] Vieira R G, Guerra F K, Vale M R and Araujo M M 2016 Comparative performance analysis between static solar panels and single-axis tracking system on a hot climate region near to the equator *Renewable & Sustainable Energy Reviews* **64** pp 672-681
[10] Altuzarra O, Macho E, Aginaga J and Petuya V 2015 Design of a solar tracking parallel mechanism with low energy consumption *Proceedings of the Institution of Mechanical Engineers, Part C: Journal of Mechanical Engineering Science* **229** pp 566-579
[11] Edgar R, Stachurski Z and Cochard S 2016 Optimisation direct normal insolation of rectangular PV platforms *Solar Energy* **136** pp 166-173
[12] Mauro S, Battezzato A, Biondi G and Scarzella C 2015 Design and test of a parallel kinematic solar tracker *Advances in Mechanical Engineering* **7** pp 1-16
[13] Nsengiyumva W, Chen S G, Hu L and Chen X 2018 Recent advancements and challenges in solar tracking systems (STS): A review *Renewable & Sustainable Energy Reviews* **81** pp 250-279
[14] Seme S, Stumberger B, Hadziselimovic M 2016 A novel prediction algorithm for solar angles using second derivative of the energy for photovoltaic sun tracking purposes *Solar Energy* **137** pp 201-211
[15] Sharaf Eldin S A, Abd-Elhady M S and Kandil H A 2016 Feasibility of solar tracking systems for PV panels in hot and cold regions *Renewable Energy* **85** pp 228-233
[16] Sinha S and Chandel S S 2016 Analysis of fixed tilt and sun tracking photovoltaic-micro wind based hybrid power systems *Energy Conversion and Management* **115** pp 265-275
[17] Alexandru C and Pozna, C 2009 Dynamic modeling and control of the windshield wiper mechanisms *WSEAS Transactions on Systems* **8** pp 825-834
[18] Alexandru C and Alexandru, P 2010 The virtual prototype of a mechatronic suspension system with active force control *WSEAS Transactions on Systems* **9** pp 927-936
[19] Alexandru C and Alexandru, P 2011 A comparative analysis between the vehicles' passive and active suspensions *International Journal of Mechanics* **5** pp 371-378
[20] Alexandru P, Macaveiu D and Alexandru C 2012 Design and simulation of a steering gearbox with variable transmission ratio *Proceedings of the Institution of Mechanical Engineers, Part C: Journal of Mechanical Engineering Science* **226** pp 2538-2548
[21] Bernard A 2005 Virtual engineering: methods and tools *Journal of Engineering Manufacture* **219** pp 413-421
[22] Tarniță D, Marghitu D 2013 Analysis of a hand arm system *Robotics and Computer-Integrated Manufacturing* **29**, pp 493-501
[23] Tarniță D, Catană M and Tarniță D N 2014 Contributions on the modeling and simulation of the human knee joint with applications to the robotic structures *Mechanisms and Machine Science* **20** pp 283-297
[24] Marcu A, Alexandru C and Barbu I 2018 Modeling and simulation of a dual-axis solar tracker for PV modules *Proceedings of the Xth Product Design, Robotics, Advanced Mechanical & Mechatronic Systems and Innovation Conference PRASIC 2018*

IOP Conf. Series: Materials Science and Engineering **514** (2019) 012038 doi:10.1088/1757-899X/514/1/012038

Research concerning the achievement of eco-innovation virtual hub tools to increase Romanian economic environment competitiveness

I Radulescu[1]

[1]University POLITEHNICA Bucharest, Romania

E-mail: irina.radulescu@upb.ro

Abstract. The paper presents the virtual hub experimental model achievement consisting of an innovative tools series in order to facilitate innovation, knowledge transfer and waste recycling efficiency of electrical and electronic equipment.

The Hub includes a tools set to increase recycling efficiency: the passport for electrical and electronic products recycling, the multi-criteria analysis method for the product, pointing relevant features for recycling, the implementation model of eco-innovative technologies at company level, the recyclability evaluation method based on idealism. It also includes an eco-innovation library, an eco-innovation and training laboratory, a Forum, as well as User Guides for each tool and a cost-benefit analysis guide for e-waste recycling.

The obtained tools, in their form, are an absolute novelty in the field of WEEE recycling and they address a new way of thinking and a new philosophy about material recovery and recycling and resource economy.

1. Introduction

Actual development directions to put in practice the Europe 2020 strategy are focused to an intelligent and sustainable growth based on eco-innovation, that will reduce the impacts on the environment, by "increasing resilience to environmental pressures or using natural resources more efficiently and responsibly", [1].

It is the 7th Environment Action Programme guides the European environment policy until 2020, being focused to reduce the pressure over environment and to eliminate the gap between innovation and market. The eco-innovation and green technologies represent the key to a competitive and prosperous Europe, creating new jobs in this field, by linking the natural environment to economy, the global demand for renewable energy to resource-efficient solutions, [1]. It is an important way to do research in studies and policy documents about how to develop eco – innovation capabilities and practices aiming "significant commercial potential across all economic sectors". Market developments focused on investment increase and "accelerate innovation in environmental technologies, products and services".

Helping companies and innovation service providers by delivering integrated information on eco-innovation represent useful tools for a "solid decision-making basis for policy development".

A 3-year initiative financed by the European Commission's Directorate-General for the Environment from the Competitiveness and Innovation framework Programme (CIP) is represented by the Eco-Innovation Observatory. It "is developing an integrated information source and a series of analyses on eco-innovation trends and markets, targeting business, innovation service providers, policy makers as well as researchers and analysts" by directly information on two major EU initiatives: the Environmental Technologies Action Plan (ETAP) and Europe INNOVA. The

Observatory emphasizes that eco-innovation represents a phenomenon present in all economic sectors, being relevant for all types of innovation: "Eco-innovation is the introduction of any new or significantly improved product (good or service), process, organizational change or marketing solution that reduces the use of natural resources (including materials, energy, water and land) and decreases the release of harmful substances across the whole life-cycle", [2].

The key-role of Eco-Innovation Observatory is to publish studies and reports of European Commission that broadens the area for "green sectors and activities, by collecting information on firm-level strategies". So, "eco-innovations are most widespread in the manufacturing industries, such as: electricity, gas, stream and air conditioning supply; and water supply, sewerage, waste management and remediation activities", [3]. New directions to improve resource efficiency are represented by the development of resource efficient products and services and their introduction in production and on the market. The Framework Program Horizon 2020 and technology and innovation platforms must disseminate the eco – innovative results in the European Union, by using instruments as trade fairs, market information, technology platforms and innovation partnerships.

At national level, one of the instruments that supply enterprises and SMEs to increase their competitiveness is represented by COSME Program, with its 2.5 billion euros budget for the period 2014 – 2020. Being a financial instrument, it promotes and sustains the access to finance, being a support to encourage the "entrepreneurial culture and the creation of new companies", [2].

Romanian SMEs have to face many difficulties to enrol in a sustainable development: poor information on environmental impacts and risks, lack of knowledge regarding the potential benefits of environmental management and eco-innovation, insufficient access to information or appropriate training, [3].

Taking all this into account, the paper presents the achievement of an experimental model for a virtual hub, which consists in innovative tools series in order to facilitate innovation, knowledge transfer and waste recycling efficiency of electrical and electronic equipment.

2. Provided and achieved objectives of the eco – innovation virtual hub

Research made in the national project EcoInnEWaste: „Virtual hub for eco-innovation to increase competitiveness in the field of electrical and electronic equipment waste recycling" were focused on objectives responding to economic, social and environmental challenges, in order to increase Romanian organizational competitiveness in the electrical and electronic equipment waste recycling and the involvement of public and private entities in promoting eco-innovation and development of green economy, [5].

The main provided and achieved objectives are:

- increasing the competitiveness of economic agents that are active in the field of WEEE recycling, by providing innovative tools to facilitate collaboration and knowledge transfer on eco-innovation, between universities - research institutes - authorities - business environment,
- creating a multidisciplinary research network and a sustainable partnership to lead toward a short term increasing of the institutional capacity and human resource skills devoted to applied research and innovative technologies development,
- increasing the visibility of Romanian and European research by knowledge creation and its dissemination, which will allow the creation and strengthening of a field of competence in the field of sustainable e-waste recycling, [5].

Business environment faces a lot of challenges concering the solving of main issues related to e-waste recycling, in order to recover recyclable materials, to eliminate hazardous components and to preserve in safe conditions the environment and human health.

Developing modern tools in e-waste recycling domain is a must for entrepreneurs, at cooperation level and in suppliers area. Information related to this topic must circulate and be disseminated and the network universities - research institutes - authorities - business environment should be strengthened.

Encouraging the eco – innovation and the innovators involves sustaining the idea, finding support at authorities level and in research field, implementing and disseminating obtained results, esspecially the ecological ones.

3. Development of the eco-innovation virtual hub tools. Results and discussions

In order to support SMEs to increase their management potential toward eco-friendly directions, a national research project focused on eco – innovation hub achievement, designed to offer useful business tools and to allow users access to relevant information resources. WEEE recycling and eco – innovation are up to date targets of SMEs management, being a modern way to increase their competitiveness, to facilitate the collaboration and knowledge transfer between key- actors.

During 3 years project, it was developed an experimental model of a virtual hub consisting of a series of innovative tools facilitating innovation, knowledge transfer, efficiency of the general waste recycling and electrical and electronic equipment waste (e-waste, WEEE) - in particular.

The working space of the Hub consists of four platforms, their application is done by the access of a common interface (www.hubecoinnewaste.ase.ro) – Figure 1, being possible to have a separate access, [5].

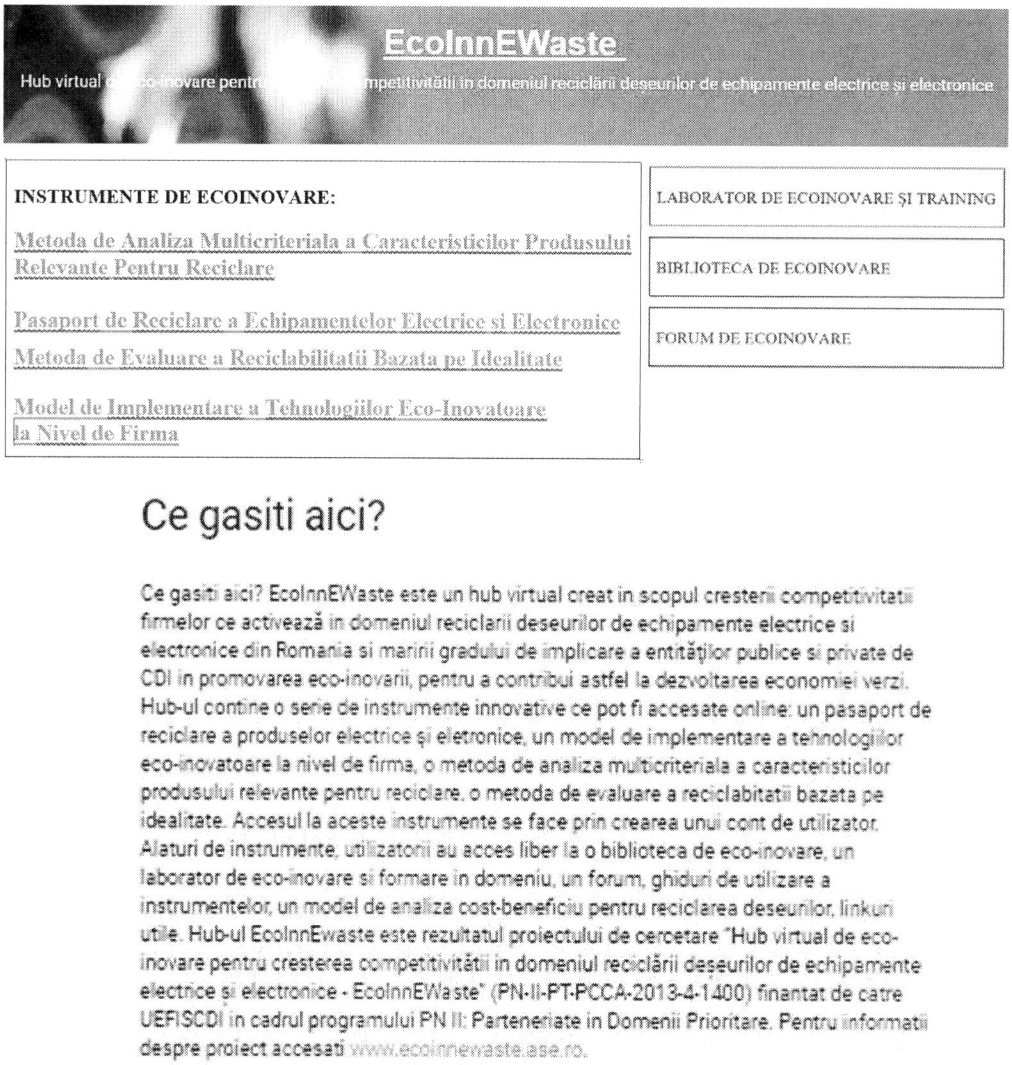

Figure 1. Common hub interface

So, there are:

- a range of tools designed to improve the efficiency of recycling and located on the platform: www.instrumente.ecoinnewaste.ase.ro – Figure 2:
  - recycling passport for electrical and electronic products, with separate sections for the manufacturer and the recycler;
  - multicriteria analysis method of product characteristics relevant for recycling;
  - implementation model of eco-innovative technologies at company level;
  - assessing method of recyclability based on ideality.
- an eco-innovation library, at the address: www.biblioteca.ecoinnewaste.ase.ro;
- an eco-innovation and training laboratory, at the address: www.portal.ecoinnewaste.ase.ro;
- a Forum, at the address: www.forum.ecoinnewaste.ase.ro.

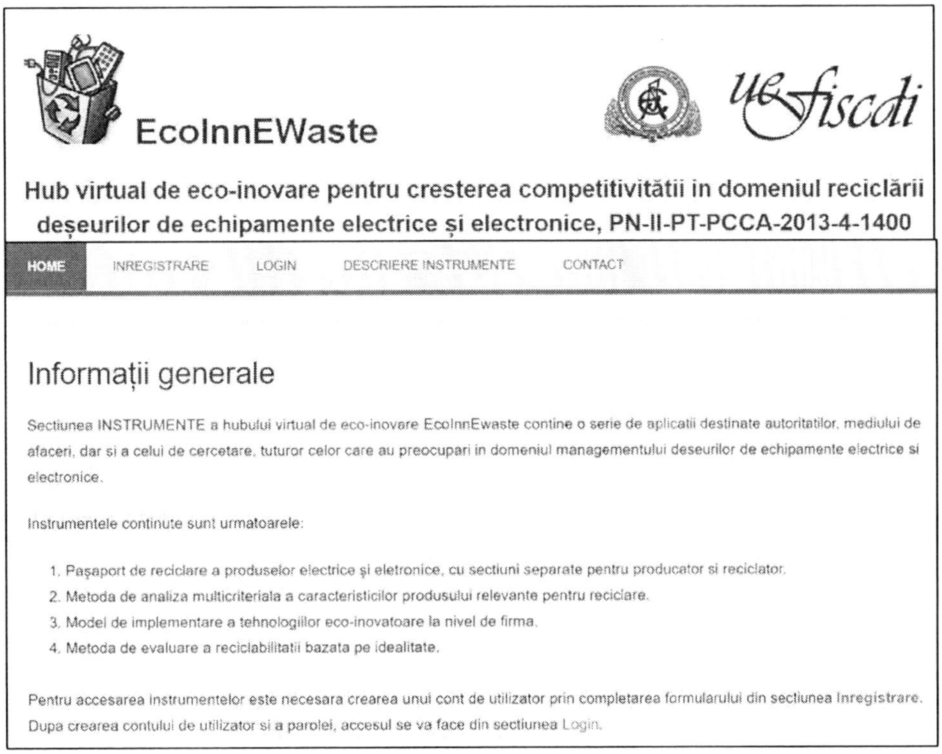

Figure 2. Instruments

To help the user on the common interface there are user guides for each tool and a Cost-Benefit Analysis Guide for e-waste recycling.

The particularity of these instruments consists of the novelty in WEEE recycling domain, bridging the innovative conceptual thinking with the innovative applicative plan, which offers detailed design of these tools, including associated software applications.

By using the *Recycling Passport for electrical and electronic products* many actors involved in recycling are aware of the materials value that can be recovered from WEEE and the resulting economic effects. The recycling passport offers the possibility to connect the user, the recycler and the manufacturer and to know information about product embedded materials and their market value. The structure of this instrument is designed to give briefly and clearly the most important and relevant information for recycling.

The recycling passport represents a special environment, linking the manufacturer and the recycler and providing to the recycling departments employees easy and fast access to necessary information for professional dismantling, re-use and sorting, [5].

Analyzing the *Multi-Criteria Method for Product Recycling Characteristics* it must be mentioned that the proposed method permits simultaneously:

- the association of some important coefficients to the classified elements;
- easy application for a big amount of elements to be classified, with minimal use of time and human involvement,
- easy application of evaluation criteria, using also their analytical hierarchy.

The method works with a a set of 19 criteria created for WEEE analysis relevant to Romanian recycling, taking into account the basic characteristics of the product categories. These criteria are proposed for traders that treat, reuse and recycle the e-waste and they have been structured into three areas: economic, environmental and technical.

Because WEEE are very diferent, there were divided into categories mentioned in the European and Romanian legislation (10 categories up to 2018 and 6 categories in 2018) and they are grouped according to the recycling technology or treatment. Users can also choose a product-level analysis, which offers the selection from a predefined list or the user may bring in products to be compared, [5].

The eco-innovative technology implementation model represents a path to eco-innovation offered to firms, helping to develop and/or bring to market new ecological technologies, products and services that reduce the overall environmental impact. Influence factors analysis can lead to the creation of sustainable solutions, by the economic use of resources and negative environmental impacts reduction.

Familiarizing the applicant with the model is done by an 6435 method example, a relatively new method, which is proposed and used in innovative product and service management, representing the basis for using the implementation model for eco - innovative technologies. It was developed an Excel application to understand how to use the method in an application example, then the user can analyze the options and chances of success, by completing with own data the proposed implementation model. The possibilities of analysis and control for the technology and for the product are the same, being specific for each one the parameters and factors involved in finding the optimal version, [5].

Another instrument is represented by the *methodology for the recyclability assessment based on idealism.* The central element of innovation is based on translating the theory of evolution into idealism in the WEEE products recycling. According to this philosophy, all systems /products evolve in the sense of materials substitution with fields, which are immaterial elements. A product is the easier to recycle as the material is smaller part. If the material is very small, even if the product is not recycled, the loss of potential reusable material is very low. An ideal recyclable product is a product made up of fields only that no longer requires recycling. This ideality theory can be used as a target in product development, [5].

The structure of *the eco-innovation library* proves its usefulness by its multi-level documentation components that enable users to access relevant information resources concerning materials recovery, WEEE recycling and eco-innovation areas. The development of eco-innovation library components were focused on all business issues, being useful for producers, users, suppliers or companies interested in WEEE issues.

The library content is divided into five modules concerning WEEE management and eco-innovation topics:

- Module 1 - Legislation, with sub-modules: National Legislation, European Union Legislation, Legislation of Other Countries;
- Module 2 - Stakeholders, with sub-modules: Companies involved in WEEE management, Organizations and NGOs, Authorities and relevant information;
- Module 3 - Studies, guides, methodologies in WEEE and eco-innovation domain, with sub-modules: Doctoral theses, dissertations; books; Guides, reports and specialist studies; Journals and conferences papers;
- Module 4 - Examples of good practices in eco-innovative technologies;
- Module 5 – Statistics.

Each module and submodule are filled with materials, their purpose is to be updated constantly from databases and to offer permanent new information. Modules interfaces have been designed to become a user-friendly tool.

Considering the descriptive documentation module side, it is designed to provide users a complex key-role template in the area of eco-innovation and WEEE management and to facilitate the inclusion in virtual learning communities that have been set up at national and international level in the e-waste management field: non-governmental organizations, forums, institutions, associations, agencies, etc. Considering the formative training module side, it is designed to ensure the learning process dynamics and training, skills development and skill accumulation for all actors of the target groups, [5].

Another direction of the project research is to create a multidisciplinary research network and a sustainable partnership, by using instruments as eco – innovation and training laboratory and the project forum. The aim is to increase the institutional capacity and human resource skills devoted to applied research, also the development of innovative technologies.

By participating in scientific and technical events in Romania and abroad there were established a series of contacts and best practices exchanges between academic and scientific environment and business environment: the University of Bucharest, Babes Bolyai University from Cluj, GreenWEEE, Remat Holding, Ecotic, Environ, Rorec. Specialists from Spain, Italy, Czech Republic, Turkey, the Netherlands, Republic of Moldova, Croatia brought knowledge in this research field, [5].

4. Conclusions

The project aims are to design useful instruments for many entities involved in e-waste recycling and to offer information to be aware of the materials value that can be recovered from WEEE, also economic effects of the recovery activity. Also, the WEEE platform of the project is a relationship tool for all those involved in recycling, bridging the users, the recyclers and the manufacturers in order to work toward sustainable development of their products, technologies and services, meaning their firms activities.

Project team partners have a good cooperation with Romanian and abroad specialists and their participation in various national and international events developed a system of relationships that is reflected in the institutional capacity increasing and human resources competencies dedicated to applied research and innovative technologies development.

By increasing the visibility of Romanian and European research by the recycling knowledge progress and its dissemination there are created and consolidated new and modern fields of interest in sustainability area.

The experimental hub and its instruments meet the EU's 7th Environment Action Programme objectives that require new technologies and approaches to business, by using innovative and sustainable ideas and contribute to transform firms in European companies more competitive, [1].

The modern thinking of WEEE platform involves legislative, utilitarian, informative, associative and formative aspects in this domain, offering easy access, interaction and information to a large number of users: citizens, potential investors, local governments, representatives of regulatory bodies, SMEs - as beneficiaries, RDI entities (universities, research institutes) or service providers on WEEE recycling.

5. References

[1] http://ec.europa.eu/environment/index_en.htm, accessed on 26.06.2018
[2] http://www.eco-innovation.eu/index.php/, accessed on 26.06.2018
[3] http://eco-inovare.ro/en/eco-inovarea-si-imm-urile, accessed on 27.06.2018
[4] European Commission, Eco-innovation observatory, Annual Report 2012, January 2013, "*Europe in transition. Paving the way to a green economy through eco-innovation*".
[5] http://www.ecoinnewaste.ase.ro/, accessed on 27.06.2018.

IOP Conf. Series: Materials Science and Engineering **514** (2019) 012039 doi:10.1088/1757-899X/514/1/012039

Command and Control vs self Management

D Nicula[1] and S S Ghimişi[2]

[1]The Bucharest University of Economic Studies, România
[2]Constantin Brâncuşi University of Târgu Jiu, România

E-mail: nicula.dana@gmail.com

Abstract. It's estimated that in 2017 there were 111 billion lines of new software code generated by developers and most lines were written within a team. Producing software is a creative activity and therefore very different from mass-production lines, because each algorithm tries to solve a different problem. For many years, traditional system development life cycle has been used as the main algorithm for how programmer teams should work to deliver software products. Waterfall methodology is successfully used in simple, unchanging projects but its main drawback is how it handles change. Scrum encourages the creativity of each member of the team, allowing all of them to take on the role they want during the project. Each person can be a developer, a tester, a designer, depending on the responsibilities he takes between two sprints. Considering the diversity of IT projects, it makes sense to understand in what kind of projects we should use the Scrum methodology and where there is a need for more explicit management and control.

1. Introduction

Organizations are trying to innovate rapidly in response to changing market conditions, customer needs, and emerging technologies. Companies live and die on their ability to discover new businesses and create ongoing value for customers. Competitive pressure is increasing, fuelled by rapid changes in technology and society. The business world is moving from treating IT as a utility that improves internal operations to using rapid software and technology-powered innovation cycles as a competitive advantage.

The traditional program and project management models' companies have used for IT are unsuited to rapid innovation cycles. However, they are deeply embedded in the way we manage everything from operations and customer service to budgeting, governance, and strategy, because most companies avoid taking risks and enjoy micromanagement and control. A software process (also known as software methodology) is a set of related activities that leads to the production of the software.

These activities may involve the development of the software from the scratch or modifying an existing system. All companies aim to optimize the process of transforming business ideas and requirements into an application that includes all the necessary functionalities, achieves the business needs and is useful and easy to use by end users.

Selecting a Software Development Life Cycle (SDLC) methodology is a challenging task for many organizations and software engineers. What tends to make it challenging is the fact that few organizations know what criteria to use in selecting a methodology to add value to the organization. Fewer still understand that a methodology might apply to more than one life cycle model. Before considering a framework for selecting a given SDLC methodology, we need to define the different types and illustrate the advantages and disadvantages of those models.

2. Traditional and Agile Development

In this article, we will explore two of the software development models and the advantages and disadvantages of each one in order to give a guidance on when and why to use them. To begin with, the waterfall approach is the earliest one and most widely known method that was used for software development.

The Waterfall Model is a linear sequential flow, in which progress is seen as flowing steadily downwards through the phases of software implementation. This means that any phase in the development process begins only if the previous phase is complete.

The waterfall approach does not define the process to go back to the previous phase to handle changes in requirement. As shown in the picture below, the sequential phases in Waterfall model are requirements analysis, system design, implementation, testing, deployment and maintenance. All possible requirements of the system that should be developed must be captured in the first phase and listed in a requirement specification document.

The requirement specifications from first phase are studied in the second phase and based on them, the system design is prepared. This system design helps in specifying hardware and system requirements and helps in defining the overall system architecture.

With inputs from the system design, the system is first developed (implemented) in small programs called units, which are integrated in the next phase. Each unit is developed and tested for its functionality, referred to as unit testing. All the units developed in the implementation phase are integrated into a system after the individual testing of each unit. Post integration the entire system is tested for any faults and failures.

Once the functional and non-functional testing is done, the product is deployed in the customer environment or released into the market. There are some issues which might come up in the client environment. To fix those issues, several patches can be released. Also, to enhance the product some newer versions with added functionalities are released.

Maintenance is done to deliver these changes in the customer environment. All these phases are cascaded to each other in which progress is seen as flowing steadily downwards through the phases. The next phase is started only after the defined set of goals are achieved for previous phase and it is signed off, so the name "Waterfall Model"(Figure 1).

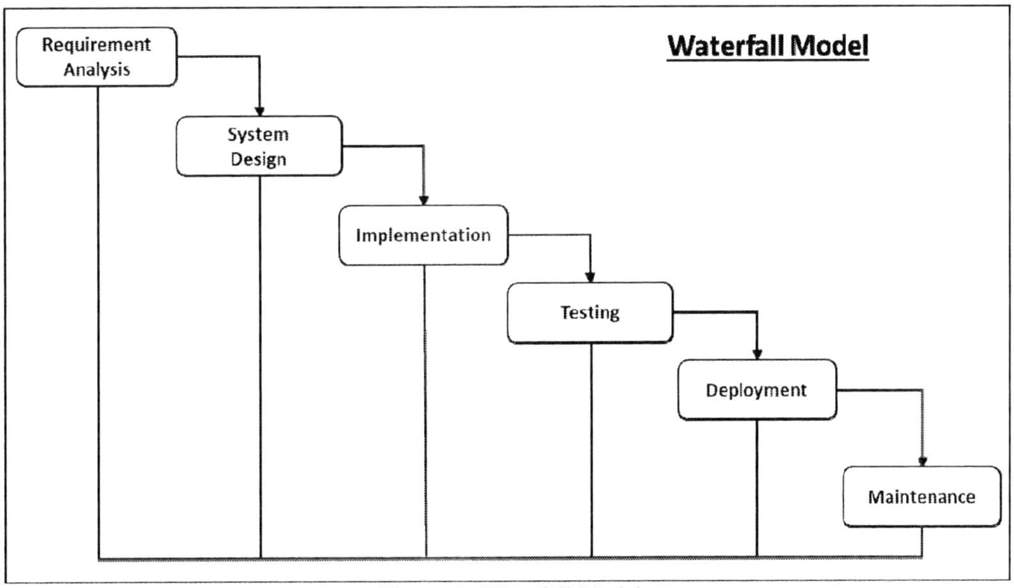

Figure 1. Waterfall Model

Since Waterfall has been widely used for software development in the last decades, we would like to emphasize that there are some situations where the use of this model might still be the most appropriate one.

For example, in short projects with stable product definition, where ample human resources with required expertise are available and where requirements are very well documented, clear and fixed, waterfall still turns out to be the appropriate methodology to use.

The disadvantage of waterfall development is that it does not allow much reflection or revision. Once an application is in the testing stage, it is very difficult to go back and change something that was not well-documented or thought upon in the concept stage. While the tools of project management, such as plans, budgets, and schedules, are necessary to control the complexity of IT initiatives, no team member was ever inspired by a Gantt chart. When projects travel beyond the realm of the known and require innovative ideas and deliverables, planning becomes less relevant and inspirational and enabling leadership becomes a key success factor.

In the past few years, the diversity and complexity of the projects has greatly increased, and a large implementation time is no longer serving in too many business situations. If development takes too long, the final solution may be outdated. Over the past 10 years, the outsourcing of software development services has been the norm. Contracts between developers and beneficiaries contained very tough clauses on how to intervene with changes to the initial requirements. Changing requirements has become a challenge both in terms of associated costs and sometimes in prolonged delivery times. For this reason, the current trend of large organizations is to give up outsourcing of project development and to hire their own developers.

Certainly, hiring developers and maintaining previous workflows could only solve one of the above-mentioned issues, namely to reduce the costs of changing requirements. The delivery term of the application can be the same regardless of whether working with a team of internal programmers or a partner firm. To reduce project risks, agile development methodologies have been adopted.

A good software engineer should have enough knowledge on how to choose the SDLC model based on the project context and the business requirements. According to Mike Cohn [1] the most suited projects for the agile process are the ones with challenging deadlines, a high degree of complexity, and a high degree of novelty for the team. If it's something the team has done before over and over then they don't need to use an agile approach. On average 65% of requirements change during software development causing waterfall projects to have an 14% worldwide success rate during the past decade. (Jim Johnson, Standish Group, 2011).

The agile methodology is a combination of iterative and incremental process models with focus on process adaptability and customer satisfaction by rapid delivery of working software product. Agile Methods break the product into small incremental builds.

These builds are provided in iterations. Each iteration typically lasts from about one to three weeks. Every iteration involves cross functional teams working simultaneously on various areas like planning, requirements analysis, design, coding, unit testing and acceptance testing. At the end of the iteration, a working product is displayed to the customer and important stakeholders. Agile is based on the adaptive software development methods, whereas the traditional SDLC models like the waterfall model is based on a predictive approach.

Predictive teams in the traditional SDLC models usually work with detailed planning and have a complete forecast of the exact tasks and features to be delivered in the next few months or during the product life cycle. Predictive methods entirely depend on the requirement analysis and planning done in the beginning of cycle. Any changes to be incorporated go through a strict change control management and prioritization. Agile uses an adaptive approach where there is no detailed planning and there is clarity on future tasks only in respect of what features need to be developed.

There is feature driven development and the team adapts to the changing product requirements dynamically. The product is tested very frequently, through the release iterations, minimizing the risk of any major failures in future. Customer Interaction is the backbone of this Agile methodology, and open communication with minimum documentation are the typical features of Agile development environment. The agile teams work in close collaboration with each other and are most often located in the same geographical location [2].

Scrum is currently one of the most studied and used agile methods, because of both its novelty and the assumption that this method is able to improve project productivity. The method uses an empirical process based on flexibility, adaptability and productivity, different than the rigid and very focused on planning way of developing with waterfall. Scrum puts an emphasis on communication and collaboration, functioning software, team self-organization, and the flexibility to adapt to emerging business realities.

It sets high-frequency management activities in order to track real-time problems during a project development process. In traditional management, there is a heavy command-and-control mentality with clearly defined roles and responsibilities for every member of the team, whereas in many agile methodologies, especially Scrum, there is no real leader. Basic collaboration mechanisms are required in order to make it work. The environment must be set up so that people have an incentive to help one another in order to optimize the whole, rather than optimizing their individual niche at the expense of the larger community. Failure to do this will prevent individuals and companies achieving the full benefits of Scrum [3].

Seen as a process, Scrum (Figure 2) takes a project and breaks it down into regular cycles referred to as Sprints. Sprints generally last between 1 and 2 weeks and involve functional teams of 5 to 9 people. Each sprint encompasses some segment of work that the overall team has agreed to accomplish. The teams that work together on each sprint are generally cross-functional and self-managing. They take a certain task that has been established in the "product backlog", similar to a list of goals for the product, and take it to a deliverable phase.

A crucial aspect of Scrum is that each sprint ends in some sort of deliverable, whether that is a formulated idea, a useful simulation, or all the way to a fully designed component. This enables iterative feedback loops in the design process and keeps progress from being made in the wrong direction.

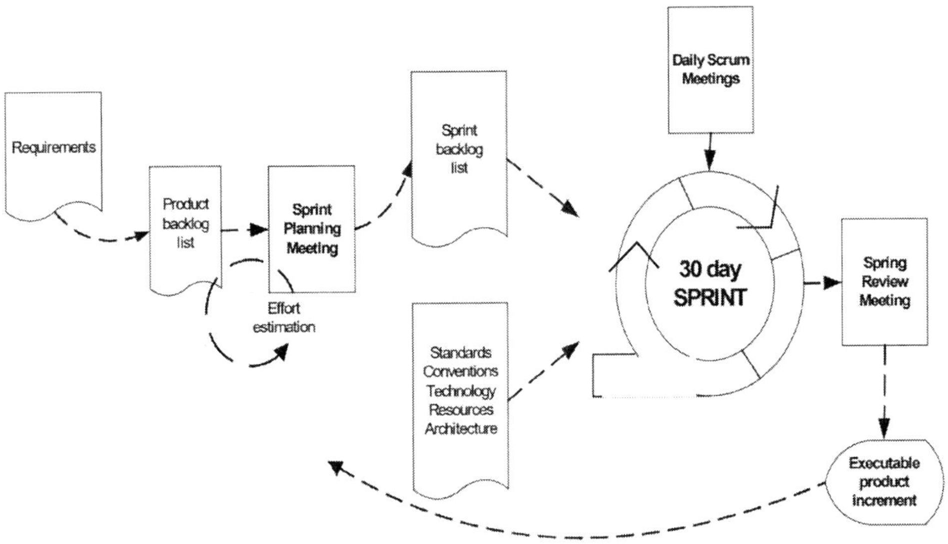

Figure 2. Scrum Model

Inside each Sprint there is a mini waterfall project since each executable product increment needs minimal requirements, design, implementation and testing. What makes Scrum different than any other methodology is the combination between the principles on which all agile methodologies are based, the limited number of roles within the team and all the ceremonies that a scrum master makes sure the team applies rigorously.

There is a document called Manifesto for Agile Software Development where twelve principles are listed. One of them says that "The best architectures, requirements, and designs emerge from self-organizing teams" [4].

Therefore, Scrum teams have the freedom to choose how to accomplish their work, rather than being directed by other stakeholders or managers. Members are empowered by the organization to structure and manage their own work. If they have the knowledge to do so, each person can be a developer, a tester, a designer, depending on the responsibilities he takes between two sprints. No one tells the development team how to turn Product Backlog into increments of potentially releasable functionality. Self-managing teams offer potential advantages over traditionally managed teams because they bring decision-making authority to the level of operational problems and increase the speed and accuracy of problem solving (Moe et al. 2009) [5].

A risk factor not to be neglected when applying the scrum methodology is related to staff fluctuations. Given that the requirements are poorly documented, changes within the team may cause the project to be affected.

Also, for the good progress of the project, it is important for team members to work well together and to respect each other. There can be a conflict between the traditional, tool-and-process focused style of project managers and the agile approach. Many project managers have difficulty stepping away from the classical techniques, such as scope, budget, and schedule planning, and migrating into a world of changing requirements, evolving expectations, and self-directed teams.

One of the success stories for the implementation of Scrum, comes from Spotify. They have grown from 30 to 250 employees three years after adopting the agile methodology, reporting a huge increase in employee satisfaction. According to Statista [6], their revenue also increased from 1 billion euros in 2013 to 4 billion euros in 2017. The engineering team from Spotify took advantage of the flexibility and experimentation that Agile encourages and decided to rewrite some rules that apparently prevented them from operating more effectively. They have created their own agile methodology, which was strongly influenced by the values of their company. One of the first changes they implemented was in the terminology used.

The terms Scrum Master and Teams were replaced with Agile Coach and Squads, which consisted of no more than eight people who were given full autonomy to operate as a cross-functional and self-organising unit, having the ability to decide what to build, how to build it and how to work together while building it.

People at Spotify tried to create a culture where employees mutually respect themselves and share credit for their achievements so that each person can feel like a valued member of the team.

Finally, and maybe the most important change of all, is the fact that Spotify encourages finding ways to ensure fast failure recovery rather than simply avoid failing at all. Some squads even use a fail wall where they publicly share their failures and whenever there is a failure, the squad conducts retrospectives which allows it to capture any learnings from the experience. Besides removing fear, experimentation also promotes a data-driven approach to taking decisions, rather than having decisions taken on the basis of personal opinion, ego or authority.

In the authors' opinion, Scrum application brings a large number of benefits to employees, managing to get them to leave the paradigm of executants and to become idea generators. It most likely encourages people to learn new things and allows them to have less aversion to risk since failure is seen as a learning experience. If the business and technical requirements received from hierarchical superiors are incomplete or even erroneous, the agile workflow allows the team to challenge the proposed solutions to a greater extent and even to create alternative ones. When applying Waterfall, the large number of requirements received simultaneously is overwhelming and the focus of the development team can be to write the code rather than to consider end user experience. A smaller number of requirements can be examined more carefully compared to hundreds of requirements that are simultaneously sent on the way to development.

The evolution from traditional project management to agile techniques requires an evolution for the project management as well, from authoritarian, top-down management to collaborative, influence-based leadership. The ability to migrate from a world of tightly-bounded scope and schedules to a world of change, vagueness, and uncertainty is the key distinguishing factor for successful agile projects and project leaders.

Every agile expert agrees that agile projects are not for novices. Process has its place in bringing order from the chaos of complex IT projects but creating the right atmosphere and staffing it with the right contributors is the foundation of a high-achieving agile team [7].

3. Conclusion

This article is only a starting point for our study. Anecdotal evidence suggests that agile methods are effective and suitable for many situations and environments. However, at present very few empirically validated studies can be found to support these claims. The existing evidence consists mainly of success stories from practicing professionals. Although these provide valuable information about practical applications, empirical studies are urgently needed for evaluating the effectiveness and the possibilities of using agile software development methods, especially in very large projects.

The goal of this type of research is to enable software professionals, projects and organizations to choose and to apply the right method at the right time and to take into consideration human motivation as well.

4. References

[1] Cohn M, 2011Deciding what kind of projects are most suited for agile *Montain goat software* https://www.mountaingoatsoftware.com/blog/deciding-what-kind-of-projects-are-most-suited-for-agile.

[2] Dingsøyr T, Moe N B, Fægri, T E and Seim E A 2018 Exploring software development at the very large-scale: a revelatory case study and research agenda for agile method adaptation *Empirical Software Engineering* **23** pp 490-520 https://doi.org/10.1007/s10664-017-9524-2

[3] James M, Walter L 2010 *Scrum Reference Card*, http://scrummethodology.com/

[4] http://agilemanifesto.org/

[5] Moe N B, Dingsøyr T, Dybå T 2009 Overcoming barriers to self-Management in Software Teams. *IEEE Softw* **26** pp 20–26

[6] Felix Richter 2018 - Spotify's Losses Widen as Royalty Costs Pile Up https://www.statista.com/chart/4894/spotify-revenue-vs-costs/

[7] Freedman R 2009 Agile teams focus on the people rather than the process *Tech decision market* https://www.techrepublic.com/blog/tech-decision-maker/agile-teams-focus-on-the-people-rather-than-the-process/

Some aspects of quality and risk management for natural gaz meters

P E Ilea[1], D Săvescu[1] and A Stoica[2]

[1]Product Design, Mechatronics and Environment Department, Transilvania University of Braşov, Romania
[2]Engineering Department, Lucian Blaga University of Sibiu, Romania

E-mail: paul.ilea@unitbv.ro

Abstract. The following paper refers to some general aspects regarding the measurement of natural gas. There are also presented the unauthorized intervention methods in natural gas distribution system and implicitly the fraudulent consumption of natural gas. It has been proposed to identify, analyze and respond to the risks involved in the measurement process, and to create a new natural gas meter by proposing an internal procedure set to prevent and reduce as much as possible unauthorized distribution. The measuring device is capable of preventing and detecting in real time unauthorized use of the distribution system.All these actions lead to changes in the competitive management process of the trading company involved in the distribution of natural gas

1. Introduction

Natural gas is a flammable gas, usually insipid and odorless, which can be found in the form of a deposit in the depths of the earth.

The science of measurements is based on the fact that the process of knowing the surrounding world relies on the experiment, that is, qualitative and / or quantitative assessment.

The strong development of all branches of science and technology has made the creation of new means and methods of measurement necessary.

Measurement is at present an essential component of any human activity, as it provides the quantitative information needed to make decisions about what needs to be done, as well as information on the results of the work done.

In industrial activity, ensuring quality control of products requires permanent measurements, both on all materials used in production and on the benchmarks following each operation and on the finished products themselves.

In the operation of exploring complex installations, for monitoring of operating parameters and for the inspection of performance maintenance, as well as to ensure the safety of the installations and the protection of the work, permanent or periodical repairs are undertaken.

Tracking consumption of raw materials, energy and fuel and taking saving measures requires the use of measuring instruments. Trade towards the population, business-to-business transactions and management of stocks can only be achieved on the basis of increasingly complex measurements.

Analyzing the process of distribution and measurement of natural gas consumption results in unauthorized, fraudulent interventions of some people with training, these interventions presenting special risks, some related to the safety of the functioning of the installations the health of people exposed to intoxications and / or explosions.

That's why we need to develop high performance equipment to minimize these risks and also through a high performance management, to elaborate a set of procedures involving all the factors, all of which lead, first of all, to the safety of the functioning of natural gas installations and, implicitly, of the people.

2. Unauthorized intervention methods on gas meters

During the inspection of the maintenance and operation of the distribution system, various types of fraudulent intervention methods have been identified: damage to the meter mechanism or its replacement, blocking of the recording mechanism , punching through the outlet of the metering chamber, measuring through the input device of the meter, opening in the cover of the recording mechanism, reversing of the meter device (counter-clockwise), counterfeit metrological marks, damage to the metrology marks, etc.

In an absolutely natural way, only a few of these interventions are presented in the paper, taking into account the ingenuity of those who, consciously, bring "contributions" to the subsequent development of the meters. This way, the closed loop of the supplier and the consumer is dynamic, the supplier must be constantly alert to unauthorized interventions. This is done by creating simple procedures, some of a general nature, others more specific, capable of being constantly in line with market requirements. Therefore, not only the performance design of gas meters should concern the attention of natural gas distributors, but also the creation of a legal, rapid intervention systems for the detection of some offenses.

2.1. Unauthorized intervention by locking the gear lever of the recording mechanism (meter)

A frequent intervention at the consumer is made by inserting some foreign bodies through the inlet of the measuring device to lock the lever that engages the recording mechanism.

Throughout the process, the natural gas meter has been completely blocked.

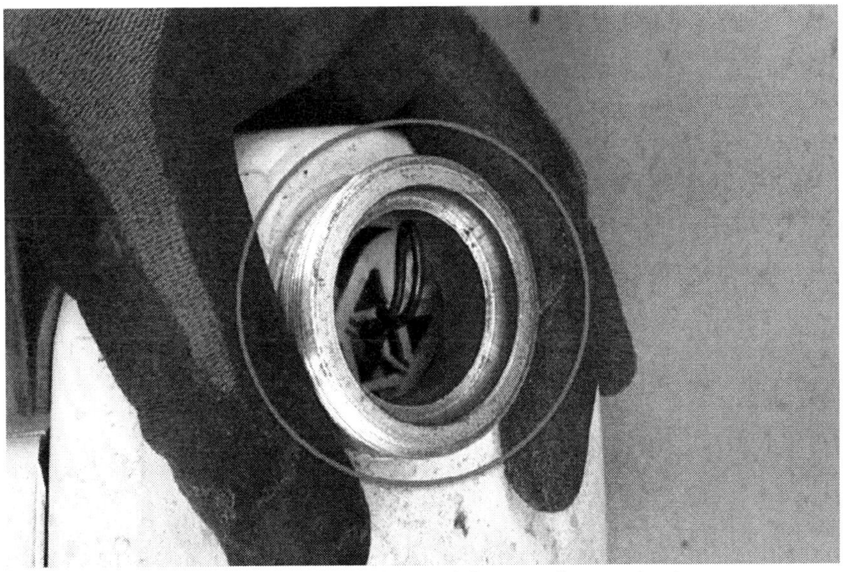

Figure 1. Locking the engagement lever of the recording mechanism

This is visible at present only when reading consumption, nonexistent to the consumer, although there is physical activity. Hence the need to create a system that quickly highlights any blocking of the recorder, when the device is in use and the recorder is stationary.

2.2. Unauthorized intervention by reverse installation of the meter device

Another unauthorized intervention commonly encountered in "practice" is the reverse installation of the meter, facing the wall (the inlet / outlet connections are reversed), having the effect of reversing the gas in the measuring apparatus and, implicitly, a negative recording of the consumption.

Figure 2. Reverse installation of the meter device

Such an intervention, as shown in Figure 2, is visible, so through a visual inspection can be detected and take counter-measures. Here, in this case, procedures are required for inspecting and ascertaining the state of functioning of the consumers. But also by designing a system for measuring pressure and direction of gas through pipelines and connecting it to a centralized recording system, fraud can quickly be detected.

2.3. Unauthorized intervention by executing a hole into the measuring chamber through the inlet of the gas meter

This method influences the correct measurement of natural gas consumption and implicitly causes a jerky / uneven operation of the measuring apparatus. It consists in making a hole in the measuring chamber through the inlet of the measuring tool. And in this case, a procedure leading to the monitoring of consumption at the point of distribution, which would allow a quick intervention, would lead to a reduction of the breaches due to fraud or the losses of the natural gas supplier by recording the correct consumption.

IOP Conf. Series: Materials Science and Engineering **514** (2019) 012040 doi:10.1088/1757-899X/514/1/012040

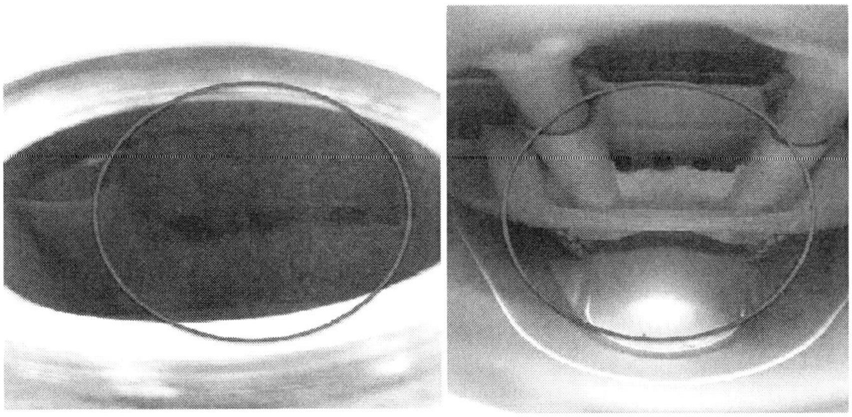

Figure 3. Hole in the measuring chamber through the inlet connection

2.4. Unauthorized intervention on the integrator caps and fasteners

Removing the protective plugs and mechanical fastener bolts is a frequent intervention.

By doing this, the measurement becomes uneven, due to piling, cutting of the pins or due to the replacement of some parts of the integrator.

The mechanical index of the measuring device can be changed (backwards),through direct access inside the meter.

Figure 4. Detach the integrator

In this case, the intervention of the consumer is easily visible, and through the periodic inspection the losses can be reduced.

3.Procedures

Natural gas metering devices are built on quality and technical norms that must allow accurate measurement of natural gas quantities.

From the above, it can be seen that there were interventions of all types on the measuring instruments, this being forbidden by law.

Natural gas theft is a crime and is punished according to law number 123 of 2012.

The desired quality of the metering device requires a risk analysis to identify existing or potential problems.

Risk management aims to manage threats that could have a negative impact on the company.

Thus, it is proposed to create a measuring apparatus that highlights the actual consumption and alerts the supplier quickly about unauthorized interventions on the measuring devices.

In order to improve the maintenance and designed activity of natural gas metering devices, the organizational management proposes to develop a set of internal procedures that prevent and limit as much as possible the unauthorized consumption of the distribution system, as follows:

3.1.For unauthorized interference by locking the engagement lever, by execution of an orifice in the measuring chamber through the inlet / outlet connection, by-pass is proposed:

- Installing sensors on the inlet and outlet connections so that a silent alarm is triggered when the meter is removed from the installation;
- The possibility of real-time detection of unauthorized use of natural gas and its forwarding to the reading compartment;
- The possibility of fast intervention in the place where the unauthorized consumption occurs;
- Development of a procedure where communication between departments is optimal and fast;
- Developing a collaboration procedure between the company and the Police.

3.2.For unauthorized intervention by reverse installation of the meter device, it is proposed:

- Mounting a gasket locking system to the reverse flow of the gas together with a throttle;
- Installing sensors on the inlet and outlet connections so that when the measuring device is being dismantled from the installation, a silent alarm will be triggered and the alarm can only be stopped by authorized distribution company personnel;
- The possibility of real-time detection of unauthorized use of natural gas and its transmission to the reading compartment;
- The possibility of rapid intervention in the place where the unauthorized consumption occurs;
- Development of a procedure where communication between departments is optimal and fast;
- Developing a collaboration procedure between the company and the Police.

3.3.For unauthorized intervention on protective plugs and fixing bolts, falsification / deterioration of metrological batches, it is proposed:

- Construction of a natural gas meter that has the integrated recording mechanism in a metal housing/casing, without the possibility to reach the internal elements of the integrator, by falsifying / damaging the metrology marks;
- The possibility of real-time detection of unauthorized use of natural gas and transmission to the reading compartment;
- The possibility of rapid intervention at the place where unauthorized consumption occurs;
- Development of a procedure where communication between departments is optimal and fast;
- Developing a collaboration procedure between the company and the Police.

3.4.For unauthorized intervention by the creating of an opening in the cover of the recorder mechanism and the insertion of a needle / wire, aiming to stop the measuring process, it is proposed to:

- Installing a pressure sensor inside the integrator that triggers a silent alarm when the hole is being drilled;

- Construction of a natural gas meter that has the integrated recording mechanism in a metal housing/casing, without the possibility to reach the internal elements of the integrator, by falsifying / damaging the metrology marks;
- The possibility of real-time detection of unauthorized use of natural gas and transmission to the reading compartment;
- The possibility of rapid intervention at the place where unauthorized consumption occurs;
- Development of a procedure where communication between departments is optimal and fast;
- Developing a collaboration procedure between the company and the Police.

4.Conclusions

As a result of this analyses , one can notice that there are many methods of unauthorized intervention on natural gas meters.

Thus, a process of identifying, analyzing and responding to the risks in the natural gas measurement process has been proposed.

It is proposed to innovate a new gas meter, focusing on the quality of the new product and its ability to prevent unauthorized use.

The new measuring device is capable of preventing and detecting in real time unauthorized use of the gas distribution system.

The advantages of improving the measuring device are as follows: reducing natural gas losses, reducing unauthorized consumption, an easier controll of the measurement / reading / billing process, increasing the revenue of the distribution company, credibility, improving the image of the company producing measuring devices, differentiation from competition, facilitation of bidding and last but not least safe exploitation and increased customer satisfaction.

5.References

[1] Stoica A, Feier D, *Aparate de măsurăşi control înindustriagazelornaturale* 2016 Editura„Universității Lucian Blaga" din Sibiu, Sibiu, România

[2] Neagu I, Constantin M, Ciocirlea-Vasilescu A 2007 *Măsuratori ş ilegislaţie metrologică.* Editura Cvasidocumentaţia Proser & Printech, Bucureşti, România

[3] Ilea P, Săvescu D, Stoica A,Unele aspecte privind măsurarea şi reglarea corectă a consumului de gaze naturale, Buletinul AGIR, An XXIII, Nr.3/2018, În curs de publicare.

[4] Săvescu D, *Some aspects regarding the concept „Research and Business".* Journal of Industrial Engineering and Management, JIEM, 2010, Vol. 3 nr.2, pag.337-352.

[5] Săvescu D, *Aspects of Risks Issues in Entrepreneurship. Part I and II,* 6th International Conference in Applied Social Science, ICASS 2017, 7-8 May, Singapore, vol. 97, pp. 217-228, IERI&PRESS USA

[6] Săvescu D, Managementul proprietăţii intelectuale 2017 Ed. LUXLIBRIS, Braşov

IOP Conf. Series: Materials Science and Engineering **514** (2019) 012041 doi:10.1088/1757-899X/514/1/012041

Aspects of the technological transfer management for capitalizing on the results obtained through scientific research

C Torcătoru[1]

[1]Product Design, Mechatronics and Environment Department, Transilvania University of Braşov, Romania

E-mail: constantin.torcatoru@unitbv.ro

Abstract. This paper attempts to provide answers to the problems currently facing higher technical education and research, focusing on technology transfer system. To this end, is defined the concept of innovation and technology transfer and are presented some aspects of the technology transfer mechanisms. Some visions on transfer management are detailed, and several procedures are propose to improve technology transfer between educational institutions and socio-economic environments. It should be borne in mind that there are problems with the students' novelty, diploma and dissertation papers. Therefore, university policies on intellectual property knowledge should include the acquisition of knowledge in the last years of study.

1. Introduction

The paper attempts to provide an answer regarding the amount of the results obtained from scientific research, except for those from the Research Centres. Those results stemming from the "creative fruit" of students who receive various topics and projects to solve during the study years, especially the results of the final years, these being on the studied or researched specialty. Depending on the theme received or chosen by the students themselves, in most cases, their results may have a market applicability, more or less with a major impact on the studied / researched field.

Often, the creator of this result is not aware of or knows the value that those results might give it by implementing the solution or idea in everyday life.

Therefore, the present paper tries to find an answer in capitalizing these results, focusing on the technology transfer system.

2. Technology transfer (TT)

People with innovative spirit are the determining factor in hiring technology for development. They are the ones who can facilitate both technical and social development, are those who can provide the relation between research and development, industry and governmental decision makers. Engineers, as well as engineering research, can make a significant contribution to providing benefits, assume and address development risks, design technological solutions to support public policy oriented to general development.

Innovation is the process of creation (generating new ideas) followed by making changes generated by it [1].

Innovation is define almost uniquely by literature [2] as an idea, a concept, a process for designing, operationalizing and experimenting with a new product model, process or new functional structure for industrial application. So, innovation compared with the invention, is not necessarily an absolute novelty, but the scope and speed of implementation are more accessible.

TT is the introduction or acquisition in the economic circuit of specific technologies and equipment, equipment and facilities resulting from research, in order to obtain new or improved processes, products or services demanded by the market, or by adopting innovative behaviour , to disseminate information, to explain, to transfer knowledge, to advise and to communicate with people who are not experts on the fundamental, applicative and pre-competitive research results, provided there is an owner of the results [1].

It can be said that the technological transfer under the transfer of manufacturing technology involves a series of formal and informal relations between R & D units and public and private economic sectors.

Thus, we can mention two general acceptance of this technology transfer:

- The assignment of an intellectual property that has been generated and developed in one place by legal means such as licensing technology;
- The process by which scientific and technological breakthroughs are converted into marketable goods and services;

The purpose of technology transfer is to strengthen the economy on a territory by accelerating the application of new technologies and resources to meet the needs and opportunities of the public and private sector.

Technology transfer is the process by which knowledge, facility or capacity of existing publicly funded research - development, are use in order to meet the needs of public and private [2].

In general, the technology transfer process involves three entities (Figure 1), which may be in a cooperative relationship.

Figure 1. Process of technology transfer

These three entities involved in technology transfer have as their primary objective the acceleration of the economic use of research results, involving the conversion from invention to innovation and successful dissemination on the market, adding value.

3. Mechanisms of technology transfer

These technology transfer mechanisms are those operations that ensure the dissemination of a particular technology from a supplier to a beneficiary. Typically, these processes can be in various forms, such as financial, technological or human, and consist of various procedures, ranging from active forms to passive forms.

These technological transfer mechanisms can be divided in different categories, as shown in Table 1.

Table 1. Technology transfer mechanisms

No.	Category	Subcategory
1	Consultancy groups	Review groups of the end user
		Technical review groups
2	Collaboration with participation for costs	Industrial consortium
		Research – Development in partnership
		Demonstration projects
		User facilities

IOP Conf. Series: Materials Science and Engineering **514** (2019) 012041 doi:10.1088/1757-899X/514/1/012041

3	Collaboration without participation for costs	Research – Development contracts
4	Staff exchanges	Works for others
		Personal Consulting
		Personally invited
		Transfer of personal
5	Licenses/spin off	Licensing
		Companies spin-off
		Scientific and technological parks
		Technological business incubators
6	Active dissemination of information	Brokerage organizations
		Conferences, seminars, workshops
		Technology Transfer Centres
		Education
7	Passive dissemination of information	Post correspondence, inclusive e-mail
		Technical reports
		Journal articles
		Video – records
		Newsletters

4. The vision of technology transfer management

Technologies are moving and evolving continuously in a wide variety of ways. A "steady" technology is worthless. Unused technologies cannot meet needs and can not produce benefits. Technological transfer is fundamental to the maturation and growth of many types of social institutions, including business, academia, military, government, etc. Without the technological transfer, any institution would not be able to sustain the pace of change.

A broader vision of technology transfer is based on seven analysis directions [3].

In this sense, each analysis directive can be associated with a question:

- Why technology transfer?

As with any other investment in the economic environment, the transfer must meet corporate objectives.

Lundquist has identified a number of technological transfer motives [4], among which we can mention: setting up new companies, new products, leadership, advanced technology, economic development, new business promotion, reputation / image, risk reduction, etc.

From a practical perspective, technologies are being transferred to solve some problems and to create well-being.

- Who are transferring?

The source and the adapter are the ones who carry out the transfer, sometimes the sources call the transfer offices to facilitate the process, sometimes adopters designate a specialist to look for useful technologies. The efficiency of technology transfer is determined by the abilities of change agents and staff.

- Where does the technology transfer happen?
  Technological transfer takes place across the value chain of product development.
- When does the transfer take place?

Transfer takes place when the technology moves along the steps in the value chain direction. In terms of source adopter, transfer takes place when certain conditions are met such as property supply, preparation adopter connection, agreement, trust, etc.

- What is the significance of technology?

The integrative vision is grounded in a collaborative process involving decisions on operations, product development, brand development and sales.

- What is the justification for the costs of technology development and technology transfer?

For and between each stage of the value chain, analysis of opportunities for investment in technology. The management of companies, laboratories and funding organizations has financial responsibilities for the use of resources in accordance with the intended purposes.

- How does technology transfer work?

The transfer can take place at the moment of its justification, being involved in three distinct stages: contacting the relationship between the source and the adopter, the agreement documented through a contract that ensures reciprocal gains and the effective movement of technology from source to adopter.

5. Technological transfer, from research institutions to socio-economic environments

Universities, colleges and vocational training centres are key points in the innovation system to attract and produce the human capital needed for innovation.

Barriers and regulations that would limit effective interaction between universities and firms should be improved as collaborative arrangements that facilitate networking. It is essential to ensure that higher education institutions offer opportunities and incentives to collaborate with each other and later with industry.

Criteria for the assessment of the performance of the study should be adjusted to reflect the multiple missions of the educational institutions, including the transfer of knowledge. To achieve this goal are essential expectations well defined and highly qualified personnel for technology transfer.

It can be said that there is an openness and a proactive attitude with concrete positive results in the development and enhancement of partnerships between universities, research and decision-making environment. At the same time, the results obtained are relatively few, compared with the needs and especially the existing opportunities [6].

There is an increase in the number of higher education graduates in Romania, but it has failed to generate a significant difference in innovation among the local economic agents. This can lead us to the conclusion that higher education in our country suffers from several fundamental chapters.

One of the key issues in this direction is the absence of extensive and profound collaboration between academia and industry.

Developing partnerships with external partners, professionalizing knowledge transfer processes and seeking synergies between regional, national, and European policy initiatives lead to promoting the capacity of universities to innovate and to improve research and learning within them.

Also, education and training play a very important role in innovation, helping companies to make various changes in production processes and adopt new technologies by improving their innovation capacity. Skills acquisition is a long process that does not end with formal education. Educational institutions provide the basis for continuous learning, but continuing skills acquisition must be encouraged, and this may involve the recognition of all forms of learning, including those through workplace qualification systems [7]. Therefore, rewarding lifelong learning and its attractiveness can help increase participation, with entrepreneurs playing an important role in innovation by helping them to transform ideas into commercial applications.

Should be implemented and carried out, both by researchers and teachers from universities, as many activities start-up and spin-off, the idea of ensuring access to research equipment in environments assistance and support with specialized staff, their partnerships with the university, in order to determine and capitalize on know - how.

We must put more emphasis on developing entrepreneurial skills and technology transfer of educational institutions have in defining their purpose in the near future as entrepreneurial universities.

It should be taken into account that to construct and operationalize a communication platform between the two areas, namely academic economics, which accelerate interaction between the two areas, including the establishment of joint teams for research, development, innovation and technology transfer [8].

The challenge for Romania is that universities and research and development units to act as key generators of IP assets.

Factors involved in this IP activity:

- Teachers and researchers;
- Doctoral students, students in their final years;
- Intellectual property offices, regional offices /area;
- National Council of Small and Medium Enterprises (CNIMMC) and businesses;
- Chambers of Commerce and Industry;
- Units of technology transfer (TT);
- Sponsors etc.

To achieve the purpose of creating capitalization should its updated operating systems, such as:

- Systems that provide industrial property culture (education, best practices etc)
- Evaluation systems (attributions, licenses, procedures, methods etc)
- Capitalization systems (rules, patent department etc) etc.

6. Conclusions

Can be considered that the relationship between the university and the socio-economic environment should be closely correlated so that the valorisation of the results obtained by the students has a maximum percentage, and the applicability of these results contributes to the improvement and evolution of the society. The interface between the university and the socio-economic environment needs to be updated periodically, so that this technology transfer process is known and understood by those who bring the results.

Increasing product competitiveness can be achieved by:

- Transferable technologies.
- Promoting innovation culture.
- Patents.
- Research results transferred to industry.
- Personnel trained and accredited (TT and innovation, technology broker).
- Innovative SMEs.
- Promoting an innovation culture etc.

Universities need to become more active actors in the knowledge economy, capable of responding effectively to market demands. The themes and topics discussed within these projects must be topical, so that the applicability percentage is as large as possible later.

The proposal to invite students to take part in the problems encountered in industrial environments, would make them look for and create solutions to solve those problems they have noticed, then having a fascinating result of the project, worthy to be rewarded and redeemed much easier.

The research is a priority for universities and the innovative research and development institutions, therefore, the human factor in these organizations should be trainee in knowing how to capitalize on the results obtained through scientific research.

To improve the education, stimulation and awareness of university studies, on the IP area, more emphasis should be place on seminars, courses, IP-themes, invitations of recognized personalities, etc.

7. References

[1] Savescu, D., *Proprietate Intelectuala,* Editura Risoprint, ISBN 978-973-53-1527-6, Cluj-Napoca, 2015.

[2] Pleşa, O., Ciotea, Fl., Naum, N. *Inovarea şi sfidările schimbării.* Editura Multimedia, ISBN 973-96197-8-9, Târgu Mureş, 1996.

[3] Rânea, C., Badea C. Dan (2003) - *Bazele Inovării şi Transferului de Tehnologie,* Editura ELECTRA, Bucureşti, 2003, ISBN 973-806780-4. Editura ELECTRA este recunoscută CNCSIS – Lista 2002, poz. 54

[4] Gary Lundquist, *A Rich Vision of Technology Transfer Technology Value Management,* Journal of Technology Transfer, 28, 265–284, 2003, Kluwer Academic Publishers]

[5] Kerzner, H., Project Management: *A Systems Approach to Planning, Scheduling, and Controlling,* Eighth Edition, John Wiley&Sons Inc., USA, 2003

[6] Tran, T.A. & Kocaoglu, D.F. (2009). *Literature Review on Technology Transfer from Government Laboratories to Industry.* Proceedings: Portland International Conference on Management of Engineering &Technology (PICMET 2009), pp 2771 – 2782

[7] Jonas Mitasiunas (2010), Innovation and Technology Transfer, *BONITA -Baltic Sea Region Programme 2007-2013,* Vilnius University.

[8] SAVESCU, D., *Aspects Regarding the Relation Between Intellectual Property and Technological Transfer,* 11th Edition of International Conference of the Carpathian Euro Region Specialists in Industrial Systems, CEurSIS 2016, 2-4 June, 2016, Baia Mare, pag. 144-149, ISBN 978-606-737-166-6

[9] * * United Kingdom Patent and Trademark Office – *Intellectual Property Patents; Trademarks; Designs; Copyright,* http://www.ipo.gov.uk.

[10] * * Portal Enterprise Europe Network – *Technology Transfer.* http://portal. enterpriseeuropenetwork, ec.europa.eu/services/technologytransfer

Author addresses

[1]TORCĂTORU, Constantin, Ph.D. Eng., University Transilvania of Braşov, B-dul Eroilor, Nr. 29, 0268412921 – ext.182, constantin.torcatoru@unitbv.ro

IOP Conf. Series: Materials Science and Engineering **514** (2019) 012042 doi:10.1088/1757-899X/514/1/012042

Some aspects regarding negotiation in business

D Săvescu[1]

[1]Product Design, Mechatronics and Environment Department, Transilvania University of Braşov, Romania

E-mail: dsavescu@unitbv.ro

Abstract. Paper presents some aspects regarding negotiation process used in technological transfer, negotiations strategies, partners orientation in negotiation time, WIN-WIN, or WIN-LOOSE, or LOOSE-LOOSE position, tactics and technics used in negotiation.

1. Introduction

Negotiation is the most important part of the technology transfer process. Negotiations between partners to grant a technology license are usually complex because it involves many basic conditions and each of them can generate different positions from those more advantageous to the most uninteresting [3].

The negotiator has the difficult task of keeping in mind the basic conditions and all the different positions, to deal with the technical aspects of the project and to constantly assess how the basic conditions can affect the commercial objectives of the license. As a rule, the negotiator has to fight for the most advantageous position, with the two components: offensive or compromise position.

It is preferable to draw up a list of internal conditions, questions on the subject and the conditions in which the negotiations are to be negotiated. When starting a negotiation, it always has to start from the idea that the other team is just as determined and competent.

The ability to analyze and appeal permanently to the relationship between the basic terms of the contract and the objectives proposed is the one that will determine the success of the negotiation for three reasons:

- The position is known; the permissible bargaining margin and the possibilities of compromise are known.
- Success in negotiations means being able to explain to the other side what its needs and objectives are in a clear and convincing way and, last but not least, at the right time.
- Discussions are conducted in a calm tone, with competence and self-confidence. Aggressive behavior clearly leads to failure and not to top-level discussion.

2. Negotiation strategies

Concerning the negotiation strategy, the negotiator has several choices:

- Integrative or distributive orientation;
- offensive or defensive orientation;
- taxing or adaptation behavior;
- between the widening of the bargaining field and its narrowing;
- between a total agreement and a partial agreement;

- between an immediate one or after a certain amount of time.

Choosing an integrative or distributive strategy is of the utmost importance, as the others are subsumed.

Gain: those who adopt such a mentality will not necessarily lose others; they do not follow what happens to the other side, this is secondary, irrelevant. It is essential for them to achieve their purpose, to achieve what they are pursuing. In the absence of rivalry and competition, gain orientation is probably the most common approach in everyday negotiations. The person with this mentality is thinking of securing their own interests and letting others take care of their own.

Integrative orientation has the following variants:

- win - win,
- winning or non-trading technique.

Win-win is a state that the benefit in any mutual interrelation is mutual. Understanding and solutions are mutually beneficial, mutually satisfactory. In this type of negotiation, the negotiator recognizes the other party as a legitimate partner, who is not challenged the right to defend their own interests, even if they are contrary to their own. An essential interest in win-win integrative negotiation is that the agreement, once found and concluded, is likely to be respected by the parties because it was obtained without resorting to the pressure of the force ratio [1].

Winning or non-trading technique is sometimes considered to be superior to win-win. Non-marketing essentially refers to the case where, in the absence of mutually beneficial solutions, it is mutually agreed not to conclude a business. No expectations have been created or contractual conditions have been established. The parties do not engage in collaboration because their values and objectives are divergent. The win-win or non-trading approach appears to be realistic at the start of a business relationship, but in the long run, it could become a non-viable option, creating difficulties in affiliated businesses because in the effort to maintain friendship relationships, many make a compromise. This is how serious problems can arise, especially if the competition operates on the win-win principle.

Distributive orientation variants are:

- win - loss;
- loss - win;
- loss - loss.

The *win-loss* orientation has the slogan: if I win, you lose. Those who adopt such guidance prevail in order to impose their will on their influential, power, accreditation, wealth, personal ascendant position, etc. For the attitude of such an attitude, conflict is not a possible source of progress, but an obstacle to be removed from the path they have drawn. In terms of game theory, the gain-loss negotiation corresponds to a zero-sum game, that is, everything that someone gains, loses someone else. The negotiator who opts for this orientation has the objective of finding or even imposing a solution, be favorable, without taking into account the interests of the adverse party (sometimes they are even denied).

Loss-win orientation is considered surrender, manifested in the form of yield or renunciation. Rarely, this orientation can be deliberately taken in order to find information about someone who might later be used in the relationship with someone else. For example, the false testimony tactic, in which you engage in a negotiation to make a preliminary (trial) order asking for a lower price and delivery facilities, justified by the high costs of introducing, launching and market promotion and the hope of subsequent massive orders. In fact, what is of interest is in fact only the bill (which is paid, even if we are not interested in the goods bought - loss-earn), which can be used at any time in negotiations with another competing supplier as proof of the existence of a favorable variants.

The *loss-loss* orientation is specific to the gain-loss negotiators, that is, 2 determined, stubborn, selfish, individuals, have something to divide, the result leads to loss-loss. Both parties behave demandingly, wanting everything or wanting to act on any side of the side. Loss-loss attitude is based on the philosophy of conflict, adversity, is the philosophy of war.

The most common variants in practice are win-win and win-loss negotiation, the remainder occurring in circumstantial situations or during a negotiation stage for certain criteria.

Choosing a particular type of bargaining is not simple, because in practice, situations such as those described above are rarely encountered. The dilemma comes from the fact that the followers of these different, even contradictory orientations are often face to face in the same negotiation and interact at their attitudes. Certain situations, by their nature, allow employment in an integrative bargaining style, and others do not. Choosing the type of negotiation is also based on personal adherence to a particular ethic [2].

Attitude in the negotiation process is permanently marked by the so-called "home account".

3. Negotiation tactics

3.1. The YES...BUT tactics

It is the kind of tactic that makes us more agreeable for the negotiating partner. "NO" is a direct and categorical negation that cuts, breaks and hits. It poses the risk of offending the partner and blocking the discussion. "NO" irritates and reverses. It is devoid of delicacy. People with tact avoid it with great care. A formula like "YES ... BUT" can be used in the sense of negation, while retaining two other options. It has three possible nuances: one that means "YES", one that means "maybe" and one that even means "NO". Anytime you can continue on the desired option. The secret of "YES ... BUT" is that it allows the formulation of your own opinion as a continuation of what the partner said and not as a direct contradiction of his opinion.

3.2. The tactics of the false offer

In short, it can be characterized as "a bargaining tricks with ... little theater". Negotiating the price is always a zero-sum game in which one can not win without the other losing. As far as possible, opponents manipulate each other, at least to the limit of loyalty and morality. One of the somewhat unfair tactics, rarely seen in manuals and often in practice, is where the buyer makes the seller an attractive price offer to eliminate the competition and motivate him to conduct the transaction. Once he has this, he finds a reason to change his initial offer. Then begins the "bargaining" that convinces the seller to accept the new offer, which is usually much more moderate. As far as possible, the seller is in the situation to have no choice.

3.3. The tactics of stress and traction

In a word, it weakens the opponent's physical and psychological resistance. As an exception and as rare as possible, when negotiating with a difficult, inconceivable and disagreeable adversary, willing to engage in useless in tough and prolonged negotiations, it is recommended to use some tricks and tactics of stress and traction. Some sort of lateral maneuvers can be used, although they are not directly offensive and humiliating, have the role of sacrificing and disturbing the opponent, prompting him to rush the end of the negotiations. The opponent can be worn vigorously through the manufacturing halls and the company's warehouses. He can be accommodated in a room exposed to infernal noises that prevent him from sleeping. At the table, it can be placed in the sun or another source of irritating light. He can be seated with his back to a door that spins and someone closes it and pushes it openly, as if by chance, can be placed on a seemingly luxurious, but uncomfortable, screaming, the only available one, unfortunately. He stays in the heap and gets tired quickly. When the long-term relationship does not interest us and we are proposing the use of such means of pressure, we must do this under the guise of the most perfect innocence and courtesy, apologizing and pretending to be victims with our opponent.

3.4. Tactics of time pressure

This tactic is based on the simple idea that, after all, there is a negotiation agenda and a negotiation agenda for the negotiators. These elements can be organized and manipulated so that the delicate problem remains at the expiration of the time spent on the negotiation process. For this purpose, any tertip and maneuvers of delay, bypass, and postponement may be used. Towards the end of the negotiations, things usually begin to precipitate. One of the discussion partners must catch the plane or the train, a strike is about to burst, etc. Under such conditions, the pace of the negotiations must be speeded up and the opponent can easily make mistakes. One of the simple but effective maneuvers for delaying the end is the recapitulation and the detachment of intermediate conclusions. Other delays may be based on a lack of documentation, one's health, departure, need for urgent travel, etc.

3.5. Tactics of "salami slice"

Also called "small steps" or "salami" tactics, it is based on the simple idea that it is easier to get the salami slice by slice, than all the sudden. When we ask too much, too quickly, the opponent can be overwhelmed for the moment and tends to resist. It seems to me much easier to respond by a refusal. It becomes increasingly difficult for him to continue playing under pressure. In return, by getting repeated partial benefits, with more time and patience, it can be easier to achieve a total win in the end. Little successes can go unnoticed, but there can be a lot of little and no success, to consolidate your position and achieve great achievements. The "step by step" tactic is exactly the opposite of "bottom-up" dealings.

3.6. Alternate tactics of negotiators

The basic idea is that when the partner changes the negotiator, you have to take everything from the beginning. A first version of this tactic makes the head of the negotiating team look really gentle and reasonable, but utterly helpless in the face of pressure from the team's specialists. Deliberately and forever premeditated and simulated, the rest of the team is tough, stubborn and seemingly irresponsible. During the negotiation process are introduced engineers, merceologists, lawyers, accountants, etc. which displays a tough and intransigent position. In this way, they create a psychological pressure to which the partner prefers to work only with the team leader and accept his more reasonable proposals. He, the faces, disagrees completely with his teammates, but he cannot go too far on them. A second version is the actual change of the negotiator. This means that during the negotiations, just when you expect less, the adversary replaces the negotiator. It can be a hard blow, difficult to deal with, because it's less pleasant to take everything from the beginning. The new negotiator has the possibility to invoke new arguments, to revoke some of the agreements already made or even to withdraw some of the concessions granted by his predecessor. The new negotiator is usually the top man who takes you away when his predecessor has already exhausted you. In diplomacy, this tactic gives excellent results.

4. Technics of negotiations

Here are the main techniques used in a negotiation. These are useful to have an overview, each technique requiring a specific, detailed and exemplary approach in order to be able to apply it effectively. Technics such as:

- *"Uly and Dove"* or *"Good Boy - Bad Boy"*: In a team of two negotiators, one of them will lead the negotiation at the beginning, and after lead the other colleague for the final stages.
- *Eludation technics*: the negotiator, in the desire to exert pressure on the partner, seeks to talk with the superiors or even with the colleagues of this important person; in this way he seeks to undermine his position, isolate him.
- *"It is important to me"*: it reflects how to act on issues that are not essential to us in order to satisfy our partners.
- *Gambit technics*: the experienced negotiator launches, besides real demand, another application that is not necessarily exaggerated and does not need it in particular.

- *Lack of empowerment*: the negotiator realizes that he is forced to yield more than he would want. One can argue that it does not have the power to sign an agreement in terms that have been discussed.
- *Dominating discussions*: A partner continues to dominate the talk initiative by holding his partners under pressure from his demands while seeking to unbalance their position.
- *Making or avoiding the first offer*: two extremes must be avoided: a) Not to make a very high bid, which would cause the other party to interpret it as a bluff; b) Do not make an offer too low because it could be interpreted as a weakness and ignorance of reality.
- *Use of time*: the request is launched at the last moment, leaving to the negotiating partner a very short time to take a decision.
- *Delay technics*: the negotiator seeks to avoid making a decision motivating lack of documents, health, holidays, travel, urgency, etc.
- *Paraverbal*: we have to talk in the same manner with the negotiating partner because only so he can understand better us, and we can enter the same wavelength.
- *Sell the Senses*: call for previous collaboration, personal or national pride, to remembrance of pleasant moments together.
- *Stimulation and follow-up*: negotiating partner is asked a series of questions about situations that have already happened and which are going to happen being stimulated memory and imagination.
- *"That's all I have"*: a buyer says "I like your product but I do not have that much money".
- *Colombo's technics*: when we do the unconscious.
- *Escalation*: the two parties (seller and buyer) have set a certain price, after which the seller raises the price, placing the buyer in an unpleasant situation.
- *Exaggerated politeness*: to a polite, friendly partner, no desire can be denied.
- *Introductory questions*: a partner will seek to gain further benefits by gathering information about partners and will build them a picture in which they can find the weaker, vulnerable points.
- *"Get-yes" technics*: a negotiator will make a small concession after the small concession made by opponents.
- *Red and Black Technics*: before the interlocutor decides, you suggest him to choose between two variants.
- *"Pocker-like"*: the negotiator will have a hard position, impenetrable; will not show anything by expression, tone, or gesture.
- *The negotiator's change*: used by warrior negotiators and difficult to cope with.
- *Application of exaggerated requests*: the possibility of later making a series of concessions
- *Partner's engagement*: Negotiation is a tiring process.
- *Estimation Technique*: A correct presentation of the product will make the partner want it more while the product is better off than your hesitation to sell it.
- *The positive association technique*: we will especially remember the positive and negative feelings, and whenever we have to deal with an event or similar action, the sentiment.
- *Technics of the assumption*: the premise that one can and supposes that everything is possible.
- *Toes Technique*: The golden rule of negotiation is never to take the word when you can very well keep silence.
- *The tender offer*: a buyer enters into a bargain with a bid high enough to remove the competition from the start. Once this has been achieved, the presumptive buyer withdraws his initial offer and thus the initial negotiation loses its validity.
- *The negotiation tactic with the proud people*: do not negotiate a price, do not negotiate services, negotiate with the interlocutor's pride. You have to beat it with your own weapons.

- *Technics of Voice Involvement*: In order to gain access to a person's unconscious, a change of speech rhythm must be made, most often in lowering the tone of the voice and its intensity.
- *Technics of using the impasse*: to request either termination of the negotiations or additional concessions
- *Moving in time and space*: placing the negotiation in another context / space or moving it over time.
- *"I do not sell to anyone" technology*: you sell nothing, you help your partner to grow, you do not lack customers.
- *Identification of the decision-maker*: identification of the decency, the user, the beneficiary and the elimination of: the clerks (secretaries, porters, service women, assistant managers etc
- *"You could do more than that" Technique*: confuses the seller, usually taking advantage of the buyer.

The above techniques find their effectiveness in using them. They must be adapted to context, to the environment in which negotiations are held. So you have to adapt the tactics to your potential to use it. Not all of the above techniques can be used by anyone and at any time. Depending on your personality, some of these techniques will be easier to use with better results. Therefore, we need to use those techniques that lead to maximizing results. But we must not forget about a principle of modern negotiation, namely WIN - WIN (in a negotiation it must win both sides). It is also very important that these tactical techniques are adapted to the personality of our interlocutor.

5. Conclusions

Negotiation is much easier when you find out what the needs and fears of the other are, when you know, beyond your expected result to put yourself in its place, try to understand empathy with your point of view. If you manage to make your negotiating partner easy to say "yes" then the solution is somewhere close. To avoid a Defendant - Winner situation, this is one of the best approaches for a win-win partnership on long-term. Whenever the solution presented by the other party is not convenient, try not to look at things personally, not to confuse the person with what is being done.

Negotiation means beyond a positive attitude, a winner and a lot of patience, pertinent insistence as well as profound engagement.

A solid basis in the negotiation process is trust. The more honesty, integrity and credibility, the more trust your business partner has, the more successful the negotiation is. If for some reason your partner considers you inferior, unreliable, it is difficult to obtain even minor concessions. When it comes to gaining someone's trust, actions say more than words. Here is a succession of things on which to build trust in the negotiating partner: demonstrate your competence; make sure that your non-verbal signals fit with your statements; maintain a professional image; communicate your good intentions; do what you say you want to do; go beyond the conventional relationship; listen; communicate as much as possible; discuss what is not to be discussed; provides accurate information without hidden thoughts; be honest, even if it costs you something; be patient; attention to fairness; negotiate for abundance and not for poverty; assume calculated risks.

"Leaders should not finish first. They have to make sure that everyone else crosses the finish line".

6. References

[1] Lempereur A, Colson A 2004 Méthode de négociation. On ne naît pas bon négociateur, on le devient. Ed. Dunod, Paris
[2] Lewiscki R, Litterrer J, Minton J and Saunders D 2004 Negotiation. Burr Ridge, Illinois
[3] Săvescu D 2016 Managementul proprietății intelectuale, Ed. LUXLIBRIS, Brașov

IOP Conf. Series: Materials Science and Engineering **514** (2019) 012043 doi:10.1088/1757-899X/514/1/012043

The components and vectors of excellence

C Puiu[1] and V Puiu[2]

[1]Travel Department, Economic College Ion Ghica of Bacău, România
[2]Engineering and Management of Mechanical System Department, Vasile Alecsandri University, of Bacău, România

E-mail: vasilepuiu@ub.ro

Abstract. This work aims at identifying the components of excellence (market, products / service, development) through a new contextual approach. An interdisciplinary scientific instrumentation is utilized, by placing, into the space of the vector theory in physics, the economical sizes that, by summing, can provide a real quantification (concrete, precise) of the excellence in business. The work proposes a new quantification modality of the excellence level through the method of vectorial composition, taken over from physics (mechanics). After the components of excellence (market, product / service, development) have been established and represented as vectors, the calculation of the resulting vector (excellence) can be done, through the classical vectorial composition, into a system of axes, determined by the excellence components.

1. Introduction

The New Economy "refers to a set of quantitative and qualitative chances that have transformed the structure, functioning and rules of the economy." It is "based on knowledge and ideas where the key factors, such as creation of jobs and higher life standards are innovative ideas and technologies, incorporated to the manufactured products and services. It's the economy where risk, uncertainty and constant chances are rather a rule than an exception" (Ișanu, 2002).

Through an explicative construing, the New Economy means conceptual activity before the physical (mechanical) activity and communication technologies that lead to a generalized (global) concept. In other words, innovations and inventions precede the production through acquiring new concepts and their means of achievement. The organizational changes take place rapidly and constantly.

In this context, the concept of excellence has appeared, a manner of making the globalized problems within the organizations compatible. Excellence in business seems to be the highest common factor without which the sizes, of any kind, cannot be correctly divided. In other words, it is the common denominator of the fractions that build the resistance structure of the economy. Through it, the precise coordinates along the variable directions of the progress and development are obtained.

Excellence is, as such, an advantage on the market competition, determining higher income rates. By definition, if we take into account all the points established, excellence becomes "improved quality of the products and services, competitive costs and satisfied consumers". It may outline a model through consumer orientation, managerial techniques and, especially, performing results.

2. Determining Excellence

Excellence can be generated / determined / influenced through cultural differences existing between nations. These derive from traditions, habits, language, quality or services, civilization level etc. The cultural dimensions make Geert Hofstede (1980) (author of this theory along with Van Hool) state: "Culture is more often a source of conflicts than of synergy. Cultural differences are a nuisance at the best and often a disaster".

The cultural differences may have important consequences in the business context. The conduct is marked by the person's image, attitude towards action, individual or group importance. Through cultural differences, the organizational structures and the firm strategies can be marked. "Conceptual analysis of the culture" utilizes, practically, a classical grill, structured on elements that can be ranked according to importance: strong / weak context; communicative values (writing, words, gestures); humans as vision (religion, traditions); action, risk, work.

Time, as relation, imposes three milestones: mono - chronism; tradition / force - in the past, present and future; time as relative value ("time is money"). In terms of vision on humans and their destiny, it can be judged through descending ("what it is") or through becoming ("what it is doing"). In both cases the attitudes to value become important. Eventually, through action, risk, work, it is the entrepreneurs that can be appreciated, not the thinkers; the American pragmatism "to make" matters. Another aspect may be when humans surrender to their condition in search for natural harmony Formatting author affiliations

3. Coordinates of Excellence

3.1. Components

Excellence, as a result of the processes of successive changes, may be derived through precise coordinates. By considering that innovation leads, by default, to changes, three steps, not four, are proposed for the coordinates of excellence, respectively: consumer, human resource and innovation - change. With these notations, the resulting vector of excellence (ē) appears, in a geometrical system, tri-axial, as shown in Fig. 1.

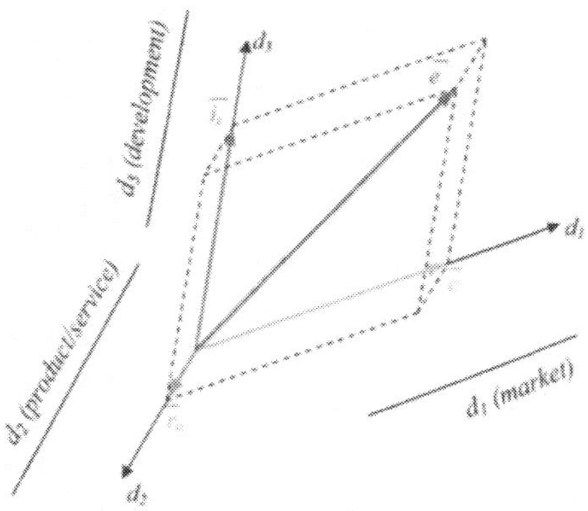

Figure 1. The components and vectors of excellence
Notations: d_1 *- market;* d_2 *- product / service;* d_3 *– development; c – consumer;* r_u *– human resource* i_s *- innovation-change.*

The motivation of the three sides (axes) of the orientation to quality (attention to consumer, personnel and innovation belongs to Tom Peters (1982), the "father" of excellence in business, who, in the eighties, along with Robert Waterman, published "In Search of Excellence". The work proposes 8 themes (bases, principles) for searching excellence. These are:

- Action ("A bias for action - getting on with it");

- Consumer (“Close to the customer - learning from the customer”);
- Entrepreneur (“Autonomy and entrepreneurship”);
- Productivity through personnel motivation (“Productivity through people”);
- Relating to fundamental values (“Hand-on, value-driven-management, walk the talk”);
- Respect for those who know to do (‘Stick to the knitting – do what you know”);
- Constructions with simple and adaptive structures (‘Simple form, lean staff”);
- Efficient results (“Simultaneous loose-tight properties – have autonomy in some areas, central ideas / values in others”).

The formula for Excellence is summarized thus:

$$\text{Excellence} = \text{efficiency} + \text{effectiveness} + \text{stimulating environment} \quad (1)$$

By corroborating this relation to the above context, excellence, according to Tom Peters, seems to become the sum of the eight themes, each of them keeping, in its contents, a discreet, but suggestively outlined secret. Such secrets might be the ascent in action, capacity of listening to the consumer, favouring the autonomy and innovating spirit, motivating the personnel for productivity, mobilizing fundamental values, appreciation of those who know how to do, simple and, especially adaptable constructive forms and combining rigor with flexibility (Marian, 2007).

By reverting to Fig.1, it may be asserted that excellence, thus represented, as touchable size, leads to success. It can be maintained through strategies (at the level of each coordinate).

3.1.1. Consumer

For surviving, the firm is aware that the final of the product / service quality is the consumer. This awareness seems to be suggestively given by the direction (d_1) indicating the market that provokes the motion needed by the system.

All aspects of perspective regarding the value addition for consumer are considered. This perceives a value prospectively evaluated. It includes, through summing, benefits and costs from the alternant comparative offers that are known. Satisfaction is associated, as size, to the manner of perception of the level being achieved on the product / service characteristics related to expectances. If the created value is important, it may lead to the consumer's loyalty and, eventually, through extension, to his fidelity. The firm is obliged to differentiate, as much as possible, its products / services from the competition, but it has to increase the innovative degree with which the consumer will be satisfied, both through expectances and the qualitative surprise. The strategies of the company are expressed through competence, flexibility, added value and new opportunities.

3.1.2. Human Resource

The direction (d_2) — product / services - that means action, is the human resource, determined at the excellence level. Through their scientific contents, from strategic planning to the personnel's satisfaction, all the concerned aspects are considered for calculation, respectively:

- Recruiting, selection, integration and vocational training, career development;
- Evaluations (jobs, performance of the personnel);
- Rewards, equity, equal chances;
- Rights, obligations;
- Labour protection and safety;
- Conflict management, communication and relations with the trade unions;
- Efficiency of the human resources usage;
- Total quality management in the field of human resources;
- Orientations, trends, objectives.

"Human is the centre of all business. All other resources, grounds, building, mechanisms, equipment, vehicles or money are or secondary importance only. Without people business cannot be done. Business has no other purpose but to serve the people's needs" (Professor Dr. Michael C. Wilkinson, dean of the Faculty of Management, Derby University, UK) (Mathis, Nica, Rusu, 1997).

The management of the human resources requires an updated approach of the contents and objectives (concepts, politics and management responsibilities), types of specialists in the fields and activity organizing. From this perspective, the company is outlined as a subsystem of the human resource and is involved in the internal organization of the human resources department, whose attributions and types of organizational structures are settled.

We can quote again Prof. Wilkinson "(. . .) people have various roles within a business and have many connections with other business. People are committed to a business. In some cases they are its managers, in others they are suppliers, customers, creditors, debtors and often they are shareholders of an initiative. However, frequently they may have all these attributes at a time" (Mathis, Nica, Rusu, 1997).

The inter-functional collaboration inside a company eventually determines excellence. Among the structural organization, processes, managerial functions and aptitudes, the suitable lubricant needs to be inserted in order to provide an optimal running of the entire assembly. The human resource puts together knowledge, talent, imagination, creation (Mathis, Nica, Rusu, 1997).

3.1.3. Innovation – Change

The differentiation of the products / services from the competition is performed through innovation that, in all cases, leads to change (Fig.1. - direction (d_3) development). As a result of a creative activity, innovation requires superior engineering / managerial solutions, able to change a compromise / conflict (technical, economical or mixed) among at least two systems involved in business. It directs the company towards new performances (Stegăroiu, Niculescu, 2000). Innovating and innovation can be explained separately, but independently. If "innovating" means a process of renewing, innovation is its result.

In dictionaries the term is explained as: "new, change, transformation or solution to a technical problem or a problem of work organizing intended for improving the work productivity, technical perfection or rationalization of the solutions being applied" [21].

In this entire explicative introduction "innovation is considered, in general, the main motor of the economical increase of the nowadays global economy. By implementing innovations into practice products can be obtained with improved quality features, service of superior quality, new production processes, more efficient and cleaner (ecologic), enhanced models of the business management system, modem methods of workmanship management etc. There are multiple motivations for the companies and organizations to innovate, such as: increasing their market quota, conquering new markets, improving the product quality, widening the product range, replacing the obsolete products", diminishing the impact on environment etc. Innovation is indissolubly created by creativity (...) [18].

Innovation has been classified into four types: product innovation, process innovation, marketing innovation and organizational innovation [17]. The OSLO manual, version 3 (2005) recommends the following definition: "An innovation is the implementation of a new product or significantly improved (a product or a service) or of a process, a new marketing method, or a new organizational method in the business practice, work place organizing or foreign relationship". The condition is for the product, process and marketing or organizational methods to represent novelties (or significant improvements) for the company [18].

Innovation leads to change. According to the extent of change needed for implementation and to the technological intensity, innovations are radical *(or breakthrough innovations)* and *incremental.*

The term of product is used for both goods and services [18]. In virtue of this, radical innovations are considered dimensions of newness and produce absolutely new solutions in systems, processes, goods / services. They get developed in new business that will lead to important chances. These affect branches of the existing industries or the bases of other prospective industries and create new markets. Here come, as examples, the optical fibers, lasers and computer industry [18]. They assure a much higher value to consumers. The radical innovations to product distinguish between the old and the new product through the description of its usage as a result of several performance or material features and superior components. These innovations induce new technologies (in exclusivity) or combinations usable at a superior level compared to the existing ones. However, a definition of radical innovations does not meet an acceptable consent of the researchers yet (Mc. Dermott, 2002).

Incremental innovations "adapt, simplify, refine and improve product and / or extant production / distribution systems" [18]. Definitions of incremental innovations were formulated by CM Banbuy and W Mitchell in 1995. They present them as "refinements and expansions of the settled projects that lead to substantial price or functional benefits to the users" (Banburry, 2005) [18]. It has also been asserted that "incremental innovations are, usually, minor improvements of the existing products and processes" [18, 20]. When the performances of an existing product / service get improved during the life cycle through successive versions, the incremental innovation comes up. Products can be simple or complex. Their improvements consist of components or materials with superior performances (simple products) or partial changes brought to the structural subsystems (complex products). The existing production limits can be expanded or additional (complementary) functions can be added to the existing functions. If the radical innovations are discontinuous, the incremental ones are continuous, of a higher frequency, motivated by the needs for market satisfaction. The former bring a much higher profit to the companies, compared to the latter category.

There are "imitated" innovations, as well. These are "new for the company, not for the market'. They are marked by low technological innovating extent and market (Garcia, 2002).

Another notion that has to be introduced to the context is invention. If invention means discovery through creation, innovation is its trading, so, the confusion between them has to be avoided. "An invention consists of creating a new configuration, composition or material, device or process" [19].

In addition "an invention is the first occurrence of a new idea (concept) for a new product (process), whilst innovation is the first trading of the new idea". The invention becomes innovation after processing through activities of production and marketing and traded on the market [19].

Edward B. Roberts (2007) considers that innovation is a process composed of:

- Generation of an idea or invention;
- Transforming that idea into a business or other useful application.

The innovating activities have a scientific, technological, organizational and economical (financial, commercial) content. These actions materialize and implement the innovations [17]. These activities are correlated with the research - development domain where the technical – scientific novelties are categorized.

The innovating activities are protected by the curtain of company secrets (scientific, technological and economical).

The innovation sources can be internal and external (Drunker, 2002). The internal ones are: *the unexpected event* (a surprise commercial success must be exploited); *non-concordances* (between expectances and results or suppositions); *process necessities* (creating opportunities through the improvement of processes); *changes in industry and on the market* (increases of the request on the market or increase of the industry are favorable for new innovative processes). The external sources of the company may be: *demographic changes, perceptive* (understanding the consumer's perceptions) and *new knowledge* (patents, know-how) [18].

The innovative action produces changes. Changes require "replacement, modification, transformation or changing into shape and / or content of an object, product, work, service, activity or process" (Burduș, 2000). By generalizing, "the change represents a permanence of our life, finding it in the evolution of all states and structures", the organization is (...) an entity in change, obliged to permanently enhance and adapt its level, structure and quality of the transformations that it produces on several input elements in order to meet the needs of the market, personnel and administration and to earn on its own in terms of profit, image, market quota, expansion capacity, resistance to competition etc." (Rusu, 2003).

The management of change approaches specific aspects within the organizational area and outside it, too. Among these, the following nay be emphasized: imperative of the change (development of the organizations, organization based on knowledge, implementation of the permanent learning concept); the organizational structure and change (classical and modern structures, remodeling of several structural elements); organizational culture in front of the change (element of strong impact in change); diagnostic; planning of the change process (strategies, politics, stages, models, their implementation into the organization); provision of the conditions for implementing the change (the consolidation of the changes to be efficient); strategic and leadership changes (negotiation elements, styles of managers and leaders, risk and complexity in change); instruments of change (procedures, techniques) (Rusu, 2003).

4. Vectors of excellence

Vectors are physical symbols that represent a size in physics, such as speed, acceleration, force etc. They are characterized by size (module) and orientation (direction and sense).

Vectors can operate through addition, subtraction and product (scalar and vectorial). Related to the application point, its orientation in space is given by direction and angle with the reference axes (e.g. X, Y and Z).

In a plane, vectors get composed by the parallelogram rule, and in space it is the parallelepiped rule that applies. The resulting vector will be the diagonal of the parallelogram (in a plane) or parallelepiped (in space).

In the case of any disposition, in space, the three directions (d_1 - market, d_2 — product / service, d_3 - development), can be translated parallel up to a point of intersection (application point). Through vectorial composition, the three components, the vectors $\underline{c}$ (consumer), $\underline{r_u}$ (human resource) and $\underline{i_s}$ give the resultant that in this case is ē - excellence, as shown in Fig.1 and 2.

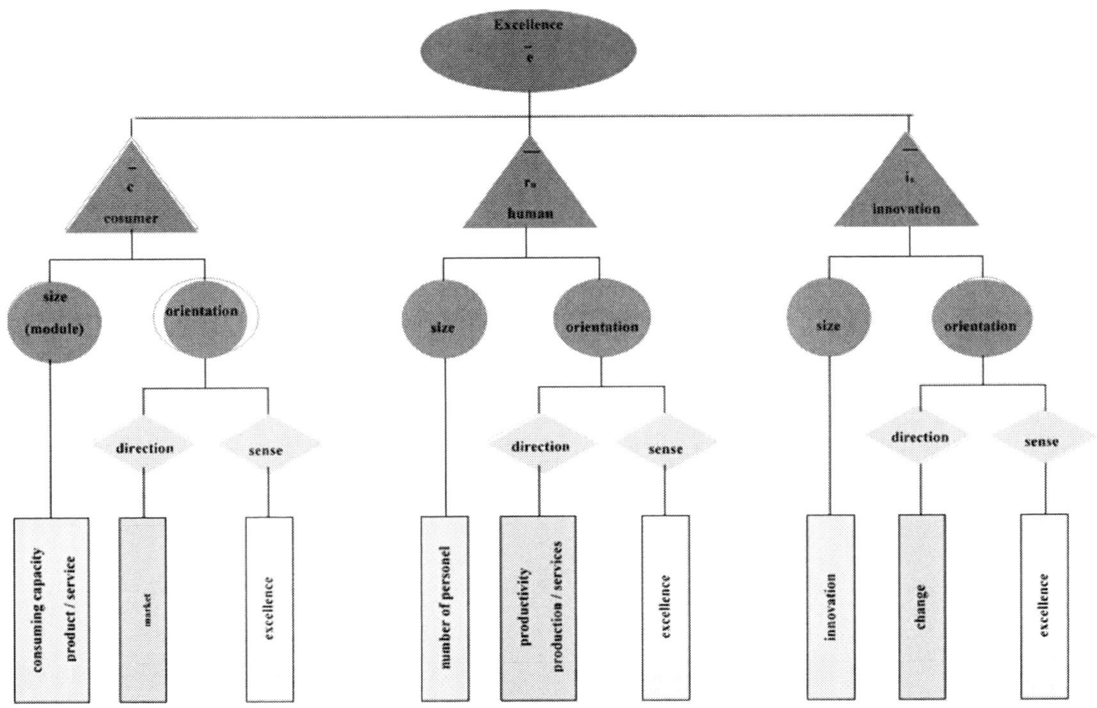

Figure 2. Vectours of excellence

IOP Conf. Series: Materials Science and Engineering **514** (2019) 012043 doi:10.1088/1757-899X/514/1/012043

5. Calculation Model

Based on the motivations and considerations presented so for and reverting to Fig.1, a model of estimation of excellence through a vectorial method is proposed. By intersecting parallels to the three directions (d_1 - market, d_2 - production / services and d_3 - development) a generalized system of axes will be obtained along which the excellence vectors can move: consumer (c), human resource (r_u) and innovation-change (i_s). By composing them, the resulting vector will be excellence, ($\bar{e}$).

Vectors, in physics, are represented by size (module) and orientation (direction and sense). These, transposed to our context, are shown in Fig. 2.

The vector consumer (c) has as size (module) the product / service consuming capacity related to the degree of satisfaction; as direction tire market appears and the sense is given by excellence.

The vector human resource (r_u) is characterized by the number of personnel members (as size or module), production / services as direction and excellence as sense. The vector innovation-change (i_s) is defined in size (module) as innovation, direction means change and sense is excellence as well.

Through their composition the vector excellence ($\bar{e}$) will result, characterized by a size (module) expressed in units of excellence and direction and sense that depend on the angles with the system axes. This vector is the diagonal of the oblique parallelepiped formed by the vectors (c), (r_u) and (i_s). The vector excellence ($\bar{e}$) is expressed through a relation in the form of:

$$\bar{e} = \bar{c} + \bar{r_u} + i_s \quad (2)$$

If the three directions are particularized to three orthogonal axes Ox (d_1), Oy (d_2), Oz (d_3), the configuration illustrated in Fig. 3 will result.

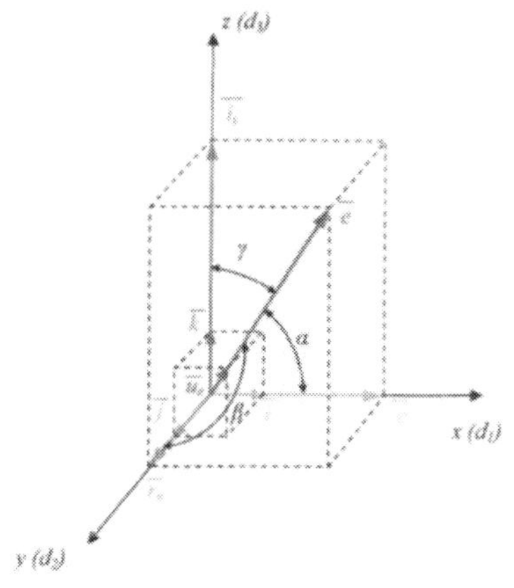

Figure 3. Versors of excellence

The versors (unit vectors) (i), (j), (k) are considered. Their module (size) is equal to 1 (unit of excellence). In this case, the projections of the vectors (c), (r_u) and (i_s) along the three coordinates will become:

$$\bar{c} = c_x \bar{i}$$
$$\bar{r_u} = r_{uy} \bar{j} \quad (3)$$
$$\bar{i_s} = i_{sz} \bar{k}$$

The sizes (modules) of the three vectors (components of excellence) will become:

$$c_x = pr_{OX}\, \bar{e} = e \cos\alpha$$

$$r_{uy} = pr_{OY}\, \bar{e} = e \cos\beta \tag{4}$$
$$i_s = pr_{OZ}\, \bar{e} = e \cos\gamma$$

The angles α, β and γ are position angles of the vector ($\bar{e}$) towards the axes x (d_1), y (d_2) and z (d_3):

$\alpha = <(\bar{e}, \bar{i})$, $\beta = <(\bar{e}, \bar{j})$; $\gamma = <(\bar{e}, \bar{k})$. Further, the size of excellence can be established through the relation:

$$e = (c_x^2 + r_u^2 + i_s^2)^{1/2} \tag{5}$$

The expression of the vector ($\bar{e}$) is:

$$\bar{e} = c_x \bar{i} + r_u \bar{j} + i_s \bar{k} \tag{6}$$

Of course, among the director cosines there is the relation:

$$\cos^2\alpha + \cos^2\beta + \cos^2\gamma = 1 \tag{7}$$

The relations (4) give the proportionality among the sizes of the component vectors of excellence and the position angles of the vector of excellence:

$$\frac{c_x}{\cos\alpha} = \frac{r_u}{\cos\beta} = \frac{i_s}{cos\gamma} \tag{8}$$

If the sign of the director cosines is changed, the orientation direction will also change. Thus, the following will result:

$$cos\,\alpha = \frac{c_x}{\pm\sqrt{c_x^2 + r_u^2 + i_s^2}} = \frac{c_x}{\pm e}$$

$$cos\,\beta = \frac{r_u}{\pm\sqrt{c_x^2 + r_u^2 + i_s^2}} = \frac{r_u}{\pm e} \tag{9}$$

$$cos\,\gamma = \frac{i_{xs}}{\pm\sqrt{c_x^2 + r_u^2 + i_s^2}} = \frac{i_s}{\pm e}$$

By knowing the position angles of the vector of excellence towards the three directions the components of excellence (projections) can be determined. This affirmation is reciprocally valid, i.e. if the projections of the vector of excellence are known, the angles, α, β and γ can be found depending on the size of the excellence vector.

If, to the initial system (x, y, z) a rotation determined by the cosines of the angles made by rotated axes is applied, a new system (x', y', z') will be obtained, as shown below (Table 1):

Table 1. Correspondence between system (x, y, z) and (x', y', z').

	O_X	O_Y	O_Z
O_X'	$\cos\alpha_1$	$\cos\beta_1$	$\cos\gamma_1$

O_Y'	$\cos \alpha_2$	$\cos \beta_2$	$\cos \gamma_2$
O_Z'	$\cos \alpha_3$	$\cos \beta_3$	$\cos \gamma_3$

The new coordinates can be calculated by means of the formulae:

$$x' = x\cos\alpha_1 + y\cos\beta_1 + z\cos\gamma_1$$
$$y' = x\cos\alpha_2 + y\cos\beta_2 + z\cos\gamma_2 \quad (10)$$
$$z' = x\cos\alpha_3 + y\cos\beta_3 + z\cos\gamma_3$$

Where the symbols mean:

- α_1, β_1 and γ_1 - the angles of the Ox' axis with O_x, O_y, O_z and $\cos\alpha_1$, $\cos\beta_1$, $\cos\gamma_1$, are the director cosines of the new axis Ox'
- α_2, β_2 and γ_2 - the angles of the Oy' axis with Ox, Oy, Oz and $\cos\alpha_2$ $\cos\beta_2$ $\cos\gamma_2$ are the director cosines of the new axis Oy'
- α_2, β_2 and γ_2- the angles of the Oz' axis with Ox, Oy, Oz and $\cos\alpha_3$ $\cos\beta_3$ $\cos\gamma_1$, are the director cosines of the new axis Oz'.

The demonstration of the formula (10) is immediate, by considering the projections of the excellence vector $\bar{e}$ *on* the three axes Ox', Oy', Oz' and the scalar products between the versors.

$$\bar{e} = c_x\bar{i} + r_u\bar{j} + i_s\bar{k} = x\bar{i} + y\bar{j} + z\bar{k} \quad (11)$$

And the versors of the new axes i', j', k' with:

$$\bar{i}' = \cos\alpha_1\, \bar{i} + \cos\beta_1\, \bar{j} + \cos\gamma_1\, \bar{k}$$
$$\bar{j}' = \cos\alpha_2\, \bar{i} + \cos\beta_2\, \bar{j} + \cos\gamma_2\, \bar{k} \quad (12)$$
$$\bar{k}' = \cos\alpha_3\, \bar{i} + \cos\beta_3\, \bar{j} + \cos\gamma_3\, \bar{k}$$

The relations (10) will result.
Analogically, it will result:

$$x = x'\cos\alpha_1 + y'\cos\alpha_2 + z'\cos\alpha_3$$
$$y = x'\cos\beta_1 + y'\cos\alpha\beta_2 + z'\cos\beta_3 \quad (13)$$
$$x = x'\cos\gamma_1 + y'\cos\gamma_2 + z'\cos\gamma_3$$

Thus, in the rotation of the coordinate axes in space nine angles come up, three for each new axis related to each of the three initial axes. These angles are not independent among them, but they are linked by the fundamental relations that are known:

$$\cos^2\alpha_1 + \cos^2\beta_1 + \cos^2\gamma_1 = 1 \qquad i = 1,3 \quad (14)$$

and

$$\sum_{i-1}^{3}\cos^2\alpha_i = 1, \qquad \sum_{i=1}^{3}\cos^2\gamma_i = 1 \qquad \sum_{i=1}^{3}\cos^2\beta_i = 1$$

A general change of the axes can be done as well, i.e. a translation and a rotation, first there will be the

IOP Conf. Series: Materials Science and Engineering **514** (2019) 012043 doi:10.1088/1757-899X/514/1/012043

translation given by the coordinates x_0, y_0, z_0 of the new origin O' towards the initial system (O, x, y,

$$x = x_0 + x'\cos\alpha_1 + y'\cos\alpha_2 + z'\cos\alpha_3$$
$$y = y_0 + x'\cos\beta_1 + y'\cos\beta_2 + z'\cos\beta_3 \quad (15)$$
$$z = z_0 + z'\cos\gamma_1 + y'\cos\gamma_2 + z'\cos\gamma_3$$

z), thus obtaining the system (O, x'', y'', z''). Afterwards, by rotating this system, the translated and rotated system (O, x', y', z') will be obtained, where:
are the formulae giving the general transformation of the axes of coordinates. The axis rotation supposes orientation modifications of the excellence vector, through the angles of excellence (α, β and γ). Translations produce reorientations, through size, in the modules of the excellence vectors. Eventually, the general change (translation plus rotation) means reorientation and relation to new positioning.

The relations (16) that give the general transformation of the axes are a starting point in the mathematical approach (model) of the vectors of excellence. From this point, the graphical - vectorial model presented penetrates the perimeter of the economical superior mathematics, offering the basis of several calculation programs. The problems of maximum / minimum should be remodelled as well as, especially, the optimal conditions.

The last stage is represented by the usage of these programs by the operators specialized in quantifications of excellence in business.

6. Conclusions

The concept of *excellence* appeared as a modality of making the globalized problems within the organizations compatible.

Excellence in business is the common denominator of the fractious that build the resistance structure of economy. Through in the precise coordinates along variable directions of progress and development are obtained.

Innovating and innovation can be explained separately, but independently. If innovating means a process of renewal innovation is its result. The innovating activities have a scientific, technological, organizational (financial, commercial) content, materialize the innovations and implement them. These activities are correlated with the research - development domain, where the technical - scientific novelties are categorized. Innovative activities are protected by the curtain of the company secret (scientific, technological and economical).

A model of estimating business excellence through a vectorial method is proposed. By intersecting parallels to the three directions *(d_1 - market, d_2 - production / services and d_3 - development)* a generalized system of axes will be obtained along which the excellence vectors can move: consumer $(\bar{c})$, human resource $(\overline{r_u})$ and innovation - change $(\overline{i_s})$. By composing them, the resulting vector will be excellence, $(\bar{e})$. This vector is the diagonal of the oblique parallelepiped formed by the vectors $(\bar{c})$, $(\overline{r_u})$ and $(\overline{i_s})$. The vector of excellence $(\bar{e})$ is expressed through a relation of the form $(\bar{e}) = (\bar{c}) + (\overline{r_u}) + (\overline{i_s})$; $\bar{i}$, $\bar{j}$ and $\bar{k}$ are versors (unit vectors); their resultant is the unit of excellence. The size of excellence can be determiend through the relation $e = (c_x^2 + r_u^2 + i_z^2)^{1/2}$.

The analytic expression of the vector $(\bar{e})$ is $\bar{e} = c_x\bar{i} + r_u\bar{j} + i_s\bar{k}$.

By knowing the position angles of the excellence vector, towards the three directions, the excellence components (projections) can be easily determined. This affirmation is reciprocally valid as well, i.e. if the projections of the excellence vector are known, the angles α, β and γ can be found, depending on the size of the excellence vector.

The rotation of the axes supposes orientation modifications of the excellence vector, through the excellence angles α, β and γ .

Translations produce reorientations, by size, in the modules of the excellence vectors.

Eventually, the general change (translation plus rotation) means reorientation and relation to new positioning.

The relations that give the general transformation of the axes are a starting point in the mathematical approach (model) of the vectors of excellence.

From this point, the graphical model – vectorial – being presented penetrates the perimeter of the economical superior mathematics, offering the basis of several calculation programs.

The problems of maximum / minimum should be remodeled as well as, especially, the optimal conditions.

The last stage is represented by the usage of these programs by the operators specialized in quantifications of excellence in business.

7. References

[1] Banbury, C., Mitchell, W. (1995), "The effect of introducing important incremental innovations on market share and business survival", Strategic management journal, 16, 161-182.

[2] Burduş, E. (coord.) (2003), "Managementul schimbării organizaţionale", Editura Economică, Bucureşti, p.13.

[3] Drucker, P. (2002), "The Discipline of Innovation", Harvard Business Review (HBR).

[4] Fagerberg, J. (2004), "Innovation: A guide to the literature", The Oxford Handbook of Innovation, Oxford University Press, 1-26.

[5] Garcia, R., Calantone, R. (2002), "A critical look at technological innovation tipology and innovativeness terminology: a literature review", Journal of Product Innovation Management, 19, 110-133.

[6] Hofstede, G. (1980), "Culture's consequences. International differences in work — related values", SAGE Publications.

[7] Işanu, V. (2002), "Noua Economie: alegaţii şi evidenţe", Revista Economică, 2 (22).

[8] Marian, L. (2007), "Antreprenorii şi antreprenoriatul", Editura EFI ROM, p.63.

[9] Mathis, L.R., Nica, C.P., Rusu, C. (1997), "Managementul resurselor umane", Editura Economică, 1997.

[10] Mc.Derinott, C.M., O'Connor, G.C. (2002), "Managing Radical Innovation: An Overview of Emergent Strategy Issues". Joumal of Product Innovation Management, 19 (6), 424-438.

[11] Paraschîveseu, A.O. (2009), "Managementul excelenţei", Editura Ştiinţa - Tehnopress, Chişinău-Iaşi, p.236.

[12] Peters, T., Waterman, R.H. Jr. (1982), "In search of excellence: lessons from America's best-run companies", Harperd & Row.

[13] Puiu, C. (2010), "Satisfacţia consumatoiului şi excelenţa în afaceri", Teză de doctorat, Universitatea Tehnică „Gh. Asachi", Iaşi.

[14] Roberts, E.B. (2007), "Managing Invention and Innovation", Research – Technology Management, 50, 35 – 54.

[15] Rusu, C., (2003), "Managementul schimbării", Colecţie CETEX, Programul MBA, Editura Economică, Bucureşti.

[16] Stegăroiu, I., Niculescu, Cr. (2000), "Excelenţa în management", Editura Niculescu, Bucureşti.

[17] *** (2005) — OSLO Manual Guidelines for colecting and interpreting innovation data 3rd ed. OELD / European Communities.

[18] http://ro.wikipedia.org/wiki/Inova %c8%9 Bie

[19] http://wapemedia.mobi/eu/invention

[20] http: //www.economicswebrininstitute.org/glossary/inovate.htm

[21] http://www. merriam-webster.com/dictionary/innovation

Assisted quality inspection for hip prostheses components

B C Braun[1]

[1] Product Design, Mechatronics and Environment Department, Transilvania University, Brasov, Romania

E-mail: braun@unitbv.ro

Abstract.. In the paper is presented a new method for quality analysis of some hip prosthetic elements, surface quality standpoint. First of all it is about the head, as spherical joint for the hip prosthesis. For best bio-compatibility is absolutely necessary that its surface to be well processed standpoint roughness, for lowest friction forces, as much as possible those in the case of the natural hip joint. For this reason, within our current research activities, we have developed a virtual instrument, as software interface for efficient assisted roughness deviations for spherical joints of the hip prostheses, reported to a reference hip prosthesis. This application could be very practical and useful for the stage of adjustment and grinding prostheses heads, increasing the productivity and reducing the number of scraps cases in production

1. Introduction

Prototyping and orthotic elements in the context of aged or post-operatory persons are increasingly more used, especially due to the fact that the trend is to continuously improve performance in terms of robotic or non robotic biomechanical devices. Not only the prostheses or orthoses themselves are currently the subject of extensive studies and research, but also robotic systems, which can train certain muscle groups to stimulate and rehabilitate locomotor function, are of very high relevance [1], [2], [3], [4].

2. Importance of prosthesis in the lower limbs

Walking and running are daily activities for anyone, so that a proper locomotion system at the lower limbs is crucial. Related to this, adequate body equilibrium and balance could be strictly dependent on prostheses and orthoses used for joints and lower limbs [5], [6], [7].

Besides, it is well known that, the need for prosthetic implants in the joints increases sharply with age, especially for persons who in their youth practiced performance sports. Unfortunately, in the present there are many worrying statistics concerning the number of hospital admissions and operative interventions in this respect [8]. At the lower limbs joints, the most problems occur at the hip joints, so that the hip prostheses have seen a great deal in the last 20 years [9]. Considering this, the issue of ensuring the quality of the prosthetic elements of the joints, not only standpoint reliability but also regarding the biocompatibility is more and more addressed [10], [11], [12]. Lubrication and surface quality are both fundamental aspects in terms of functional prosthetic components due to the fact that these could influence not only the prosthesis reliability, but also the patient's comfort. For instance, in case of hip prostheses, the aspect of surface quality of the head (with the third degree joint) is very important to ensure low friction, so that small wear [10], [11]. As a result, from this point of view, surface analysis and, more exactly surfaces roughness measuring could be essential for functional

IOP Conf. Series: Materials Science and Engineering **514** (2019) 012044 doi:10.1088/1757-899X/514/1/012044

quality evaluation.

3. Developed and proposed method for hip prostheses surface quality analysis

If, initially, the basic used method for a surface roughness determination consisted generally to use as main measuring device of some dedicated roughness apparatus, in the present performance digital microscopes are more and more wide-spread. These can be equipped with specific software environments that allow images and 3D charts acquiring, but also the reconstitution of all coordinates for roughness determination using dedicated software algorithms [13]. However, the main inconvenience of these methods is that not in all situations the expenses with the equipment and software licences are justified, so that not in any cases the solution being sustainable. As a result, the researches focused first of all to develop a new method that could lead to a considerable cost reduction without losing much of the efficiency and measuring precision, method that could be applied especially for quality insurance in prosthetic elements. Concretely, it is about using of some simple and low cost measuring devices, associated to flexible software interface which role is to generate and to process the measuring results.

The method was applied for two hip prostheses providing as rejected, different head dimensions. It has been used measuring device an incremental displacement transducer, which measuring resolution is 0.2 μm. For each prosthesis, one standard was used (with the surface of the corresponding prosthesis head of the roughness) the process and scanning conditions being the same between the standard and the studied models. As analysis method, it was taken the problem to analyze a surface element providing from the active area of the hip's head prostheses. The picked area to be scan was about 3 x 3 mm, the length of each scanning vector being of 2.4 mm, according to the in force standards regarding the reference length of scanning, considering 10 equidistant scanning vectors (v_j), each one being analyzed in 10 equidistant points.

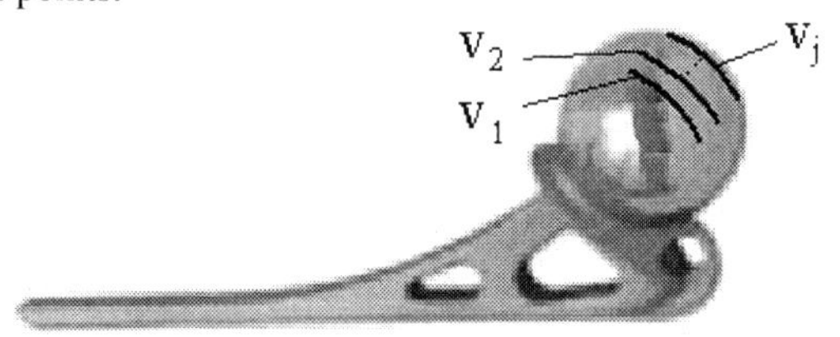

Figure 1. Vectors disposing for defining the scanning surface area [14]

To ensure a high precision on the surface quality evaluation, to the touching rod of the displacement transducer it was adapted a special element, having a a special peak, which curvature radius is up to 0.5 μm. In this way it was possible to have a proper reproduction, by touching of the scanned item profile. Due to the roughness measuring, along the 10 scanning vectors, there were processed the results taking into account 10 equidistant measuring points for each per each vector [15], [16], the data being saved into EXCEL files, to be processed (figure 2).

D
R1
-26.6
29.2
-25.4
51.4
-0.8
12.0
-42.8
33.4
-27.4
0.6

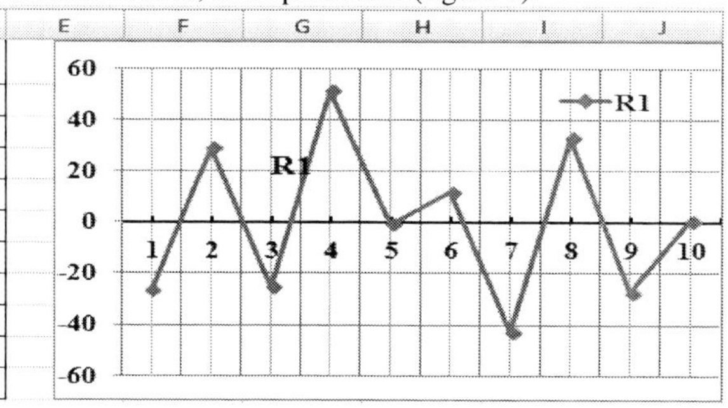

Figure 2. Example of roughness measured values in 10 equidistant points for the 1st scanning vector

For a fair, complete and effective interpretation of the results, in the research, there it was taken the problem to develop a software interface (in Lab VIEW) as virtual instrument for determining and automatic displaying of all information referring to the scanned surface's roughness. More exactly, the interface allow an immediate determination of all roughness parameters based on the values generated in EXCEL files: R_a – averaged value of the roughness, R_z – roughness averaged depth in 10 points, R_t – total roughness value, R_{max} – maximum roughness value [15], [16]. For each determined roughness, parameter on each scanning vector, it has been taken into account of the relationships (1) ÷ (4).

$$R_{a\_vj} = \frac{1}{n} \cdot \sum_{i=1}^{n} R_i \quad [15] \tag{1}$$

where:
R_{a_vj} – averaged roughness for each scanning vector ($j = 1 \div 10$);
R_i – roughness measured for all point of the current scanning vector (n = 10 – number of scanning points / each vector);

$$R_{z\_vj} = \frac{1}{5} \cdot \left(\sum_{i=1}^{n} R_{2k} - \sum_{i=1}^{n-1} R_{2k+1}\right) \quad [15] \tag{2}$$

where:
R_{z_vj} – roughness in all measured points for each scanning vector;
R_{2k} – positive measured roughness in all point of the current scanning vector;
R_{2k+1} – negative measured roughness in all point of the current scanning vector;

$$R_{t\_vj} = \mathrm{Max}(R_{2k}) + \mathrm{Max}(R_{2k+1}) \quad [15] \tag{3}$$

where:
R_{t_vj} – total roughness in all measured points for each scanning vector;

$$R_{max\_vj} = \mathrm{Max}(R_i) \quad [15] \tag{4}$$

where:
R_{max_vj} – maximum roughness in all measured points for each scanning vector;

For the calculus algorithm programming as a part of the software interface, the relationships (1) ÷ (4) have been used, also so that all roughness parameters can be automatically determined referring to the entire scanned surface, it was applied the equations (5).

$$R_a = \frac{1}{m} \cdot \sum_{j=1}^{m} R_{a\_vj}; \quad R_z = \frac{1}{m} \cdot \sum_{j=1}^{m} R_{z\_vj}; \quad R_t = \frac{1}{m} \cdot \sum_{j=1}^{m} R_{t\_vj}; \quad R_{max} = \frac{1}{m} \cdot \sum_{j=1}^{m} R_{max\_vj} \quad [15] \tag{5}$$

where:
R_a / R_z / R_t; R_{max} – averaged / all points / total / maximum roughness for entire scanned surface.

3.1 Software interface presenting

For an efficient data processing and all necessary information on the verdict on the evaluated head hop prosthesis, surface quality standpoint, the user has to run the software interface, following the next steps: first of all a control LED of state allow to set the data reading from files, if necessary (figure 3) (green means that data reading option was chosen and gray means no data reading. As programming, a case structure was associated to the Boolean LED of state.

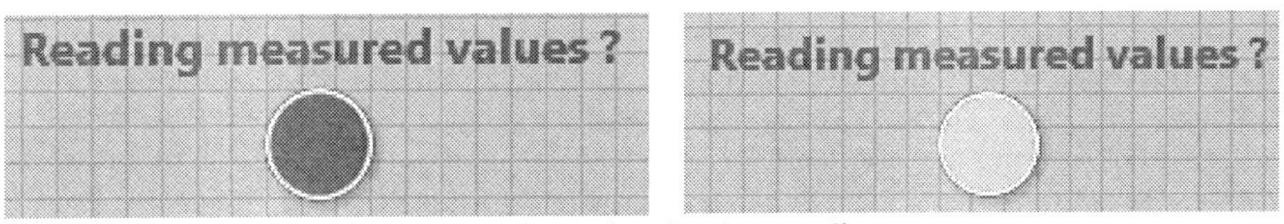

Figure 3. Option for data reading

In case of reading from files, during running the interface, the user has to select all files from which the data must to be processed.

Another aspect related to the user is to specify the dimension of the hip prosthesis's head dimension, depending on which the limit value of the tolerance for roughness deviation is automatically determined (figure 4).

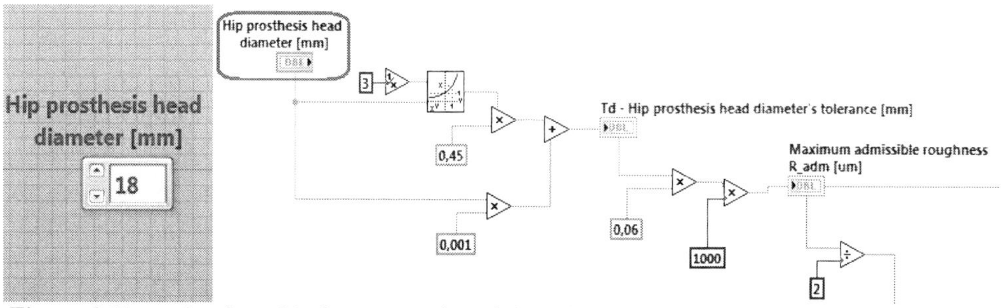

Figure 4. Example of graphical programming of the calculus algorithm for maximum admissible roughness depending on the hip prosthesis's head dimension

In red, there are all text boxes and dialog referring to the inputs (settings, numeric values etc.) and in blue are shown all generated information after software running. The fist obtained information are the following: maximum admissible roughness, averaged value of the roughness, roughness averaged depth in 10 points, total roughness value and maximum roughness value for each scanning vector. Besides is displayed a chart roughness distribution for each vector, together with the limits of roughness high and low tolerances (figure 5).

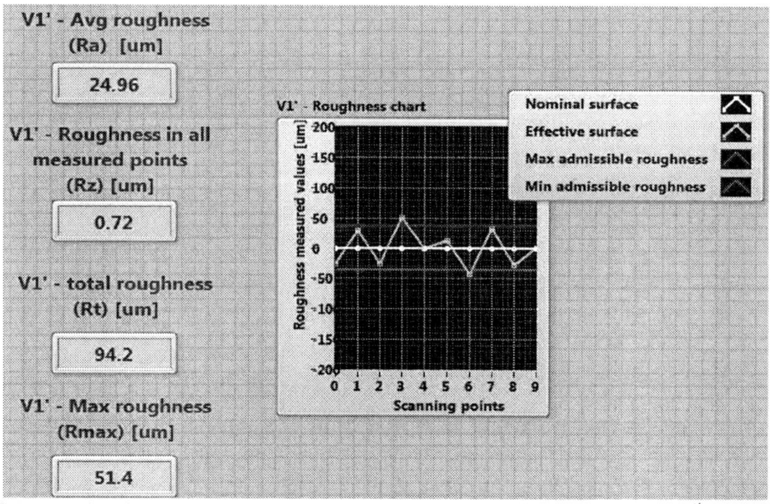

Figure 5. Example of generated information on roughness parameters for the 1st scanned vector

Applying the equation (5), the programming addressed the sequence of algorithm for roughness parameters concerning the entire scanned surface (figure 6).

A 3D chart allow also allows a spatial view of the roughness distribution chart of the scanned surface (figure 7). In terms of programming, a 3D Chart function was defined, being addressed as inputs the index of the scanned point per each vector, the index of the scanned vector and all roughness values for all scanning points.

IOP Conf. Series: Materials Science and Engineering **514** (2019) 012044 doi:10.1088/1757-899X/514/1/012044

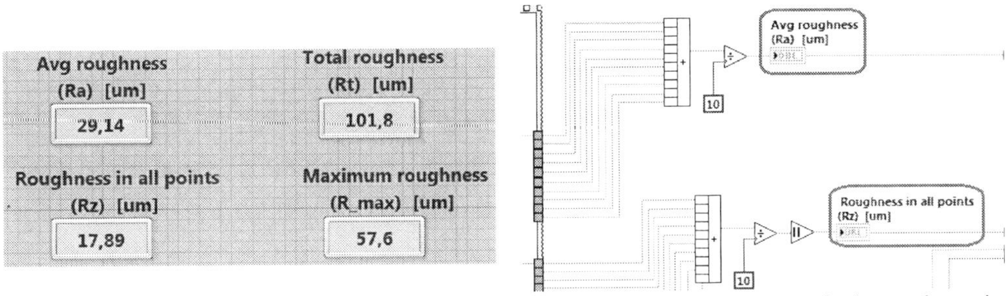

Figure 6. Programming routine for displaying the roughness parameters values referring to the entire scanned surface

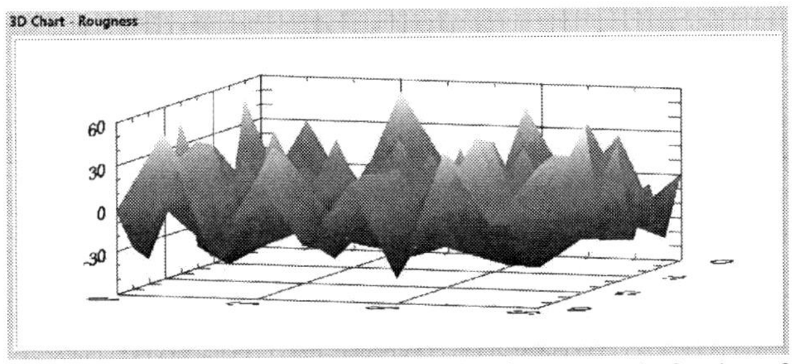

Figure 7. Example of 3D roughness chart obtained generated after interface running

However the most important information is on the verdict on the hip prosthesis surface quality standpoint (figure 8). Two LEDs of state and a selective text message inform the user if the hip prosthesis corresponds or not. Both LEDs are in gray if turned off, while the first one is green if turned on and the second is red if turned on. The first LED turns green together with the message *Hip prosthesis head surface corresponds* if all determined roughness values are lower than the imposed limits. The second LED turns in red together with the message *Hip prosthesis head surface does not correspond* if at least one of the roughness parameters is greater than the imposed limit (figure 8). To transpose this algorithm, when programming there was used a Case structure, containing a string text control with the positive message (case True) and another Case and a string text control with the negative message (False case). As output string variable, a text indicator was defined to receive either the positive, either the negative text, besides both LEDs (figure 8).

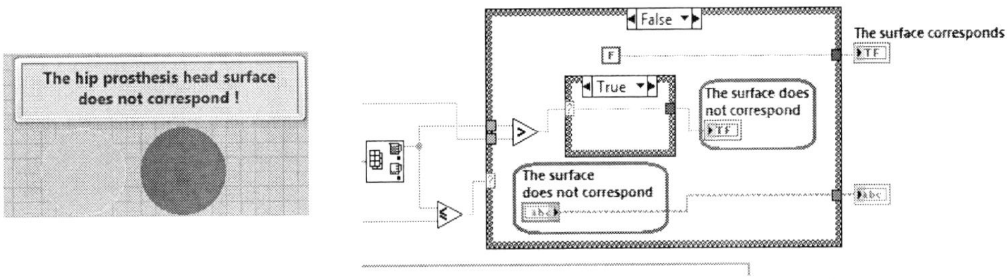

Figure 8. Example of negative verdict on a hip prosthesis evaluation

4. Results and conclusion

The method, together with the software interface has been successfully used for two hip prostheses, one for children (which head is 15 mm in diameter) and another for adults (which head is 18 mm in diameter). Both hip prostheses providing as rejected from an orthopaedic clinic, these proved to be non adequate neither surface quality standpoint. However, from the point of view didactic and research, the hip prostheses proved in all circumstances to be very useful.

Regarding the software interface, it proved to be very efficient and useful for a critical and objective analysis, providing in a very short time all the necessary information. As a result it could be widely used even in the stage of manufacturing. Besides, the proposed method, together with the interface could be used for any type of biomechanical joint prosthesis, not only for lower limbs.

5. References

[1] Steven A Gard 2003 Overview of Lower Limb Prosthetics, in the United States, Chicago, Illinois, *WRAMC and the VA Orthopedic & Prosthetic Workshop*

[2] Wilson C, Marco, A, Jean-Jacques, B, 2015, Conception, design and development of a low-cost intelligent prosthesis for one-sided transfemoral amputees, in *Reserch of Biomedical Engineering*, **31**, pp. 64-67

[3] Bar A, Ishai G, Meretsky P, Koren Y 1983 Adaptive microcomputer control of an artificial knee in level walking, in *Journal of Biomedical Engineering*

[4] Jingguo, W, Yangmin, L, Changwu, Q 2008 Analysis of dynamic stability constraints for a mobile humanoid robot, in Thailand, *Robotics and Biomimetics, ROBIO, IEEE International Conference*, pp. 377-382, ISBN: 978-1-4244-2678-2

[5] Baritz, M 2010 Correlated and interconnected analyses for human walking and standing biomechanical behavior, in *Proceedings of the 9th Wseas International Conference on Signal Processing, Robotics and Automation ISPRA, Book Series: Mathematics and Computers in Science and Engineering*, **10**, pp: 236-243, ISBN: 978-960-474-157-1

[6] Jose, M, Gomez, R 2013 Evaluation of Idiopathic Scoliosis based on Alignment, Equilibrium and Stability in *Gomez Orthotic Spine System, Torres, J Spine*, 2:5, **2**, ISSN: 2165-7939

[7] Brian, J., Ignacio, A., Sara, J., Dagmar, A. 2017 Construct Validity of the Prosthetic Limb Users Survey of Mobility (PLUS-M) in Adults With Lower Limb Amputation, in *Archives of Physical Medicine and Rehabilitation*, **98**, ELSEVIER, pp. 277-285

[8] AJRR 2016 Improving Orthopaedic Care Through Data. Annual Report, in *American Joint Replacement Registry*, pp. 15-20

[9] Ostendorf, M, Johnell, O, Malchau, H et al 2002 The epidemiology of total hip replacement in the Netherlands and Sweden, in *Acta Orthopaedica Scandinavica* 73, pp. 50 – 54

[10] Drugă, C 2011 Contribuţii la studiul comportării în exploatare a elementelor de protezare, *PhD Thesis*, scientific coordinator: Prof. Dr. Eng. Ileana Ro ca, pp 60 – 77

[11] Furey, M. 2003 Joint Lubrication, in *Biomechanics. Principles and Applications, CRC PRESS LLC, Washington D.C*, pp.75-79.

[12] Mathieu, L., Bernard, C., Felix, M., S., Linares, J., M., Jacob, J., C., Decherchi, P., Dousset, E. 2017 Biocompatibility of Four Common Orthopedic Biomaterials Following a High-Salt Diet: An In Vivo Study International, in *Journal of Molecular Sciences*, **18**

[13] Baritz M. 2015 Structural analysis by image processing of the multilayer ophthalmic polycarbonate lenses behavior during mechanical factors aggression, *Metalurgia International*, pp. 121-126

[14] Hip prostheses 2018 *Images for hip prostheses*, https://www.google.ro/search?q=hip+prostheses accesed in June 2018

[15] Rosca, I 2003 Tehnici i sisteme de măsurare, Noti e de curs, pp 25 – 31

[16] ISO 4287 1997 Terms. Definition and surface texture parameters

Acknowledgments

The researches were held in the Mechanical Engineering and Measurements Laboratories of the Product Design Faculty of Transylvania Brasov.

IOP Conf. Series: Materials Science and Engineering **514** (2019) 012045 doi:10.1088/1757-899X/514/1/012045

Virtual instrument for technical and economic standpoint evaluation of dimensional measurement methods and devices for the machined components finished products

B C Braun[1]

[1] Product Design, Mechatronics and Environment Department, Transilvania University, Brasov, Romania

E-mail: braun@unitbv.ro

Abstract. The paper presents a research stage, within PhD Thesis, focused on critical analysis of the solutions for quality insurance in the production activity, standpoint dimensional inspection. The key was the programming of a flexible and efficient software environment, which will help the operator to make the proper decision on the best method and optimal means of dimensional measurement of various categories of fabricated components. It is practical one and it could be very important for productivity increasing, for low costs and also for finite products high quality. To accomplish this software application it were defined as inputs several criteria, like: inspection process timing, measuring precision, devices costs, process costs and energetic consumption. The most part of these criteria (process duration, measuring precision, devices costs) was possible to be established very properly, due to some previous analytical and empiric results, associated to several tested measuring devices

1. Introduction

In the last decades, the exigencies on product quality have experienced a continued increasing. Engineering standpoint, for a product to be reliable and safe, it must respect serious of strict functional requirements (e.g. automotive, aeronautics, robotics, biomechanics (prosthetic and orthotic), medical equipments etc.) [1], [2], [3]. For this reason, the dimensional and form inspection and surface quality for some functional key parts in the structure of the final products play a crucial key [4], [5], [6].

Surface quality inspection standpoint, for some parts there it can be successfully used, specific measuring devices (roughness apparatus, distance or displacement transducers). However, actually, in a large range of applications, especially in case of deformable, tactile sensible parts or micro-probes, it was proved to be very useful and efficient to use digital high magnitude microscopes or image based dimensional equipments [7], [8], [9].

On the dimensional and form inspection, especially for functional role, a growing scale has the use of CMM and 3D portable scanners, for digitization and even for form modelling (if necessary).

Of the series of functional probes is the category of those who perform translational and especially rotational movements. In their case, in many situations, for dimensional inspection, especially for functional parameters control (e.g. axial or radial beating, misalignments etc.) as measuring devices are recommended the displacement or distance transducers, coupled to a PC, laptop, smart phone or I-phone. These have the main advantage of a low cost comparing with CMMs or 3D portable scanners, however there are a lot of situations in which, despite the high cost, it is recommended to use the latter [10], [11].

IOP Conf. Series: Materials Science and Engineering **514** (2019) 012045 doi:10.1088/1757-899X/514/1/012045

2. Principle of measuring devices choice for dimensional and functional inspection

Regarding the used devices and methods for quality inspection, is absolutely necessary to consider a range of essential aspects, like productivity, measuring accuracy, measuring range but also the costs concerning the acquisition and exploitation of the measuring device or equipment [12]. As a result, taking into account the productivity ask, the primordial criteria obviously is the productivity, meaning the efficiency standpoint measuring and inspection process. Besides this aspect, always must to consider the quality of the products, invoking certainly the measuring precision criteria. The measuring range could also become extremely important when relating to the nature and geometry of the parts composing the final products. Unfortunately, too high costs due to the acquisition, maintaining and exploitation of the measuring devices can often hinder obstacles in desired results obtaining standpoint quality insurance, especially via dimensional inspection [13], [14].

3. Software interface for optimal choice of measurement device

In the researches accomplished during the elaboration of the Doctoral Thesis, it has addressed, among others, the issue of critical analysis concerning the proper measuring device, starting from a concrete case, with the possibility to extrapolate for general cases. Starting from this idea, it was proposed a new solution for critical analysis to find the optimal device for dimensional inspection in case of revolution and translation parts [15].

To make it as flexible and efficient as possible, this critical analysis invoked the programming and developing of a software interface in Lab VIEW, that can be successfully used in a very large range of applications [15], [16].

3.1. Software interface description

The interface (O.E.M.A.vi), addresses, first of all, the above specified criteria. Thus means that, to have a conclusive information standpoint each criteria, when its running, the user has to complete, one by one several groups of text boxes, including all necessary technical details meaning using each one of the considered measuring device. Thus means that for all existing measuring systems, the input data can be easily and quickly be introduced by the user, similar to completing an on-line form. The groups to be completed all contain queries meaning technical and managerial data, which the operator can easily complete them. The first group of queries (figure 1) contain exclusive technical aspects on the inspection process:

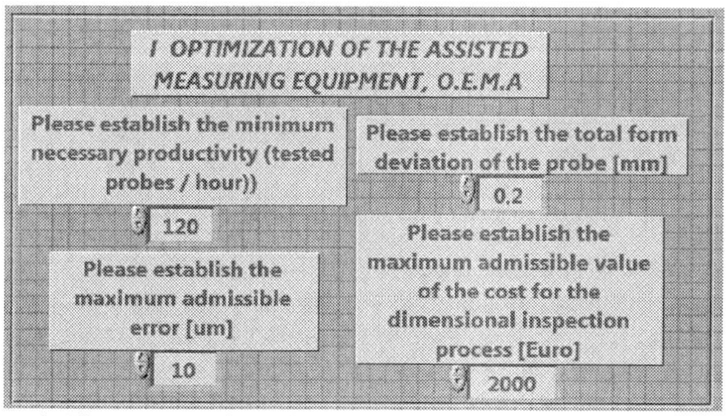

Figure 1. First group of queries as input information to be completed

Each query, as text-box, meaning input information was defined in the interface to be related to boolean operators, which role is to process the information to obtain, finally the verdict. An example in this regard can be observed in figure 2, one of the input query (as text- box input string) being related to some Boolean operators. In this case, the necessary productivity (known by the operator to be ensured) must to be compared with the automatic determined productivity (as output information), the last information being automatically generation also due to the interface running. In this way, the interface, after its running, can provide conclusive and very useful information regarding the used

IOP Conf. Series: Materials Science and Engineering **514** (2019) 012045 doi:10.1088/1757-899X/514/1/012045

device. In this case it is about the productivity standpoint (figure 2). The programming algorithm for the other input – output relationships is similar.

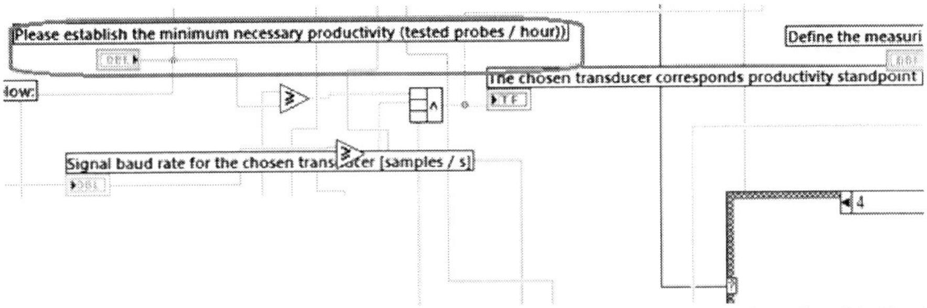

Figure 2. Necessary productivity, meaning an example of an input query, to be related in Boolean programming algorithm

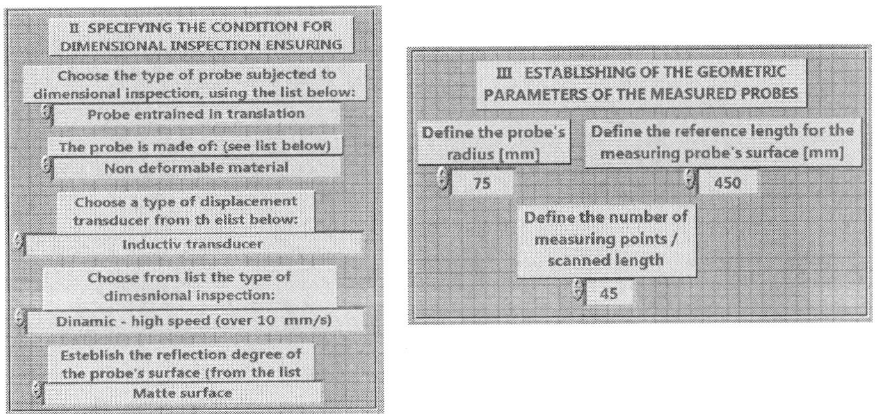

Figure 3. The 2nd and 3rd steps when input data introducing, meaning the second group of technical queries

The second group of queries, meaning the second step to be accomplished by the operator (figure 3) refers exclusively to the measuring technical conditions. All asked aspects are known by the operator, he knowing all process details, so that it will be easy to be completed. In this group, the main aspect is that each input means that the operator has to choose an option from a predefined list. For instance, referring the query on the displacement transducer to be used, it is possible to choose one from an existing list, mentioning 4 types of transducers (e.g inductive, incremental, pneumatic or with LASER transducer). The input information on geometric parameters of the probes to be inspected mean the third group of queries to be completed before interface running (figure 3).

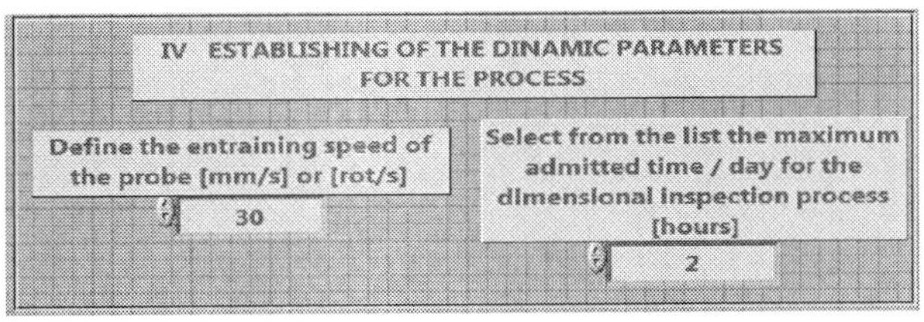

Figure 4. Specifying the process's dynamic parameters in terms of productivity

Also this category of information is very well known by the operator. After that, establishing the dynamic parameters for the inspection process is also very important, to have the information on productivity ensuring if using the selected measuring displacement transducer (figure 4). The cost criteria, one of the most important, must to be addressed as input queries in the 5th step, meaning costs information for each one of the existing measuring devices and also for the entraining system during inspection process is presented in figures 5 and 6.

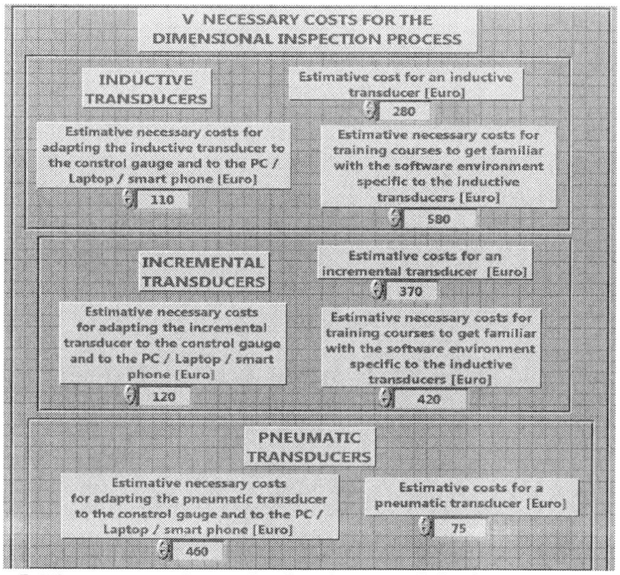

Figure 5. The costs queries on the first 3 types of displacement transducers

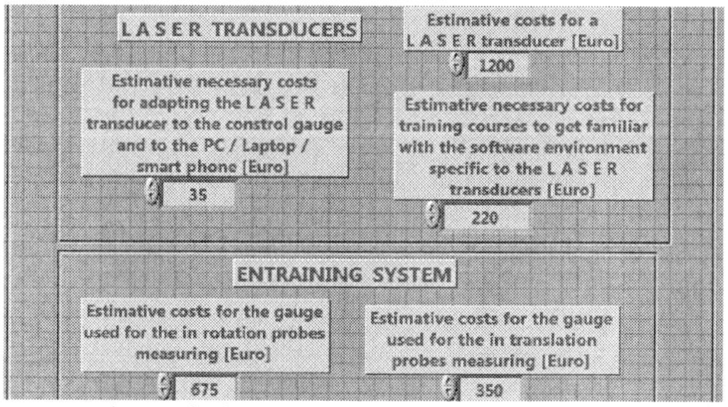

Figure 6. The costs queries on the last type of displacement transducers and entraining system

Once all group of input queries were completed, the next step is the interface running, a group of outputs (marked in blues) providing all useful and conclusive information for a verdict regarding the type of measuring device, in the previously specified technical and economical conditions (figure 7). The illustrated example presents a case in which the selected device (inductive displacement transducer) would satisfy 3 from 4 criteria (productivity, measuring range and costs, but not the measuring accuracy). Thus means that the in discussion transducer would not be the optimal one for the necessary inspection process. The rectangular state LED displays the final verdict on the analyzed measuring device to be used. In the presented case, the LED is turned off, meaning a negative verdict for the in case chosen transducer.

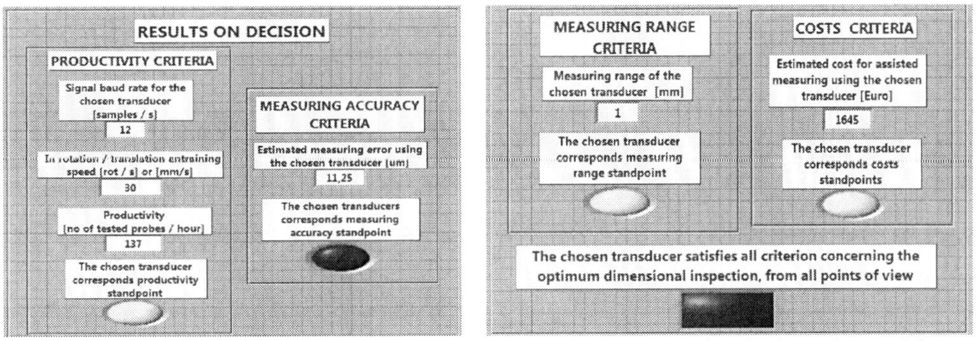

Figure 7. Example results on all criteria for one of the chosen measuring device to be used for a dimensional inspection process, due to the interface running

In terms of programming, in figure 10 there is presented an example of a sequence referring to the algorithm which uses Boolean algebra for data processing to obtain the final complete information.

The most important used function is a multiplexor AND type, having 4 Boolean inputs, being related to the 4 criteria specified in figures 8 and 9. Each one is shown as a Boolean output item (in green) (e.g. "The chosen transducer corresponds, costs standpoint"). The rectangular LED (as final verdict) shown in figure 9 corresponds to a Boolean item (framed in red in figure 8).

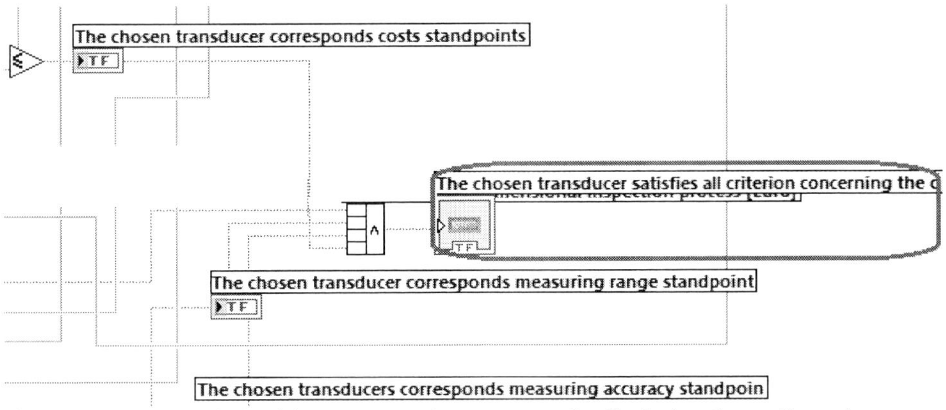

Figure 8. Sequence of algorithm programming to ensure the displaying the verdict and necessary information for the in case measuring device

4. Conclusions

In the Doctoral researches, all types of measuring devices were used, being known all technical and economic specifications. In these conditions, the presented software interface was programmed also as an important research activity. It was successfully tested for all used measuring devices. Finally, due to the use of the interface it was found that, for rotational and in translation functional probes (with application in aeronautics and automotive) the incremental proved to be the optimal one standpoint all criteria.

It was demonstrated that such of software interface could be successfully and easily used by almost any operator, specialized in metrology, on production lines. The most important aspect is that the interface, being user friendly, thus means that the staffs in field do not require any specific IT knowledge or skills.

It's using could lead to the manufacturing processes improving, especially productivity and costs standpoints, having a positive influence managerial standpoint.

5. References

[1] Kishu, M 2011 Quality assurance: Importance of systems and standard operating procedures, Perspectives in Clinical research, pp 34,35

[2] Cristea, L 2008 Research concerning the improvement of bearings quality and performances, using automatic systems, *The 19th INTERNATIONAL DAAAM SYMPOSIUM "Intelligent Manufacturing & Automation: Focus on Next Generation of Intelligent Systems and Solutions"*, DAAAM International Vienna, 22-25th October 2008, ISSN 1726-9679

[3] Giuseppe Di Leo, Liguori, C., Pietrosanto, A., Sommella, P. 2017 A vision system for the online quality monitoring of industrial manufacturing, Optics and LASERS in Engineering, *ELSEVIER*, **89**, pp. 162 - 168

[4] Micu C., Dodoc, P. 1980 Aparate şi sisteme de măsurare în construcţia de maşini, *Editura Tehnică Bucureşti*

[5] Militaru, C., Moldoveanu, M. 1991 Controlul dimensiunilor mari în construcţia de maşini, *Editura TEHNICĂ, Bucureşti*, ISBN 973-31-0305-5

[6] Cristea, L, Baritz, M 2008 Research on the improvement of bearings quality using automatic systems, *Proceedings of the 9th International Conference on Mechatronics and Precision Engineering, COMEFIM '9*

[7] Barreiro, J.; Alaiz, R.; Alegre, E. and Ablanedo, D. 2008 Surface finish control in machining processes using textural descriptors based on moments, *6th International DAAAM Baltic Conference "INDUSTRIAL ENGINEERING 24-26 April 2008*, Tallinn, Estonia

[8] Baritz M. 2015 Structural analysis by image processing of the multilayer ophthalmic polycarbonate lenses behavior during mechanical factors aggression, *Metalurgia International*, pp. 121-125

[9] Chenlong, L., Yongbo, Y., Mingyuan, Z. 2016 Uncertainty analysis of displacement measurement with Imetrum Video Gauge, ISA Transactions, *ELSEVIER,* **65**, pp. 547 - 555

[10] Ponsar, M., Chicot, J. et al 2008 Three - dimensional measurement with LASER, *The Romanian Review Precision Mechanics, Optics & Mechatronics*, **18**, pp 91 - 94

[11] COORD 3 Metrology 2015 Benefits of CMM Laser Scanning, *Metrology in Focus* http://www.coord3-cmm.com/benefits-of-cmm-laser-scanning/

[12] Braun B 2009 Contribuţii privind optimizarea echipamentului de măsurare de la automatele de control, Teză de Doctorat, conducător ştiinţific: Prof. Dr. ing. Ciprian OLTEANU, pp 86÷93

[13] QM for Windows, 2008 Modelarea şi simularea proceselor economice. Manualul utilizatorului

[14] Cristea, L, Baritz, M, Roşca, I 2006 Modular reorganisation of the automatic system for the micro-bearings dimensional inspection, *International Workshop "Advanced Researches in Computational Mechanics and Virtual Engineering", COMAT*, , ISBN 973-635-823-2

[15] Braun, B., Drugă, C.; Costaş, C 2008 The experimental determining of the dynamic dimensional control about the measuring precision, *6th International DAAAM Baltic Conference "INDUSTRIAL ENGINEERING 24-26 April 2008, Tallinn, Estonia*

[16] Braun, B., Drugă, C., Olteanu, C 2007 The statistical evaluation of the aided by pc measuring results, via Lab VIEW, *The 2nd International Conference "Computational Mechanics and Virtual Engineering" COMEC 2007*, ISBN 978 -598 –117-4

Acknowledgments

The research presented in the paper was accomplished in the Doctoral stage, due to the CNCSIS national grant, code 293, awarded to young PhD students, in the period of years 2005 and 2006.

IOP Conf. Series: Materials Science and Engineering **514** (2019) 012046 doi:10.1088/1757-899X/514/1/012046

Method to model the spinal column shape

A M Vutan[1], C M Gruescu[1], E C Lovasz[1] and V Ciupe[1]

[1] Mechatronics Department, University Politehnica of Timisoara, Romania.

E-mail: corina.gruescu@upt.ro

Abstract. Correct evaluation of the spine is important in detecting the deficiencies that may occur at this level, especially in case of scoliosis. It is proposed a new method of acquiring the coordinates of the column in order to generate a model of it. A number of eight coordinates has been chosen to use, a big enough number to assure a true model of the spine and a wide area of possible mathematical approaches. The acquiring system consists of eight accelerometers, from which is collected the pitch angle. This angle, as the angle of inclination in the frontal plan is used to compute the pairs (x, y), considering known the distance between the sensors. The accuracy of the model is validated by comparison with images taken during the measurement.

1. Introduction

Spinal deformities occur frequently in children during the period of growth most of them not having a precise cause but can also occur in adult population, due to lack of physical activity and long-term sitting position [1, 2, 3]. Researches in this field are widely spread and continuously developing, as the current schemes of treatment are long-lasting, costly and not very effective.

In order to asses a spinal deformity it is necessary to evaluate the shape of the column. In this regard, the most commonly used method of investigation is the X-Ray image. However, due to its invasive nature and limitation in providing information, a wide range of alternative methods of investigation have developed during the last decades. The purpose of seeking for new methods of investigation is to assure a non-invasive character, the electronic storage of data and richer information.

The shape of the column and different parameters associated with posture can be obtained through different methods, among which the most common are raster stereography [4], ultrasonic mapping [5, 6], thermography, image processing [7, 8], mapping based on information provided by different sensors [9, 10]. The researches in the field of designing new methods to generate mathematical models of the spine are in progress, as none of the existing ones provides enough efficiency, accuracy and realiability.

The authors of the present paper propose a method of investigation based on the use of accelerometers, which provide data for computing a set of coordinates of the column and allow the generation of a mathematical model.

2. Equipment and processing of data

The method proposed to acquire data is based on the use of a set of accelerometers attached to the back of the patient (figure 1 a). The system consists of a set of 8 MEMS IMUs connected via their I^2C bus to an I^2C multiplexer and finally to a microcontroller board (figure 1 b) [11]. The accelerometers provide the angles relative to the vertical axis, within the frontal plan (pitch angles).

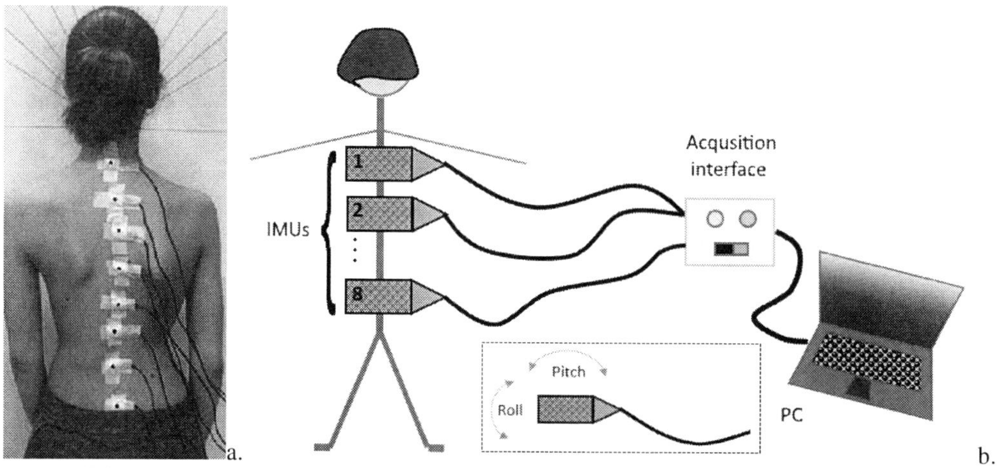

Figure 1. Scheme of the sensors applied on the column and acquisition system. [11]

The angles φ_i (i= 1... 8) and the distance l_i between the sensors (fig. 2) allow the computation of their coordinates (x_i, y_i):

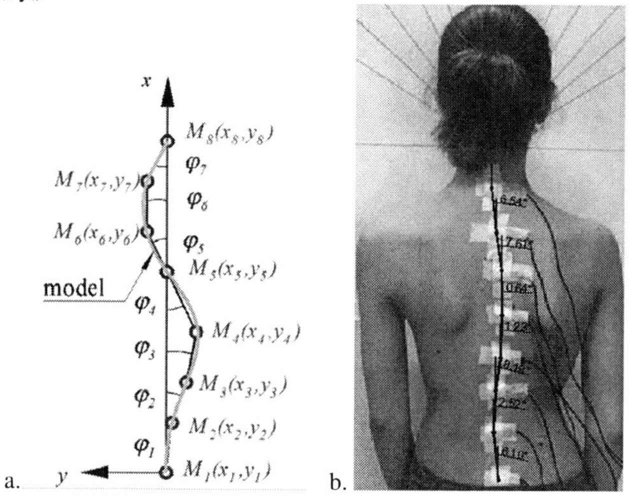

Figure 2. Points and associated parameters. a – theoretical model; b – illustrative example.

$$\begin{cases} x_{M1} = 0 \\ y_{M1} = 0 \\ x_{Mi+1} = x_{Mi} + l_i \cos\varphi_i & i = \overline{1,8} \\ y_{Mi+1} = y_{Mi} + l_i \sin\varphi_i & i = \overline{1,8} \end{cases}. \quad (1)$$

The reference system attached to the application originates in the first point. The coordinates so obtained are used to get an approximation function, which was chosen to be a fifth degree polynomial. In order to further process the shape of the column it is necessary to validate the accuracy of the acquired data. For this purpose, during the measurement were taken photos of the patient with the sensors attached. The centers of the sensors are emphasized with black point markers (fig. 1a). The images were processed so that the angles φ_i resulted distinctively. They were used to trace the shape of the column through a mathematical approach similar to the one applied to the experimental results.

3. Experimental results

The experimental program is based on measurements, which were performed with the consent of patients and their parents.

The patient with a medium "S"-shaped scoliosis was asked to tilt starting from the orthostatic posture, to the right and then to the left in progressive tilting positions with the step of 10 degrees between positions. In each position, a photo has been taken as a reference needed to validate the further mathematically generated shape.

Figure 3 explains the steps of the comparative procedure.

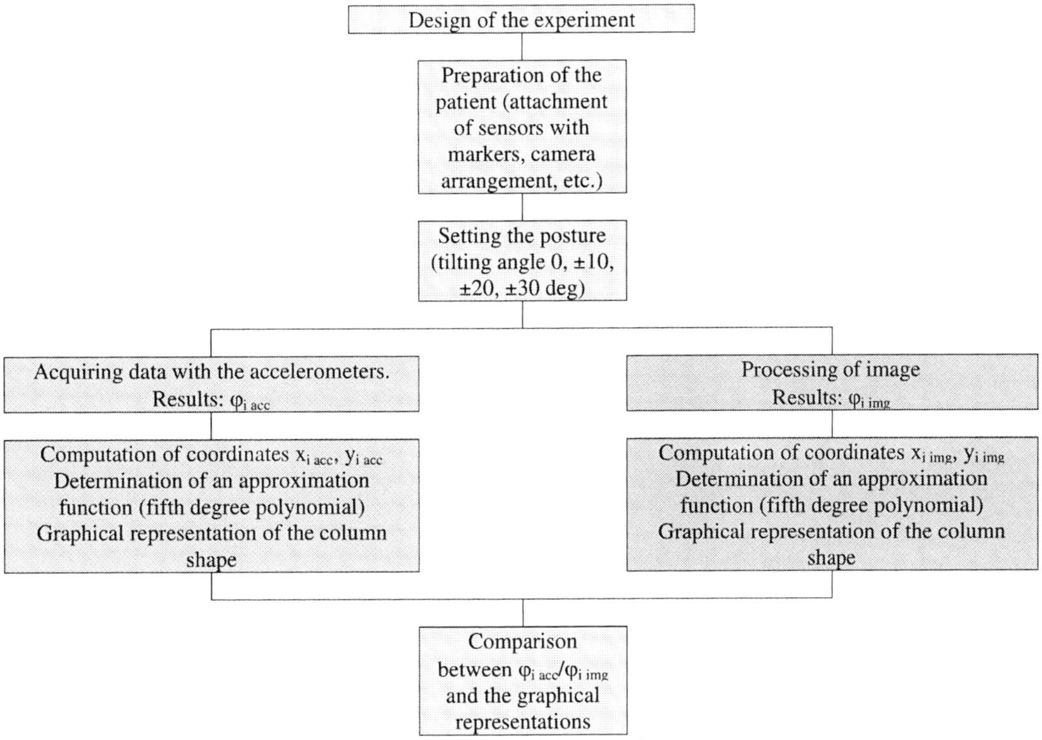

Figure 3. Design of the experiment.

The results of the procedure above are shown in figures 4 … 8, for the orthostatic posture (0 deg) and four tilted positions (± 10 deg, ± 30 deg). For each position are placed side by side the processed image, the graphical representation of the column got with the sensors (blue) and from the image (red) and the pitch angles resulted from the two approaches.

Comparing the values of the angles and the shapes of the column got through processing data from accelerometers and from direct images respectively, one can make the following notices:

- The reference standard is considered to be the image, which provides sure information regarding the shape of the column.
- The values of the angles are very similar. It is important to observe that the signs of the angles are entirely the same, which is determinant for the shape of the column. Therefore, both methods provide similar concavities, which is relevant for the qualitative and quantitative evaluation of the deformity.

- The difference between the angles is reasonable considering that the measurements are performed onto a biological system, which can not be considered rigid and still. Furthermore, the image sequence is taken at a single moment during a period of several seconds assigned to data acquisition with the accelerometers. The angles from the accelerometers are average values resulted from files recorded in an interval of five seconds, at a frequency of 2Hz.

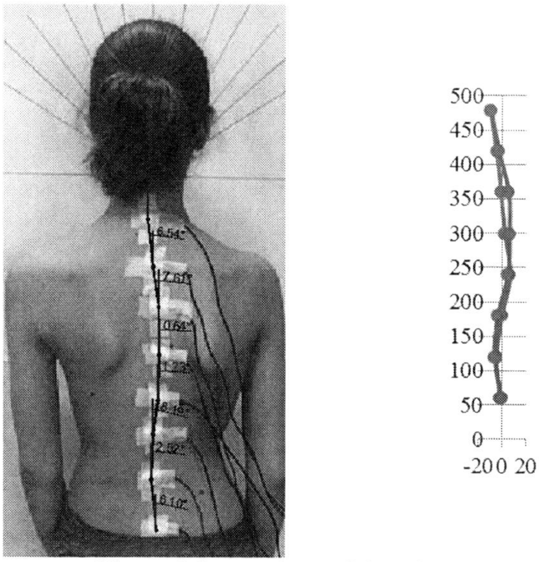

φ [deg]	
Sensors	Image
-2.77	-6.1
4.46	2.52
6.54	8.19
-4.32	1.23
-3.00	-0.64
-3.70	-7.61
-5.11	-6.54

Figure 4. Image, shape of the column and values of pitch angles (0 deg).

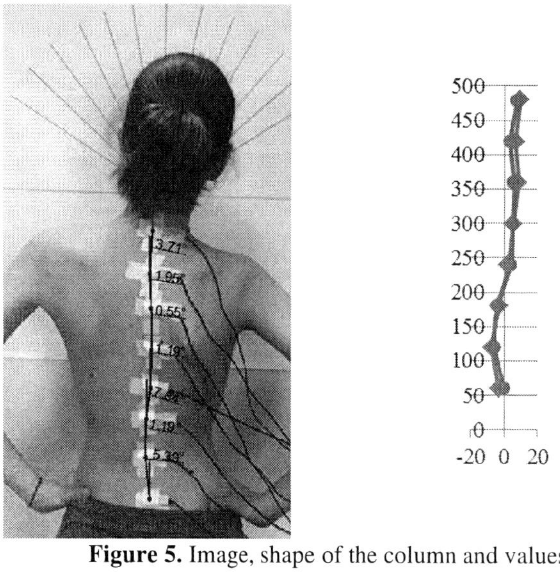

φ [deg]	
Sensors	Image
-3.62	-5.39
4	1.19
5.2	7.84
2.97	1.19
2.73	0.55
-1.26	-1.95
2.5	3.71

Figure 5. Image, shape of the column and values of pitch angles (+10 deg).

- The shapes of the column, represented on the same graph, are very similar. The differences of the coordinates (x, y) are within the range of units (millimetres), which is considered acceptable.

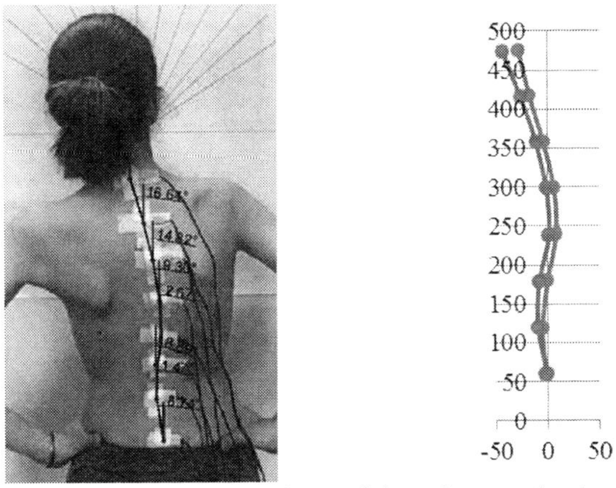

φ [deg]	
Sensors	Image
-2.65	-8.74
4.93	1.47
7.78	8.89
-2.06	-2.67
-9.73	-9.30
-12.93	-14.82
-10.58	-16.64

Figure 6. Image, shape of the column and values of pitch angles (-10 deg).

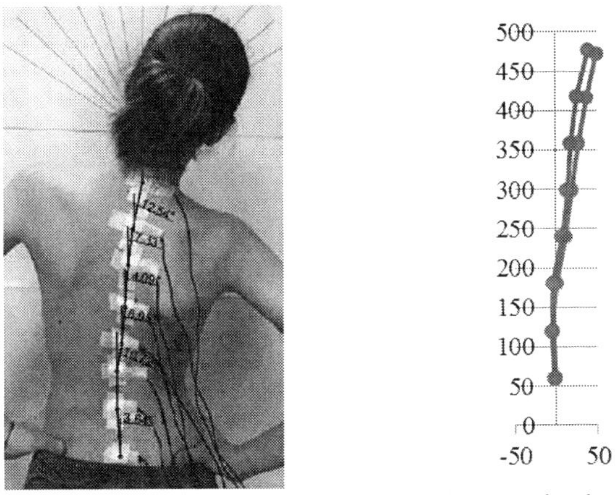

φ [deg]	
Sensors	Image
-2.49	-3.64
6	0
9.6	10.72
7.71	5.95
7.55	4.09
9.5	7.13
11.5	12.54

Figure 7. Image, shape of the column and values of pitch angles (+30 deg).

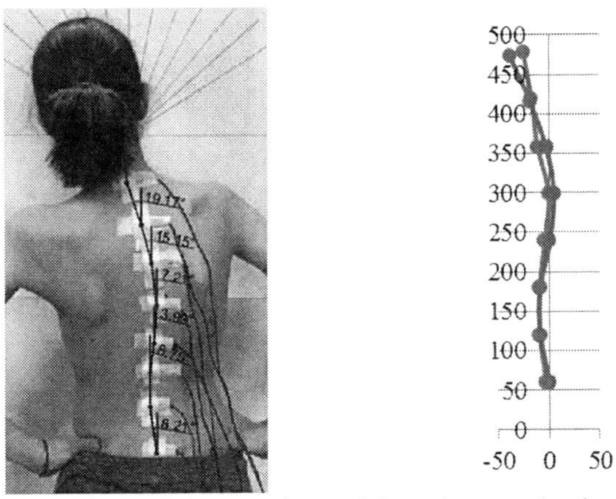

φ [deg]	
Sensors	Image
-5.75	-8.21
0	0
4.74	8.75
3.73	3.99
-10.22	-7.21
-5.6	-15.15
-7.69	-19.15

Figure 8. Image, shape of the column and values of pitch angles (-30 deg).

The discussion above allows the conclusion that the method based on acquiring data with the accelerometers is valid. It is considered reliable to be used in further measurements, planned to develop clinical studies regarding vertebral column deformities, such as early diagnosis of scoliosis and subsequent controls but also in assessing the effectiveness of exercises performed in scoliosis rehabilitation programs.

4. Conclusions

Beside the existing methods of investigation of the spinal column, alternative to the traditional X-Ray image, the authors propose a new method based on data acquired from a set of accelerometers, attached on the skin of the patient's back.

As new method, it needs validation by comparison of its results with reference data provided by a sure method. The reference is considered the image taken during acquiring data with the accelerometers.

The results obtained with the new method and the reference data are very similar in regard of pitch angle values and graphical shape. Graphical similarity also validates the mathematical approach, consisting in determination of approximation functions, which were chosen as fifth degree polynomials.

The results of the present work are intended to be used in a study regarding the change of curvatures during exercises conducted by a kinetotherapist within the treatment of scoliosis, with school-age children.

References

[1] Juskelene V, Magnus P, Bakketeig LS and Dailidiene N 1996 Prevalence and risk factors for asymmetric posturein preschool children aged 6-7 years *Int. J. of Epidemiology* **5** 1053-59

[2] Linton SJ, Hellsing AL, Halme T and Akerstedt K 1994 The effects of ergonomically designed school furniture on pupils' attitudes, symptoms and behavior *Applied Ergonomics* **5** 299-304

[3] Jianu M 2010 *Pediatric scoliosis* (Bucharest: Pro)

[4] Terheyden JH, Wetterkamp M, Gosheger G, Lange T, Schulze Bövingloh A and Schulte TL 2018 Rasterstereography versus radiography for assessing shoulder balance in idiopathic scoliosis: A validation study relative to patients' self-image *Journal of Back and Musculoskeletal Rehabilitation* 1-9 DOI: 10.3233/BMR-170867

[5] Zheng R, Hill D, Hedden D, Moreau M, Southon S and Lou E 2018 Assessment of curve progression on children with idiopathic scoliosis using ultrasound imaging method *Eur Spine J.* 1-6 doi: 10.1007/s00586-017-5457-0.

[6] Lovasz EC, Mateas M and Vilmos F 2003 Geometry determination of the Human Spine in 3D *Proc. Inter-Ing. Conference, Tg- Mures, Romania* 131-137

[7] Gruescu CM, Garaiman A and Lovasz EC 2015 Modeling of human spinal column and simulation of spinal deformities *Mechanika* **21** 214-219

[8] Mateas M, Vacarescu V, Lovasz EC, Nemes IDA and Mircea M 2007 About a new method for the 3D determination of the human spine geometry *Proc. of The 12th IFToMM International Congress, Besanson, France* 300-305

[9] Ciobanu O, Ciobanu G and Aiello A 2017 An application of Kinect sensor for scoliosis and kyphosis screening *Proc. of the International Conference on Medical and Biological Engineering (CMBEBIH)* 713-717

[10] Voinea GD, Butnariu S and Mogan G 2017 Measurement and Geometric Modelling of Human Spine Posture for Medical Rehabilitation Purposes Using a Wearable Monitoring System Based on Inertial Sensors *Sensors* **17**(1) 3

[11] Vutan AM, Ciupe V, Gruescu CM and Lovasz E-C 2017 Experimental method for dynamic evaluation of spinal column deformation exercises *New Advances in Mechanism and Machine Science, Mechanisms and Machine Science* **57** 137-146 https://doi.org/10.1007/978-3-319-79111-1_13

IOP Conf. Series: Materials Science and Engineering **514** (2019) 012047 doi:10.1088/1757-899X/514/1/012047

Novel solutions for actuation of an above elbow prosthesis

EC Lovasz[1], CM Gruescu[1], C Sticlaru[1], CE Moldovan[1] and M Stoian[1]

[1]Mechatronics Department, University Politehnica Timisoara, Romania

Email:corina.gruescu@upt.ro

Abstract. The above elbow prosthesis must provide the movement of the elbow, which naturally is composed by a flexion – extension movement. The mechanical system must fulfil a set of conditions regarding kinematics, dynamics, precision, size and load. The paper describes two novel solutions to generate the flexion – extension movement of the elbow. The description contains the kinematic scheme, elements of calculus and the CAD generated constructive solution.

1. Introduction

The limbs of the human bodies, whose functions have been attenuated or cancelled by traumas or accidents, can be partially or totally replaced with proper prosthesis. Regarding the upper limb and according to the level of amputation, the prosthesis can be [1]:

- From shoulder
- Above elbow
- Below elbow.

The above elbow prosthesis can be a very complex technical assembly because of its specific mobility within an open kinematic chain. The movement of the natural elbow is composed by a wide amplitude flexion – extension movement (~150 deg).

The market offers a large range of products, which embody different solutions regarding the actuation, the control, mechanical system etc. [2 ... 6].

The control of prosthesis recorded an obvious progress from command operating with myoelectric or piezoelectric signals to direct cortical signals [7, 8]. The trend is to get more and more sophisticated ways of control with the purpose of making the prosthesis easier to handle by its user.

The actuation system consists of miniature electric/hydraulic motors, according to the technological evolution, which made possible smaller and smaller components.

The mechanical system which conveys the energy into movements with given trajectory and amplitude, alike the natural behaviour is still subject to improvement.

These systems are kinematic chains [9], on which the following minimum set of initial conditions is imposed:

- Trajectory of the final point (and, eventually, of other points on the chain)
- Velocity of displacement
- Precision of displacement
- Maximal handling load
- Overall dimensions
- Overall weight.

The existing solutions are based on complex linkages [10], gears [11], wires and pulleys or combinations of these. The present paper proposes two novel solutions to generate the flexion –

extension movement of the elbow:

- Mechanism with bevel gears and wires
- Mechanism with linkages and gears.

2. Elbow prosthesis driven by a mechanism with bevel gears and wires

The prosthesis consists of two jointed parts corresponding to the arm and the forearm. The mechanism contains a bevel gear train and a wire transmission (fig. 1).

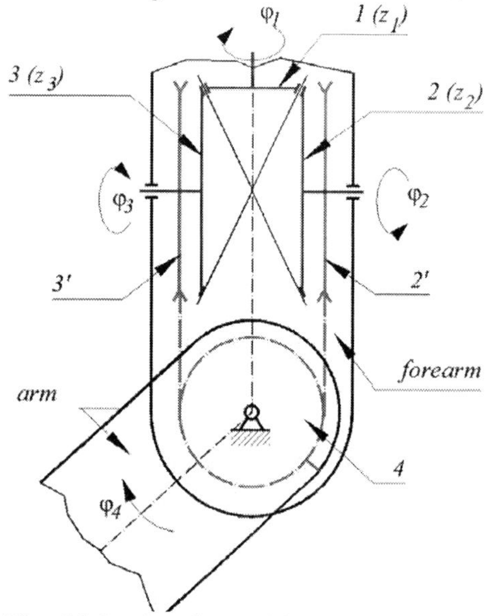

Figure 1. Kinematic scheme of the Mechanism with bevel gears and wires

The driving bevel gear (1) meshes with the driven identical bevel gears (2) and (3), placed in symmetrical positions in order to generate identical antagonistic angular displacement φ_2=-φ_3. The shafts of the bevel gears (2) and (3) are mounted in bearings held in the housing of the forearm. On the shafts are fastened the identical pulleys (2') and (3') with antagonistic movement and working in a wire transmission that also includes the pulley (4). The pulley is jointed in the centre of the elbow joint, which is fixed in the housing of the arm allowing the rotation of the forearm in regard with the arm. The value of the rotation angle in the elbow is limited by attaching the wire to the pulley (4).

By actuating of the drive bevel gear the wire is developed from the pulley (2) drive the pulley (4) and enveloped around the pulley (3). The computation of the transmission function of the actuating mechanism is given by considering the transmission ratios with the relationship:

$$\varphi_4 = \frac{z_1 \cdot r_{2'}}{z_2 \cdot r_{4'}} \cdot \varphi_1 , \qquad (1)$$

where: φ_1, φ_4 is the angular displacement of the drive gear and pulley transmission,
z_1, z_2=z_3 - the number of teeth of bevel gears,
r_2=r_3, r_4 - the radii of the pulleys.

Considering the entrance data:

- $z_1 = 16$
- $z_2 = z_3 = 32$
- $r_{2'} = 8\,mm$
- $r_{4'} = 16\,mm$,

the transmission function is linear as it is given in figure 2.

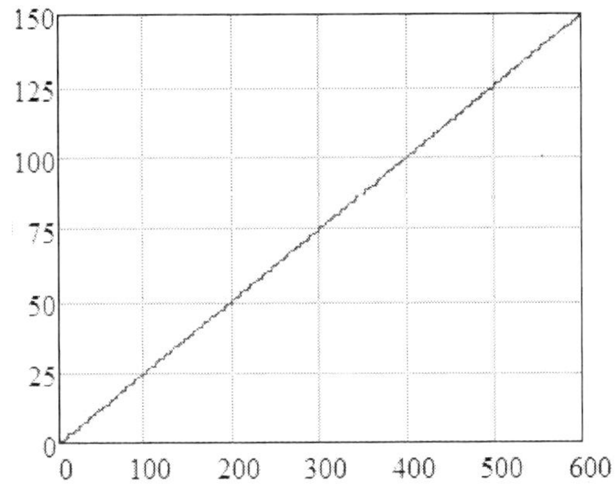

Figure 2. Transmission function of the mechanism with bevel gears and wires

The actuation of the mechanism is assured by a stepper motor series. The mechanism in extreme positions, corresponding to maximum flexion and extension, is represented in figure 3 a, b as CAD generated models.

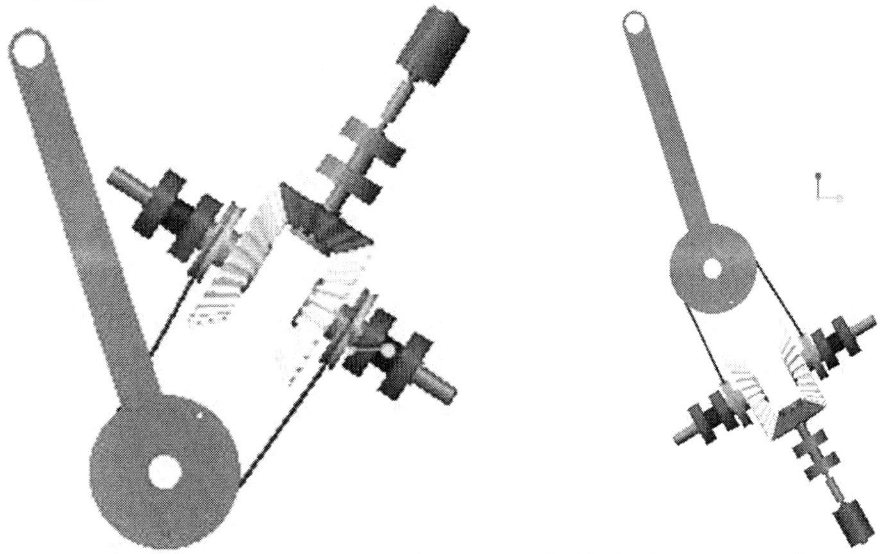

a. Mechanism corresponding to maximum flexion

b. Mechanism corresponding to maximum extension

Figure 3. CAD model of the mechanism with bevel gears and wires

The system is designed to handle on object of 2.5 kg along an extension of 150 deg by an angular displacement of the stepper motor of 600 deg. The maximum angle of rotation of the driving element (1) is 1200 deg. Considering the length of the forearm of 250 mm, the torque needed to handle the imposed mass is 6.25 Nm.

3. Elbow prosthesis driven by a mechanism with geared linkages

The mechanism is conceived so that the linear actuator (2 and 3) is attached to the forearm and the driven element – spur gear (5) is fixed with the housing of the arm (fig. 4 a and b). The joint denoted B_0 corresponds to the elbow joint.

A gear segment is attached to the slider (3), which meshes with the driven spur gear (5).

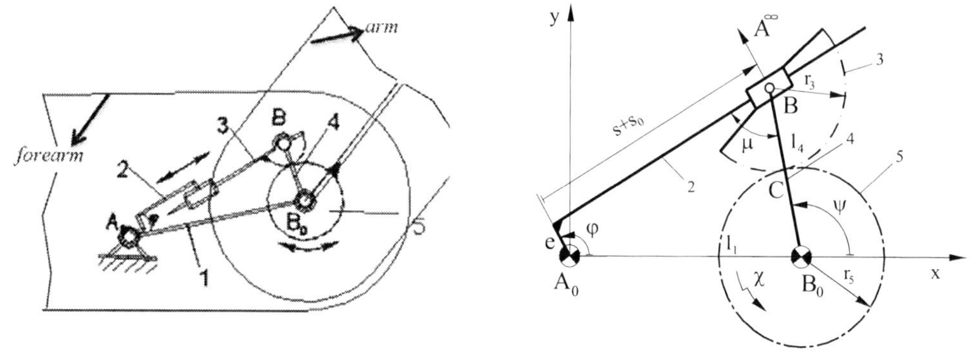

a. Overall scheme b. Kinematic schema

Figure 4. Scheme of the mechanism with linkages and gears

The mechanism is actuated with the linear servo motor. The transmission function of the geared mechanism [12, 13] can be computed based on the scheme in figure 4b as follows:

$$\chi(s) = (1-\rho)\cdot\psi(s) + \rho\cdot\varphi(s), \tag{2}$$

where: $\chi(s)$ - output angle or transmission function,

$\rho = \pm r_3 / r_5$ - gear ratio

$\varphi(s), \psi(s)$ - transmission functions of the base four-bar linkage:

$$\varphi(s) = 2\cdot\mathbf{arctan}\left((B(s) \pm \sqrt{A(s)^2 + B(s)^2 - C(s)^2})/(A(s) - C(s))\right), \tag{3}$$

with:

$$\begin{aligned} A(s) &= 2\cdot l_1 \cdot e_S, \\ B(s) &= 2\cdot l_1\cdot(s_0 + s), \\ C(s) &= -l_1^2 - e_S^2 - (s_0+s)^2 + l_4^2. \end{aligned} \tag{4}$$

and

$$\psi(s) = \mathbf{arccos}\left((e_S^2 + (s_0+s) - l_1^2 - l_4^2)/2l_1l_4\right). \tag{5}$$

The synthetized mechanism considered the lengths of the linkages (l_1 = 100 mm; l_4 = 21 mm) and the transmission ratio ($i = r_5/r_3 = 0.5$) known and provided:

- minimum transmission angle of the base mechanism: μ_{min} = 35 deg
- stroke of the slider: s = 32.3 mm (for the initial position s_0 = 84 mm).

The transmission functions (fig. 5, 6 and 7) shows that the angular displacement of the forearm is about 150 deg according to the imposed design parameters.

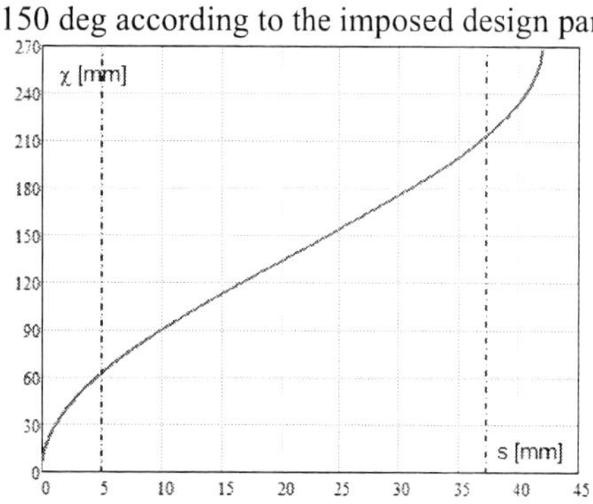

Figure 5. Transmission function 0-ord

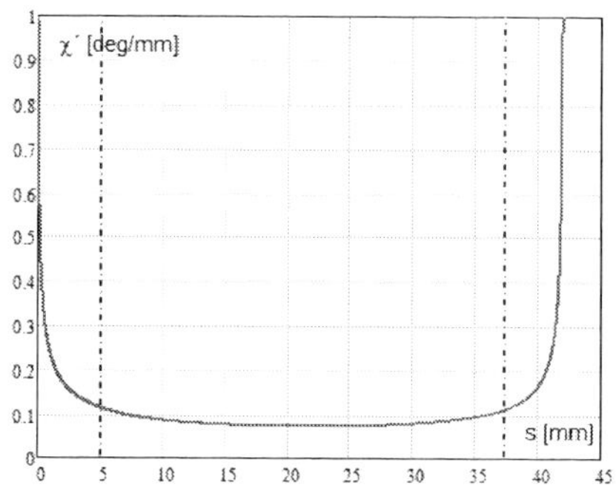

Figure 6. Transmission function 1-ord

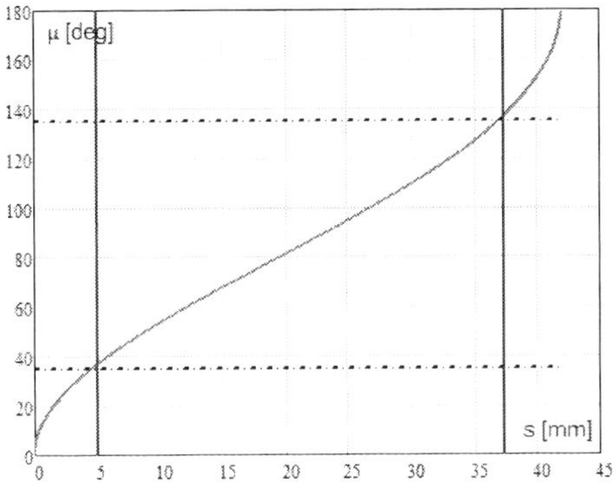

Figure 7. Transmission angle of the base mechanism

The mechanism in extreme positions, corresponding to maximum flexion and extension, is represented in figures 4 and 5 as CAD generated models.

Figure 8. CAD model of the mechanism with linkages and gears – maximum extension

Figure 9. CAD model of the mechanism with linkages and gears – maximum flexion

4. Conclusions

Two solutions of mechanisms to generate the flexion – extension movement of the elbow are proposed. Both solutions fulfil the conditions regarding the amplitude of the angular movement. Furthermore, they are compact in construction and small in size. The elements do not imply special geometry, so that they are easy to machine at low cost. The actuation force or torque is small for usually tasks of the human hand.

The both mechanism structure generate linear or approximately linear transmission functions, which are favourable to be controlled and actuated, in order to generate the oscillating motion of ~150 deg.

5. References

1. Kulkarni T, Uddanwadiker R 2016 Overview: Mechanism and Control of a Prosthetic Arm *Molecular & cellular biomechanics: MCB* **12**(3) 1-49
2. http://www.ottobock.ro/proteze/extremitatea-superioară/prezentare-generală/sistemul-protetic-transhumeral-dynamicarm/
3. http://www.tehnortopro.ro/produs/proteza-de-brat-actionata-prin-cablu/
4. http://theranova.ro/products/proteze-membru-superior/
5. http://ortoprofil.ro/produse/proteza-de-brat-functionala-actionata-prin-cablu/
6. http://www.rosal-ortopedic.ro/catalog/produse/proteze/proteze-membrul-superior/proteza-de-dezarticulatie-de-cot/
7. Velliste M, Perel S, Spalding M, Whitford AS, Schwartz A.B. 2008 Cortical control of a prosthetic arm for self-feeding *Nature* **453** 1098–1101
8. Gilja V et al. 201) A high-performance neural prosthesis enabled by control algorithm design *Nature Neuroscience* **15** 1752–1757
9. Matsyuk I N, Shlyahov EM, Yehurnov OI 2018 Some aspects of synthesis of linkage of complex structures *Geotechnical and Mining Mechanical Engineering, Machine Building* **3** 57-63
10. Escudero AZ, Álvarez J, Leija L 2001 Inverse kinematics for a parallel myoelectric elbow. Conf. Engineering in Medicine and Biology Society. *Proc. of the 23rd Annual International Conference of the IEEE* **22**(2) 1427 – 1430
11. Casolo F, Cinquemania S, Giberti H 2017 Optimization of Myoelectric Elbow Prosthesis Transmission *International Journal of Applied Engineering Research* **12**(21) 10909-10920
12. Lovasz E-C, Modler K-H, Pop C, Margineanu D, Maniu I 2018 Type synthesis and analysis of geared linkages with linear actuation *Mechanika* **24**(1) 108-114
13. Lovasz E-C, Mărgineanu D, Ciupe V, Maniu I, Gruescu CM, Zăbavă ES, Stan SD 2017 Design and control solutions for haptic elbow exoskeleton module used in space telerobotics *Mechanism and Machine Theory* **107** 384–398

IOP Conf. Series: Materials Science and Engineering **514** (2019) 012048 doi:10.1088/1757-899X/514/1/012048

Analysis of postural balancing movements in case of gait cycle in people with visual dysfunctions

M I Baritz[1]

[1]Product Design, Mechatronics and Environment Department, Transilvania University of Braşov, Romania

E-mail: mbaritz@unitbv.ro

Abstract. In the literature, it is often mentioned that uncorrected visual dysfunctions can influence the behaviour and motor activities of individuals, thus causing a limitation or diminution in the amplitude or frequency of their actions. The effects of visual dysfunctions of any kind may be amplified and can vary in time and in varying degrees both the posture, balance, stability and the type of gait of the patients. In this paper are presented the general aspects related to the identification of the parameters that affect the posture and the stability through the visual system dysfunctions and also the ways of compensating some of the causes that determine the postural imbalances. In the second part of the paper the experimental system and the recording methodology that emphasizes the effect of limitations in the functioning of the visual system, on the stability and the cycle of gait is presented. In this regard, a sample of 5 subjects with an average age of 25.1 years, with a visual emmetropic system, were induced a visual dysfunction related to the erroneous perception of the ambient colours and the postural equilibrium during gait cycle time. In the third part of the paper are presented the results and the conclusions obtained by applying the experimental methodology developed in this research.

1. Introduction

As shown in various papers presenting definitions and general aspects of the biomechanics of the human body in a classic way," ideal postural alignment (viewed from the side) is defined as a straight line that passes through the: ear lobe, bodies of the cervical vertebrae, acromion, bodies of the lumbar vertebrae, slightly posterior to the hip joint, slightly anterior to the axis of the knee joint, just anterior to the lateral malleolus". [1] A correct posture is represented by the normal state of balance between the muscular and skeletal system and as such posture is a continuous action to change the ratio between the muscular effect and the positioning of the skeleton, depending on the activity of the subject. At the same time, the fatigue state installed after a period of motor activity causes a decrease in muscle stability, an excessive movement of the spine accompanied by a stress sustained in the skeletal structure or an increase in pain and pressure in the nervous system. The normal posture kept for a long time brings a number of benefits including: less muscle fatigue, better health, better functioning of internal organs, and generally a well-being and self-confidence. These manifestations are identified in human subjects both, standing and sitting or lying, the different being related to the supporting surface and to the anthropometric dimensions of the human body. If the subject's activity, also includes complex movements of the human body (movement of the arms, legs or neck / head and of course walking on different surfaces), then the musculoskeletal coordination and posture with the sensory systems are the most important neuro-motor actions, controlled by the "central unit" - the brain. The position as sitting position but on the same place requires a more detailed analysis of the use of the upper part of the human body, predominantly of the hands, trunk or head and eyes. Therefore, the proprioceptive and vestibular sensory systems along with the visual and auditory ones will substantially influence the static and dynamic postural equilibrium level.

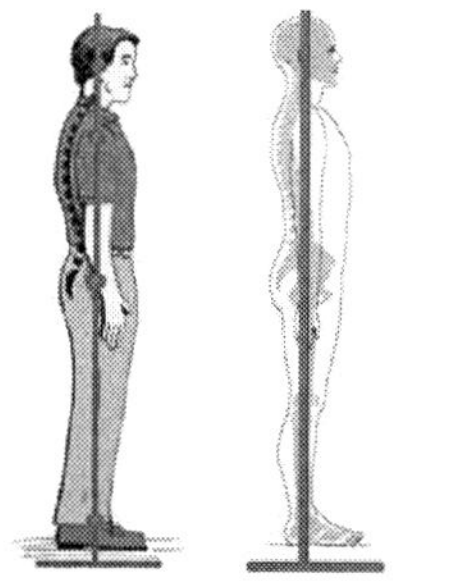

Figure 1. Standing normal posture of human body

Another important element is the working surface and the difficulties the human subject manages, since any overthrow of the anthropometric limits (forcing the worker to perform a certain manoeuvre) can lead to a change in the postural balance, the balance on the period of movement or the induction of musculoskeletal problems with pathological effects.

Various ergonomic and biomechanical researches [2,3,4] have highlighted that incorrect manipulation of objects, weights (volumes) too large, displacement of the human subject on horizontal or vertical distances with these objects can trigger the spine (the most affected by these activities), a series of pathologies requiring long recovery periods, complex methods and means of rehabilitation of dysfunctions and temporary or even permanent incapacity to work. Sometimes these common problems are also associated with visual dysfunctions, thus increasing the difficulty of action of the human subject. This is all the more important as the movements are mainly controlled and coordinated by the visual system and as such it is not only used to form images of objects but also to provide information on the visual cortex about the position and orientation of the movement of the body of the human subject. The central area of the field of vision perceived by the eyeball determines that the subject can clearly distinguish small objects with very good spatial resolution and optimal control over colour perception and contrast differences. The peripheral area of the visual field is "responsible" for identifying the direction and the sense of movement of the objects.

As shown in the literature "for gathering details from a moving object, the fovea has to perform pursuits by coordinated eye and head movements. Eye following movements are used to register floating objects in the fovea with a velocity up to 50–100°/s. If there is a higher velocity, a retinal slip is inserted resulting in saccades". [4]

Another important factor by which the postural balance in both static and dynamic position may deteriorate is identified by the age of the subjects. [5] The higher the age, the greater the likelihood of falling (to lose the postural balance) and all aspects of the resources that ensure the existence and preservation of the posture are altered (cognitive processing, biomechanical constraints, movement strategies, control of dynamics, orientation in space and sensory strategies).[6,7,8]

2. Experimental setup

To carry out the research, a methodology has been developed to analyze the effect of a chromatic visual dysfunction on posture and trajectory of the gait cycle. The experimental installation consists of a *footscan* pressure plate connected to a laptop with dedicated software for recording planting pressures along with all the gait cycle parameters (the centre of force trajectory, time, speed etc.), a set of special glasses with a filter (red and green), white light source with continuous and intermittent emission, dimensional measurement equipment for evaluating and framing the sample of subjects in anthropometric standards (Fig. 2). The sample of subjects consists of five emmetropic subjects, with an average age of 25.1 years old and no loco-motor dysfunction. They were tested on an identical procedure for all, and the evaluation of the influence of chromatic stress and luminous intensity applied to the visual system, the gait cycle and the postural behaviour of the subjects was followed. In order to have a system that only indicates the influence of visual stress, a normal ambient noise level, no vibrations or other sources of distraction, with constant environmental conditions (temperature 22^0C, humidity 40%, pressure 750 mmHg) was ensuring. The subjects were informed about the stages

IOP Conf. Series: Materials Science and Engineering **514** (2019) 012048 doi:10.1088/1757-899X/514/1/012048

of the procedures and were trained to participate effectively and affectively (concentration ability) during the experiment.

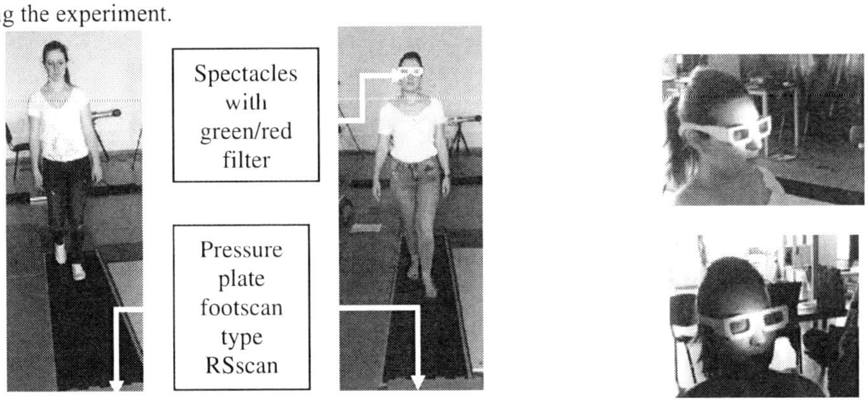

Figure 2. Experimental setup

For the evaluation process to be correlated between all the subjects of the sample, at the initial stage there were recorded the cycle of these subjects, in optimal conditions, without any visual stress or of any other nature, wearing all the same type of footwear. Monocular and binocular visual acuity was tested before starting the recordings and was ultimately compared with the visual acuity of the subjects measured while wearing coloured glasses and after inducing stress with compound white light. Each time, it was intended to obtain visual acuity values from subjects under the same environmental conditions and immediately after induction of intermittent light.

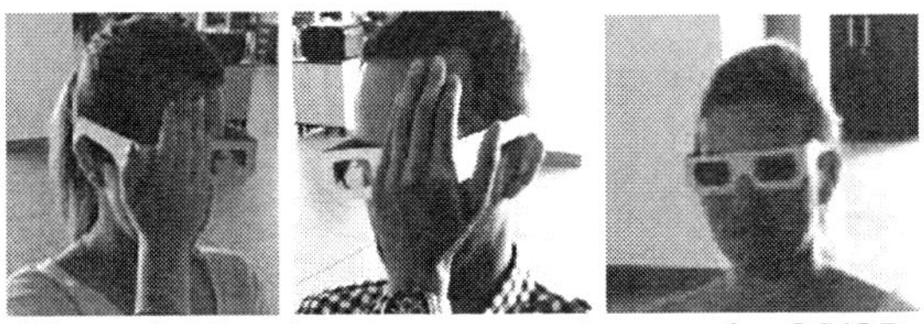

Figure 3. Visual acuity measure (monocular OS/OD and binocular) with chromatic filter spectacles

3. Results and conclusions

Following this test, constant visual acuity values were identified prior to recording, with all subjects having visual acuity (VA) monocular (OD-right eye, OS-left eye) and binocular 1/1 for both eyes (emmetropic subjects) and different values when using coloured filter spectacles (fig.4).

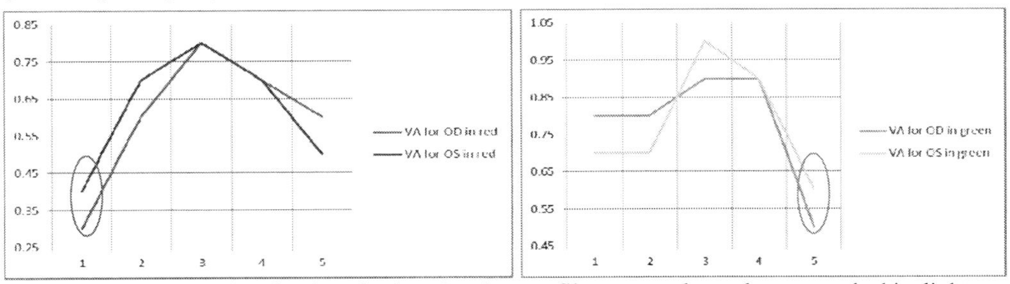

Figure 4. Monocular visual acuity in red and green filter spectacles and compound white light

Therefore, a visual diminution of the visual acuity of the subjects was induced temporarily, and there were at that moment recordings of the leg centre trajectory of the left leg force, respectively for the right foot. In Fig.5, the recordings for the subject No. 1 having the lowest visual acuity for the red filter simulation spectacles and the subject No. 5 having the smallest visual acuity for the green filter spectacles are presented.

(In Fig. 5, the black line represents the initial trajectory without the filters spectacles, and the coloured ones the trajectories corresponding to the use of the spectacles with red or green filters).

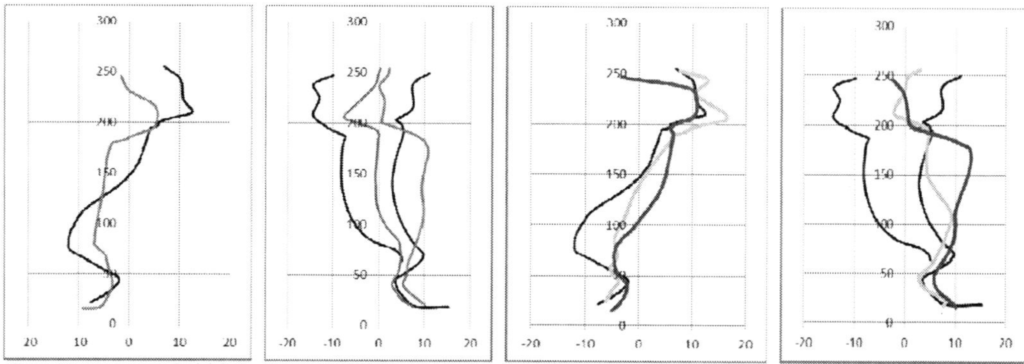

Force center trajectories for left foot, subject no.1. in red filter spectacles | *Force center trajectories for right foot, subject no.1 in red filter spectacles* | *Force center trajectories for left foot, subject no.5. in green filter spectacles* | *Force center trajectories for right foot, subject no.5. in green filter spectacles*

Figure 5. Force centre trajectories (mm x mm) for subject no.1/no.5 with red/green filter spectacles

From the analysis of these records (Fig. 5), there was a change for the left / right leg (in the sense of diminishing) of about **50%** of the maximum and minimum positioning points of the red filter spectacle of the original ones (black line) when it went normal and relaxed. This aspect indicates an additional contraction in the loco-motor system that causes the subject to go much tense and strained when wearing the red filter spectacles. These aspects are corroborated by the change in the level of visual acuity for both the right eye and the left eye under the same conditions. In the case of the subject No. 5 it changes the maximum and minimum limits, with the preponderance to the right foot, in percentage of about **67%** indicating a bigger imbalance on the right side of the left posture.

Due to the fact that the force centre variation traces recorded by the *RSscan pressure plate* for each left / right foot contact are analyzed as polynomial functions it is possible to use the **ARMAX** model to estimate the temporal evolution of these trajectories.

As a future research direction, we can identify the possibility of using this ARMAX model in recording situations with multiple visual stress variants to model the behaviour of subjects in relation to temporal evolution.

4. References

[1] http://www.mccc.edu/~behrensb/documents/PostureBIG.pdf accessed September 2018;

[2] Fay B. Horak, 2006, Postural orientation and equilibrium: what do we need to know about neural control of balance to prevent falls? *Age and Ageing (*2006); 35-S2;

[3] Fay B. Horak et al., 1997, Postural Perturbations: New Insights for Treatment of Balance Disorders, *Physical Therapy*, June (1997);

[4] Michaela Friedrich et al., 2008, InXuence of pathologic and simulated visual dysfunctions on the postural system *Exp Brain Res* (2008), 186:305–314;

[5] P.A. Fransson et al., 2004, Balance control and adaptation during vibratory perturbations in middle-aged and elderly humans, *Eur J Appl Physiol* (2004) 91: 595–603;

[6] A. Manso et al., 2016, Vestibular rehabilitation with visual stimuli in peripheral vestibular disorders, *Braz J Otorhinolaryngol.*,(2016), 82(2):232-241;

[7] P.A. Winkler, 2009, Ocular fixation, vestibular dysfunction, and visual motion hypersensitivity, *Optometry* (2009) 80, 502-512;

[8] Kathrine Jáuregui-Renaud, 2009, Postural Balance and Peripheral Neuropathy, InTech book, ch.5, pag 125-142, http://dx.doi.org/10.5772/55344, (2009);

Acknowledgments

In these experiments we've developed the investigations with equipment from "Advanced Mechatronic Systems Research Center - C04" and Applied optometric Laboratory at University Transylvania of Brasov, with the students help from Optometric study Program and Master.